APPLIED MICROBIAL SYSTEMATICS

Applied Microbial Systematics

Edited by

Fergus G. Priest

Heriot-Watt University,
Edinburgh

and

Michael Goodfellow

University of Newcastle,
Newcastle upon Tyne

KLUWER ACADEMIC PUBLISHERS
DORDRECHT / BOSTON / LONDON

Library of Congress Cataloging-in-Publication Data

ISBN 0-4127-1660-7 (HB)
ISBN 0-7923-6518-6 (PB)

Published by Kluwer Academic Publishers,
P.O. Box 17, 3300 AA Dordrecht, The Netherlands.

Sold and distributed in North, Central and South America
by Kluwer Academic Publishers,
101 Philip Drive, Norwell, MA 02061, U.S.A.

In all other countries, sold and distributed
by Kluwer Academic Publishers,
P.O. Box 322, 3300 AH Dordrecht, The Netherlands.

Printed on acid-free paper

Printed in the Netherlands

TABLE OF CONTENTS

ADDRESSES OF THE CONTRIBUTORS

Siv Ahrné Department of Food Technology, University of Lund, PO Box 124, S-221 00 Lund, Sweden. (e-mail: Siv.Ahrne@livsteki.lth.se)

Lars Axelsson MATFORSK, Norwegian Food Research Institute, Oslovein 1, N-1430 AS, Norway. (e-mail: lars.axelsson@matforsk.nlh.no)

James R. Brown Department of Bioinformatics, SmithKline Beecham Pharmaceuticals, 1250 S. Collegeville Road, PO Box 5089, Collegeville, PA 19426-0989, USA. (e-mail: James_R_Brown@sbphrd.com)

Nicolas P. Burton Department of Biological Sciences, University of Warwick, Coventry, CV4 7AL,UK. (e-mail: ny@dna.bio.warwick.ac.uk)

Vanderlei P. Canhos Faculdade de Engenharia de Alimentos (FEA) - Universidade Estadual de Campinas; Rua Professor Zeferino Vaz, s/n; Cidade Universitária, Barão Geraldo; CEP 13081-970; Campinas (SP), Brazil. (e-mail: vcanhos@bdt.org.br)

Heitor L. C. Coutinho Embrapa Solos, Rua Jardim Botânico 1024, Rio de Janeiro, RJ, 22460-000, Brazil. (e-mail: heitor@cnps.embrapa.br)

Ronald L. Crawford Environmental Biotechnology Institute, University of Idaho, Moscow, ID#83844-1052, USA. (e-mail: crawford@uidaho.edu)

Susan J. Dewar Department of Biological Sciences, Heriot-Watt University, Edinburgh EH14 4AS, Scotland UK. (e-mail: S.J.Dewar@hw.ac.uk)

John C. Dodd International Institute of Biotechnology, Biotechnology MIRCEN, Department of Biosciences, University of Kent, Canterbury, Kent CT2 7NJ, UK. (e-mail: J.C.Dodd@ukc.ac.uk)

Martina M. Ederer Environmental Biotechnology Institute, University of Idaho, Moscow, ID 83844-1052, USA. (e-mail: ederer@uidaho.edu)

Dagmar Fritze DSMZ, Mascheroder Weg 1b, D-38124 Braunschweig, Germany. (e-mail: dfr@dsmz.de)

Brett M. Goebel Australian Magnesium Corporation, Demonstration Plant, PO Box 488, Gladstone, Queensland 4680, Australia. (e-mail: bmg@amc.aust.com)

Michael Goodfellow Department of Agricultural and Environmental Science, University of Newcastle, Newcastle upon Tyne, NE1 7RU, UK. (e-mail: m.goodfellow@ncl.ac.uk)

William D. Grant Department of Microbiology, University of Leicester, PO Box 138, Leicester, LE1 9HN, UK. (e-mail: WDG1@le.ac.uk)

Makiko Hamamoto Japan Collection of Microorganisms, RIKEN, Wako-shi, Saitama, 351-0198, Japan. (e-mail: hamamoto@jcm.riken.go.jp)

Gudmundur O. Hreggvidsson Institute, Biology University of Iceland, Liftaeknihus, Keldnaholt, IS-112 Reykjavic, Iceland. (e-mail: gudmundo@iti.is)

Richard A. Humber USDA-ARS Plant Protection Research Unit, US Plant, Soil and Nutrition Laboratory, Tower Road, Ithaca, NY 14853-2901, USA. (e-mail: rah3@cornell.edu)

Peter Jeffries Department of Biosciences, University of Kent, Canterbury, Kent CT2 7NJ, UK. (e-mail: P.Jeffries@ukc.ac.uk)

Kristine K. Koretke Department of Bioinformatics, SmithKline Beecham Pharmaceuticals, 1250 S. Collegeville Road, PO Box 5089, Collegeville, PA 19426-0989, USA. (e-mail: Kristin_K_Koretke@sbphrd.com)

Jakob K. Kristjansson Institute, Biology University of Iceland, Liftaeknihus, Keldnaholt, IS-112 Reykjavic, Iceland. (e-mail: jakobk@iti.is)

Gilson P. Manfio Coleção de Culturas Tropical (CCT) - Fundação André Tosello; Rua Latino Coelho, 1301; Parque Taquaral; CEP 13087-010; Campinas (SP), Brazil. (e-mail: gilson@bdt.org.br)

Fatima M. S. Moreira Departamento de Ciencias do Solo, Universidade Federal de Lavras, Caixa Postal 37, Lavras-MG, 37200-000, Brazil. (e-mail: fmoreira@ufla.br)

Takashi Nakase Japan Collection of Microorganisms, RIKEN, Wako-shi, Saitama 351-0198, Japan. (e-mail: nakase@jcm.riken.go.jp)

Paul R. Norris Department Biological Sciences, University of Warwick, Coventry, CV4 7AL, UK. (e-mail: pn@dna.bio.warwick.ac.uk)

Valéria M. de Oliveira Fundaçâo André Tosello, R. Latino Coelho 1301, Campinas - SP, 13087-010, Brazil. (e-mail maia@bdt.org.br)

Fergus G. Priest Department of Biological Sciences, Heriot-Watt University, Edinburgh EH14 4AS, Scotland UK. (e-mail: f.g.priest@hw.ac.uk)

Gerrit Voordouw Department of Biological Sciences, University of Calgary, 2500 University Drive NW, Calgary (AB) T2N 1N4, Canada. (e-mail: voordouw@ucalgary.ca)

Lawrence G. Wayne Tuberculosis Research Laboratory, Veterans Administration Medical Center, 5901 Long Beach, Ca 90822, USA. (e-mail: waynelg@oco.net)

Vera Weihs DSMZ, Mascheroder Weg 1b, D-38124 Braunschweig, Germany. (e-mail: vew@dsmz.de)

John M. Young Landcare Research, 120 Mt Albert Road, Mt Albert, Auckland, New Zealand. (e-mail: youngj@landcare.cri.nz)

PREFACE

FERGUS G PRIEST and MICHAEL GOODFELLOW

Microbial systematics has enjoyed two major advances in the latter half of this century: the introductions of numerical phenetics and molecular techniques for direct comparisons of organismal genomes. Numerical phenetics (taxonomy) was very influential during the 1960s and 70s in providing the first objective approach to bacterial classification. Numerical taxonomy gave microbiologists the solid theoretical base for classification and identification that had been sought for so long. Indeed, the concepts and procedures are still used for routine identification in the applied microbiology laboratory in the form of kits and automated identification machines. Although numerical phenetics no longer ranks as the primary method of microbial classification, the theoretical base and need for phenotypic descriptions of new taxa will continue to make the approach important for many years.

While numerical phenetics made no claim to have a phylogenetic content, the lack of an evolutionary backbone to microbial classification was held by many to be a serious detraction. A related problem was the inability of phenotypic data to provide a hierarchical taxonomic system, species could generally be defined with accuracy but phenotypically defined genera have commonly been shown to be unstable and even the most committed numerical taxonomists realised that higher taxa (families etc.) could not be derived from phenotypic descriptions. It had been appreciated in the 1960s that nucleic acids and proteins contained the necessary information for deriving phylogenetic classifications but accessing that information was so technically difficult that evolutionary studies were restricted in scope and could only be conducted by the specialist protein chemist. However, the scene was set for the application of nucleic acid sequence analysis to microbial systematics once the analytical techniques became available, and by 1990 bacterial taxonomy was set on a molecular phylogenetic course from which it is unlikely to be moved in the foreseeable future. Indeed, the completion of more genome sequences will provide a more robust basis for molecular phylogenetics and the resultant classifications will be even more influential.

Molecular systematics has penetrated microbial classification at every level from the very top in placing prokaryotes in two of the three domains of life to the very bottom in resolving the population structures of bacterial species and recognition of clones. In so doing it has influenced the practice of microbiology in numerous ways but particularly by providing classifications which are more stable than those based on phenotypic characters and yet which are largely congruent with the latter. In general, microbiologists are appreciating the polyphasic classifications that are emerging. Phylogenetic trees, ideally supported by phenetic information, abound in the literature and are being used to address and explain various aspects of microbiology such as ecology, metabolism, pathogenicity and physiology as well as the possible evolution of microorganisms. Molecular methods are offering powerful approaches to recognizing biodiversity and achieving rapid and accurate identification. In these ways, taxonomy is becoming more widely appreciated as providing the basic working framework for microbiologists, as it does for botanists and zoologists.

The aim of the book is to illustrate the utility and practical applications of modern systematics. Little mention is made of the methods of systematics; instead the value of a taxonomic approach to assist understanding of microbiological problems is emphasized. We wanted to demonstrate how improved systematics is advancing our knowledge of related disciplines such as microbial diversity, ecology, evolution and certain areas of biotechnology.

We would like to express our sincere thanks to all the contributors for their stimulating and comprehensive chapters and in particular to Olivier Sparagano for his eagle-eyed proofreading and assistance with the indexes.

I   INTRODUCTION

1   MICROBIAL SYSTEMATICS: BACKGROUND AND USES

MICHAEL GOODFELLOW

Department of Agricultural and Environmental Science,
University of Newcastle, Newcastle upon Tyne, NE1 7RU, UK

1   Some Guiding Principles

This chapter is intended to be a brief introduction to the concepts and practices embraced by microbial systematics, the scientific study of the kinds and diversity of microorganisms and their relationships. The subject is generally divided into three related disciplines, the taxonomic trinity recognised by Cowan (1955), namely, classification, nomenclature and identification. Classification and identification are core scientific disciplines that are relevant to both pure and applied microbiologists. Nomenclature is central to all facets of the microbial sciences as microbiologists need to know what organisms they are working with before they can pass on information about them within and outwith the scientific community. In other words, the name of an organism is the key to its literature, an entry to what is known about it.

It is important to understand the basic concepts and specialist terms used in microbial systematics. The first step, microbial classification, is the ordering of microorganisms into groups or taxa on the basis of similarities and differences. The product of this exercise is an orderly arrangement or system of classification that is designed to show relationships between organisms and to serve as an information storage and retrieval system. Relationships are usually based on a combination of genotypic and phenotypic data. The genome encompasses the sum of the information on the genes (genetic information) whereas all observable information on an organism, irrespective of whether it is derived from the genotype or phenotype, constitutes the phenome (phenetic information). The term classification covers both the process of classification and the product of that process though the outcomes of classification are often referred to as taxonomies. In contrast, taxonomy can be seen as the theoretical study of classification including its bases, principles and rules (Simpson, 1961).

The primary purpose of a practical or utilitarian taxonomy is to provide classifications that are useful for varied scientific and practical purposes, including identification, and also to generate databases replete with relevant information about organisms. Such general

F. G. Priest and M. Goodfellow (eds.), Applied Microbial Systematics, 1-18
© 2000 Kluwer Academic Publishers. Printed in the Netherlands.

purpose classifications should be stable, objective and predictive. Three basic kinds of classification can be recognised, namely, artificial, phenetic and phylogenetic classifications, though in practice they overlap.

Early systems of microbial classification were, at least in part, artificial, that is they were based on a few properties that were used to sort strains, often irrespective of any true affinity. In practice, the overreliance placed on a few subjectively chosen morphological and behavioural properties led to serious misclassifications of microorganisms, including bacteria, as exemplified by comprehensive studies on aerobic, endospore-forming bacilli (Nielsen *et al.*, 1995), pseudomonads (Palleroni, 1993) and streptomycetes (Williams *et al.*, 1989), and protozoa such as the ameoboflagellates (De Jonckheere, 1998) and kinetoplastids (Philippe, 1998).

A special purpose classification is a particular type of artificial classification where one or a few properties of organisms are weighted for a particular purpose. The artificial separation of members of the *Mycobacterium tuberculosis* complex (*M. africanum, M. bovis, M. canetti, M. microti* and *M. tuberculosis*) is a case in point as members of the constituent species are difficult to distinguish from one another and which from a purely taxonomic perspective ought to be considered as a single species (Goodfellow & Magee, 1998). However, clinical imperatives currently override taxonomic interests not least because *M. bovis* and *M. tuberculosis* show marked epidemiological differences as agents of tuberculosis. The net result of the epidemiological interest was a weighting of a battery of phenotypic tests to characterise slow-growing mycobacteria using techniques which tended to exaggerate differences between *M. bovis* and *M. tuberculosis*.

Relationships that record the similarity between the present properties of organisms, without reference to how they came to possess them, are termed phenetic. Phenetic classifications are generated by assigning organisms to groups on the basis of overall similarity or resemblance based on genotypic and phenotypic characters. The resultant classifications have a high information content as they are founded on a broad range of measurable features. In contrast to special purpose classifications, phenetic classifications are predictive, that is, they permit the greatest number of predictive statements to be made about members of defined groups.

Phylogeny is the study of the evolutionary history of organisms. It usually connotes the study of long-term evolutionary change. In contrast, population genetics is the study of short-term change. The study of genealogy, the branching pattern of evolutionary lineages, is called cladistics; a clade is a monophyletic group. Phylogenetic relationships are inferred from various types of phenetic relationships that rest on assumptions on how evolution might have occurred (Gupta, 1998). However, in the absence of actual time units, relationships may be expressed in topological units (numbers of nodes and internodes) or number of inferred evolutionary changes (Sneath, 1988, 1989).

Phylogenetic classifications are not inherently more stable, objective or predictive than phenetic systems (Sokal, 1985; Sneath, 1988). Indeed, phenetic classifications may be considered to be superior on the grounds that they are more readily verifiable. The procedure used to generate a phenetic classification involves gathering data that are then analysed using established statistical techniques (Stackebrandt *et al.*, 1999). This procedure is entirely objective and can be repeated by other scientists. Phenetic classifications can also be evaluated in light of the application of independent taxonomic methods. In contrast, phylogenetic classifications are not directly verifiable; it is not possible to check them against "true phylogenies" as these are unknown and unknowable.

Sneath (1989) also pointed out that phylogenetic groups would be of little practical value were they found to have no phenetic coherence. Conversely, if two phyletic lines were to become almost identical, as a result of extreme convergence due to strong selective pressures, there would be little practical value in separating them. These observations suggest that the taxonomic importance of phylogenetic groupings is based on the implicit assumption that such groups will also be found to be phenetically coherent. If this were not so, then phylogenetic groupings would have to be seen as special purpose classifications for the study of genealogical relationships (Gilmour, 1940; Sneath, 1988).

The second step in the taxonomic trinity, nomenclature, deals with the terms used to denote taxonomic categories or ranks, such as species, genera or families, and the process of assigning correct, internationally recognised names to organisms (Ride *et al.*, 1985; Bousfield, 1993; MacAdoo, 1993; Greuter *et al.*, 1994). The scientific names of prokaryotes are regulated by successive editions of the *International Code of Bacterial Nomenclature* (Sneath, 1992). The three reforms introduced in the "Bacteriological Code" edited by Lapage *et al.* (1975) are worthy of special consideration:

- A new starting document and starting date were introduced with the publication of the *Approved Lists of Bacterial Names* on January 1, 1980 (Skerman *et al.*, 1980). Names published prior to 1980 but omitted from the Approved Lists lost their standing in nomenclature thereby clearing away thousands of useless names. Old names are available for renewal if the provisions for doing so are met.
- Names of new taxa can only be validly published in the *International Journal of Systematic Bacteriology* (IJSB) though they can be effectively published in relevant international journals then legitimised by announcement in Validation Lists published periodically in the IJSB. The IJSB is published by the UK Society for General Microbiology on behalf of the International Union of Microbiological Societies.
- For valid publication, nomenclatural types must be designated.

An historic account which led to these changes is available (Sneath, 1986). The *Approved Lists of Bacterial Names* were reprinted in 1989 together with a list of names validly published in the IJSB or cited in the Validation Lists between 1 January, 1980 and 1 January, 1989 (Moore & Moore, 1989).

Fungal nomenclature is regulated by the *International Code of Botanical Nomenclature* (Greuter *et al.*, 1994) and that of the protozoa by both the *International Code of Zoological Nomenclature* (Ride *et al.*, 1985) and by the *Botanical Code* (Greuter *et al.*, 1994). These Codes are currently undergoing extensive revision, the changes envisaged will bring them more into line with the *International Code of Bacterial Nomenclature* (Sneath, 1992). There are also moves to produce a *Universal Code of Nomenclature*.

It is important to recognise that the nomenclature of a group of organisms does not depend on the correct latinisation of words, but on the thoroughness of the preceding taxonomic work. When microorganisms have been rigorously characterised and classified it is a relatively simple matter to apply the rules of nomenclature. Once the name of a taxon has been chosen, the next step is to publish it. The paper submitted for publication should give the derivation of the new name and the name itself should carry the relevant indication of novelty (e.g. sp. nov., gen. nov.).

Identification, the final stage in the taxonomic trinity, covers both the act and result of determining whether an unknown organism belongs to a particular group in a previously made classification. It involves determining the key characters of the unknown organism and matching of these against databases containing information on validly described taxa. Identification is an important practical activity as it underpins important facets of microbiology, including microbial ecology, industrial biotechnology and medical microbiology. Some areas, such as diagnostic hospital microbiology are almost entirely devoted to identification (Magee, 1998). Developments in other areas, such as microbial ecology and microbial technology, are severely hampered by a lack of suitable identification systems (Bull *et al.*, 1992).

Classification and identification are markedly data dependent and hence are in a constant state of development as they are influenced by the application of new taxonomic concepts and methods. Nomenclature, by ensuring that the correct internationally recognised names are given to taxa (classification) and unknown strains (identification), spans both disciplines and, as with classification and identification, is constantly being adapted and refined.

2   Current Trends

Microbial systematics has undergone spectacular changes over the past forty years as systematists have revised and extended existing classifications in light of developments in chemistry, molecular biology and data handling systems. The resultant improved classifications together with the availability of representatives of well circumscribed taxa have helped promote advances in procedures used to identify microorganisms of ecological, medical and industrial importance. The beginning of this new era in microbial systematics can be traced back to developments in chemotaxonomy, molecular systematics and numerical phenetic taxonomy (Goodfellow & O'Donnell, 1993).

2.1   CHEMOTAXONOMY

Chemical data derived from the analysis of whole organisms and cell components, using physico-chemical methods such as gas chromatography, thin layer chromatography and high performance liquid chromatography, can be used to classify microorganisms according to the pattern of distribution of different components within and between members of different taxa (Goodfellow & O'Donnell, 1994). Chemotaxonomic analyses of chemical macromolecules, particularly amino acids and peptides (e.g. from peptidoglycan and pseudomurein), lipids (lipopolysaccharides, fatty acids), polysaccharides and related polymers (e.g. methanochondroitin and wall sugars), proteins (e.g. bacteriochlorophylls and whole-organism protein patterns), enzymes (e.g. hydrolases, lyases and multilocus enzyme electrophoretic patterns) and other complex polymeric compounds, such as isoprenoid quinones, all provide grist to the chemotaxonomic mill.

Chemical fingerprints of taxonomic value can be obtained by using analytical chemical techniques, such as Curie-point pyrolysis mass spectrometry (PyMS), Fourier-transform infra-red spectroscopy and dispersive Raman microscopy (Magee, 1993; Goodfellow *et al.*, 1994; Naumann *et al.*, 1994; Goodacre *et al.*, 1998). The speed and reproducibility of pyrolysis mass spectrometry and its applicability to a wide range of pathogenic microorganisms make it an attractive method for interstrain comparison and epidemiological studies (Goodfellow *et al.*, 1997a). Other whole-organism approaches which provide valuable data for delineating between microbial taxa include fatty acid (Stead *et al.*, 1992) and protein analyses (Vauterin *et al.*, 1993), and the elucidation of enzyme and volatile metabolite profiles (James, 1994; Larsen & Frisvad, 1995).

2.2   NUMERICAL PHENETIC TAXONOMY

Numerical taxonomy, the grouping by numerical methods of taxonomic units into taxa based on shared unit characters, was first applied to microbial classification by Sneath (1957a, b). The primary objective was to assign individual strains to homogeneous groups or clusters (taxospecies) using large sets of phenotypic data, and to use the resultant quantitative results for the numerically circumscribed groups to generate improved identification systems. When introduced, the numerical taxonomic procedure was in sharp contrast to the prevailing orthodoxy as taxa, notably taxospecies, were delineated using many equally weighted features, not by a few subjectively weighted morphological and behavioural properties. The theoretical basis of numerical taxonomy is well documented (Sokal, 1985; Sackin & Jones, 1993; Stackebrandt *et al.*, 1999).

The application of numerical taxonomic procedures has led to significant improvements in the classification of microorganisms, notably bacteria. The taxonomy of most established bacterial genera has benefited from a revision of their pre-1960 classification using numerical taxonomic procedures; good examples include *Bacillus* (White *et al.*, 1993), *Mycobacterium* (Wayne *et al.*, 1996) and *Rhodococcus* (Goodfellow *et al.*, 1998). In contrast, relatively few numerical phenetic surveys have been carried out on other microbial groups, including the fungi and protozoa (Goodfellow & Dickinson, 1985). The numerical taxonomic method has been less successful in circumscribing higher taxonomic ranks, but this is almost certainly due to the types of data used rather than to fundamental flaws in numerical methods.

2.3   MOLECULAR SYSTEMATICS

The most specific and informative methods for classifying microorganisms are based on the determination of precise amino acid and nucleotide sequences of individual proteins and specific regions of the chromosome. For noncoding sequences, such as various rRNAs, tRNAs and introns, phylogenetic analysis can be carried out based on nucleotide sequence data. However, for gene sequences that encode proteins, analyses can be performed based on either amino acid or nucleotide sequence data.

Nucleic acid sequencing methods have developed rapidly in recent years so that comparative sequencing of homologous genes is now a standard procedure in microbial systematics. Conserved genes, such as those coding for ribosomal RNA, are widely used to establish relationships at and above the genus level (Woese, 1987; Kurtzman, 1992; Suh & Nakase, 1995; Coombs *et al.*, 1998). Less conserved genes have been examined to distinguish between closely related species (e.g. the gyrase B genes to distinguish between closely related *Acinetobacter* species; Yamamoto & Harayama, 1996) and strains (e.g. the cholera toxin genes to type toxin-producing strains of *Vibrio cholerae*; Alm & Manning, 1990).

Nucleic acid sequences are usually deposited in sequence databases such as the DNA Data Bank of Japan (Tateno *et al.*, 1998), the European Molecular Biology Laboratories Database (Stroesser *et al.*, 1999), the GenBank Database (Benson *et al.*, 1998) and the Ribosomal Database Project (Maidak *et al.*, 1997) and can be retrieved for comparative studies. Thousands of nucleic acid sequences from eukaryotes, prokaryotes and viruses are available under unique accession numbers and may be searched for by genus and species of origin, as well as by homology with a nucleotide sequence provided by the user. Substantial databases are also available for several genes coding for widely distributed and well conserved proteins such as actin, elongation factor - 1α, glyceraldehyde-3-phosphate dehydrogenase, heat shock protein 70 and tubulins.

It is common knowledge that rDNA genes are essential for the survival of all organisms. These genes are highly conserved in eukaryotes and prokaryotes and are being used to generate the universal tree of life (Fig. 1). This approach to classification rests on two premises, namely, that lateral gene transfer has not occurred between rDNA genes and that the degree of dissimilarity of rDNA sequences between a given pair of organisms is representative of the range of variation shown by the corresponding whole genomes. The good congruence found between phylogenies based on rDNA sequences and those generated from studies on other conserved molecules, such as elongation factors, ATPase subunits and RNA polymerases lends substance to this latter point (Ludwig *et al.*, 1993; Olsen & Woese, 1993). There is some evidence that horizontal gene transfer can occur between 16S rDNA genes, as exemplified by aeromonads (Sneath, 1993) and rhizobia (Eardly *et al.*, 1996).

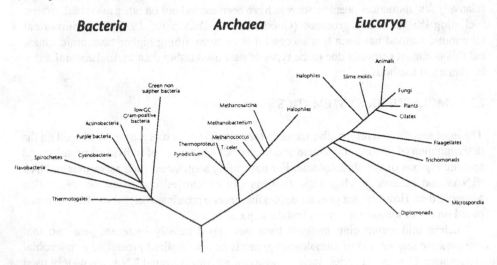

Figure 1. Universal phylogenetic tree based on ribosomal RNA sequence comparisons.

There is also evidence that 16S rDNA genes from members of closely related bacterial species may be so conserved that they cannot be used to differentiate between strains at the species level (Fox *et al.*, 1992). This is the case with some species of *Aeromonas* (Martinez-Murcia *et al.*, 1992), *Bacillus* (Fox *et al.*, 1992), *Legionella* (Fry *et al.*, 1991) and *Tsukamurella* (Yassin *et al.*, 1997). The terms rRNA species complex and rRNA superspecies have been proposed for bacteria which have virtually identical 16S rRNA sequences but can be distinguished on the basis of DNA:DNA relatedness.

Evolutionary relationships between microorganisms need to be interpreted with care as estimates of phylogeny are based on relatively simple assumptions when seen within the context of the complexities of evolutionary processes. Indeed, all methods of phylogenetic inference are based on assumptions that can be violated by data to a greater or lesser extent (Hillis *et al.*, 1993; Gupta, 1998). Potential problems in interpreting nucleotide sequence data include the reliability of sequence alignments, non-independence of sites, inequalities in base substitution frequencies at homologous sites, and lineage-dependent inequalities in rates of change, lateral gene transfer, and comparison of paralogous sequences which are the results of unidentified ancient gene duplication events.

DNA:DNA hybridisation. The complementary strands of DNA, once denatured, can, under suitable experimental conditions, reassociate to form native duplexes. The specific pairings are between adenine and thymine and guanine and cytosine; the overall pairing of the nucleic acid fragments is dependent on similar linear arrangements of these bases along the DNA. The amount of molecular hybrid and its thermal stability provide an average measurement of nucleotide sequence similarity when nucleic acids from different organisms are compared.

DNA:DNA relatedness data have been extensively used to delineate microbial species (Stackebrandt & Goebel, 1994). The *Ad Hoc Committee on Reconciliation of Approaches to Bacterial Systematics* (Wayne *et al.*, 1987) proposed a formal definition of bacterial species, namely, that a species should generally include strains showing approximately 70% or more DNA:DNA relatedness with a difference of 5°C or less in thermal stability (ΔT_m). This guideline has been used to clarify relationships between species in diverse bacterial genera, including streptomycetes (Labeda, 1998).

Molecular fingerprinting. The accurate delineation of groups within a species is assuming greater importance in all branches of microbiology, particularly diagnostic microbiology. Indeed, the control of communicable diseases, including tuberculosis, would not be possible without the use of typing methods to help locate the sources of infection in susceptible populations (Towner & Cockayne, 1993; van Belkum *et al.*, 1998). Molecular marker systems are also needed for the detection of genetically engineered microorganisms in the environment (Prosser, 1994).

A broad range of highly specific molecular methods are now available for typing microorganisms (Gürtler & Stanisich, 1996; Caetano-Anollés & Gresshoff, 1997). The various techniques have advantages and disadvantages when applied to specific situations. In addition to the ability to discriminate between strains within target species is the ease of performance and interpretation of tests and the availability of reagents. The usefulness of typing systems vary depending on the specific pathogen of interest.

3   Polyphasic Taxonomy

Microbial systematics began as a largely intuitive science but has become increasingly objective due to the introduction of the concepts and methods outlined above. The new advances, especially in molecular systematics, have stimulated the need to compare established and more recent approaches to microbial classification (Wayne *et al.*, 1987; Murray *et al.*, 1990). This exercise promoted the view that classification at all levels in the taxonomic hierarchy should be based upon the integrated use of genotypic and phenotypic data (Vandamme *et al.*, 1996; Goodfellow *et al.*, 1997b). This approach, known as polyphasic taxonomy, was introduced by Colwell (1970) to signify successive or simultaneous studies on groups of organisms using a combination of taxonomic methods designed to yield good quality genotypic and phenotypic data. Genotypic data are derived from the analysis of nucleic acids and phenotypic data from cultural, chemotaxonomic, morphological, nutritional and other expressed features.

It was only with the availability of rapid data acquisition and improved data handling systems that the polyphasic approach to the circumscription of microbial taxa became manageable (Canhos *et al.*, 1993; Vandamme *et al.*, 1996). Most descriptions of new cultivable bacteria are based on a judicious selection of genotypic and phenotypic data, as exemplified by proposals for the genera *Desulfocella* (Brandt *et al.*, 1999), *Prauserella* (Kim & Goodfellow, 1999) and *Roseivivax* (Suzuki *et al.*, 1999) and for the species *Erwinia pyrifoliae* (Kim W-S. *et al.*, 1999), *Nocardia salmonicida* (Isik *et al.*, 1999) and *Streptomyces thermoalkalitolerans* (Kim, B. *et al.*, 1999). Similar integrated approaches are being used to describe protozoal (Foissner, 1998) and fungal (Bridge *et al.*, 1995; Boysen *et al.*, 1996; Mordue *et al.*, 1996) taxa, notably yeasts (Takashima & Nakase, 1998; Rosa *et al.*, 1999).

Although polyphasic taxonomy is now widely practiced little attempt has been made to spell out the methods which should be used to generate genotypic and phenotypic information. At present, polyphasic taxonomic studies tend to reflect the interests of individual research groups and the equipment and procedures which they have at their disposal. However, it is difficult to be too prescriptive about which methods should be applied as those employed need to reflect the rank of the taxa under investigation (Table 1). Nevertheless, it is clear that rRNA sequencing is a powerful tool for establishing relationships between and within microbial groups (Woese, 1987; Cavalier-Smith, 1993; Coombs *et al.*, 1998) though it is of less value in unravelling relationships below the genus level (Goodfellow *et al.*, 1997b). In contrast, DNA:DNA relatedness, molecular fingerprinting and phenotypic studies provide valuable data for delineating groups at species and infrasubspecific levels (Stackebrandt & Goebel, 1994; Wayne *et al.*, 1996).

It is now widely accepted amongst microbiologists that nomenclature should reflect genomic relationships and that established classifications should be re-examined within this context (Wayne *et al.*, 1987; Murray *et al.*, 1990). There are no problems in naming new taxa given congruence between genotypic and phenotypic data. However, there are a number of instances where bacterial taxospecies encompass two or more genomic species. Wayne and his colleagues recommended that in such instances genomic species should not be formally named until they can be differentiated by using suitable phenotypic markers.

TABLE 1. Sources of taxonomic information*

Cell component	Analysis	Taxonomic rank		
		Genus or above	Species	Subspecies or below
Chromosomal DNA	Base composition (mol.% G+C)	√	√	
	DNA:DNA hybridisation		√	
	Restriction patterns (RFLP, ribotyping)		√	√
DNA segments	DNA probes	√	√	√
	DNA sequencing	√	√	√
	PCR based DNA fingerprinting (PCR-RFLP, RAPD)		√	√
Ribosomal RNA	DNA:rRNA hybridisation	√	√	
	Nucleotide sequences	√	√	
Proteins	Amino acid sequences	√	√	
	Electrophoretic patterns		√	√
	Multilocus enzyme electrophoresis			√
	Serological comparisons	√	√	√
Chemical markers	Peptidoglycan	√		
	Fatty acids	√	√	
	Isoprenoid quinones	√	√	
	Mycolic acids	√	√	
	Polar lipids	√	√	
	Polyamines	√		
	Polysaccharides		√	
	Teichoic acids	√	√	
Whole organisms	Pyrolysis mass-spectrometry		√	√
	Rapid enzyme tests		√	√
Expressed features	Morphology	√	√	
	Physiology	√	√	

* Modified from Priest & Austin (1993) and Vandamme *et al.* (1996).

  Abbreviations: RFLP, restriction fragment length polymorphism; PFGE, pulsed field gel electrophoresis; RAPD, randomly amplified polymorphic DNA fingerprints; PCR, polymerase chain reaction.

Such taxospecies can be considered as nomenspecies. It has been suggested that genomic species accommodated in nomenspecies should be referred to as genomovars (Rosselló et al., 1991; Ursing et al., 1995). This is a useful concept as it indicates that genomic species are an integral part of a nomenspecies and hence should not be overlooked in subsequent taxonomic work.

The polyphasic taxonomic approach to circumscribing microbial taxa can be expected to help meet many of the primary challenges facing microbial systematists, notably the need to generate well defined taxa, a stable nomenclature and improved identification schemes. However, the concept does not address the evolutionary processes that generate taxonomic diversity. This is clearly a major omission but is also a fertile area for collaboration between microbial systematists and population geneticists. However, it is debatable whether a theoretically sound species concept would prove to be as practically useful as the current operational species concept.

4   Microbial Systematics in Practice

4.1   PHYLOGENY

The application of modern taxonomic concepts and practices has had a profound effect on microbial classification at all levels in the taxonomic hierarchy. In particular, the pivotal belief in the prokaryotic - eukaryotic divide (Chatton, 1937; Stanier & van Niel, 1941) has been challenged by the work of Carl Woese and his colleagues (Fox et al., 1977; Woese & Fox, 1977; Woese, 1987). Molecular comparisons of small subunit rRNA and associated biochemical and genomic evidence led Woese et al. (1990) to formally propose the replacement of the bipartite view of life with a tripartite scheme based on three urkingdoms or domains, namely the Bacteria (formerly eubacteria), Archaea (formerly archaebacteria) and Eucarya (the eukaryotes).

The tripartite scheme has been widely accepted amongst microbiologists though several systematists continue to question the elevation of the archaebacteria (and hence eubacteria) to a taxonomic rank that corresponds to the eukaryotes (Mayr, 1990, 1998; Margulis & Guerro, 1991; Cavalier-Smith, 1992; Gupta, 1998). This on-going controversy is considered in detail by Brown and Koretke (Chapter 2) within the wider context of determining universal tree topologies. It is also clear from their contribution that phylogenetic analysis of PCR amplified rRNA molecules provides an effective way of highlighting novel clinically significant isolates of pathogenic bacteria and of unravelling the extent of microbial species diversity in different ecosystems.

A consensus is beginning to emerge about the broad outlines of fungal phylogeny, a theme developed by Hamamoto and Nakase in Chapter 3. This consensus mainly rests on persuasive molecular evidence, notably that involving ribosomal gene sequences, gathered mainly for ascomycete and basidiomycete fungi. Phylogenetic trees based on small subunit rRNA gene sequences show that the higher fungi form a monophyletic group composed of sister taxa corresponding to the phyla Ascomycota and Basidiomycota; these taxa are thought to have separated about 390 million years ago. The lower fungi, which are considered to predate the ascomycetes and basidiomycetes, also encompass two phylogenetic groups, the chytridomycetes and zygomycetes. The taxonomy of the main fungal groups is likely to remain fluid for some time to come as so many fungi remain undiscovered and undescribed.

4.2   SOIL, PLANTS AND INSECTS

The application of polyphasic taxonomy has led to significant improvements in the systematics of microorganisms of agricultural importance. This is particularly so for mycorrhizal fungi and rhizobia, organisms which play a vital role in sustaining soil fertility. The application of molecular systematic techniques is promoting an understanding of the population ecology of these organisms which in turn is driving forward important advances in mycorrhizal and rhizobial technologies. The significance of the improvements in mycorrhizal systematics are considered by Jeffries and Dodd in Chapter 4. These authors stress the importance of identifying mycorrhizas in determining inter- and intraspecific genetic variation and in mapping the ecological distribution of individual genotypes. Similarly, the practical importance of rhizobial systematics is the subject of Chapter 5 by Coutinho and his colleagues. These authors stress the importance of unravelling the full extent of rhizobial diversity and show how the application of advanced systematic methods is promoting improvements in the performance of rhizobial inoculants.

Another agriculturally important group of microorganisms the plant pathogenic bacteria. In Chapter 6, Young considers the impact of contemporary taxonomic concepts and methods on the classification and identification of phytopathogenic bacteria. He points out that plant pathologists have been especially well served by the introduction of improved identification methods, notably those based on the use of oligonucleotide probes and DNA primers which amplify specific sequences in target organisms. In contrast, Young rails against the progressive revision of genera based on phylogenetic refinements as he considers that this practice undermines the need for a stable nomenclature and hence imposes a burden on practitioners for little discernible gain.

It is also essential to define the systematics of insect/microbe interactions if the complex nature of these relationships are to be elucidated. These interactions range from mutualism through opportunistic and obligate pathogenicity to complete interdependence in endosymbiotic pairings. In Chapter 7, Priest and Dewar stress that molecular techniques are essential for determining the inferred evolution of endosymbiotic bacteria which cannot be cultivated in the laboratory, for relating virulence with certain lineages of pathogen, for analysing the genetic population structures of bacteria and their insect hosts, and for distinguishing between pathogenic and non-pathogenic strains of biocontrol agents such as *Bacillus sphaericus*. In a complementary chapter, Humber highlights the relevance of insect mycology with emphasis on the ascomycetes and hyphomycetes which constitute the largest and taxonomically most complex group of almost one thousand taxa of fungal entomopathogens.

4.3   ENVIRONMENT AND ITS EXPLOITATION

Access to well characterised microorganisms with particular properties is at a premium in the search and discovery of new commercially significant bioactive compounds, for bioprocessing and for environmental protection. The choice of isolates for industrial screening programmes is essentially a problem of distinguishing between known strains and recognising novel ones. The recognition of novel, that is, previously undescribed organisms presupposes the existence of reliable and workable classification and identification procedures.

Until recently, little attention was paid to the extremophiles but these organisms are now seen as a rich source of new natural products, notably thermostable enzymes. The special properties that allow these organisms to live in a range of harsh chemical and physical environments are often the ones which make them interesting in an industrial context. Much of the emphasis in extremophile research has been on discovering novel and unusual archaea and bacteria. In Chapter 9, Kristjannsen and his colleagues consider developments in the systematics of extremophiles and outline the importance of such studies for microbial technology.

In Chapter 10, Goebel and his colleagues review the systematics of acidophilic bacteria, organisms which are increasingly being used commercially in mineral processing. Further rapid developments can be expected in this area as the application of rRNA gene targeting has the potential to clarify questions concerning the nature of the mixed microbial populations involved in bioleaching. This approach has already been extensively used to help clarify the individual prokaryotic components of the complex microbial communities found in oil fields. The practical significance of such work is spelt out by Voordouw in Chapter 11. It seems likely that certain microorganisms are either indigenous to oil fields or are transported there from neighbouring subsurface regions. It has also been shown that the halophilic and thermophilic properties of oil field bacteria are usually absent from those found in surface water and soil. Progress made in subsurface microbiology indicates that microorganisms are able to live in rocks by maintaining themselves through chemolithotrophic metabolism.

The biodegradation of compounds of anthropogenic origin, especially xenobiotic ones alien to existing enzyme systems, has become an important area of basic and applied research. The assignment of many bacteria capable of degrading arthropogenic compounds to the genus *Sphingomonas* suggests that sphingolipids may have a role to play in the biodegradation of such compounds. The relationship between the activity and taxonomy of sphingomonads is considered by Ederer and Crawford in Chapter 12 with particular reference to the degradation of xenobiotic pollutants, notably the pesticide pentachlorophenol (PCP). The fact that PCP-degraders and PCP genes have been patented indicates that phylogenetic analyses of sphingomonads have far reaching implications.

4.4   FOOD AND MEDICINE

The production of fermented foods is assuming greater importance in the food processing industry. It is, therefore, important to predict, control and improve the activities of lactic acid bacteria (LAB) involved in the acid fermentation of food. These challenges are addressed in Chapter 13 by Axelsson and Ahrné within the context of developments of LAB systematics. Taxonomic revisions based on phenetic and phylogenetic data show that LAB can be classified into sixteen genera that belong to the low G + C subdivision of the Gram-positive bacteria. This improved classification provides a sound base for studying the diversity, ecology and distribution of specific LAB taxa in fermented products. Thus, molecular techniques, such as the use of species-specific oligonucleotide probes and dot blot hybridisation, are being used to determine the dominant LAB in diverse fermented foods. Similarly, molecular strain typing methods are being used to study the performance of LAB starter cultures and inoculants employed as additives in functional food type products.

The practice of medical microbiology is founded on the ability to distinguish between isolated microorganisms and to predict their pathogenicity and epidemiology from *in vitro* characteristics. This predictive capability is a product of systematic studies which have stretched over several years. This process is exemplified in Chapter 14 by Wayne with reference to mycobacterial systematics. Wayne shows how taxonomic methods and concepts that were evolving over the past fifty years, and the specific applied diagnostic needs that were also unfolding, came together to shape the direction of mycobacterial systematics. In particular, he shows how the introduction of specific chemotherapy for the treatment of tuberculosis had a marked effect on the systematic study of slowly-growing mycobacteria. Wayne makes it clear that the current healthy state of mycobacterial systematics is based on a polyphasic taxonomic approach which reconciles the interests of both makers and users. He also shows that the success of this approach owes much to a series of co-operative studies carried out under the auspices of the *International Working Group on Mycobacterial Taxonomy.*

4.5 MICROBIAL RESOURCE CENTRES AND LEGISLATION

The need to study and conserve microbial diversity is a major challenge for the successful development of the biotechnology industry and a key element in moving towards the goal of sustainable development. Microbial Resource Centres (MRCs) are an essential part of the infrastructure needed to promote this work. For both ecological understanding and biotechnologies to advance, ways need to be found to isolate, classify, conserve and catalogue a vastly greater array of microorganisms, including the structurally complex and fastidious. The scientific challenges facing MRCs are considered by Canhos and Manfio in Chapter 15. These authors note that those working in new age MRCs will be required not only to develop their current responsibilities but will need to expand their activities to address the challenges of biology in the era of bioinformational science. They also point out that MRCs have to operate within the Framework of the Convention on Biological Diversity. The activities of MRCs are also being influenced by changes in biosafety regulations for the transport of hazardous biological material.

In the final chapter, Fritze and Weihs illustrate how the assignment of microorganisms to hazard groupings within the context of risk assessments has an important role in various regulatory fields. In particular, they show that permits to work with certain types of organisms, to allow the import and export of microbial cultures, or the use of microorganisms for biotechnological purposes rely on a comprehensive and accurate systematic framework and the ability of practitioners to identify strains. They also stress how systematics underpin laws governing the use of genetically engineered microorganisms and patent, reference and quality control strains.

References

Alm, R.A. & Manning, P.A. (1990). Biotype-specific probe for *Vibrio cholerae* serogroup 01. *Journal of Clinical Microbiology* **28**, 823-824.

Benson, D.A., Boguski, M.S., Lipman, D.J., Ostell, J. & Ouellette, B.F. (1998). GenBank. *Nucleic Acids Research* **26**, 1-7.

Bousfield, I.J. (1993). Bacterial nomenclature and its role in systematics. In *Handbook of New Bacterial Systematics*, pp. 317-338. London, Academic Press Ltd.

Boysen, M., Skouboe, P., Frisvad, J. & Rossen, L. (1996). Reclassification of the *Penicillium roqueforti* group into three species on the basis of molecular genetic and biochemical profiles. *Microbiology* **142**, 541-549.

Brandt, K.K., Patel, B.K.C. & Ingvorsen, K. (1999). *Desulfocella halophila* gen. nov., sp. nov., a halophilic, fatty-acid oxidising, sulfate-reducing bacterium isolated from sediments of the Great Salt Lake. *International Journal of Systematic Bacteriology* **49**, 193-200.

Bridge, P.D., Holderness, M., Paterson, R.R.M. & Rutherford, M. (1995). Multidisciplinary characterization of fungal plant pathogens. *Bulletin OEPP/EPPO Bulletin* **25**, 125-131.

Bull, A.T., Goodfellow, M. & Slater, J.H. (1992). Biodiversity as a source of innovation in biotechnology. *Annual Review of Microbiology* **42**, 219-257.

Caetano-Anollés, G. & Gresshoff, P.M., eds. (1997). *DNA Markers : Protocols, Applications and Overviews.* New York, Wiley

Canhos, V.P., Manfio, G.P. & Blaine, L.D. (1993). Software tools and databases for bacterial systematics and their dissemination *via* global networks. *Antonie van Leeuwenhoek* **64**, 205-229.

Cavalier-Smith, T. (1992). Origins of secondary metabolism. *Ciba Foundation Symposium* **171**, 64-80.

Cavalier-Smith, T. (1993). Kingdom Protozoa and its 18 phyla. *Microbiological Reviews* **57**, 953-994.

Chatton, E. (1937). *Titres et travaux scientifiques (1906-1937) de Edouard Chatton.* Sette, Italy, E. Sottano.

Colwell, R.R. (1970). Polyphasic taxonomy of bacteria. In *Culture Collections of Microorganisms*, pp. 421-436. Edited by H. Iizuka & T. Hasegawa. Tokyo, University of Tokyo Press.

Coombs, G.H., Vickerman, K., Sleigh, M.A. & Warren, A., eds. (1998). *Evolutionary Relationships among Protozoa.* Dordrecht, Kluwer Academic Publishers.

Cowan, S.T. (1955). The principles of microbial classification. Introduction: the philosophy of classification. *Journal of General Microbiology* **12**, 314-319.

De Jonckheere, J.F. (1998). Relationships between ameoboflagellates. In: *Evolutionary Relationships among Protozoa*, pp. 181-194. Edited by G.H. Coombs, K. Vickerman, M.A. Sleigh & A. Warren. Dordrecht, Kluwer Academic Publishers.

Eardly, B.D., Wang, F.S. & van Berkum, P. (1996). Corresponding 16S rRNA gene segments in *Rhizobiaceae* and *Aeromonas* yield discordant phylogenies. *Plant and Soil* **186**, 69-74.

Foissner, W. (1998). The karyorelictids (Protozoa:Ciliophora), a unique and enigmatic assemblage of marine, intestinal ciliates: a review emphasizing ciliary patterns and evolution. In *Evolutionary Relationships among Protozoa*, pp. 304-325. Edited by G.H. Coombs, K. Vickerman, M.A. Sleigh & A. Warren. London, Chapman and Hall.

Fox, G.E., Pechman, K.G. & Woese, C.R. (1977). Comparative cataloguing of 16S ribosomal ribonucleic molecular approach to prokaryotic systematics. *International Journal of Systematic Bacteriology* **27**, 44-57.

Fox, G.E., Wisotzkey, J.D. & Jurtshuk, Jr. P. (1992). How close is close: 16S rRNA sequence identity may not be sufficient to guarantee species identity. *International Journal of Systematic Bacteriology* **42**, 166-170.

Fry, N.R., Saunders, N.A., Warwick, S. & Embley, T.M. (1991). The use of 16S ribosomal RNA analyses to investigate the phylogeny of the family *Legionellaceae. Journal of General Microbiology* **137**, 1215-1222.

Gilmour, J.S.L. (1940). Taxonomy and philosophy. In *The New Systematics*, pp. 461-474. Oxford, Clarendon Press.

Goodacre, R., Timmins, E.M., Burton, R., Kaderbhai, N., Woodward, A.M., Kell, D.B. & Rooney, P.J. (1998). Rapid identification of urinary tract infection bacteria using hyperspectral whole-organism fingerprinting and artificial neural networks. *Microbiology* **144**, 1157-1170.

Goodfellow, M. & Dickinson, C.H. (1985). Delineation and description of microbial populations using numerical methods. In *Computer-Assisted Bacterial Systematics*, pp. 165-225. Edited by M. Goodfellow, D. Jones & F.G. Priest. London, Academic Press.

Goodfellow, M. & Magee, J.G. (1998). Taxonomy of mycobacteria. In *Mycobacteria*. I. *Basic Aspects*, pp. 1-71. Edited by P.R.J. Gangadharam & P.A. Jenkins. New York, Chapman & Hall.

Goodfellow, M. & O'Donnell, A.G. (1993). The roots of bacterial systematics. In *Handbook of New Bacterial Systematics*, pp. 3-54. Edited by M. Goodfellow & A.G. O"Donnell. London, Academic Press Ltd.

Goodfellow, M. & O'Donnell, A.G. eds. (1994). *Chemical Methods in Prokaryotic Systematics*. Chichester, John Wiley & Sons.

Goodfellow, M., Chun, J., Atalan, E. & Sanglier, J-J. (1994). Curie point pyrolysis mass spectrometry and its application to bacterial systematics. In *Bacterial Diversity and Systematics*, pp. 87-104. Edited by F.G. Priest, A. Ramos-Cormenzana & B.J. Tindall. New York, Plenum Press.

Goodfellow, M., Freeman, R. & Sisson, P.R. (1997a). Curie-point pyrolysis mass spectrometry as a tool in clinical microbiology. *Zentralblatt für Bakteriologie* **285**, 133-156.

Goodfellow, M., Manfio, G.P. & Chun, J. (1997b). Towards a practical species concept for cultivable bacteria. In *Species: The Units of Biodiversity*, pp. 25-59. Edited by M.F. Claridge, H.A. Dawah & M.R. Wilson. London, Chapman & Hall.

Goodfellow, M., Alderson, G. & Chun, J. (1998). Rhodococcal systematics: problems and developments. *Antonie van Leeuwenhoek* **74**, 1-18.

Greuter, W., Barrie, F.R., Burdet, H.M., Chaloner, W.G., Demoulin, V., Hawksworth, D.L., Jfragensen, P.M., Nicholson, D.H., Silva, D.C., Trehane, P. & McNeil, J. (1994). *International Code of Botanical Nomenclature. Renum Vegetable* **131**, 1-389.

Gupta, R.S. (1998). Protein phylogenies and signature sequences: A reappraisal of evolutionary relationships among archaebacteria, eubacteria and eukaryotes. *Microbiology and Molecular Biology Reviews* **62**, 1435-1491.

Gürtler, V. & Stanisich, V.A. (1996). New approaches to typing and identification of bacteria using the 16S-23S rDNA spacer region. *Microbiology* **142**, 3-16.

Hillis, D.M., Allard, W. & Miyamoto, M.M. (1993). Analysis of DNA sequence data : phylogenetic inference. *Methods of Enzymology* **224**, 456-487.

Isik, K., Chun, J., Hah, Y.C. & Goodfellow, M. (1999). *Nocardia salmonicida* nom. rev., a fish pathogen. *International Journal of Systematic Bacteriology* **49**, 833-837.

James, A.L. (1994). Enzymes in taxonomy and diagnostic bacteriology. In *Chemical Methods in Prokaryotic Systematics*, pp. 471-492. Edited by M. Goodfellow & A.G. O'Donnell. Chichester, John Wiley & Sons Ltd.

Kim, S.B. & Goodfellow, M. (1999). Reclassification of *Amycolatopsis rugosa* Lechevalier *et al.* 1986 as *Prauserella rugosa* gen. nov., comb. nov. *International Journal of Systematic Bacteriology* **49**, 507-512.

Kim, B., Sahin, N., Minnikin, D.E., Zakrzewska-Czerwinska, J., Mordarski, M. & Goodfellow, M. (1999). Classification of thermophilic streptomycetes, including the description of *Streptomyces thermoalkalitolerans* sp. nov. *International Journal of Systematic Bacteriology* **49**, 7-17.

Kim, W-S., Gardan, L., Rhum, S-L. & Geider, K. (1999). *Erwinia pyrifoliae* sp. nov., a novel pathogen that affects Asian pear trees (*Pyrus pyrifolia* Nakai). *International Journal of Systematic Bacteriology* **49**, 899-906.

Kurtzman, C.P. (1992). rRNA sequence comparisons for assessing phylogenetic relationships among yeasts. *International Journal of Systematic Bacteriology* **42**, 1-6.

Labeda, D.P. (1998). DNA relatedness among the *Streptomyces fulvissimus* and *Streptomyces griseoviridis* phenotypic cluster groups. *International Journal of Systematic Bacteriology* **48,** 829-832.

Lapage, S.P., Sneath, P.H.A., Lessel, E.F., Skerman, V.B.D., Seeliger, H.P.R. & Clark, W.A. (1975). *International Code of Nomenclature of Bacteria, 1975 Revision.* Washington, D.C., American Society for Microbiology.

Larsen, T.O. & Frisvad, J.C. (1995). Characterisation of volatile metabolites from 47 Penicillium taxa. *Mycological Research* **99,** 1153-1166.

Ludwig, W., Neumaier, J., Klugbauer, N., Brockmann, E., Roller, C., Jilg, S., Reetz, K., Schachtner, I., Ludvigsen, A., Bachleitner, M., Fischer, U. & Schleifer, K.H. (1993). Phylogenetic relationships of bacteria based on comparative sequence analysis of elongation factor Tu and ATP-synthase β-subunit genes. *International Journal of Systematic Bacteriology* **64,** 285-305.

MacAdoo, T.O. (1993). Nomenclatural literacy. In *Handbook of New Bacterial Systematics*, pp. 339-358. Edited by M. Goodfellow & A.G. O'Donnell. London, Academic Press.

Magee, J.T. (1993). Whole-organisms fingerprinting. In *Handbook of New Bacterial Systematics*, pp. 383-427. Edited by M. Goodfellow & A.G. O"Donnell. London, Academic Press Ltd.

Magee, J.T. (1998). Taxonomy and nomenclature of bacteria. In *Topley and Wilson's, Microbiology and Microbial Infections, Volume 2, Systematic Bacteriology*, pp. 45-64. Edited by A. Balows & B.I. Duerden. London, Arnold.

Maidak, B.L., Olsen, G.J., Larsen, N., Overbeek, R., McCaughey, M.J. & Woese, C.R. (1997). The Ribosomal Database Project (RDP). *Nucleic Acids Research* **5,** 109-111.

Margulis, L. & Guerro, R. (1991). Kingdoms in turmoil. *New Scientist* **129,** 46-50.

Martinez-Murcia, A.J., Benlloch, S. & Collins, M.D. (1992). Phylogenetic interrelationships of members of the genera *Aeromonas* and *Plesiomonas* as determined by 16S ribosomal DNA sequencing: Lack of congruence with results of DNA:DNA hybridisation. *International Journal of Systematic Bacteriology* **42,** 412-421.

Mayr, E. (1990). A natural system of organisms. *Nature* **348,** 491.

Mayr, E. (1998). Two empires or three. *Proceedings of the National Academy of Science, USA* **95,** 9720-9723.

Moore, W.E.C. & Moore, L.V.D., eds. (1989). *Index of the Bacterial and Yeast Nomenclatural Changes Published in the International Journal of Systematic Bacteriology since the 1980 Approved Lists of Bacterial Names (1 January 1980 to 1 January 1989).* Washington, D.C., American Society for Microbiology.

Mordue, J.E.M., Banniza, S., Bridge, P.D., Rutherford, M.A. & Holderness, M. (1996). Integrated biochemical, cultural and numerical methods. In: *Rhizoctonia species : Taxonomy, Molecular Biology, Ecology, Pathology and Disease Control.* pp. 87-98. Edited by B. Sneh, S. Jabaji-Hare, S. Neate & G. Dijst. London, Kluwer Academic Publishers.

Murray, R.G.E., Brenner, D.J., Colwell, R.R., De Vos, P., Goodfellow, M., Grimont, P.A.D., Pfennig, N., Stackebrandt, E. & Zavarzin, G.A. (1990). Report of the *ad hoc* committee on approaches to taxonomy within the *Proteobacteria*. *International Journal of Systematic Bacteriology* **40,** 213-215.

Naumann, D., Helm, D. & Schultz, C. (1994). Characterization and identification of micro-organisms by FT-IR spectroscopy and FT-IR microscopy. In *Bacterial Diversity and Systematics*, pp. 67-85. Edited by F.G. Priest, A. Ramos-Cormenzana & B.J. Tindall. New York, Plenum Press.

Nielsen, P., Fritze, D. & Priest, F.G. (1995). Phenetic diversity of alkaliphilic *Bacillus* strains: proposal for nine new species. *Microbiology* **141,** 1745-1761.

Olsen, G.J. & Woese, C.R. (1993). Ribosomal RNA: A key to phylogeny. *FASEB Journal* **7,** 113-123.

Palleroni, N.J. (1993). *Pseudomonas* classification. A new case history in the taxonomy of Gram-negative bacteria. *Antonie van Leeuwenhoek* **64,** 231-251.

Philippe, H. (1998). Molecular phylogeny of kinetoplastids. In *Evolutionary Relationships among Protozoa,* pp. 195-212. Edited by G.H. Coombs, K. Vickerman, M.A. Sleigh & A. Warren. Dordrecht, Kluwer Academic Publishers.

Priest, F.G. & Austin, B., eds. (1993). *Modern Bacterial Taxonomy.* London, Chapman and Hall.

Prosser, J.I. (1994). Molecular marker systems for detection of genetically engineered micro-organisms in the environment. *Microbiology* **140**, 5-17.

Ride, W.D.L., Sabrosky, C.W., Bernardi, G. & Melville, R.V., eds. (1985). *International Code of Zoological Nomencalture.* Berkeley and Los Angeles, University of California Press.

Rosa, C.A., Lachance, M-A., Starmer, W.T., Barker, J.S.F., Bowles, J.M. & Schlag-Edler, B. (1999). *Kodamaea nitidulidarum, Candida restingae* and *Kodamaea anthophila,* three new related yeast species from ephemeral flowers. *International Journal of Systematic Bacteriology* **49**, 309-318.

Rosselló, R., García-Valdéz, J., Lalucat, J. & Ursing, J. (1991). Genotypic and phenotypic diversity of *Pseudomonas stutzeri. Systematic and Applied Microbiology* **8**, 124-127.

Sackin, M.J. & Jones, D. (1993). Computer-assisted classification. In *Handbook of New Bacterial Systematics,* pp. 281-313. London, Academic Press Ltd.

Simpson, G.G. (1961). *Principles of Animal Taxonomy.* New York, Columbia University Press.

Skerman, V.B.D., McGowan, V. & Sneath, P.H.A. (1980). Approved lists of bacterial names. *International Journal of Systematic Bacteriology* **30**, 225-420.

Sneath, P.H.A. (1957a). Some thoughts on bacterial classification. *Journal of General Microbiology* **17**, 184-200.

Sneath, P.H.A. (1957b). The applications of computers to taxonomy. *Journal of General Microbiology* **17**, 201-226.

Sneath, P.H.A. (1986). Nomenclature of bacteria. In *Biological Nomenclature Today IUBS Monograph Series No. 2,* pp. 38-48. Edited by W.D.L. Ride and T. Younes. Eynsham, IRL Press.

Sneath, P.H.A. (1988). The phenetic and cladistic approaches. In *Prospects in Systematics,* pp. 252-273. Edited by D.L. Hawksworth. Oxford, Clarendon Press.

Sneath, P.H.A. (1989). Predictivity in taxonomy and the probability of a tree. *Plant Systematics and Evolution* **167**, 43-57.

Sneath, P.H.A., ed. (1992). *International Code of Nomenclature of Bacteria, 1990 Revision.* Washington, D.C., American Society for Microbiology.

Sneath, P.H.A. (1993). Evidence from *Aeromonas* for genetic crossing-over in ribosomal sequences. *International Journal of Systematic Bacteriology* **43**, 626-629.

Sokal, R.R. (1985). The principles of numerical taxonomy: Twenty-five years later. In *Computer-Assisted Bacterial Systematics,* pp. 1-20. Edited by M. Goodfellow, D. Jones & F.G. Priest. London, Academic Press Ltd.

Stackebrandt, E. & Goebel, B.M. (1994). Taxonomic note: a place for DNA:DNA reassociation and 16S rRNA sequence analysis in the present species definition in bacteriology. *International Journal of Systematic Bacteriology* **44**, 846-849.

Stackebrandt, E., Tindall, B., Ludwig, W. & Goodfellow, M. (1999). Diversity and systematics. In *Biology of the Prokaryotes,* pp. 674-720. Edited by J.W. Lengeler, G. Drews & H.G. Schlegel. Stuttgart, Thieme.

Stanier, R.Y. & van Niel, C.B. (1941). The main outlines of bacterial classification. *Journal of Bacteriology* **42**, 437-466.

Stead, D.E., Sellwood, J.E., Wilson, J. & Viney, I. (1992). Evaluation of a commercial microbial identification system based on fatty acid profiles for rapid, accurate identification of plant pathogenic bacteria. *Journal of Applied Bacteriology* **72**, 315-321.

Stroesser, G., Tuli, M.A., Lopez, R. & Sterk, P. (1999). The EMBL nucleotide sequence database. *Nucleic Acids Research* **27**, 18-24.

Suh, S-O. & Nakase, T. (1995). Phylogenetic analysis of the ballistosporous anamorphic genera *Udeniomyces* and *Bullera*, and related basidiomycetous yeasts, based on 18S rDNA sequence. *Microbiology* 141, 901-906.

Suzuki, T., Muroga, Y., Takahama, M., Shiba, T. & Nishimura, Y. (1999). *Roseovivax halodurans* gen. nov., sp. nov. and *Roseovivax halotolerans* sp. nov., aerobic bacteriochlorophyll-containing bacterium isolated from a saline lake. *International Journal of Systematic Bacteriology* 49, 629-634.

Takashima, M. & Nakase, T. (1998). *Bullera penniseticola* sp. nov. and *Kockovaella sacchari* sp. nov., two new yeast species isolated from plants in Thailand. *International Journal of Systematic Bacteriology* 48, 1025-1030.

Tateno, Y., Fukami-Kobayashe, K., Miyazaki, S., Sugawara, H. & Gojobori, T. (1998). DNA Data Bank of Japan at work on genome sequence data. *Nucleic Acids Research* 26, 16-20.

Towner, K.J. & Cockayne, A. (1993). *Molecular Methods for Microbial Identification and Typing*. London, Chapman & Hall.

Ursing, J.B., Rosselló-Mora, R.A., Garcia-Valdés, E. & Lalucat, J. (1995). Taxonomic note: a pragmatic approach to the nomenclature of phenotypically similar genomic groups. *International Journal of Systematic Bacteriology* 45, 604.

Van Belkum, A., Scherer, S., Loek, van A. & Verbrugh, H. (1998). Short-sequence DNA repeats in prokaryotic domains. *Microbiology and Molecular Biology Reviews* 62, 275-293.

Vandamme, P., Pot, B., Gillis, M., De Vos, P., Kersters, K. & Swings, J. (1996). Polyphasic taxonomy, a consensus approach to bacterial systematics. *Microbiological Reviews* 60, 407-438.

Vauterin, L., Swings, J. & Kersters, K. (1993). Protein electrophoresis and classification. In *Handbook of New Bacterial Systematics*, pp. 251-280. London, Academic Press Ltd.

Wayne, L.G., Brenner, D.J., Colwell, R.R., Grimont, P.A.D., Kandler, P., Krichevsky, M.I., Moore, L.H., Moore, W.E.C., Murray, R.G.E., Stackebrandt, E., Starr, M.P. & Trüper, H.G. (1987). Report of the *ad hoc* committee on reconciliation of approaches to bacterial systematics. *International Journal of Systematic Bacteriology* 37, 463-464.

Wayne, L.G., Good, R.C., Böttger, E.C., Butler, R., Dorsch, M., Ezaki, T., Gross, W., Jones, V., Kilburn, J., Kirschner, P., Krichevsky, M.I., Ridell, M., Shinnick, T.M., Springer, B., Stackebrandt, E., Tárnok, Z., Tasaka, H., Vincent, V., Warren, N.G., Knott, C.A. & Johnson, R. (1996). Semantide- and chemotaxonomy-based analyses of some problematic phenotypic clusters of slowly growing mycobacteria, a cooperative study of the International Working Group on Mycobacterial Taxonomy. *International Journal of Systematic Bacteriology* 46, 280-297.

White, D., Sharp, R.J. & Priest, F.G. (1993). A polyphasic taxonomic study of thermophilic bacilli from a wide geographic area. *Antonie van Leeuwenhoek* 64, 357-386.

Williams, S.T., Goodfellow, M. & Alderson, G. (1989). Genus *Streptomyces* Waksman and Henrici 1943, 339[AL]. In *Bergey's Manual of Systematic Bacteriology, Volume 4*, pp. 2452-2492. Baltimore, Williams and Wilkins.

Woese, C.R. (1987). Bacterial evolution. *Microbiological Reviews* 51, 221-271.

Woese, C.R. & Fox, G.E. (1977). Phylogenetic structure of the prokaryotic domain: the primary kingdoms. *Proceedings of the National Academy of Sciences, USA* 74, 5088-5090.

Woese, C.R., Kandler, O. & Wheelis, M.L. (1990). Towards a natural system of organisms: Proposal for the domains Archaea, Bacteria, and Eucarya. *Proceedings of the National Academy of Sciences, USA* 87, 4576-4579.

Yamamoto, S. & Harayama, S. (1996). Phylogenetic analysis of *Acinetobacter* strains based on the nucleotide sequences of *gyr* B genes and on the amino acid sequences of their products. *International Journal of Systematic Bacteriology* 46, 506-511.

Yassin, A.F., Rainey, F.A., Burghardt, J., Brzezinka, H., Schmitt, S., Seifert, P., Zimmermann, O., Mauch, H., Gierth, D., Lux, I. & Schaal, K.P. (1997). *Tsukamurella tyrinosolvens* sp. nov. *International Journal of Systematic Bacteriology* 47, 607-614.

II PHYLOGENY

2 UNIVERSAL TREES

DISCOVERING THE ARCHAEAL AND BACTERIAL LEGACIES

JAMES R. BROWN *and* KRISTIN K. KORETKE

*Microbial Bioinformatics Group, Department of Bioinformatics,
SmithKline Beecham Pharmaceuticals, 1250 S. Collegeville Road,
PO Box 5089, Collegeville PA 19426-0989, USA*

1   Introduction

The motivations behind the construction of universal trees, phylogenies which are inclusive of all living organisms, are two-fold. First, there is the rational classification of life based on evolutionary principles and relationships. Second, universal trees provide a framework to help better understand the patterns and processes involved in early cellular evolution. However, even a cursory survey of the literature reveals that the universal tree topologies are as diverse as the molecular markers used, either ribosomal RNAs (rRNAs) or proteins. Does this mean that distant evolutionary events can not be accurately reconstructed using presently available methods of phylogenetic analysis? This might be true in some cases where extreme variations in the rate of evolutionary change between different species can confound the use of certain proteins and rRNAs as molecular markers.

Conversely, the different universal tree topologies may only appear to be contradictory because we are hindered by inadequate data, an incomplete understanding of cellular processes or a combination of both. The data issue is now being dramatically addressed by the remarkable generation of complete genome sequences from many diverse groups of organisms, although ever reaching a plateau in our knowledge of genome diversity is still highly unlikely. Mining these databases in search of new evolutionary patterns will be a major challenge over the next several years, perhaps decades.

The conduct of this search will no longer be limited to specialists in evolution. Combined with the widespread use of phylogenetic tools, molecular evolutionary analysis is now being broadly applied in a number of new and diverse areas of study (as witnessed by the contributions to this volume). In nearly all environments or hosts studied, *in situ* PCR amplified rRNA sequences have revealed many new bacterial species, previously undetected using traditional *in vitro* culture methods. Phylogenetic analysis of PCR amplified

F. G. Priest and M. Goodfellow (eds.), Applied Microbial Systematics, 19-55
© 2000 *Kluwer Academic Publishers. Printed in the Netherlands.*

rRNA molecules is being used to identify new clinical isolates of pathogenic bacteria and to catalogue microbial species diversity from different environments. The annotation of new bacterial genomes will increasing rely upon phylogenetic analysis for the prediction of the structure, and ultimately biochemical function, of novel proteins encoded by potential open reading frames (ORFs). In the war against infectious diseases, phylogenetic analysis can be used to trace the evolutionary history of those genes responsible for antibiotic resistance and virulence as well as to help direct the rational design of drugs against new bacterial targets.

Underlying any evolutionary analysis is the commonality of life whereby all genetic histories are ultimately linked by descent from a common cellular ancestor which existed billions of years ago. Therefore, it is fundamental for any worker applying the principles and tools of evolutionary biology to have some exposure to the concept and controversies surrounding the construction of the universal tree and the conceptualization of the last common ancestor or cenancestor (Fitch & Upper, 1987). This is the purpose of this chapter.

2   The Archaea, Bacteria and Eucarya

Over 50 years ago, Chatton (1937) and Stanier and van Niel (1941) suggested that life could be subdivided into two fundamental cellular categories, prokaryotes and eukaryotes (summarized in Doolittle & Brown, 1994). The distinction between the two groups was subsequently refined as studies of cellular biology and genetics progressed such that prokaryotes became universally distinguishable from eukaryotes on the basis of missing internal membranes (such as the nuclear membrane and endoplasmic recticulum), nuclear division by fission rather than mitosis and the presence of a cell wall (Stanier & van Niel, 1962; Stanier, 1970). The definition of eukaryotes was broadened to include the endosymbiont hypothesis which describes how eukaryotes improved their metabolic capacity by engulfing certain prokaryotes and converting them into intracellular organelles, principally mitochondria and chloroplasts (Margulis, 1970).

In the late 1970s the fundamental belief in the prokaryote-eukaryote dichotomy was shattered by the work of Carl Woese, George Fox, and co-workers. By digesting *in vivo* labelled 16S rRNA using T1 ribonuclease then cumulating and comparing catalogues of the resultant oligonucleotide "words", Woese's group was able to derive dendrograms showing the relationships between different bacterial species. Analyses involving some unusual methanogenic bacteria revealed surprising and unique species clusterings among prokaryotes (Fox *et al.*, 1977). So deep was the split in the prokaryotes that Woese & Fox (1977a) proposed to call the methanogens and their relatives "archaebacteria", a name which relayed their distinctness from the true bacteria or "eubacteria" as well as contemporary preconceptions that these organisms might have thrived in the environmental conditions of a younger Earth.

In 1990, Woese, Kandler & Wheelis formally proposed the replacement of the bipartite view of life with a new tripartite scheme based on three urkingdoms or domains; the Bacteria (formally eubacteria), Archaea (formally archaebacteria) and Eucarya (eukaryotes, still the more often used name) [Fig. 1]. The rationale behind this revision came from a growing body of biochemical, genomic and phylogenetic evidence which, when viewed collectively, suggested that the archaebacteria were worthy of a taxonomic status equal to that of eukaryotes and eubacteria. While there was wide acceptance of this re-classification by most archaebacteriologists, there was serious dissent, most notably expressed by Mayr (1990), Margulis & Guerrereo (1991), and Cavalier-Smith (1992) who challenged the elevation of the archaebacteria (and hence the eubacteria) to a taxonomic rank equal that of eukaryotes (reviewed in Doolittle & Brown 1994; Brown & Doolittle, 1997). At the centre of the controversy surrounding the concept of the three domains, are the archaebacteria and their degree of uniqueness from the eubacteria. Therefore, it is appropriate to briefly review the biology of these interesting organisms.

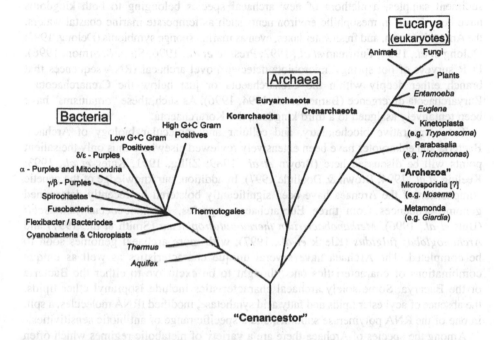

Figure 1. Schematic drawing of a universal rRNA tree showing the relative positions of evolutionary pivotal groups in the domains Bacteria, Archaea, and Eucarya (eukaryotes). In this tree, branching orders are based on Cavalier-Smith (1983) and Olsen et al. (1994) but branch lengths are not meaningful. The location of the root (the cenancestor) corresponds with that proposed by reciprocally rooted gene phylogenies (Gogarten et al., 1989; Iwabe et al., 1989; Brown & Doolittle, 1995). The question mark beside the archezoan group microsporidia denotes recent suggestions that it might branch higher in the eukaryotic portion of the tree, perhaps with the fungi (Keeling & McFadden 1998).

3   What are the Archaea?

3.1   THE ARCHAEA ALONE

According to rRNA trees, there are two groups within the Archaea: the kingdoms Crenarchaeota and Euryarchaeota (Woese, 1987). The Crenarchaeota are generally hyperthermophiles or thermoacidophiles (some genera are *Desulfurococcus, Pyrodictium, Sulfolobus, Thermofilum* and *Thermoproteus*). The Euryarchaeota span a broader ecological range and include hyperthermophiles (some genera are *Pyrococcus* and *Thermococcus*), methanogens (one genus is *Methanosarcina*), halophiles (some genera are *Halobacterium* and *Haloferax*), and even thermophilic-methanogens (some genera are *Methanobacterium, Methanococcus* and *Methanothermus*). However, it is important to note that microbial species assemblages in extreme environments are not exclusively archaeal as bacteria-specific rRNA signatures can also be amplified from such sites. In addition, through polymerase chain reaction (PCR) amplification of rRNA sequences from water and sediment samples, a plethora of new archaeal species belonging to both kingdoms have been found in mesophilic environments such as temperate marine coastal waters, the Antarctic Ocean, and freshwater lakes, even as marine sponge symbionts (Delong, 1992; Delong *et al.*, 1994; Fuhrman *et al.*, 1992; Preston *et al.*, 1996; Stein & Simon, 1996). PCR surveys of hot springs' microbiota detected novel archaeal rRNA sequences that branch either deeply within the Crenarchaeota or just below the Crenarchaeota - Euryarchaeota divergence (Barnes *et al.*, 1994, 1996). As such, these "organisms" have been tentatively assigned to a third kingdom, the Korarchaeota.

The comparative biochemistry, and cellular and molecular biology of Archaea, Bacteria and eukaryotes have been extensively reviewed elsewhere thus only the salient points will be discussed here (Brown *et al.*, 1989; Zillig, 1991, Kates *et al.*, 1993; Keeling *et al.*, 1994; Brown & Doolittle 1997). In addition, our knowledge of the genetic composition of the Archaea have been significantly bolstered by recently determined genome sequences from three Euryarchaeota species, *Methanococcus jannaschii* (Bult *et al.*, 1996), *Methanobacterium thermoautotrophicum* (Smith *et al.*, 1997) and *Archaeoglobus fulgidus* (Klenk *et al.*, 1997), with more archaeal genomes soon to be completed. The Archaea have several unique characteristics as well as unique combinations of characteristics once thought to be exclusive to either the Bacteria or the Eucarya. Some solely archaeal characteristics include isopranyl ether lipids, the absence of acyl ester lipids and fatty acid synthetase, modified tRNA molecules, a split in one of the RNA polymerase subunits, and a specific range of antibiotic sensitivities.

Among the species of Archaea there are a variety of metabolic regimes which often differ greatly from the better known metabolic pathways of Bacteria and eukaryotes (reviewed in Danson, 1993). For example, both ATP-dependent and pyrophosphate dependent phosphofructokinases can occur in Bacteria and eukaryotes while Archaea use either ADP-dependent or pyrophosphate linked kinases. Hexokinase is ATP-dependent in Bacteria, eukaryotes and the archaeon *Thermoproteus* but ADP-linked in *Pyrococcus* (Kengen *et al.*, 1994). The conversion of pyruvate to acetate, a reaction which bridges glycolytic and citric acid cycles, is catalysed by a pyruvate dehydrogenase multienzyme complex in Bacteria and eukaryotes whereas Archaea and some anaerobic eukaryotes (*Entamoeba, Giardia* and *Trichomonas*) appear to employ pyruvate:ferredoxin oxidore-ductase (Danson, 1993; Schönheit & Schäfer, 1995). However, one component of the pyruvate dehydrogenase complex, dihydrolipoamide dehydrogenase, has been detected in halophilic Archaea although its function is unclear (Danson *et al.*, 1984).

3.2   ARCHAEA AND BACTERIA

Archaea and Bacteria are united in the "realm of prokaryotes" by generally similar cell sizes, a lack of a nuclear membrane and organelles, and the presence of a large circular chromosome occasionally accompanied by one or more smaller circular DNA plasmids (Charlebois *et al.*, 1991). Although the origin of DNA replication of any archaeal large chromosome has yet to be confirmed experimentally, there is little suggestion for a significant departure from the bacterial model of a single replication initiation site (Forterre & Elie, 1993; Edgell & Doolittle, 1997). The close sequence similarity between archaeal and bacterial topoisomerases and gyrases provides further, if indirect, evidence of comparable chromosome structure among the two groups.

Many archaeal genes appear to be organized into Bacteria-like operons. Furthermore, many archaeal operons and gene clusters are arranged in a similar fashion to those of Bacteria (reviewed Ramirez *et al.*, 1993; Keeling *et al.*, 1994). As an example, rRNA operons in Bacteria and chloroplasts are arranged in the order 16S-23S-5S. Archaea have the same operon organization although there is some variation, such as the distal location of the 5S rRNA gene in some thermoacidophiles and methanogens (reviewed in Brown *et al.*, 1989). Of the 11 genes in the *E. coli* spectinomycin (*spc*) operon, the same order occurs for nine and eleven genes in *Sulfolobus acidocaldarius* and *Methanococcus vanniellii*, respectively (Auer *et al.*, 1989; Ramirez *et al.*, 1993). Both species have three small additional open reading frames (ORFs) within their *spc* operons.

Like Bacteria, archaeal mRNAs do not have 5' end caps and often have Shine-Dalgarno ribosome binding sites (reviewed in Amils *et al.*, 1993). However, the locations of putative Shine-Dalgarno sequences relative to the translational initiation codon are more variable in Archaea, and the upstream sequences of several highly expressed genes bear little resemblance to Shine-Dalgarno motifs (see Figure 6 in Ramirez *et al.*, 1993).

Other features shared between the Archaea and Bacteria include Type II restriction enzyme systems (Nolling & Vos, 1992), the absence of splicesomal introns, and the presence of homing endonucleases typical of group I introns found in mitochondria and bacteriophages (Brown *et al.*, 1989; Belfort *et al.*, 1995). Inteins are unusual introns spliced at the protein rather than the mRNA level. They were first reported in the catalytic subunit of yeast vacuolar type ATPase (Hirata *et al.*, 1990; Kane *et al.*, 1990). The first archaeal intein was found in the DNA polymerase gene of *Thermococcus litoralis* (Perler *et al.*, 1992), and several novel intein insertions were reported in the whole genome sequence of *Methanococcus jannaschii* (Bult *et al.*, 1996).

The bacterial cell division protein FtsZ has been discovered in several species of Archaea (Baumann & Jackson 1996; Wang & Lutkenhaus, 1996). FtsZ is thought to be a distant homologue to eukaryotic tubulins since both proteins are GTPases that polymerize into filaments in the presence of GTP (Erickson, 1995). The plant, *Arabidopsis thaliana*, has a nuclear-encoded FtsZ homolog which is directed to the chloroplast (Osteryoung & Vierling, 1995). Phylogenetic analysis also supports the notion that FtsZ is a distant homologue to tubulin, with the FtsZ of Archaea and Bacteria forming distinct yet closely related groups (Baumann & Jackson, 1996). Several other cell division proteins were identified in the *M. jannaschii* genome but not the entire suite of genes known to function in cell septation or chromosome partitioning in Bacteria (Bult *et al.*, 1996).

3.3   ARCHAEA AND EUKARYOTES

While cell division in the Archaea might function as in Bacteria, many components of DNA replication, transcription and translation are definitely more eukaryote-like. Hints of genetic homology among Archaea and eukaryotes were found in early studies using antibiotics (summarized in Amils *et al.*, 1993). Archaea and eukaryotes are both sensitive to aphidicolin, an inhibitor of DNA polymerase, to which Bacteria are refractory (reviewed in Forterre & Elie, 1993). Bacteria are sensitive to streptomycin, an anti-70S ribosome directed inhibitor. Archaea and eukaryotes are both refractory to streptomycin but are sensitive to certain anti-80S ribosome directed inhibitors (such as anisomycin).

Later studies provided proof of significant similarities between archaeal and eukaryotic DNA replication, transcriptional, and translational components. All known archaeal DNA polymerases belong to the eukaryotic family B type which are absent from bacterial species (Edgell & Doolittle, 1997). In fact, suprisingly few DNA replication proteins are homologous across all three domains although the process of DNA strand elongation is functionally similar. Many DNA replication/repair proteins which are homologous among Archaea and eukaryotes are absent in Bacteria. The *Methanococcus jannaschii* genome revealed several putative homologues to proteins involved with the replication factor complex (rfc) in eukaryotes (Bult *et al.*, 1996). Flap endonuclease I (FEN-1) and Rad2 are DNA repair enzymes encoded by duplicate genes in eukaryotes to which there are single gene homologues in Archaea but not in Bacteria (DiRuggiero *et al.*, 1999). Archaea have close counterparts to eukaryotic Dmc1 and Rad51 proteins which, in turn, are distantly related to bacterial RecA proteins and serve similar functions in homologous recombination, DNA repair and SOS response (Sandler *et al.*, 1996; Brendel *et al.*, 1997; Rashid *et al.*, 1997; DiRuggiero *et al.*, 1999). Another highly divergent, yet universal, gene family encodes for NTP binding proteins involved in chromosome condensation and DNA recombination and repair (Elie *et al.*, 1997). In *M. jannaschii* and *S. solfataricus*, there is a homologue to the product *pelota* which controls early stages of meiosis and mitosis in the fruitfly, *Drosophila melanogaster* (Ragan *et al.*, 1996).

Transcriptional components are also strikingly similar between eukaryotes and the Archaea. Archaeal RNA polymerases (RNAP) are evolutionary closer to those of eukaryotes (Pühler *et al.*, 1989; Zillig *et al.*, 1993). Bacterial RNAPs have a simple structure comprised of only four major subunits while eukaryotic and archaeal RNAPs have a minimum of seven homologous subunits (Gropp *et al.*, 1986; Zillig *et al.*, 1993; Langer *et al.*, 1995). Phylogenetic trees using the sequences for either large RNAP subunits strongly show eukaryotic and archaeal genes as close relatives (Pühler *et al.*, 1989; Zillig *et al.*, 1991; Klenk *et al.* 1993).

Archaea and eukaryotes also share other features of transcription that are apparently absent in Bacteria. The typical eukaryotic promoter consists of a TATA box sequence located about -30 bp upstream of the transcriptional start nucleotide. The eukaryotic core RNA polymerase does not contact the DNA template strand directly at this site, rather the enzyme attaches to a specific transcription factor, TFIID, bound to the DNA strand. The fully assembled initiation complex of RNA polymerase II consists of at least five transcription factors, TFIIA, TFIIB, TFIID, TFIIE and TFIIF, with a sixth factor, TFIIS, binding to the RNA polymerase once strand elongation commences (RNA polymerases I, II and III have specific suites of TFs that are numbered accordingly). TFIID is a multimeric protein which includes the important TATA-binding protein or TBP. TBP is a general transcription factor insofar that it appears to be required for the initiation of transcription of all RNA polymerase II transcribed genes, including those without a recognizable TATA box, as well as genes transcribed by RNA polymerases I and III.

Reiter *et al.* (1990) demonstrated the importance of TATA box-like upstream sequences in the transcription of the 16S/23S rRNA gene in *Sulfolobus*. Later, archaeal homologues of eukaryotic TFIIB, TFIIS and TBP were identified. A TFIIB homologue in *Pyrococcus woesei* was discovered via a database search (Ouzounis & Sander, 1992) and was subsequently cloned from *Sulfolobus shibatae* (Qureshi *et al.*, 1995). The TBP major component of TFIID was found in *Thermococcus celer* (Marsh *et al.*, 1994) and *P. woesei*, where it was elegantly demonstrated to be functional in transcription (Rowlands *et al.*, 1994). Later, it was demonstrated that yeast and human TBPs can substitute for native transcription factors in cell-free archaeal transcription systems (Wettach *et al.*, 1995; Thomm, 1996). However, 5' capping, persistent poly-A tailing, monocistronic messages and splicesomal introns are features unique to eukaryotic mRNA, and were likely derived after any eukaryote / Archaea divergence.

Similar to the situation in transcription, the Archaea appear to have several eukaryote-like translation initiation factors (IF). In *S. acidocaldarius* (Keeling *et al.*, 1996), and the three completed archaeal genomes, an IF-2 homolog has been found which is more similar to eukaryotic than bacterial IF-2 proteins. All IF-2 proteins appear to belong to a single gene family which like EF-G/2, self-recycle GTP (Keeling & Doolittle, 1995; Keeling *et al.*, 1996; Cousineau *et al.*, 1997). All the self-recycling GTP proteins belong to a larger multi-gene family which includes the EF-1α/Tu and eIF-2α; proteins that depend upon an external guanine nucleotide exchange factor for GTP recycling. At least 11 different translation initiation factor proteins have been identified from *M. jannaschii*, of which three match eukaryotic homologues (Bult *et al.*, 1996). A hypusine-containing protein which is similar to eukaryotic translation initiation factor eIF-5a (Bartig *et al.*, 1992) was also found in *S. acidocaldarius* and other Archaea. Subsequent sequence analyses suggested that eIF-5a is related to bacterial elongation factor, EF-P, which lacks hypusine, and that two other eukaryotic translation factors, eIF-1A and SUI1, are also universally distributed (Kyrpides & Woese, 1998). The close relationship between Archaea and eukaryotes is supported by the phylogenetic analyses of several other translation proteins including elongation factors (Iwabe *et al.*, 1989; Balduaf *et al.*, 1996), ribosomal proteins (Brown & Doolittle, 1997; Müller & Wittmann-Liebold, 1997); aminoacyl-tRNA synthetases (Brown & Doolittle, 1995; Brown *et al.*, 1997) and methionine aminopeptidase (Keeling & Doolittle, 1996).

Comparative analyses of translational components have also fueled several evolutionary controversies. Lake and co-workers argued for the eocyte universal tree topology based on similarities in ribosome shapes among Crenarchaeota and eukaryotes (Henderson *et al.*, 1984; Lake, 1988; Lake *et al*, 1984, 1985). However, the different lobes and protuberances supposedly diagnostic of eocyte ribosomes, were later found in species of halophiles and thermophilic methanogens (Stöffler-Meilicke *et al.*, 1986; Stöffler & Stöffler-Meilicke, 1986). The ribosomes of eukaryotes and Crenarchaeota are also more protein-rich compared to those of other Archaea and *E. coli*. However, the bulking-up of ribosomes with proteins is more likely a general adaptation to high temperature environments, since several thermophile-methanogenic Archaea, as well as a hyperther-mophilic bacterium, *Aquifex pyrophilus*, have high ratios of protein to rRNA in their ribosomes (Acca *et al.*, 1993).

Certain DNA-binding proteins have a strong resemblance to eukaryotic histones, in terms of both primary sequence (Sandman *et al.*, 1990; Grayling *et al.*, 1994), and three-dimensional structure (Starich *et al.*, 1996). Initially isolated from the methanogen *Methanothermus fervidus*, these histone-like proteins are called HMf. Histone encoding genes have now been sequenced from different members of the Euryarchaeota, where their number can vary between species (Reeves *et al.*, 1997). Putative archaeal nucleosomes differ from eukaryotic nucleosomes in that DNA strands are constrained about HMf particles as positive supercoils rather than the conventional negative supercoil conformation imposed by histones (Musgrave *et al.*, 1991; Reeves *et al.*, 1997). While eukaryotic histones form only H2A / H2B and H3 / H4 heterodimers, archaeal histones can assemble as both homodimers and heterodimers. Archaea also have proteins similar to bacterial DNA-binding proteins, known as HU, which are not evolutionary linked to histones, although they perform similar functions (Bianchi, 1994).

Pre-rRNA processing in eukaryotes and archaea involves many different small nucleolar rRNAs (snoRNAs) associating with the protein fibrillarin. The gene for fibrillarin has been cloned and sequenced from methanogenic Archaea (Amiri, 1994; Bult *et al.*, 1996) and antibodies against fibrillarin have precipitated snoRNAs in *Sulfolobus* (Dennis, 1997). Although a report of U3-like RNA cloned from *Sulfolobus acidocaldarius* (Potter *et al.*, 1995) was subsequently determined to be an error (Russell *et al.*, 1997), it is still possible that related snRNAs are involved in the processing of archaeal pre-rRNA (Durovic & Dennis, 1994). Introns in tRNA genes of Archaea and eukaryotes are of similar size, and occur in mostly the same positions although some archaeal tRNA introns have shifted locations (Belfort & Weiner, 1997). Excision of tRNA introns in eukaryotes involves a site-specific endonuclease which is comprised of two subunits (Kleman-Leyer *et al.*, 1997; Trotta *et al.*, 1997). These subunits are similar to each other as well as being homologous to archaeal tRNA endonucleases.

Several metabolic processes are also uniquely found in the Archaea and eukaryotes. In eukaryotes, isoprenoid biosynthesis proceeds with the formation of mevalonate from 3-hydroxy-3-methylglutaryl coenzyme A (HMGCoA) which is catalyzed by an NADPH-dependent HMGCoA reductase. The archaeal species, *Haloferax volcanii* and *Sulfolobus solfataricus*, have biosynthetic HMGCoA reductase genes similar to those of eukaryotes (Lam & Doolittle, 1992; Bochar *et al.*, 1997). While most bacteria do not have biosynthetic HMGCoA reductases, the bacterium *Pseudomonas mevalonii* does have a highly diverged biodegradative NAD^+-dependent HMGCoA reductase. The evolutionary picture of archaeal HMGCoA reductases has become more convoluted with the discovery that *Archaeoglobus fulgidus* has an HMGCoA reductase gene which is more similar to bacterial NAD^+-dependent, rather than eukaryotic NADPH-dependent, HMGCoA reductases (Doolittle & Logsdon 1998).

Conversion of acetyl-CoA to acetate in bacteria requires the coordinated activities of two enzymes, phosphate acetyltransferase and acetate kinase while hyperthermophilic Archaea employ a single enzyme, an ADP-forming acetyl-CoA synthetase (Schönheit & Schäfer, 1995), which is also found in two anaerobic, amitochondrial eukaryotes,

Entamoeba histolytica (Reeves *et al.*, 1977) and *Giardia lamblia* (Lindmark, 1980). Archaea have a simplified, and possibly ancestral, version of the 26S proteasome, a multimeric complex responsible for ATP-dependent proteolysis in eukaryotes (Zwickl *et al.*, 1992; Wenzel & Baumeister, 1993).

In summary, there are some features that distinguish the Archaea from the Bacteria and eukaryotes, most notably the structure and composition of their membranes. Primarily, the members of the Archaea are unique in having a combination of traits which, until now, were believed to be exclusive to either Bacteria or eukaryotes. However, the distribution of any gene throughout the Bacteria, Archaea and eukaryotes needs to be constantly evaluated as new genomic sequences become available.

4   Rooting the Universal Tree

There are three possible topologies for any universal tree:
(i) Bacteria diverged first from a lineage producing Archaea and eukaryotes,
(ii) a proto-eukaryotic lineage diverged from a fully prokaryotic (Bacteria and Archaea) lineage, or
(iii) Archaea diverged from a lineage leading to eukaryotes and Bacteria. However, on the basis of a solitary gene, it is impossible to derive an objective rooting for the universal tree. Typically, the rooting for a particular organismal tree, for example all mammalian species, would be determined by including sequence data from a known outgroup species, such as cold-blooded vertebrates. However, outgroup species are not available for a gene tree consisting of all living organisms unless specific assumptions are made such as the progression of life from a prokaryotic to a eukaryotic cell. Therefore, the branching order of the three domains emerging from the cenancestor can only be established by some method unrelated to either outgroup organisms or theories about primitive and advanced states.

In 1989, a solution to this problem using ancient duplicated genes, was simultaneously proposed in separate papers by Gogarten *et al.* and Iwabe *et al.* Their collective reasoning was as follows: although there can be no organism which is an outgroup for a tree relating all organisms, one could root a tree based on the sequences of outgroup genes produced by an early gene duplication (Fig. 2A). Iwabe *et al.* applied this concept by reciprocal rooting trees for paralogous elongation factor genes. Elongation factors (EF) are a family of GTP-binding proteins which facilitate the binding of aminoacylated tRNA molecules to the ribosome (EF-Tu in Bacteria and EF-1α in eukaryotes and Archaea) and the translocation of peptidyl-tRNA (EF-G in Bacteria and EF-2 in eukaryotes and Archaea). There were five conserved regions which could be aligned between EF-Tu/1α and EF-G/2 sequences which, at the time, were available from one Archaeon, *Methanococcus vanniellii*, and several species of Bacteria and eukaryotes. According to protein sequence similarity and neighbor-joining trees, both EF-1α and EF-2 of Archaea were more similar to their respective eukaryotic, rather than bacterial, homologues.

Gogarten *et al.* (1989) developed composite trees based on a second gene duplication, that of the V-type (found in Archaea and eukaryotes) and F-type (found in Bacteria) ATPase subunits. The catalytic β subunit of F-type ATPases is most similar to the A or

70-kDa subunit of V-type ATPases, while the α subunit of F-type ATPases is most similar to the B or 60-kDa subunit of V-type ATPases. In agreement with the elongation factor rooting, reciprocally rooted ATPase subunits trees showed the Archaeon, in this case *Sulfolobus acidocaldarius*, to be closer to eukaryotes than to Bacteria.

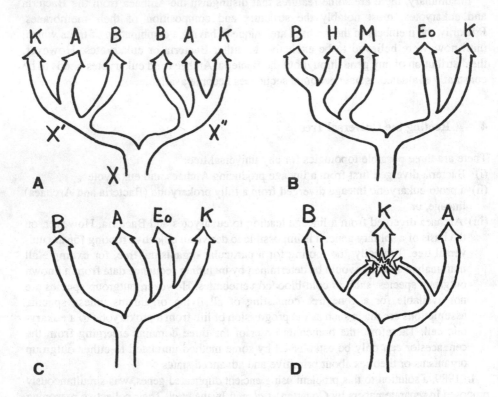

Figure 2. Alternative scenarios about early cellular evolution. A) Conceptual rooting of the universal tree using paralogous genes. Suppose that some gene (X) was duplicated (X' and X") in the cenancestor such that all extant organisms have both genes. Provided that some sequence similarity still exists between genes X' and X", then reciprocally rooted gene trees could be constructed. The positioning of Archaea (A) and eukaryotes (K) as sister groups with the Bacteria (B) as the outgroup, has been consistently supported by such rootings (Gogarten *et al.*, 1989; Iwabe *et al.*, 1989; Brown & Doolittle, 1995). B) The 1988 eocyte tree as proposed by Lake (1998). The Crenarchaeota or eocytes (Eo) form a clade with eukaryotes while the Euryarchaeota, namely halophiles (H) and methanogens (M) cluster with the Bacteria. C) The most recent eocyte tree advocated by Rivera & Lake (1992). The eocytes are the closest group of Archaea to eukaryotes with the Euryarchaeota (A) being more distantly related although no longer branching with the Bacteria. D) The "chimeric" or "fusion" hypothesis which suggests that eukaryotes arose when a "Gram-negative" bacterium engulfed an archaeon and components of their genomes fused (Zillig *et al.*, 1989; Zillig, 1991; Gupta & Singh, 1994; Lake & Rivera, 1994; Martin & Müller, 1998).

5   Testing the Rooting

Archaebacteriologists (or "archaeologists") were already primed to accept the conclusions of these duplicated gene rootings of the universal tree, since early on there was a general feeling that the Archaea were somehow "missing links" between Bacteria and eukaryotes. At the time, eukaryote-like functional and structural characteristics of archaeal RNA and DNA polymerases and some ribosomal proteins were known. Woese, Kandler & Wheelis (1990) incorporated the protein rooting in their formulation of the three domains, Archaea, Bacteria and Eucarya. Although archaeal and bacterial rRNA sequences are slightly more similar, Woese and colleagues placed the root of the ribosomal tree such that Archaea and Eucarya were sister groups. Here, it is important to emphasize that their "archaeal" universal tree was formulated from three different data analyses. The separate monophyly of the Archaea, Bacteria and Eucarya was suggested by rRNA gene trees (Woese, 1987) while the grouping of Archaea and Eucarya together arose from the reciprocally rooted gene trees for elongation factors and ATPase subunits (Gogarten *et al.*, 1989; Iwabe *et al.*, 1989).

However, some researchers have seriously challenged the topology of the archaeal universal tree. In his rRNA tree proposed in 1988, Lake broke up the Archaea (then archaebacteria) by placing the Crenarchaeota, which he called "eocytes", in a clade with eukaryotes, now named "Karyotes" (Fig. 2B). Methanogens, halophiles and Bacteria were in a separate group called the "Parkaryotes" which notably had Bacteria in a clade with halobacteria (Lake *et al.*, 1985). This early revisionist universal tree was based on differences in ribosome shapes (Henderson *et al.*, 1984; Lake *et al.*, 1984) and a novel phylogenetic analysis of rRNA sequences (Lake, 1988). However, as the rRNA dataset grew and more ribosome structures were determined, Lake's 1988 eocyte tree became untenable. Later, Rivera & Lake (1992) found new support for the eocyte tree, this time in the analysis of a specific 11 amino acid insertion shared in the EF-1α genes of eukaryotes and Crenarchaeota, but absent from Euryarchaeota and Bacteria. This 1992 eocyte tree resembled the 1988 tree by still having a Crenarchaeota - Eukaryote clade, but differed in reassignment of Euryarchaeota and Bacteria into individual clades (Fig. 2C).

Cammarano and co-workers (Tiboni *et al.*, 1991; Cammarano *et al.*, 1992; Creti *et al.*, 1994) added several new EF sequences from the Archaea as well as a deeply branching bacterium, *Thermotoga maritima*. Although they did not employ a reciprocal rooting, their analyses of EF-G/2 sequences showed strong support for a monophyletic clade of the Archaea, subdivided into the kingdoms Crenarchaeota and Euryarchaeota. However, another analysis of elongation factor genes by Baldauf *et al.* (1996), including many deep-branching species from all domains, found support, albeit statistically weak, for the divergence of eukaryotes within the Archaea as a sister group to the Crenarchaeota which bolsters the eocyte tree of Rivera & Lake (1992).

Forterre *et al.* (1993) argued that neither the elongation factor nor ATPase datasets can settle the issue of rooting the universal tree. Their major criticisms concerned the paucity of taxa (which was largely addressed by Baldauf *et al.* [1996]) and the fact that only 120 amino acids could be aligned with confidence between EF-Tu/1α and EF-G/2 genes which are 390 to 460 and 700 to 860 amino acids long, respectively.

ATPase subunit gene phylogenies were more problematic. Based on greater similarities between archaeal and eukaryotic V-type ATPases over bacterial F_0F_1 type ATPases, earlier analyses placed the root of the universal tree in the Bacteria (Gogarten et al., 1989; Iwabe et al., 1989). At the time, known bacterial ATPases were of the F_0F_1 type, while V-type ATPases were exclusive to Archaea and eukaryotes. It was proposed that a gene duplication in the cenancestor, resulted in the F-type β / V-type A (or 70-kDa) subunit, on the one hand, and the F-type α / V-type B (or 60-kDa) subunit on the other.

Subsequently, archaeal V-type ATPases were reported for two bacterial species, *Thermus thermophilus* (Tsutsumi et al., 1991) and *Enterococcus hiraea* (Kakinuma et al., 1991), and a F_1- ATPase β subunit gene was found in the Archaeon, *Methanosacrina barkeri* (Sumi et al., 1992). Forterre et al. (1993) suggested that the ATPase subunit gene family had not been fully determined, and that other paralogous genes might exist. However, Hilario & Gogarten (1993) suggest that the observed distribution of ATPase subunits is the result of a few lateral gene transfers between species of Archaea and Bacteria. In support of their view, broader surveys have failed to detect archaeal V-type ATPases in other bacterial species (Gogarten et al., 1996).

More recently, the rooting of the universal tree was attempted using another duplicated gene family, the aminoacyl-tRNA synthetases (Brown & Doolittle, 1995). Novel isoleucyl-tRNA synthetase sequences from species belonging to deep evolutionary lineages of Bacteria, Archaea and eukaryotes were used to construct a universal gene tree which was rooted by valyl- and leucyl-tRNA synthetases. The sisterhood of eukaryotes and Archaea, as well as the separate monophyly of all three domains, were strongly supported by this analysis. A similar conclusion was reached in the phylogenetic analysis of tryptophanyl- and tyrosyl-tRNA synthetase (Brown et al., 1997; Fig. 3). However, new genomic sequence data reveals that several bacteria have isoleucyl-tRNA synthetases highly similar to those of eukaryotes (Brown et al., 1998). It has been proposed that horizontal gene transfer of an isoleucyl-tRNA synthetase gene from a eukaryote to certain bacterial species might have been favoured as a defense against a naturally occurring anti-bacterial agent. Now that more archaeal and bacterial aminoacyl-tRNA synthetases are known, further opportunities exist to derive multiple rooted universal trees although the emerging picture is that several aminoacyl-tRNA synthetase gene trees are inconsistent with the archaeal tree (Brown, 1998).

6   Genomic Chimerism and the Origin of Eukaryotes

Several different paralogous gene phylogenies provide a general consensus that the root of the universal tree lies somewhere in the Bacteria thus positioning Archaea and eukaryotes as sister groups (also see Lawson et al., 1996). However, there is still uncertainty about this rooting since each duplicated gene dataset has its own particular, and significant, shortcomings. Furthermore, three or four genes spanning a few thousand base pairs may not be representative of entire genomes with over 1000 genes and, at least, several million base pairs.

There are many properties that make rRNA a suitable molecular marker for phylogenetic reconstruction: it occurs in all living organisms, its sequences are highly conserved, and there is no compelling evidence for inter-specific transfers of rRNA genes. However, even the monophyly of Archaea, Bacteria and eukaryotes, as strongly suggested by rRNA trees, is open to challenge. The higher G-C content of rRNA genes of certain organisms, such as thermophilic Archaea and Bacteria, could be biasing phylogenetic reconstruction, and new

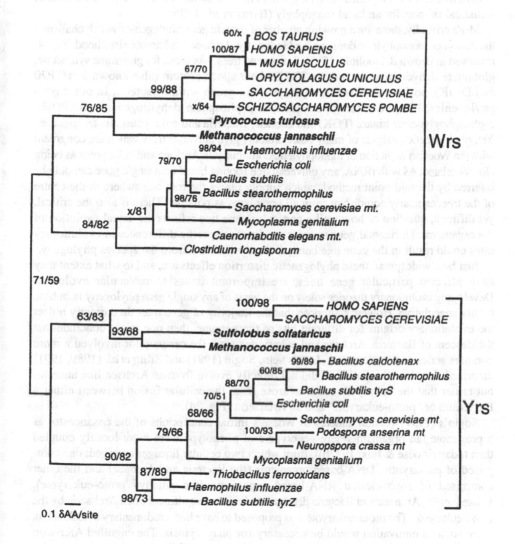

Figure 3. Reciprocally rooted universal trees for tryptophanyl- and tyrosyl-tRNA synthetases (Wrs and Yrs, respectively). All phylogenies were constructed using the neighbour-joining method as implemented by the program NEIGHBOR of the PHYLIP 3.57c package (Felsenstein, 1993). The scale bar represents 0.1 expected amino acid replacements per site as estimated by the program PROTDIST using the Dayhoff PAM substitution matrix (Dayhoff *et al.*, 1972). In this figure, the numbers are the percent occurrence of nodes in 500 bootstrap re-samplings of maximum parsimony (PAUP; Swofford, 1993) and neighbour-joining analyses, respectively. Values less than 50% are not shown. Species names of Bacteria are in plain-face, those of Eucarya or eukaryotes in uppercase, and those of Archaea in bold-face. Also indicated are eukaryotic nuclear-encoded mitochondria (mito. or mt) or chloroplast (chlo.) targeted isoforms.

environmental PCR amplified rRNA sequences have in some instances reduced the overall statistical support for archaeal monophyly (Barns et al., 1996).

More critically, there are a growing number of protein gene phylogenies which challenge the notion of monophyletic domains, and indirectly, Archaea - eukaryote sisterhood (Fig. 4; reviewed in Brown & Doolittle, 1997). Some gene trees, like those for glutamine synthetase, glutamate dehydrogenase, and the 70-kDa heat shock protein (also known as HSP70 and DnaK), position Archaea as a paraphyletic group within Bacteria. In other gene phylogenies, like those for glyceraldehyde-3-phosphate dehydrogenase (GAPDH), 3-phosphoglycerate kinase (PGK) and enolase, Bacteria and eukaryotes cluster together. Yet, phylogenetic analyses of many other proteins give universal trees which are congruent with the Woesian depiction of monophyletic domains with Archaea and eukaryotes as being closely related. As with rRNA, any universal tree rooting based on a single gene can only be inferred by the mid-point method, which simply places the root somewhere in the centre of the tree, as nearly equidistant from all organisms as possible. There is also the critical, yet difficult, question of how well a particular gene tree reflects the actual evolution of the organisms. Horizontal gene transfer and lineage specific differences in evolutionary rates could result in the gene tree being radically different from the species phylogeny.

Just how widespread these phylogenetic distortion effects are, and to what extent they have affected particular gene trees, are important issues to molecular evolution. Developing evolutionary theories solely on the basis of any single gene phylogeny is, at best, highly speculative. However, suppose that the majority of gene trees do correctly reflect the evolutionary origins for different bits of the genome, then one might conclude that the descent of Bacteria, Archaea and eukaryotes from the cenancestor involved a more complex series of genetic events. In this vein, Sogin (1991) and Zillig et al. (1989, 1991) theorized that the eukaryotic cell did not directly evolve from an Archaea-like ancestor but rather that the eukaryotic nucleus arose from the cellular fusion between either a Bacterium or "proto-eukaryote" and an Archaeon (Fig. 2d).

Sogin's (1991) version adheres to Woese's initial conceptions of the cenancestor as a progenote [an organism where genotype and phenotype were more loosely coupled than today (Woese & Fox, 1977b)], from which two cellular lineages emerged; one comprised of prokaryotic, DNA-based organisms (the Bacteria and Archaea) and the other comprised of sophisticated RNA-based organisms (the putative proto-eukaryote). Subsequently, Archaea and Bacteria diverged before the engulfment of an Archaeon by the proto-eukaryote. The proto-eukaryote was proposed to have had a rudimentary cytoskeleton, since such an innovation would be necessary for phagocytosis. The engulfed Archaeon formed the cell nucleus which led to the replacement of the host RNA genome by a DNA-based one. Sogin suggested that contradictions between rRNA phylogenies, which show contemporary Archaea and Bacteria as most similar, and paralogous protein gene trees, which show Archaea and eukaryotes as sister groups, exist because eukaryotic rRNA is a remnant of the proto-eukaryote genome.

Zillig et al. (1991) similarly proposed that separate lineages of Archaea and Bacteria descended from the cenancestor, but that eukaryotes did not exist until a cellular fusion occurred between species from the two prokaryotic groups. The model of Zillig et al., unlike that of Sogin, has the cenancestor as a genote, a prokaryote with a fully functional genome, rather than a progenote. Gupta & Golding (1993) have elaborated upon Zillig et al.'s

Argininosuccinate synthetase
Aspartyl-tRNA synthetase
ATPase α subunit
ATPase β subunit
DNA polymerase B
Ef1α/Tu
Ef-G/2
HSP60
Isoleucyl-tRNA synthetase
Ribosomal proteins (18)
RNA polymerase subunit A
RNA polymerase subunit B
SecY
Tryptophanyl-tRNA synthetase
Tyrosyl-tRNA synthetase

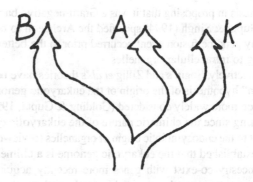

ALADH
Citrate synthase
FGARAT
Glutamate dehydrogenase II
Glutamine synthetase I
Gyrase B
HisA, HisC, HisF, HisG, HisIE
HSP70
IMPDH
Ribosomal Proteins (3)
TrpD

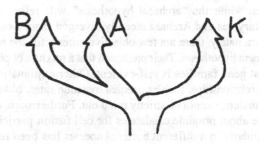

Enolase
FeMn superoxide dismutase
GAPDH
HisB
PGK
ProC
TrpB

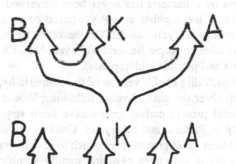

Acetyl-coenzyme A synthetase
Glu-tRNA reductase
Dihydrofolate reductase
HisD, HisH
Photolyase
TrpA, TrpC, TrpE, TrpG

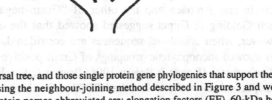

Figure 4. Alternative rootings of the universal tree, and those single protein gene phylogenies that support them. Individual gene trees were constructed using the neighbour-joining method described in Figure 3 and were reported in Brown & Doolittle (1997). Protein names abbreviated are: elongation factors (EF), 60-kDa heat shock protein (HSP60), 5-aminolevulinic acid dehydratase (ALADH), phosphoribosylformylglycinamidine synthetase (FGARAT), 70-kDa heat shock protein (HSP70), inosine monophosphate dehydrogenase (IMPDH), glyceraldehyde 3-phosphate dehydrogenase (GAPDH) and 3-phosphoglycerate kinase (PGK).

hypothesis in proposing that it was a Gram-negative bacterium that engulfed an Archaeon; later Gupta & Singh (1994) specified the Archaeon to be an eocyte. Both theories suggest that any genome fusion event occurred prior to the better-accepted bacterial endosymbiosis leading to intracellular organelles.

Collectively, Sogin's and Zillig *et al.*'s theories have been referred to as the "chimeric" or "fusion" hypothesis of the origin of the eukaryotic genome, although Zillig *et al.*'s version has been more widely considered (Golding & Gupta, 1995). This terminology is somewhat confusing since the chimeric nature of the eukaryotic cell has been long recognized with respect to the endosymbiotic origin of organelles (reviewed in Gray, 1992). In addition, it has been established that the eukaryotic genome is a chimera where genes of ancient eukaryotic ancestry co-exist with genes more recently acquired from bacterial endosymbionts. In the context here, the term "chimera hypothesis" will be applied to suggestions that the eukaryotic genome originated from a fusion between two independent, non-eukaryotic genomes, while the "archaeal hypothesis" will refer to the more conventional view, that eukaryotes and Archaea recently diverged from a common ancestor.

Unfortunately, there are few objective criteria for the rejection or acceptance of any of the chimera hypotheses. Their prediction that a mixture of phyletic relationships exists among different gene families is self-evident. Other explanations for the observed mixing of domain relationships, such as unequal mutation rates, hidden gene paralogy and horizontal gene transfers, cannot be strictly ruled out. Furthermore, only very broad speculation can be made about possible candidates for cell fusion participants. Although bacteria living intracellularly in a different bacterial species has been reported (Larkin & Henk, 1996), phagocytosis by a bacteria has never been observed. Nor is there any evidence for the existence of the sophisticated RNA-based organisms with cytoskeletons integral to Sogin's proto-eukaryotic model. On the other hand, the endosymbiosis hypothesis, clearly establishes α-purple Bacteria and cyanobacteria as the respective progenitors of mitochondria and plastids (chloroplasts).

Nonetheless, Zillig *et al.*'s version of the chimeric hypothesis has found some interest and support. Over the past 20 years following Woese's initial universal rRNA trees, many archaeal protein coding genes have been sequenced that are homologous to counterparts in Bacteria and eukaryotes. Golding & Gupta (1995) constructed unrooted phylogenetic trees for 24 protein genes which were common to all three domains. They found that nine of the protein trees gave statistically significant support for the monophyly of domains (*sensu* Woese's rRNA trees) as well as the closer association of Archaea and eukaryotes (by mid-point rooting). Eight protein trees were not statistically significant. Seven protein trees supported an alternative topology with two clades, one of Gram-positive Bacteria and Archaea and the other of "Gram-negative" Bacteria and eukaryotes, which Golding & Gupta suggested showed that the eukaryotic genome is a chimera. However, when additional sequences are considered, none of the seven protein gene trees showed monophyletic groupings of Gram-positives and Archaea or Gram-negatives and eukaryotes (Roger & Brown 1996; Brown & Doolittle, 1997).

Gupta and co-workers view the HSP70 or DnaK gene phylogeny among the strongest evidence supporting the chimeric origin for eukaryotes hypothesis (Gupta & Singh, 1992, 1994; Gupta and Golding, 1993, 1996; Gupta *et al.*, 1994). HSP70 genes have been found in all species of Bacteria and eukaryotes investigated, thus far. However, the distribution of HSP70 genes in archaeal genomes is less than universal. Although found in a few methanogens and halophiles, no HSP70 genes were detected in the completely sequenced genomes of either *M. jannaschii* (Bult *et al.*, 1996) or *A. fulgidus* (Klenk *et al.*, 1997). Further, there is no genetic or biochemical evidence for the existence of HSP70 in a species of Crenarchaeota. HSP70 genes from Archaea and Gram-positive Bacteria are clearly more similar, both with respect to phylogeny (Fig. 5), and the absence of a specific 25 amino acid long sequence found in HSP70 genes of other Bacteria and eukaryotes. However, the HSP70 phylogeny suffers the same drawbacks of any single protein tree, in that it cannot be uniquely rooted. Since the rooting is subjective, either clustering possibility for Gram-negative Bacteria, with eukaryotes or with Gram-positive Bacteria and Archaea, have equal validity.

If Gram-positive Bacteria and Archaea did diverge first from the cenancestor, then it would be reasonable to assume that these groups should be monophyletic. However, the HSP70 tree shows these groups as paraphyletic rather than monophyletic, whereas eukaryotes and all other Bacteria can be resolved as separate monophyletic groups. In order to reconcile the observed HSP70 gene phylogeny, one would have to postulate the unlikely scenario of a rapid rate acceleration in eukaryotes, while Gram-positive bacterial and archaeal proteins evolved so slowly as to appear similar to each other (Roger & Brown, 1996).

7   Alternatives to Chimerism

There is little doubt that different protein phylogenies can project widely different evolutionary scenarios. Feng, Cho & R.F. Doolittle (1997) tried to estimate the timing of ancient divergence events by calibrating the rates of amino acid substitution in 64 different enzyme sets to the vertebrate fossil record. In their dataset, 38 proteins had sufficient representation from species across all three domains to be used to evaluate the universal tree (Feng *et al.* were also interested in more recent divergence events, such as those separating plants, animals and fungi, thus not all of their datasets needed to be universal in species spectrum). They found that Archaea clustered with eukaryotes (AK cluster), positioning Bacteria as the outgroup, in 8 protein trees; Archaea and Bacteria were the closest domains (AB cluster) in 11 protein trees and; Bacteria and eukaryotes clustered together (BK cluster) in 15 protein trees. Four protein trees were anomalies where two different archaeal species, in the same phylogeny, grouped with either Bacteria or eukaryotes.

Interestingly, when the Archaea was the outlier, eukaryotes usually clustered with Gram-negative bacteria rather than with Gram-positive bacteria or cyanobacteria. When eukaryotes were the outgroup, then the Archaea would tend to cluster with the Gram-positive bacteria. In estimating divergence times, which was the principal motive behind their analysis, Feng *et al.* reverted to the archaeal tree stating that the first split occurred 3200-3800 Mya between the Archaea and Bacteria, then eukaryotes diverged from the Archaea about 2300 Mya. This dating of the Archaea-Bacteria split was earlier than a previous estimate by R.F. Doolittle's group (Doolittle *et al.*, 1996), however, the timing is still problematic if cyanobacteria existed at least 3450 Mya, as suggested by microfossils (Schopf, 1993).

Brown & W.F. Doolittle (1997), constructed 66 universal protein trees and catalogued the frequency of nearest pairs of domains as well as the occurrence of domain monophyly

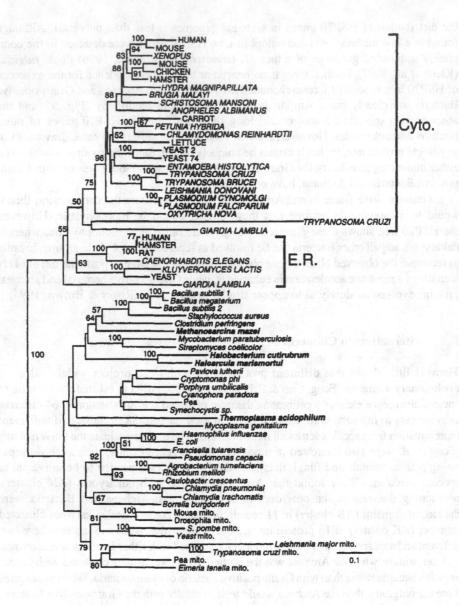

Figure 5. Phylogeny of 70-kDa heat shock proteins (HSP70 or DnaK). Cytostolic (Cyto.) and endoplasmic reticulum (E.R.) eukaryotic isoforms are indicated. Phylogeny was constructed using the neighbour-joining method described in Figure 3. Numbers show the percent occurrence of nodes in 100 bootstrap replications, and only values greater than 50% are shown.

or integrity. Included in the study were several protein enzymes that either participated in the same biochemical pathway, such as histidine biosynthesis, or physically interacted with each other, such as ribosomal proteins. Mean interdomain distances for each protein (Archaea to eukaryotes, Archaea to Bacteria, and eukaryotes to Bacteria) were determined by averaging the pairwise distances between all available sequences from the different domains and the differences between mean pairwise distances were statistically compared (Table 1). In a total of 56 protein trees, the mean interdomain distance scores of one of the three possible domain clusters (AK, AB or AK) was significantly lower ($P < 0.05$) than the other two possibilities. The clusterings AK, AB and BK occurred in 31, 18 and 7 of the comparisons, respectively.

Several possible trends in the clustering of domains could be seen in these multiple universal trees. First, there was a rough partitioning of gene tree topologies on the basis of biochemical function. Most of the proteins involved in either DNA replication, transcription or translation (including most ribosomal proteins), so-called information roles, seem to favour the AK grouping. A notable exception is gyrase B where Archaea and Bacteria appear highly similar. That Bacteria and Archaea might have similar gyrase B / topoisomerase II molecules is not surprising given that the chromosome configurations for both groups appear to be comparable.

Interdomain relationships inferred from metabolic enzymes are less clear. Proteins of the glycolytic pathway strongly supported the BK cluster while chaperones, membrane bound proteins involved in secretion, and ATP-proton pumps favoured the AK cluster. Other biosynthetic and chaperone proteins supported the AB grouping. Protein trees derived from the enzymes of the tryptophan and histidine biosynthetic pathway showed little favour towards any particular rooting, and the differences between clusters were statistically non-significant.

Domain integrity considers whether Archaea, Bacteria and eukaryotes are resolved as separate monophyletic groups [*sensu* Woese's rRNA tree (Woese *et al.*, 1990)] or whether one or more domains are paraphyletic [*sensu* the Archaea in Lake's eocyte tree (Rivera & Lake, 1992)]. Paraphyletic clusterings would suggest lateral gene transfers between particular species belonging to different domains (Smith *et al.*, 1992). Considering only those gene trees with two or more species from each domain, universal phylogenies with one or more paraphyletic domains were found to be nearly twice as frequent as phylogenies with three monophyletic domains (33 versus 17 protein trees; Brown & Doolittle 1997). All trees with three monophyletic domains supported the AK grouping, and never the BK or AB clustering. The occurrence of monophyletic domain trees was not restricted by protein function since examples could be found among informational pathway proteins as well as metabolic enzymes.

In protein trees where either cluster AB or BK are most strongly supported, species associations could provide clues about the nature of potential participants in lateral gene transfers. Discernible trends are usually not very apparent since the closest domains are not monophyletic, the major branch points are often ill-resolved, and species diversity might be low. In phylogenies supporting an AB grouping, archaeal species often branch among Gram-positive Bacteria (Gupta & Singh, 1992; Brown *et al.*, 1994; Brown & Doolittle 1997). As an example, the HSP70 phylogeny positions archaeal species amongst the Gram-positive Bacteria although the exact branching order among these groups and the cyanobacteria are poorly resolved (Fig. 5).

TABLE 1.　Summary of protein gene phylogenies (taken from Brown & Doolittle, 1997). Interdomain distance was the expected number of amino acid replacements per site averaged over all pairwise comparisons between species of Archaea (A), eukaryotes (K) and Bacteria (B). Distances were calculated as described in Figure 3. The lowest mean value (closest pair of domains) for a particular protein are in bold-face. Also indicated are whether the means are significantly (ANOVA, p ≤ 0.05) different (*) or not (ns). Domain integrity was determined by inspection of neighbour-joining trees for either monophyletic (■) or para/polyphyletic (☆) domains. If only a single species from a particular domain was known, the box for that domain is marked with □. Definitions of abbreviated protein names are given in Figure 4

Function	Gene Product	Inter-Domain Distance			ANOVA	Domain Integrity		
		A-K	A-B	K-B	α0.05	K	A	B
DNA Repair	DNA polymerase II	**1.231**	1.910	2.047	*	■	■	□
& Replication	Gyrase B	2.406	**0.706**	2.327	*	■	□	☆
	Photolyase Class I	1.318	**1.188**	1.310	ns	■	□	☆
Transcription	RNA polymerase subunit A	**1.138**	1.803	2.126	*	■	■	■
	RNA polymerase subunit B	**1.187**	1.612	2.010	*	■	■	■
Translation	Elongation factor G/2	**1.153**	1.337	1.475	*	■	■	■
	Elongation factor Tu/1α	**0.638**	1.152	1.289	*	■	☆	■
	Isoleucyl-tRNA synthetase	1.272	**1.188**	1.613	*	■	■	■
	Aspartyl-tRNA synthetase	**1.330**	4.469	4.276	*	■	□	■
	Tryptophanyl-tRNA synthetase	**1.849**	2.665	3.297	*	■	☆	■
	Tyrosyl-tRNA synthetase	**1.283**	2.564	3.048	*	■	■	■
	Ribosomal protein L2	**1.031**	1.558	1.989	*	■	☆	■
	Ribosomal protein L3	**1.209**	2.176	2.700	*	■	□	■
	Ribosomal protein L5	**1.066**	1.686	1.990	*	■	☆	■
	Ribosomal protein L6	**1.267**	2.271	2.969	*	■	☆	■
	Ribosomal protein L10	**1.752**	3.599	4.600	*	■	☆	■
	Ribosomal protein L11	1.941	**1.367**	2.591	*	■	☆	■
	Ribosomal protein L14	**0.775**	1.141	1.279	*	■	☆	■
	Ribosomal protein L15	**2.012**	2.910	3.194	*	■	■	■
	Ribosomal protein L22	**1.374**	3.186	2.769	*	■	■	■
	Ribosomal protein L23	**0.852**	1.685	1.656	*	■	☆	■
	Ribosomal protein L30	**1.799**	1.811	2.582	*	■	■	■
	Ribosomal protein S5	**0.950**	1.411	1.650	*	■	■	■
	Ribosomal protein S7	**0.788**	2.215	2.400	*	□	☆	■
	Ribosomal protein S8	4.049	**2.257**	3.978	*	■	■	■
	Ribosomal protein S9	**1.238**	2.097	2.022	*	■	■	■
	Ribosomal protein S10	**1.188**	1.230	1.564	*	■	■	■
	Ribosomal protein S11	**0.712**	1.213	1.492	*	■	☆	■
	Ribosomal protein S12	**0.684**	2.995	3.425	*	■	☆	■
	Ribosomal protein S15	**1.629**	2.728	3.594	*	■	□	■
	Ribosomal protein S17	4.505	**1.435**	4.135	*	■	■	☆
	Ribosomal protein S19	**0.976**	1.916	1.864	*	■	■	■

TABLE 1. continued

Function	Gene Product	Inter-Domain Distance			ANOVA	Domain Integrity		
		A-K	A-B	K-B	α0.05	K	A	B
Central metabolism	GAPDH	2.760	2.699	0.684	*	☆	☆	☆
	PGK	1.481	1.490	0.808	*	☆	■	■
	Enolase	1.095	0.900	0.665	*	■	□	☆
	Acetyl-coenzyme A synthetase	1.065	0.992	1.119	ns	■	□	☆
	Citrate synthase	1.910	1.231	2.047	*	■	□	☆
Amino acid biosynthesis	Argininosuccinate synthetase	1.141	2.115	2.003	*	■	■	■
	Glutamate dehydrogenase II	1.200	0.866	1.209	*	☆	☆	☆
	Glutamine synthetase I	2.899	0.977	2.826	*	■	☆	☆
	HisA	3.274	1.477	3.980	*	■	☆	☆
	HisB	0.931	1.025	0.872	*	☆	☆	☆
	HisC	2.175	1.815	2.008	*	■	☆	☆
	HisD	1.121	1.082	0.945	ns	■	☆	☆
	HisF	1.072	0.859	1.335	*	□	☆	☆
	HisG	1.607	1.431	1.498	*	■	■	■
	HisH	1.326	1.282	1.345	ns	□	☆	☆
	HisIE	1.696	1.191	1.597	*	■	☆	☆
	ProC	2.235	2.408	1.418	*	☆	□	☆
	TrpA	1.454	1.406	1.332	ns	■	☆	☆
	TrpB	1.087	1.093	0.638	*	■	☆	☆
	TrpC	1.296	1.373	1.338	ns	■	☆	☆
	TrpD	1.629	1.180	1.579	*	☆	☆	☆
	TrpE	1.196	1.271	1.358	*	☆	☆	☆
	TrpG	0.858	1.013	0.870	*	☆	☆	☆
Cofactors	Dihydrofolate reductase	2.060	1.939	1.810	ns	■	□	☆
Purine biosynthesis	IMPDH	1.377	0.767	1.231	*	■	□	☆
	FGARAT	2.282	1.084	1.827	*	■	☆	☆
Respiration	FeMn superoxide dismutase	1.138	1.032	0.810	*	☆	☆	☆
Porphyrins	ALADH	1.128	0.851	1.255	*	☆	□	☆
	Glu-tRNA reductase	1.539	1.591	1.500	ns	■	□	☆
Chaperones	70-kDa heat shock protein	0.867	0.658	0.816	*	■	☆	☆
	60-kDa heat shock protein	1.113	2.342	2.687	*	☆	■	☆
Membrane	SecY - protein secretion	1.434	2.942	3.119	*	■	☆	■
ATP-proton	ATP synthase F1 α subunit	0.704	1.888	1.909	*	■	☆	☆
	ATP synthase F1 β subunit	0.636	2.114	2.061	*	■	☆	☆

In the glutamine synthetase tree, an essential enzyme in ammonia assimilation and glutamine biosynthesis, Archaea and Gram-positive bacteria also cluster together (Tiboni et al., 1993; Brown et al., 1994). The glutamine synthetase phylogeny shown in Figure 6, which has been updated from Brown et al. (1994) with new archaeal, bacterial and eukaryotic sequences, displays the subdivisions within this protein family. Glutamine synthetase type I (GSI) is widely found in the Bacteria and Archaea but not in eukaryotes while glutamine synthetase type II (GSII) occurs in eukaryotes, and a few species of bacteria. Bacteria with a GSII type synthetase usually have a GSI type as well.

Within the GSI type there are two further subdivisions: GSI-α type proteins occur in the Archaea and low G+C Gram-positive bacteria (Bacillus cereus, B. subtilis, Clostridium acetobutylicum, Lactobacillus delbrueckii, and Staphylococcus aureus), Thermatogales (Thermotoga maritima) and the green non-sulfur bacteria (Deinococcus radiodurans); GSI-β type enzymes occur in all other bacteria, including cyanobacteria, proteobacteria and high G+C Gram positives. These subdivisions are also supported by data on gene structure and function. GSI-β type proteins have a 24-28 amino acid long insertion which is absent from GSII and GSI-α type proteins, with the exception of Archeoglobus fulgidus and Sulfolobus solfataricus which have a 22 amino acid long insertion in the same location as those characteristic of GSI-β type synthetases. However, the archaeal and bacterial inserted regions are highly divergent from one another. Unique to nearly all GSI-β enzymes is post-translational regulation of expression by reversible deadenylylation and adenylylation (reviewed in Brown et al., 1994). Only cyanobacteria lack this control mechanism but sequence analysis suggests that this might be due to a secondary loss in function.

The inclusion of new GSI sequences lends further insight into the evolution of glutamine synthetase. Within the GSI-α clade, the Archaea form the lowest branches, except for the halophile, Haloferax volcanii which clusters with the low G+C Gram-positive bacteria. Sulfolobus solfataricus, a member of the Crenarchaeota, and A. fulgidus, a member of the Euryarchaeota, branch first and second lowest in the GSI-α clade, respectively. Three species of methanogens branch higher and as a separate clade. The intermixing of H. volcanii with bacterial species would suggest that bacterial GSI-α might have originated from an archaeon closely related to contemporary halophilic lineages.

The present phylogeny extends the occurrence of GSI-α type synthetase to another bacterial species which is low branching in rRNA trees, Deinococcus radiodurans, an organism tolerant of high radiation levels and a close relative to the thermophile, Thermotoga maritima (Olsen et al., 1994). The occurrence of GSI-α type proteins in two lower branching bacteria as well as more recently diverged low G+C Gram positive organisms, would suggest either that gene transfer from an archaeon to a bacterium occurred at least twice. Alternatively, there was only a single gene transfer because Deinococcus radiodurans and Thermotoga maritima are actually closely related to low G+C Gram-positive bacteria.

Interestingly, the extreme thermophile Aquifex aeolicus, the most deeply branching bacterium in the rRNA tree (Burggraf et al., 1992; Deckert et al., 1998), does not cluster with either Deinococcus or Thermotoga rather it has a GSI-β type enzyme (including its characteristic amino acid sequence insertion). The branching order within the GSI-β cluster would suggest two further subgroups, one of the proteobacteria (except Helicobacter pylori)

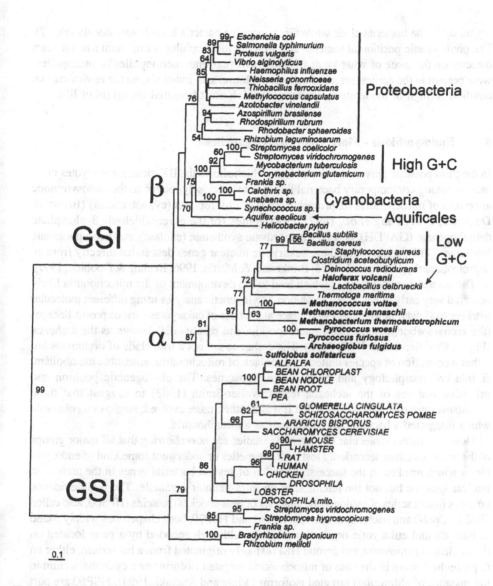

Figure 6. Neighbour-joining tree of glutamine synthetase. Clusters of GSI and GSII type synthetases are shown as well as the a and b subgroups within the GSI type. The positions of various bacterial groups are shown, including the high G+C and low G+C Gram-positives. Numbers show the percentage occurrence of nodes in 100 bootstrap replications, and only values greater than 50% are shown.

and the other of *A. aeolicus*, cyanobacteria and low G+C Gram-positive bacteria. Although the precise branching order of GSI-β sequences is somewhat indeterminate, it is clear that *Aquifex* is not the earliest bacterial lineage as proposed in rRNA phylogenies. The phylogeny for HSP60, discussed in the next section, similarly places *A. aeolicus*, higher up in the bacterial clade while Gram-positive bacteria branch very deeply (Fig. 7). The phylogenetic position of bacterial and archaeal thermophiles is important since it bears directly on the issue of what kinds of organisms (extreme thermophiles or mesophiles) were present in the earliest stages of cellular evolution and, indirectly, on the environmental conditions (high or moderate temperatures) which most favoured the origin of life.

8   Endosymbiosis - More Than Just Organelles

In the third possible universal tree topology, which shows the BK cluster, eukaryotes often branch among contemporary bacterial lineages most closely related to the endosymbiotic ancestors of mitochondria (α-proteobacteria) or chloroplasts (cyanobacteria) (Brown & Doolittle, 1997; Feng et al., 1997). Phylogenies for the glyceraldehyde 3-phosphate dehydrogenase (GAPDH), phosphoenolpyruvate synthetase (enolase), and triosephosphate isomerase (TPI), suggest that these eukaryotic nuclear genes descended directly from an α-proteobacteria (Henze et al., 1995; Brinkmann & Martin, 1996; Keeling & Doolittle, 1997).

The bacterial endosymbiosis which lead to the development of the mitochondria likely occurred very early in eukaryotic evolution. Phylogenetic analyses using different molecular markers (reviewed in Keeling, 1998), place at the base of eukaryotes various protist lineages (the archamoeboe, metamonads, microsporidia and parabasalia) known as the archezoa (Fig. 1; Cavalier-Smith 1983, 1993). The archezoa are not a true clade of organisms but rather a collection of species united by their lack of mitochondria, anaerobic metabolism, simple cell morphology and bacteria-like ribosomes. The phylogenetic position and primitive features of the archezoa lead Cavalier-Smith (1983) to suggest that these organisms were relic lineages, and that it was another, more evolved, single-cell eukaryote which integrated a bacterial endoysmbiont as a mitochondria.

However, recent molecular phylogenetic studies are now showing that all major groups of the archezoa either secondarily lost their organelles or underwent some kind of endosymbiosis which resulted in the successful fixation of several bacterial genes in the archezoan nuclear genome but not the retention of an intra-cellular organelle. The main evidence emerges from a series of phylogenetic analyses of heat-shock 60 proteins (HPS60, also called GroEL, Cpn60 and bacterial common antigen) and HSP70, both chaperones widely found in bacteria and eukaryotic organelles (Fig. 7). Although encoded by a gene located on the nuclear chromosome, eukaryotic HSP60 likely originated from a bacterium, either an α-proteobacterium in the case of mitochondria targeted isoforms or a cyanobacterium in the instance of chloroplast targeted isoforms (Viale and Arakaki, 1994). HSP60 are part of a larger protein family which includes another eukaryotic chaperone, TCP-1 (or CCT), and TCP-1's closest prokaryotic relative, a widely found archaeal chaperone called the thermosome (see Brown & Doolittle, 1997).

Trichononas vaginalis, a parabaslium, has mitochondria-specific HSP60 (Roger et al., 1996) and HSP70 genes (Germot et al., 1996). A mitochondria-specific HSP60 homologue has been found in Giardia lamblia, a diplomonad which is among the deepest branching archezoan (and eukaryotic) lineage (Roger et al., 1998). Clark & Roger (1995) describe HSP60 and another mitochondrion specific gene, pyridine nucleotide transhydrogenase from the protist Entamoeba histolytica. Although lacking mitochondria, Entamoeba is

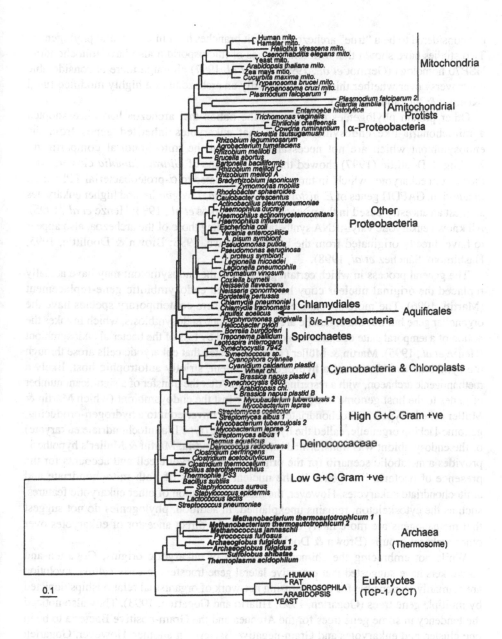

Figure 7. Neighbour-joining tree of bacterial and organellar 60-kDa heat shock proteins (HSP60 or Cpn60) rooted by archaeal thermosome and eukaryotic TCP-1 proteins. Note that HSP60 of amitochondrial protists clusters with similar proteins from eukaryotes targeted to the mitochondria (mito.) and α-proteobacteria, the putative bacterial symbiont ancestor of the mitochondria.

not considered to be a "true" archezoan since it branches high in eukaryotic phylogenies. Two studies have shown that different species of microsporidia also have mitochondrial HSP70 homologs (Germot *et al.*, 1997; Hirt *et al.* 1997) although there is considerable controversy over whether this group is actually an archezoan or a highly modified fungi (Keeling & McFadden, 1998).

Other protein phylogenies also support the notion that archezoa had once adopted a mitochondria-like endosymbiont, and that eukaryotes inherited genes from the endosymbiont which are not necessarily active in the mitochondrial compartment. Keeling & Doolittle (1997) showed that the TPI gene of *Giardia lamblia* clusters with those from eukaryotes which, in turn, are closely related to α-proteobacterial TPI genes. In addition, GAPDH genes of *E. histolytica*, *G. lamblia*, *T. vaginalis*, and higher eukaryotes suggest an ancestry based in the proteobacteria (Markos *et al.*, 1993; Henze *et al.*, 1995). All known eukaryotic valyl-tRNA synthetases, including those of the archezoa, also appear to have directly originated from the bacteria (Brown, 1998; Brown & Doolittle, 1995; Hashimoto, Sánchez *et al*, 1998).

The general process in which certain genes from the endosymbiont may have actually replaced the original nuclear copy has been called endosymbiotic gene-replacement (Martin, 1996). The more specialized instances where contemporary species have the organellar gene but not the organelle are termed cryptic endosymbiosis, which invokes the notion of a temporal state of endosymbiosis, followed by loss of the bacterial endosymbiont (Henze *et al.*, 1995). Martin & Müller (1998) suggested that eukaryotic cells arose through the symbiotic association of a hydrogen-dependent, strictly autotrophic host, likely a methanogenic archaeon, with a respiring bacterium. After the transfer of a significant number of genes to the host genome, the host cell either lost the endoysmbiont (which Martin & Müller called a type I amitochondriate eukaryote), converted it to a hydrogen-producing, genome-lacking organelle, called the hydrogenosome (type II amitochondriate eukaryote) or the endosymbiont was transformed into a mitochondria. Martin & Muller's hypothesis provides a metabolic scenario for the origin of the eukaryotic cell and accounts for the presence of bacteria-like genes in the nuclear genomes of both mitochondriate and amitochondriate eukaryotes. However, the mode of evolution of other eukaryotic features, such as the cytoskeleton, remains unexplained. In addition, phylogenies do not suggest that methanogens are more likely candidates for the direct ancestor of eukaryotes over other archaeal groups (Brown & Doolittle, 1997).

While not embracing the chimera hypothesis of eukaryotic origins, Gogarten and co-workers have suggested that extensive lateral gene transfers in early cellular evolution are primarily responsible for the confusing net-work of organismal relationships depicted by multiple gene trees (Gogarten, 1995; Hilario and Gogarten, 1993). They also noticed the tendency in some gene trees for the Archaea and the Gram-positive Bacteria to be in one cluster, and eukaryotes and Gram-negative Bacteria in another. However, Gogarten *et al.* (1996) suggested that any transfer of genes from Bacteria to eukaryotes had occurred earlier than organellar endosymbiosis.

Aside from the possibility of variable rates of evolution or gene convergence, two basic evolutionary scenarios might be behind the existence of so many conflicting protein gene trees. First, the eukaryotic nuclear genome is a chimera which resulted from some past cellular and genome fusion event involving a bacterium and an archaeon. As such, one subset of eukaryotic genes would be similar to archaeal homologues while a different subset would look most like bacterial counterparts. The most ancient divergence event would be that separating the Bacteria from the Archaea. However, protein trees where the Archaea and Bacteria cluster together, the two domains are never depicted as monophyletic groups which suggests more recent gene transfers are responsible for such tree topologies.

The second scenario depicts the archaeal tree as being essentially correct - that is, from the cenancestor, two lineages emerged, one leading to the Bacteria with the other splitting later into the Archaea and eukaryotes. Accordingly, many gene trees should position eukaryotes and the Archaea as sister domains, and each as separate mono-phyletic groups, which is often the case. Protein gene trees incongruent with the archaeal tree must be explained in terms of either lateral gene transfers or specific gene losses which occurred post-emergence of the three domains. Collectively, phylogenetic evidence from both duplicated (hence rooted) and unrooted, gene trees would seem generally, but not absolutely, consistent with the latter scenario. Radical differences in evolutionary rates, which simultaneously occurred throughout all species in one domain but not the remaining two domains, would also produce incongruent protein trees. However, preliminary analyses provide little evidence for such extreme evolutionary rate differentials at the domain level (Brown & Doolittle, 1997).

9   Evolutionary Biology and Genomics

As genomic sequencing projects hit their stride over the coming years, placing this new information into a meaningful evolutionary context will be a major challenge to evolutionary biologists. A new dimension to evolutionary analysis will be added as we move into the era of post-genomics where expression and functional data are also being collected across the entire genome of an organism (greatly facilitated by the so-called DNA chip technology). Thus, genomic information from a wide variety of Archaea, Bacteria and eukaryotes presents a new opportunity to synthesize a grander view of the prokaryote-eukaryote transition.

In some respects, additional genomic sequence data has complicated our perceptions about cellular biology - it was easier to pose plausible hypotheses when, in our ignorance, we could speculate about the contents of a genome. Perhaps, the most notable features about the first bacterial genomes sequenced were those genes that were missing rather than those that were present. Decades of genetic experiments and partial genome sequences of the mainstays of microbiology, *Escherichia coli* and *Bacillus subtilis* (which are now completed), perhaps lifted our expectations about the contents of these first bacterial genomes. If we are surprised by the absence of even a few types of genes from different organisms, then the ubiquity of any gene family is open to question. In this sense, the development of more robust universal trees is even more urgent in order to determine where genes appear or disappear - in the twigs or the load-bearing branches of the tree of life?

It is also evident that the Bacteria, Archaea and eukaryotes are much too diverse to be characterized by just one complete genome apiece. Fortunately, genomes sequences are becoming available from many examples of Bacteria and Archaea. A thorough sampling of the genome diversity of eukaryotes will naturally take longer due to their more complex genomes. Although the complete genomic sequence of *Saccharomyces cerevisiae* is known, fungi are unique among eukaryotes in having several bacterial and archaeal metabolic pathways such as aromatic amino acid biosynthesis. As such, comparsions between many different genomes will be important in discerning pan-domain characteristics from limited, intradomain or intrakingdom variation. If some consensus could be reached regarding the basic gene composition of bacterial, archaeal and eukaryotic genomes then the evolutionary origins of specific cellular characters might become clearer, although exceptions will always be found. Functional genomics holds the promise of allowing comparisons of gene regulatory mechanisms across diverse groups of organisms as well.

In some instances of phylogenetic reconstruction, we are perhaps pushing methodologies too far in trying to resolve the very deepest evolutionary branches. At other times, the gene trees seem to be highly robust, yet "wrong" from the perspective of organismal evolution. Most likely, there are many correct phylogenetic reconstructions of *gene* evolution but relatively fewer of *organismal* evolution. This statement implies that the evolution of genomes, or parts thereof, is sometimes decoupled from that of the host organism. Lateral gene transfer, is one example of a decoupled process. Now, what is needed is the development of new evolutionary paradigms where genomes, biochemistry and organisms are all considered, in concert.

Acknowledgments

We thank Andrei N Lupas for his comments on the nature of the HSP60 protein family.

References

Acca, M., Bocchetta, M., Ceccarelli, E., Creti, R., Stetter, K.O. & Cammarano, P. (1993). Updating mass and composition of archaeal and bacterial ribosomes, archaeal-like features of ribosomes from the deep-branching bacterium *Aquifex pyrophilus*. *Systematic and Applied Microbiology* 16, 129-137.

Amils, R., Cammarano, P. & Londei, P. (1993). Translation in Archaea. In *The Biochemistry of Archaea (Archaebacteria)*, pp. 393-438. Edited by M. Kates, D. J. Kushner, & A. T. Matheson). Elsevier; Amsterdam.

Amiri, K.A. (1994). Fibrillarin-like proteins occur in the domain Archaea. *Journal of Bacteriology* 176, 2124-2127.

Auer, J., Spicker, G. & Böck, A. (1989). Organization and structure of the *Methanococcus* transcriptional unit homologous to the *Escherichia coli* "spectinomycin operon", implications for the evolutionary relationship of 70S and 80S ribosomes. *Journal of Molecular Biology* 209, 21-36.

Baldauf, S.L., Palmer, J.D. & Doolittle, W.F. (1996). The root of the universal tree and the origin of eukaryotes based on elongation factor phylogeny. *Proceedings of the National Academy of Sciences of the USA* 93, 7749-7754.

Barns, S.M., Fundyga, R.E., Jefferies, M.W. & Pace, N.R. (1994). Remarkable archaeal diversity detected in a Yellowstone National Park hot spring environment. *Proceedings of the National Academy of Sciences of the USA* 91, 1609-1613.

Barnes, S.M., Delwiche, C.F., Palmer, J.D. & Pace, N.R. (1996). Perspectives on archaeal diversity, thermophily and monophyly from environmental rRNA sequences. *Proceedings of the National Academy of Sciences of the USA* **93**, 9188-9193.

Bartig, D., Lemkemeier, K., Frank, J., Lottspeich, F. & Klink, F. (1992). The archaebacterial hypusine-containing protein: Structural features suggest common ancestry with eukaryotic translation initiation factor 5A. *European Journal of Biochemistry* **204**, 751-758.

Baumann, P. & Jackson, S.P. (1996). An archaebacterial homologue of the essential eubacterial cell division protein FtsZ. *Proceedings of the National Academy of Sciences of the USA* **93**, 6726-6730.

Belfort, M., & Weiner, A. (1997). Another bridge between kingdoms: tRNA splicing in Archaea and eukaryotes. *Cell* **89**, 1003-1006.

Belfort, M., Reaban, M.E., Coetzee, T. & Dalgaard, J.Z. (1995). Prokaryotic introns and inteins: a panoply of form and function. *Journal of Bacteriology* **177**, 3897-3903.

Bianchi, M.E. (1994). Prokaryotic HU and eukaryotic HMG1: a kinked relationship. *Molecular Microbiology* **14**, 1-5.

Bochar, D.A., Brown, J.R., Doolittle, W.F., Klenk, H-P., Lam, W., Schenk, M.E., Stauffacher, C.V. & Rodwell, V.W. (1997). 3-Hydroxy-3-methylglutaryl coenzyme reductase of *Sulfolobus solfataricus*: DNA sequence, phylogeny, expression in *Escherichia coli* of the *hmgA* gene, and purification and kinetic characterization of the gene product. *Journal of Bacteriology* **179**, 3632-3638.

Brendel, V., Brocchieri, L., Sandler, S.J., Clark, A.J. & Karlin, S. (1997). Evolutionary comparisons of RecA-like proteins across all major kingdoms of living organisms. *Journal of Molecular Evolution* **44**, 528-541.

Brinkmann, H. & Martin, W. (1996). Higher-plant chloroplast and cytosolic 3-phosphoglycerate kinases: a case of endosymbiotic gene replacement. *Plant Molecular Biology* **30**, 65-75.

Brown, J.R. (1998) Aminoacyl-tRNA synthetases: evolution of a troubled family. In *Thermophiles - the Keys to Molecular Evolution and the Origin of Life?* pp. 217-230. Edited by J. Wiegel & M. Adams. Taylor & Francis Group Ltd., London.

Brown, J. R. & Doolittle, W.F. (1995). Root of the universal tree of life based on ancient aminoacyl-tRNA synthetase gene duplications. *Proceedings of the National Academy of Sciences of the USA* **92**, 2441-2445.

Brown, J.R. & Doolittle, W.F. (1997). Archaea and the prokaryote to eukaryotes transition. *Microbiology and Molecular Biology Reviews* **61**, 456-502.

Brown, J.R., Masuchi, Y., Robb, F.T. & Doolittle, W.F. (1994). Evolutionary relationships of bacterial and archaeal glutamine synthetase genes. *Journal of Molecular Evolution* **38**, 566-576.

Brown, J.R., Robb, F.T., Weiss, R. & Doolittle, W.F. (1997). Evidence for the early divergence of tyrosyl- and tyrptophanyl-tRNA synthetases. *Journal of Molecular Evolution* **45**, 9-16.

Brown, J.R., Zhang, J. & Hodgson, J.E. (1998). A bacterial antibiotic resistance gene with eukaryotic origins. *Current Biology* **8**, R365-R367.

Brown, J.W., Daniels, C.J. & Reeve, J.N. (1989). Gene structure, organization and expression in archaebacteria. *CRC Critical Reviews in Microbiology* **16**, 287-338.

Bult, C.J., White, O., Olsen, G.J., Zhou, L., Fleischmann, R.D., Sutton, G.G., Blake, J.A., Fitzgerald, L.M., Clayton, R.A., Gocayne, J.D., Kertavage, A.R., Dougherty, B.A., Tomb, J.-F., Adams, M.D., Reich, C.I., Overbeek, R., Kirkness, E.F., Weinstock, K.G., Merrick, J.M., Glodek, A., Scott, J.L., Geoghagen, N.S.M., Weidman, J.F., Fuhrmann, J.L., Nguyen, D., Utterback, T.R., Kelley, J.M., Peterson, J.D., Sadow, P.W., Hanna, M.C., Cotton, M.D., Roberts, K.M., Hurst, M.A., Kaine, B.P., Borodovsky, M., Klenk, H.-P., Fraser, C.M., Smith, H.O., Woese, C.R. & Venter, J.C. (1996). Complete genome sequence of the methanogenic Archaeon, *Methanococcus jannaschii*. *Science* **273**, 1058-1073.

Burggraf, S., Olsen, G.J., Stetter, K.O. & Woese, C.R. (1992). A phylogenetic analysis of *Aquifex pyrophilus*. *Systematic and Applied Microbiology* **15**, 352-356.

Cammarano, P., Palm, P., Creti, R., Ceccarelli, E., Sanangelantoni, A.M. & Tiboni, O. (1992). Early evolutionary relationships among known life forms inferred from elongation factor EF-2/EF-G sequences: Phylogenetic coherence and structure of the archaeal domain. *Journal of Molecular Evolution* **34**, 396-405.

Cavalier-Smith, T. (1983). A 6-kingdom classification and a unified phylogeny In *Endocytobiology II: Intracellular space as oligogenetic*, pp. 1027-1034. Edited by H.E.A. Schenk & W.S. Schwemmler. Berlin, Walter de Gruyter & Co..

Cavalier-Smith, T. (1992). Bacteria and eukaryotes. *Nature*, **356** 570.

Cavalier-Smith, T. (1993). Kingdom protozoa and its 18 phyla. *Microbiological Reviews* **57**, 953-994.

Charlebois, R.L., Schalkwyk, L.C., Hofman, J.D. & Doolittle, W.F. (1991). Detailed physical map and set of overlapping clones covering the genome of the archaebacterium *Haloferax volcanii* DS2. *Journal of Molecular Biology* **222**, 509-524.

Chatton, E. (1937). *Titres et Travaux Scientifiques*. Setes, Sottano, Italy.

Clark, C.G. & Roger, A.J. (1995). Direct evidence for secondary loss of mitochondria in *Entamoeba histolytica*. *Proceedings of the National Academy of Sciences of the USA* **92**, 6518-6521.

Cousineau, B., Leclerc, F. & Cedergren, R. (1997). On the origin of protein synthesis factors: A gene duplication/fusion model. *Journal of Molecular Evolution* **45**, 661-670.

Creti, R., Ceccaralli, E., Bocchetta, M., Sanangelantoni, A.M., Tiboni, O., Palm, P. & Cammarano, P. (1994). Evolution of translational elongation factor (EF) sequences: Reliability of global phylogenies inferred from EF-1α(Tu) and EF-2(G) proteins. *Proceedings of the National Academy of Sciences of the USA* **91**, 3255-3259.

Danson, M.J. (1993). Central metabolism of the Archaea In *The Biochemistry of Archaea (Archaebacteria)*, pp. 1-24. Edited by M. Kates, D.J. Kushner & A.T. Matheson. Elsevier, Amsterdam.

Danson, M.J., Eisenthal, R., Hall. S., Kessell, S.R. & William, D.L. (1984). Dihydrolipoamide dehydrogenase from halophilic archaebacteria. *Biochemical Journal* **218**, 811-818.

Dayhoff, M.O., Eck, R.V. & Park, C.M. (1972). A model of evolutionary change in proteins, in *Atlas of Protein Sequence and Structure*, vol. 5 pp. 89-99. Edited by M.O. Dayhoff. Washington, D.C: National Biomedical Research Foundation.

Deckert G., Warren, P.V., Gaasterland, T., Young, W.G., Lenox, A.L., Graham, D.E., Overbeek, R., Snead, M.A., Keller, M., Aujay, M., Huber, R.A., Feldman, R., Short, J.M., Olsen, G.J. & Swanson, R.V. (1998). The complete genome sequence of the hypertherophilic bacterium *Aquifex aeolicus*. *Nature* **392**, 353-358.

Delong, E.F. (1992). Archaea in coastal marine environments. *Proceedings of the National Academy of Sciences of the USA* **89**, 5685-89.

Delong, E.F., Wu, K.Y., Prézelin, B.B. & Jovine, R.V.M. (1994). High abundance of Archaea in Antarctic marine picoplankton. *Nature* **371**, 695-97.

Dennis, P.P. (1997). Ancient ciphers: translation in Archaea. *Cell* **89**, 1007-1010.

DiRuggiero, J., Brown, J.R., Bogert, A.M. & Robb, F.T. (1999). DNA repair systems in Archaea: mechanisms from the last universal common ancestor? *Journal of Molecular Evolution* **49**, 474-494.

Doolittle, W.F. & Brown, J.R. (1994). Tempo, mode, the progenote and the universal root. *Proceedings of the National Academy of Sciences of the USA* **91**, 6721-6728.

Doolittle, W.F. & Logsdon, Jr., J.M. (1998). Archaeal genomics: Do archaea have a mixed heritage? *Current Biology* **8**, R209-R211.

Doolittle, R.F., Feng, D.-F., Tsang, S., Cho, G. & Little, E. (1996). Determining divergence times of the major kingdoms of living organisms with a protein clock. *Science* **271**, 470-477.

Durovic, P. & Dennis, P.P. (1994). Separate pathways for excision and processing of 16S and 23S rRNA from the primary rRNA operon transcript from the hyperthermophilic archaebacterium *Sulfolobus acidocal-*

darius: similarities to eukaryotic rRNA processing. *Molecular Microbiology* 13, 229-242.

Edgell, D.R. & Doolittle, W.F. (1997). Archaea and the origin(s) of DNA replication proteins. *Cell* 89, 995-998.

Elie, C., Baucher, M.-F., Fondrat, C., & Forterre, P. (1997). A protein related to eucaryal and bacterial DNA-motor proteins in the hyperthermophilic archaeon *Sulfolobus acidocaldarius*. *Journal of Molecular Evolution* 45, 107-114.

Erickson, H.P. (1995). FtsZ, a prokaryotic homolog of tubulin? *Cell* 80, 367-370.

Felsenstein, J. (1993). PHYLIP (Phylogeny Inference Package) version 3.5c. Department of Genetics, University of Washington, Seattle. Distribution: http://evolution.genetics.washington.edu/phylip.html

Feng, D-F., Cho, G. & Doolittle, R.F. (1997). Determining the divergence times with a protein clock: Update and reevaluation. *Proceedings of the National Academy of Sciences of the USA* 94, 13028-13033.

Fitch, W.M. & Upper, K. (1987). The phylogeny of tRNA sequences provides evidence for ambiguity reduction in the origin of the genetic code. *Cold Spring Harbour Symposium on Quantative Biology* 52, 759-767.

Forterre, P. & Elie, C. (1993). Chromosome structure, DNA topoisomerases and DNA polymerases in archaebacteria (Archaea). In *The Biochemistry of Archaea (Archaebacteria)*, pp. 325-365. Edited by M. Kates, D.J. Kushner, and A.T. Matheson. Amsterdam: Elsevier.

Forterre, P., Benachenhou-Lahfa, N., Confalonieri, F., Duguet, M., Elie, C. & Labedan, B. (1993). The nature of the last universal ancestor and the root of the tree of life, still open questions. *BioSystems* 28, 15-32.

Fox, G.E., Magrum, L.J., Balch, W.E., Wolfe, R.S. & Woese, C.R. (1977). Classification of methanogenic bacteria by 16S ribosomal RNA characterization. *Proceedings of the National Academy of Sciences of the USA* 74, 4537-4541.

Fuhrman, J.A., McAllum, K. & Davis, A.A. (1992). Novel major archaebacterial group from marine plankton. *Nature* 356, 148-149.

Germot, A., Philippe, H. & Le Guyader, H. (1996). Presence of a mitochondrial-type 70-kDa heat shock protein in *Trichomonas vaginalis* suggests a very early mitochondrial endosymbiosis in eukaryotes. *Proceedings of the National Academy of Sciences of the USA* 93, 14614-14617.

Germot, A., Philippe, H. & Le Guyader, H. (1997). Evidence for loss of mitochondria in microsporidia from a mitochondria-type HSP70 in *Nosema locustae*. *Molecular and Biochemical Parasitology* 87, 159-68.

Gogarten, J.P. (1995). The early evolution of cellular life. *Trends in Ecology and Evolution* 10, 147-151.

Gogarten, J.P., Hilario, E. & Olendzenski, L. (1996). Gene duplications and horizontal transfer during early evolution. In *Evolution of Microbial Life, SGM 54* pp. 267-292. Edited by D.M. Roberts, G. Alderson, P. Sharp and M. Collins. Cambridge: Cambridge University Press.

Gogarten, J.P., Kibak, H., Dittrich, P., Taiz, L., Bowman, E.J., Bowman, B.J., Manolson, N.F., Poole, R.J., Date, T., Oshima, T., Konishi, J., Denda, K. & Yoshida, M. (1989). Evolution of the vacuolar H⁺-ATPase: Implications for the origin of eukaryotes. *Proceedings of the National Academy of Sciences of the USA* 86, 6661-6665.

Golding, G.B. & Gupta, R.S. (1995). Protein-based phylogenies support a chimeric origin for the eukaryotic genome. *Molecular Biology and Evolution* 12, 1-6.

Gray, M. (1992). The endosymbiont hypothesis revisited. *International Reviews in Cytology* 141, 233-357.

Grayling, R.A., Sandman, K. & Reeve, J.N. (1994). Archaeal DNA binding proteins and chromosome structure. *Systematic and Applied Microbiology* 16, 582-590.

Gropp, F., Reiter, W-D., Sentenac, A., Zillig, W., Schnabel, R., Thomm, M. & Stetter, K.O. (1986). Homologies of components of DNA-dependent RNA polymerases of archaebacteria, eukaryotes and eubacteria. *Systematic and Applied Microbiology* 7, 95-101.

Gupta, R.S. & Golding, G.B. (1993). Evolution of HSP70 gene and its implications regarding relationships between archaebacteria, eubacteria and eukaryotes. *Journal of Molecular Evolution* 37, 573-582.

Gupta, R.S. & Golding, G.B. (1996). The origin of the eukaryotic cell. *Trends in Biochemical Sciences* 21, 166-171.

Gupta, R.S. & Singh, B. (1992). Cloning of the HSP70 gene from *Halobacterium marismortui*: Relatedness of archaebacterial HSP70 to its eubacterial homologs and a model for the evolution of the HSP70 gene. *Journal of Bacteriology* 174, 4594-4605.

Gupta, R.S. & Singh, B. (1994). Phylogenetic analysis of 70 kD heat shock protein sequences suggests a chimeric origin for the eukaryotic cell nucleus. *Current Biology* 4, 1104-1114.

Gupta, R.S., Aitken, K., Falah, M. & Singh, B. (1994). Cloning of *Giardia lamblia* heat shock protein HSP70 homologs: Implications regarding origin of eukaryotic cells and of endoplasmic reticulum. *Proceedings of the National Academy of Sciences of the USA* 91, 2895-2899.

Hasimoto, T., Sánchez, L.B., Shirakura, T., Müller, M. & Hasegawa, M. (1998). Secondary absence of mitochondria in *Giardia lamblia* and *Trichomonas vaginalis* revealed by valyl-tRNA synthetase phylogeny. *Proceedings of the National Academy of Sciences of the USA.* 95, 6860-6865.

Henderson, E., Oakes, M., Clark, M.W., Lake, J.A., Matheson, A.T. & Zillig, W. (1984). A new ribosome structure. *Science* 225, 510-512.

Henze, K.A., Badr, A., Wettern, M., Cerff, R. & Martin, W. (1995). A nuclear gene of eubacterial origin in *Euglena gracilis* reflects cryptic endosymbioses during protist evolution. *Proceedings of the National Academy of Sciences of the USA,* 92, 9122-9126.

Hilario, E. & Gogarten, J.P. (1993). Horizontal transfer of ATPase genes - the tree of life becomes the net of life. *BioSystems* 31, 111-119.

Hirata, R., Ohsumk, Y., Nakano, A., Kawasaki, H., Suzuki, K. & Anraku, Y. (1990). Molecular structure of a gene, VMA-1, encoding the catalytic subunit of H(+)-translocating adenosine triphosphatase from vacuolar membranes of *Saccharomyces cerevisiae*. *Journal of Biological Chemistry* 265, 6726-6733.

Hirt, R.P., Healy, B., Vossbrinck, C.R., Canning, E.U. & Embley T.M. (1997). A mitochondrial HSP70 orthologue in *Vairimorpha necatrix*: molecular evidence that microsporidia once contained mitochondria. *Current Biology* 7, 995-998.

Iwabe, N., Kuma, K-I., Hasegawa, M., Osawa, S. & Miyata, T. (1989). Evolutionary relationship of Archaea, Bacteria, and eukaryotes inferred from phylogenetic trees of duplicated genes. *Proceedings of the National Academy of Sciences of the USA* 86, 9355-9359.

Kakinuma, Y., Igarishi, K., Konishi, K. & Yamato, I. (1991). Primary structure of the alpha-subunit of vacuolar-type Na$^+$-ATPase in *Enterococcus hirae*, amplification of a 1000 bp fragment by polymerase chain reaction. *FEBS Letters* 292, 64-68.

Kane, P.M., Yamashiro, C.T., Wolczyk, D.F., Neff, N., Goebi, M. & Stevens, T.H. (1990). Protein splicing converts the yeast TFP1 gene product to the 69-kD subunit of the vacuolar H(+)-adenosine triphosphatase. *Science* 250, 651-657.

Kates, M., Kushner, D.J. & Matheson, A.T. (1993). *The Biochemistry of Archaea (Archaebacteria)*. Amsterdam: Elsevier.

Keeling, P.J. (1998). A kingdom's progress: Archezoa and the origin of eukaryotes. *BioEssays* 20, 87-95.

Keeling, P.J. & Doolittle, W.F. (1995). Archaea: Narrowing the gap between prokaryotes and eukaryotes. *Proceedings of the National Academy of Sciences of the USA* 92, 5761-5764.

Keeling, P.J. & Doolittle, W.F. (1996). Methionine aminopeptidase-1: the MAP of the mitochondrion. *Trends in Biochemical Sciences* 21, 285-286.

Keeling, P.J. & Doolittle, W.F. (1997). Evidence that eukaryotic triosephosphate isomerase is of alpha-proteobacterial origin. *Proceedings of the National Academy of Sciences of the USA* 94, 1270-1275.

Keeling, P.J. & McFadden, G.I. (1998). Origins of microsporidia. *Trends in Microbiology* 6, 19-23.

Keeling, P.J., Charlebois, R.L. & Doolittle, W.F. (1994). Archaebacterial genomes: eubacterial form and eukaryotic content. *Current Opinions in Genetics and Development* 4, 816-822.

Keeling, P.J., Baldauf, S.L., Doolittle, W.F., Zillig, W. & Klenk, H.-P. (1996). An infB-homolog in *Sulfolobus acidocaldarius*. *Systematic and Applied Microbiology* 19, 312-321.

Kengen, S.W.M., de Bok, F.A.M., van Loo, N.-D., Dijkema, C., Stams, A.J.M. & de Vos, W.M. (1994). Evidence of the operation of a novel Embden-Meyerhof pathway that involves ADP-dependent kinases during sugar fermentation by *Pyrococcus furiosus*. *Journal of Biological Chemistry* 269, 17537-17541.

Kleman-Leyer, K., Armbruster, D.W. & Daniels, C.J. (1997). Properties of *H. volcanii* tRNA intron endo nuclease reveal a relationship between the archaeal and eucaryal tRNA intron processing systems. *Cell* 89, 839-847.

Klenk, H-P., Palm, P. & Zillig, W. (1993). DNA-dependent RNA polymerases as phylogenetic marker molecules. *Systematic and Applied Microbiology* 16, 138-147.

Klenk, H.-P., Clayton, R.A., Tomb, J.-F., White, O., Nelson, K.E., Ketchum, K.A., Dodson, R.J., Gwinn, M., Hickey, E.K., Peterson, J.D., Richardson, D.L., Kerlavage, A.R., Graham, D.E., Kyrpides, N.C., Fleischmann, R.D., Quackenbush, J., Lee, N.H., Sutton, G.G., Gill, S., Kirkness, E.F., Dougherty, B.A., McKenny, K., Adams, M.D., Loftus, B., Peterson, S., Reich, C.I., McNeil, L.K., Badger, J.H., Glodek, A., Zhou, L., Overbeek, R., Gocayne, J. D., Weidman, J. F., McDonald, L., Utterback, T., Cotton, M. D., Spriggs, T., Artiach, P., Kaine, B.P., Sykes, S.M., Sadow, P.W., D'Andrea, K.P., Bowman, C., Fujii, C., Garland, S.A., Mason, T.M., Olsen, G.J., Fraser, C.M., Smith, H.O., Woese, C.R. and Venter, J.C. (1997). The complete genome sequence of the hyperthermophilic, sulphate-reducing archaeon *Archaeoglobus fulgidus*. *Nature* 390, 364-370.

Kyrpides, N.C. & Woese, C.R. (1998). Universally conserved translation initiation factors. *Proceedings of the National Academy of Sciences of the USA* 95, 224-228.

Lake, J.A. (1988). Origin of the eukaryotic nucleus determined by rate-invariant analysis of rRNA sequences. *Nature*, 331, 184-86.

Lake, J.A. & Rivera, M.C. (1994). Was the nucleus the first endosymbiont? *Proceedings of the National Academy of Sciences of the USA* 91, 2880-2881.

Lake, J.A., Henderson, E., Oakes, M. & Clark, M.W. (1984). Eocytes: a new ribosome structure indicates a kingdom with a close relationship to eukaryotes. *Proceedings of the National Academy of Sciences of the USA* 81, 3786-3790.

Lake, J.A., Clark, M.W., Henderson, E., Fay, S.P., Oakes, M., Scheinman, A., Thornber, J.P. & Mah, R.A. (1985). Eubacteria, halobacteria, and the origin of photosynthesis: The photocytes. *Proceedings of the National Academy of Sciences of the USA* 82, 3716-3720.

Lam, W.L. & Doolittle, W.F. (1992). Mevinolin-resistant mutations identify a promoter and the gene for a eukaryote-like 3-hydroxy-3-methylglutaryl-coenzyme A reductase in the archaebacterium *Haloferax volcanii*. *Journal of Biological Chemistry* 267, 5829-5834.

Langer, D., Hain, J., Thuriaux, P. & Zillig, W. (1995). Transcription in Archaea: similarity to that in Eucarya. *Proceedings of the National Academy of Sciences of the USA* 92, 5768-5772.

Larkin, J.M. & Henk, M.C. (1996). Filamentous sulfide-oxidizing bacteria at hydrocarbon seeps of the Gulf of Mexico. *Microscopy Research Technique* 33, 23-31.

Lawson, F.S., Charlebois, R.L. & Dillon, J.-A.R. (1996). Phylogenetic analysis of carbamoylphosphate synthetase genes: evolution involving multiple gene duplications, gene fusions, and insertions and deletions of surrounding sequences. *Molecular Biology and Evolution* 13, 970-977.

Lindmark, D.G. (1980). Energy metabolism of the anaerobic protozoon *Giardia lamblia*. *Molecular and Biochemical Parasitology* 1, 1-12.

Margulis, L. (1970). *Origin of Eukaryotic Cells*.New Haven CT, Yale University Press.

Margulis, L. & Guerreo, R. (1991). Kingdoms in turmoil. *New Scientist* 129, 46-50.

Markos, A., Miretsky, A. & Müller, M. (1993). A glyceraldehyde-3-phosphate dehydrogenase with eubacterial features in the amitochondriate eukaryote, *Trichomonas vaginalis*. *Journal of Molecular Evolution* 37, 631-643.

Marsh, C.L., Reich, C.I., Whitlock, R.B. & Olsen, G.J. (1994). Transcription factor IID in the Archaea: sequences in the *Thermococcus celer* genome would encode a product closely related to the TATA-binding protein of eukaryotes. *Proceedings of the National Academy of Sciences of the USA* 91, 4180-4184.

Martin, W. (1996). Is something wrong with the tree of life? *BioEssays* 18, 523-527.

Martin, W. & Müller, M. (1998). The hydrogen hypothesis for the first eukaryote. *Nature* 392, 37-41.

Mayr, E. (1990). A natural system of organisms. *Nature* 348, 491.

Müller, E.-C. & Wittmann-Liebold, B. (1997). Phylogenetic relationship of organisms obtained by ribosomal protein comparison. *Cellular and Molecular Life Sciences* 53, 34-50.

Musgrave, D.R., Sandman, K.M. & Reeve, J.N. (1991). DNA binding by the archaeal histone HMf in positive supercoiling. *Proceedings of the National Academy of Sciences of the USA* 88, 10397-10401.

Nolling, J. & Vos, W.M. (1992). Characterization of the archaeal, plasmid-encoded type II restriction-modification system MthTI from *Methanobacterium thermoformicicum* THF: homology to the bacterial NgoPII system from *Neisseria gonorrhoeae*. *Journal of Bacteriology*, 174, 5719-5726.

Olsen, G.J., Woese, C.R. & Overbeek, R. (1994). The winds of (evolutionary) change: breathing new life into microbiology. *Journal of Bacteriology* 176, 1-6.

Osteryoung, K.W. & Vierling, E. (1995). Conserved cell and organelle division. *Nature* 376, 473-474.

Ouzounis, C. & Sander, C. (1992). TFIIB, an evolutionary link between the transcription machineries of archaebacteria and eukaryotes. *Cell* 71, 189-190.

Perler, F.B., Comb, D.G., Jack, W.E., Moran, L.S., Qiang, B., Kucera, R.B., Benner, J., Slatko, B.E., Nwankwo, D.O., Hempstead, S.K., Carlow, C.K.S. & Jannasch, H. (1992). Intervening sequences in an Archaea DNA polymerase gene. *Proceedings of the National Academy of Sciences of the USA* 89, 5577-5581.

Potter, S., Durovic, P. & Dennis, P.P. (1995). Ribosomal RNA precursor processing by a eukaryotic U3 small nucleolar RNA-like molecule in an Archaeon. *Science* 268, 1056-1060.

Preston, C.M., Wu, K.Y., Molinski, T.F. & DeLong, E.F. (1996). A psychrophilic crenarchaeon inhabits a marine sponge: *Crenarchaeum symbiosum* gen. nov., sp. nov. *Proceedings of the National Academy of Sciences of the USA* 93, 6241-6246.

Pühler, G., Leffers, H., Gropp, F., Palm, P., Klenk, H-P., Lottspeich, F., Garrett, R.A. & Zillig, W. (1989). Archaebacterial DNA-dependent RNA polymerases testify to the evolution of the eukaryotic nuclear genome. *Proceedings of the National Academy of Sciences of the USA* 86, 4569-4573.

Qureshi, S.A., Khoo, B., Baumann, P. & Jackson, S.P. (1995). Molecular cloning of the transcription factor TFIIB homolog from *Sulfolobus shibatae*. *Proceedings of the National Academy of Sciences of the USA* 92, 6077-6081.

Ragan, M.A., Logsdon Jr., J.M., Sensen, C.W., Charlebois, R.L. & Doolittle, W.F. (1996). An archaebacterial homolog of pelota, a meiotic cell division protein in eukaryotes. *FEMS Microbiology Letters* 144, 151-155.

Ramirez, C., Köpke, A.K.E., Yang, D-C., Boeckh, T. & Matheson, A.T. (1993). The structure, function and evolution of archaeal ribosomes. In *The Biochemistry of Archaea (Archaebacteria)*. pp. 439-66. Edited by M. Kates, D.J. Kushner, and A.T. Matheson. Amsterdam:Elsevier.

Rashid, N., Morikawa, M., Nagahisa, K., Kanaya, S. & Imanaka, T. (1997). Characterization of a RecA/Rad51 homologue from the hyperthermophilic archaeon *Pyrococcus* sp. KOD1. *Nucleic Acids Research* 25, 719-726.

Reeves, J.N., Sandman, K. & Daniels, C.J. (1997). Archaeal histones, nucleosomes and transcription initiation. *Cell* 89, 999-1002.

Reeves, R.E., Warren, L.G., Suskind, B. & Lo, H.S. (1977). An energy-conserving pyruvate-to-acetate pathway in *Entamoeba histolytica*: Pyruvate synthase and a new acetate thiokinase. *Journal of Biological Chemistry* 252, 726-731.

Reiter, W-D., Hüdepohl, U. & Zillig, W. (1990). Mutational analysis of an archaebacterial promoter: essential role of a TATA box for transcription efficiency and start-site selection in vitro. *Proceedings of the National Academy of Sciences of the USA* 87, 9509-9513.

Rivera, M.C. & Lake, J.A. (1992). Evidence that eukaryotes and eocyte prokaryotes are immediate relatives. *Science* 257, 74-76.

Roger, A.J. & Brown, J.R. (1996). A chimeric origin for eukaryotes re-examined. *Trends in Biochemical Sciences* 21, 370-371.

Roger, A.J., Clark, C.G. & Doolittle, W.F. (1996). A possible mitochondrial gene in the early-branching amitochondriate protist *Trichomonas vaginalis*. *Proceedings of the National Academy of Sciences of the USA* 93, 14618-14622.

Roger, A.J., Svard, S.G., Tovar, J., Clark, C.G., Smith, M.W., Gillin, F.D. & Sogin, M.L. (1998). A mitochondrial-like chaperonin 60 gene in *Giardia lamblia*: evidence that diplomonads once harbored an endosymbiont related to the progenitor of mitochondria. *Proceedings of the National Academy of Sciences of the USA* 95, 229-234.

Rowlands, T., Baumann, P. & Jackson, S.P. (1994). The TATA-binding protein: a general transcription factor in eukaryotes and archaebacteria. *Science* 264, 1326-1329.

Russell, A.G., Moniz de Sa, M. & Dennis, P. (1997). A U3-like small nucleolar RNA in Archaea: retraction. *Science* 277, 1189.

Sandler, S.J., Satin, L.H., Samra, H.S. & Clark, A.J. (1996). RecA-like genes from three archaean species with putative protein products similar to Rad51 and Dmc1 proteins of the yeast *Saccharomyces cerevisiae*. *Nucleic Acids Research* 24, 2125-2132.

Sandman, K., Krzycki, J.A., Dobrinski, B., Lurz, R. & Reeve, J.N. (1990). HMf, a DNA-binding protein isolated from the hyperthermophilic Archaeon *Methanothermus fervidus*, is most closely related to histones. *Proceedings of the National Academy of Sciences of the USA* 87, 5788-5791.

Schönheit, P. & Schäfer, T. (1995). Metabolism of hyperthermophiles. *World Journal of Microbiology and Biotechnology* 11, 26-75.

Schopf, J.W. (1993). Microfossils of the early Archaen apex chart: new evidence of the antiquity of life. *Science* 260, 640-646.

Smith, D.R., Doucette-Stamm, L.A., Deloughery, C., Lee, H., Dubois, J., Aldredge, T., Bashirzadeh, R., Blakely, D., Cook, R., Gilbert, K., Harrison, D., Hoang, L., Keagle, P., Lumm, W., Pothier, B., Qiu, D., Spadafora, R., Vicaire, R., Wang, Y., Wierzbowski, J., Gibson, R., Jiwani, N., Caruso, A., Bush, D., Safer, H., Patwell, D., Prabhakar, S., McDougall, S., Shimer, G., Goyal, A., Pietrokovski, S., Church, G.M., Daniels, C.J., Mao, J.-I., Rice, P., Nölling, J. & Reeve, J.N. (1997). Complete genome sequence of *Methanobacterium thermoautotrophicum* ΔH: Functional analysis and comparative genomics. *Journal of Bacteriology* 179, 7135-7155.

Smith, M.W., Feng, D-F. & Doolittle, R. F. (1992). Evolution by acquisition: the case for horizontal gene transfers. *Trends in Biochemical Sciences* 17, 489-493.

Sogin, M.L. (1991). Early evolution and the origin of eukaryotes. *Current Opinion in Genetics and Development* 1, 457-463.

Stanier, R.Y. (1970). Some aspects of the biology of cells and their possible evolutionary significance.

Symposium of the Society of General Microbiology, **20**, 1-38.

Stanier, R.Y. & van Niel, C. B. (1941). The main outlines of bacterial classification. *Journal of Bacteriology* **42**, 437-466.

Stanier, R.Y. & van Niel, C.B. (1962). The concept of a bacterium. *Archives für Mikrobiologie* **42**, 17-35.

Starich, M.R., Sandman, K., Reeve, J.N. & Summers, M.F. (1996). NMR structure of HMfB from the hyperthermophile, *Methanothermus fervidus*, confirms that this archaeal protein is a histone. *Journal of Molecular Biology* **255**, 187-203.

Stein, J.L. & Simon, M.I. (1996). Archaeal ubiquity. *Proceedings of the National Academy of Sciences of the USA* **93**, 6228-6230.

Stöffler, G. & Stöffler-Meilicke, M. (1986). Electron microscopy of archaebacterial ribosomes. *Systematic and Applied Microbiology*, **7**, 123-30.

Stöffler-Meilicke, M., Böhme, C.,Strobel, O., Böck, A. & Stöffler, G. (1986). Structure of ribosomal subunits of *M. vanniellii*: Ribosomal morphology as a phylogenetic marker. *Science* **231**, 1306-1308.

Sumi, M., Sato, M.H., Denda, K., Date, T. & Yoshida, M. (1992). A DNA fragment homologous to F_1-ATPase β subunit amplified from genomic DNA of *Methanosarcina barkeri*: Indication of an archaebacterial F-type ATPase. *FEBS Letters* **314**, 207-210.

Swofford, D.L. (1993). PAUP: phylogenetic analysis using parsimony, version 3.1.1 computer program. Distributed by the Illinois Natural History Survey. Champaign, IL, USA

Thomm, M. (1996). Archaeal transcription factors and their role in transcription initiation. *FEMS Microbiology Reviews* **18**, 159-171.

Tiboni, O., Cammarano, P. & Sanangelantoni, A.M. (1993). Cloning and sequencing of the gene encoding glutamine synthetase I from the archaeon *Pyrococcus woesei*: anomalous phylogenies inferred from analysis of archaeal and bacterial glutamine synthetase I sequences. *Journal of Bacteriology* **175**, 2961-2969.

Tiboni, O., Cantoni, R., Creti, R., Cammarano, P. & Sanangelantoni, A.M. (1991). Phylogenetic depth of *Thermotoga maritima* inferred from analysis of the *fus* gene: Amino acid sequence of elongation factor G and organization of the *Thermotoga str* operon. *Journal of Molecular Evolution* **33**, 142-151.

Trotta, C.R., Miao, F., Arn, E.A., Stevens, S.W., Ho, C.K., Rauhut, R. & Abelson, J.N. (1997). The yeast tRNA splicing endonuclease: A tetrameric enyzme with two active site subunits homologous to the archaeal tRNA endonucleases. *Cell* **89**, 849-858.

Tsutsumi, S., Denda, K., Yokoyama, K., Oshima, T., Date, T. & Yoshida, M. (1991). Molecular cloning of genes encoding major subunits of a eubacterial V-type ATPase from *Thermus thermophilus*. *Biochimica Biophysica Acta* **1098**, 13-20.

Viale, A.M. & Arakaki, A.K. (1994). The chaperone connection to the origins of eukaryotic organelles. *FEBS Letters* **341**, 146-51.

Wang, X. & Lutkenhaus, J. (1996). FtsZ ring: the eubacterial division apparatus conserved in archaebacteria. *Molecular Microbiology* **21**, 313-319.

Wenzel, T. & Baumeister, W. (1993). *Thermoplasma acidophilum* proteasomes degrades partially unfolded and ubiquitin-associated proteins. *FEBS Letters* **326**, 215-218.

Wettach, J., Gohl, H.P., Tschochner, H., & Thomm, M. (1995). Functional interaction of yeast and human TATA-binding proteins with an archaeal RNA polymerase and promoter. *Proceedings of the National Academy of Sciences of the USA* **92**, 472-476.

Woese, C.R. (1987). Bacterial evolution. *Microbiological Reviews* **51**, 221-271.

Woese, C.R. & Fox, G.E. (1977a). Phylogenetic structure of the prokaryotic domain: The primary kingdoms. *Proceedings of the National Academy of Sciences of the USA* **51**, 221-271.

Woese, C.R. & Fox, G.E. (1977b). The concept of cellular evolution. *Journal of Molecular Evolution* **10**, 1-6.

Woese, C.R., Kandler, O. & Wheelis, M.L. (1990). Towards a natural system of organisms: Proposal for the domains Archaea, Bacteria and Eucarya. *Proceedings of the National Academy of Sciences of the USA* **87**, 4576-4579.

Zillig, W. (1991). Comparative biochemistry of Archaea and Bacteria. *Current Opinion Genetics Development* **1**, 457-463.

Zillig, W., Klenk, H-P., Palm, P., Leffers, H., Pühler, G., & Garrett, R.A. (1989). Did eukaryotes originate by a fusion event? *Endocytobiosis Cell Research* **6**, 1-25.

Zillig, W., Palm, P., Klenk, H-P., Pühler, G., Gropp, F. & Schleper, C. (1991). Phylogeny of DNA-dependent RNA polymerases: testimony for the origin of eukaryotes. In *General and Applied Aspects of Halophilic Microorganisms*, pp. 321-322. Edited by F. Rodriguez-Valera. New York: Plenum Press.

Zillig, W., Palm, P., Klenk, H-P., Langer, D., Hüdepohl, U., Hain, J., Lanzendörfer, M. & Holz, I. (1993). Transcription in Archaea. In *The Biochemistry of Archaea (Archaebacteria)*. pp. 367-91. Edited by M. Kates, D.J. Kushner, & A.T. Matheson. Amsterdam: Elsevier.

Zwickl, P., Grizwa, A., Pühler, G., Dahlmann, B., Lottspeich, F., & Baumeister, W. (1992). Primary structure of the *Thermoplasma* proteasome and its implications for the structure, function, and evolution of the multicatalytic proteinase. *Biochemistry* **31**, 964-972.

3   PHYLOGENETIC RELATIONSHIPS AMONG FUNGI INFERRED FROM SMALL SUBUNIT RIBOSOMAL RNA GENE SEQUENCES

MAKIKO HAMAMOTO *and* TAKASHI NAKASE

Japan Collection of Microorganisms, RIKEN
Hirosawa 2-1, Wako-shi, Saitama 351-0198, Japan

1   Introduction

The recent focus on microbial biodiversity has demonstrated that a wide variety of unknown fungi are present in the natural world. Microorganisms are indispensable to the global environment and to human welfare as they are essential for the turnover of organic matter and produce important bioactive compounds such as antibiotics. It is estimated that less than 10% of microorganisms have been described with those found in subtropical to tropical regions receiving scant attention. The number of microbial species, including unculturable microorganisms, is believed to be especially high in these regions which contain perhaps more than half of the species of plants and animals living on the Earth.

It is becoming increasingly clear that many of the recently described fungi do not fit into conventional taxonomic groups, established mainly on the basis of phenotypic characteristics. How should such fungi be classified? One approach which can be expected to be useful is to help establish relationships using information derived from the analysis of small subunit ribosomal RNA (SSU rRNA) gene sequences. However, it is necessary to evaluate the significance of both conventional taxonomic criteria and molecular criteria such as those based on sequencing SSU rRNA genes in order to obtain a balanced, polyphasic classification of these organisms.

This chapter is focused on the taxonomy of eukaryotic microorganisms belonging to the fungal kingdom. The relationships between fungi inferred from SSU rRNA gene sequences are outlined with special reference to the basidiomycetes.

2   Kingdom Fungi and the Problems with Deuteromycetes

Fungi form one of the major kingdoms of the domain Eucarya which is generally considered monophyletic according to the phylogenetic tree constructed from nucleotide sequences of the SSU rRNA gene (Bruns *et al.*, 1992). The fungal kingdom consists of four taxa, namely, the Ascomycota, Basidiomycota, Chytridiomycota and Zygomycota. The former two taxa are generally called higher fungi reflecting the dikaryotic stage in

F. G. Priest and M. Goodfellow (eds.), Applied Microbial Systematics, 57-71

their life cycles and the formation of mycelia with septa. The latter two, the lower fungi, lack septa and form multinuclear mycelia. Figure 1 illustrates the evolutionary relationships of the major lineages in the fungal kingdom inferred from SSU rRNA gene sequences. It is apparent from the phylogenetic tree that the higher fungi and lower fungi are monophyletic. Relationships within each taxon will be discussed in Sections 3 and 4 of this chapter.

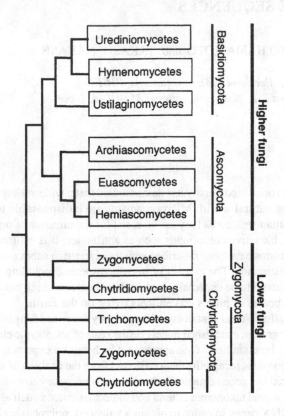

Figure 1. Evolutionary relationships between the major lineages in the fungal kingdom inferred from small subunit ribosomal RNA gene sequences. The lengths of the branches do not reflect the phylogenetic distances. The chronological order of the branches within the lower fungi is ambiguous.

The higher fungi have been classified into ascomycetes, basidiomycetes and deuteromycetes using conventional criteria based on the life cycle and the pleomorphy of the various recognized states produced by these fungi. The two former classes are called teleomorphs because they possess sexual states, while the latter are referred to as anamorphic because the members lack a sexual state. Above the generic level, teleomorphic species are classified by their sexual reproductive patterns and the morphology of their sexual organs. In contrast, anamorphic species are grouped according to the patterns of conidiogenesis and the shape of conidia. However, recent systematic analyses based on rRNA gene sequences demonstrate that the anamorphic fungi, the deuteromycetes, cannot be assigned to a distinct taxonomic group (Hori & Osawa, 1987; Guého *et al.*, 1989; Bruns *et al.*, 1991; Fell *et al.*, 1992; Berbee & Taylor, 1993; Nishida *et al.*, 1993; Suh & Sugiyama, 1993). The concept of the deuteromycetes has been questioned by the new systematic analyses, which assign the fungi in the deuteromycetes to either the ascomycetes or the basidiomycetes.

A similar situation exists within the yeasts. In former editions of "The Yeast, a Taxonomic Study", the imperfect yeasts were included as a distinct line from the ascomycetous and the basidiomycetous yeasts. However, in the fourth edition of this reference work the imperfect yeasts, whose taxonomy suffers the same conceptual problem as the deuteromycetes, were assigned as appropriate into ascomycetous and basidiomycetous taxa (Kurtzman & Blanz, 1998). The new edition therefore adopts the taxonomic idea that each taxon contains both teleomorphic and anamorphic genera and species while the deuteromycetes as a conventional form taxon with a dual taxonomy is denied. This concept for fungal systematics has the advantage of integrating the teleomorphs and anamorphs into one natural classification irrespective of sexual reproduction and the relationships between teleomorphic and anamorphic phases.

The conventional systematics, relying as it did on phenotypic characteristics, often ambiguously assigned yeast species (both anamorphic and teleomorphic) to a higher taxonomic rank than genus thereby creating considerable heterogeneity in the deuteromycetes. For example, the anamorphic yeast *Saitoella complicata* was classified initially in the genus *Rhodotorula* based on the morphology of its vegetative cells and their physiological characteristics (Goto & Sugiyama, 1970). *Saitoella complicata* was later reclassified in the family *Cryptococcaceae* on the basis of its unique ascomycetous characteristics including a negative diazonium blue B (DBB) reaction and an ascomycetous-type cell wall ultrastructure (Goto *et al.*, 1987). Nevertheless, this reclassification remained ambiguous since no teleomorph was found. Recent analysis of the organism based on the new systematics is discussed in Section 3.1.3.

The genus *Rhodosporidium*, which gave rise to the concept of basidiomycetous yeasts, is a second example of this confusion. *Rhodosporidium* had long been classified in the order Ustilaginales because of its characteristic properties of sexual life cycle (Banno, 1967) such as the formation of secondary hyphae, germination of thick-walled probasidia from teleospores and subsequent propagation of basidiospores (Kreger-van Rij, 1984). There was general agreement about this classification since mainstream yeast taxonomy had relied absolutely on form taxa based on the biological significance of the sexual stage. However, recent systematic analysis revealed that the genus *Rhodosporidium* belongs to the class Urediniomycetes, and the genus *Rhodotorula*, considered as the anamorph of the genus *Rhodosporidium*, is also included in the same class (see Section 3.2.1). This new finding highlighted the problems associated with conventional, phenotypic taxonomy. In the traditional classification, the anamorphic *Rhodotorula* was placed in the deuteromycetes simply because it had no sexual stage the close relationship with the genus *Rhodosporidium* was hidden.

Nonetheless, further consideration is needed regarding the consequences of abandoning the deuteromycetes as a higher taxon and the integration of meiotic and mitotic species, as discussed by Reynolds & Taylor (1993). For example, it would be necessary to place vast numbers of deuteromycete taxa into a phylogenetic classification to realize the integration of teleomorphs and anamorphs and it is unlikely that representative strains of many of these species will be studied, at least in the current research climate. Consequently many deuteromycete taxa could be left "in limbo".

3   Phylogenetic Relationships Among Higher Fungi

Several phylogenetic trees based on nucleotide sequences of the SSU rRNA gene show that the higher fungi are monophyletic and possess two monophyletic sister groups which correspond to the phyla Ascomycota and Basidiomycota (Bruns et al., 1992; Berbee & Taylor, 1993; Nishida & Sugiyama, 1994). The separation between ascomycetes and basidiomycetes is thought to have occurred about 390 million years ago during the early Devonian period (Berbee & Taylor, 1993).

3.1   ASCOMYCOTA

A comparison of SSU rRNA gene sequences indicates that the ascomycetes are mono-phyletic, although support for the basal branch is not strong (Bruns et al., 1992). Ascomycetes fall into three evolutionary groups, which may have been established during the coal age, that is from 310 million (230 to 390 million) to 330 million (240 to 430 million) years ago (Berbee & Taylor, 1993). It can be seen from Figure 1 that the earliest branch from the common ancestor of the ascomycetes is the Archiascomycetes (Nishida & Sugiyama, 1994). These organisms separated from the ancestor before the euascomycete and hemiascomycete branches were formed. In contrast, the euascomycetes and hemiascomycetes form monophyletic taxa which exhibit a paraphyletic relationship. The archiascomycota are estimated to have evolved about 330 million years ago (early Carboniferous period) and the euascomycetes and hemiascomycetes around 310 million years ago (mid-Carboniferous period) (Berbee & Taylor, 1993).

3.1.1 Hemiascomycetes
This phylogenetic group which contains the so-called ascomycetous yeasts is well supported as a monophyletic taxon. These fungi lack ascogenous hyphae, spend most of their life cycle as unicellular (yeast) cells without forming fruiting bodies, and are exemplified by the yeast genus Saccharomyces. They are the true yeasts which separated some 310 million years ago (230 to 390 million) from the filamentous ascomycete group with fruiting bodies. The hemiascomycetes include the asexual Candida albicans, which is considered to have followed a similar evolutionary pathway to Saccharomyces. The phylogenetic tree suggests that the first true yeast may have been a filamentous type which preceded Dipodascus uninucleata and Endomyces geotrichum (Berbee & Taylor, 1993).

The conventional taxonomy of some of the ascomycetous yeasts has been considered within the context of phylogenetic data. For example, Schwanniomyces, which can be distinguished from Debaryomyces on the basis of ascospore surface topography, was grouped on the Debaryomyces branch by analyses of three regions of the small and large rRNA subunits (about 900 bases in total) which suggested that *Schwanniomyces* should be integrated with the genus *Debaryomyces*. (Kurtzman & Robnett, 1991). In contrast, Yamada *et al.* (1991) claimed that the genus *Schwanniomyces* should be retained because of the importance of ascospore morphology and the number of substituted bases in the partial nucleotide sequences of the SSU rRNA. Such discussions raise an important question, whether to give greater weight to conventional phenotypic characteristics such as ascospore morphology or phylogenetic data. This dilemma has yet to be resolved.

3.1.2 *Euascomycetes*

The euascomycetes form a sister taxon to the hemiascomycetes and are thought to have radiated into at least three taxa approximately 280 million years ago (180 to 320 million years) (Berbee & Taylor, 1993). The euascomycetes are filamentous ascomycetes which enclose their sexual spore sacs in fruiting bodies. During their life cycle, ascogenous hyphae are formed and most of the life cycle is spent in a filamentous form. Euascomycetes without a sexual life cycle, such as the genus *Aspergillus*, are included in this phylogenetic group.

In traditional classifications, ascomycetes are often grouped on the basis of morphological similarities among ascomata that are thought to be homologous. However, the assessment of character homology has been shown to be problematic. For instance, the classes Hymenoascomycetes and Loculoascomycetes were proposed on the basis of whether the ascus wall is functionally unitunicate or bitunicate, respectively (Luttrell, 1951; Barr, 1987, 1990). However, the monophyly of the classes Hymenoascomycetes and Loculoascomycetes was rejected on the basis of SSU rRNA gene sequence data which revealed three groups, namely, the Geoglossaceae, Pezizales and Pyrenomycetes in the Hymenoascomycetes while the class Pyrenomycetes clustered with the Pleosporales (subclass Loculoascomycetes) (Spatafora, 1995). The rejection of the monophyly of the Hymenoascomycetes is consistent with the results of an independent analysis of nucleotide sequence data from chitin synthase genes (Bowen *et al.*, 1992).

Berbee (1996) also found that SSU rRNA gene sequence data did not provide strong support for the monophyly of the loculoascomycetes. *Capronia pilosella* (subclass Loculoascomycetes, Order Chaetothyriales) clusters with plectomycete members rather than with the other Loculoascomycetes, although the other two loculoascomycete members, the Dothideales and the Pleosporales, appear to be monophyletic groups (Berbee, 1996). Within the Loculoascomycetes, the Pleosporales forms a monophyletic group in 99% of bootstrap parsimony trees, while the Dothideales appears as a monophyletic group but without statistical support. Berbee deduced from the phylogenetic tree that the two-layered ascus wall must have either evolved at least twice or have been lost at least once. Spatafora (1995) suggested that further sampling of major groups of Loculoascomycetes, inoperculate discomycetes (Geoglossaceae) and lichenized forms would clarify the distribution of these fungi among ascomycetes. At present, the phylogenetic relationships among euascomycetes remain very confused.

3.1.3 *Archiascomycetes*

The Archiascomycetes consist of a wide variety of fungi, including the plant parasitic genera *Taphrina* and *Protomyces*, *Saitoella complicata*, an anamorphic yeast isolated from soil collected in the Himalayas, *Schizosaccharomyces pombe*, a fission yeast and *Pneumocystis carinii*, the mammalian lung pathogen (Nishida & Sugiyama, 1994). These species possess characteristics of both ascomycetes and basidiomycetes (Fig. 2), except for *Pneumocystis carinii* which cannot be grown in pure-culture.

Species of the genera *Taphrina* and *Protomyces* live on plants in a filamentous form, generate asci and produce ascospores. The ascospores germinate in asci; the germinable cells can be cultured artificially like yeast cells (Kramer, 1973). In addition, these fungi spend most of their life cycle in the dikaryotic phase which is more characteristic of basidiomycetous fungi.

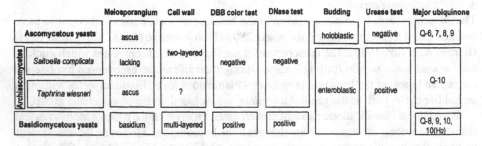

Figure 2. Representative characteristics of ascomycetous yeasts, basidiomycetous yeasts and the Archiascomycetes, *Saitoella complicata* and *Taphrina wiesneri*.

Saitoella complicata is a monotypic species which is a saprophytic anamorphic yeast (Goto *et al.*, 1987). Some of the chemotaxonomic characteristics of members of this taxon are similar to those of ascomycetous yeasts, such as the negative DBB reaction and lack of extracellular DNase activity, but some other features are similar to those of basidiomycetous yeasts, notably positive urease activity and the presence of the major ubiquinone system Q-10. As for ultrastructure, the cell wall is a typical ascomycetous two layered structure, whereas budding is typical of the basidiomycetous enteroblastic type (Goto *et al.*, 1987).

Fungi, such as the ones described above, which possess phenotypic characteristics of both ascomycetes and basidiomycetes, have been grouped in the Archiascomycetes on the basis of SSU rRNA data. This group, which was once called the basal ascomycetes (Berbee & Taylor, 1993) or early ascomycetes (Taylor *et al.*, 1994), is a phylogenetic link to the ascomycetes and basidiomycetes. It is thought that these taxa branched out from the ancestor in the chronological order of basidiomycetes, archiascomycetes and ascomycetes (Nishida & Sugiyama, 1994). Nevertheless, support for the monophyly of the archiascomycetes is not strong. The poorly resolved phylogeny provides few clues for determining which basal ascomycete most closely resembles the common ancestor to the group. Common phenotypic characteristics are needed to provide a definition for the taxon.

3.2   BASIDIOMYCOTA

Phylogenetic analyses of the SSU rRNA gene sequences of basidiomycetes indicate the existence of three major groups which correspond to the classes Hymenomycetes,

Urediniomycetes and Ustilaginomycetes (Fig. 1; Swann & Taylor, 1995). Each of these three major groups is monophyletic with a basal divergence between the Urediniomycetes and the Hymenomycetes, and subsequently between these taxa and the Ustilaginomyces. The timing of the divergence is calculated to be 340 million years ago (early Carboniferous period; Berbee & Taylor, 1993). Each group possesses unique biochemical and ultrastructural properties which allow their differentiation (Wolters & Erdman, 1986; Prillinger *et al.*, 1990, 1991a, b; Swann & Taylor, 1993, 1995; Boekhout *et al.*, 1993,1995; Wells, 1994; Suh & Sugiyama, 1994; Fell *et al.*, 1995; McLaughlin *et al.*, 1995).

3.2.1 *Urediniomycetes*
Fungi comprising this phylogenetic group are characterized by homogeneous phenotypic characteristics such as their ultrastructural and cellular carbohydrate composition. They form simple septal pores, disc-like spindle pole bodies and holo- or phragmobasidia (Wells, 1994; McLaughlin *et al.*, 1995), and commonly contain mannose as the major cellular carbohydrate, besides glucose and galactose; fucose is also present in most species (Sugiyama *et al.*, 1985; Prillinger *et al.*, 1991a). For convenience, the urediniomycetes can be divided into four statistically-supported major groups; the *Agaricostilbum* group, the *Erythrobasidium* group, the Sporidiales group and the Uredinales group (Swann & Taylor, 1995). However, it was subsequently revealed that the phylogenetic group corresponding to the class Urediniomycetes contained another group comprising three *Sporobolomyces* species (Hamamoto, unpublished data). The new phylogenetic group is supported statistically and, for the time being, will be referred to as the *Subbrunneus* group. The Urediniomycetes currently consists of five phylogenetic groups (Fig. 3) as described below.

The Sporidiales group. This group comprises *Heterogastridium pycnidioideum* (the type species), *Leucosporidium scottii* (the type species), *Mastigobasidium intermedium* (the type species) *Rhodosporidium toruloides* (the type species), 4 *Sporidiobolus* species (including the type species *S. johnsonii*), and 14 anamorphic species classified in the genera *Bensingtonia* (1 species), *Rhodotorula* (3 species including the type species *R. glutinis*) and *Sporobolomyces* (9 species including the type species *S. roseus*). The genera *Rhodosporidium* and *Sporidiobolus* have been assigned to the Ustilaginales due to the similarity of their sexual life cycles. Nucleotide sequence analysis of the SSU rRNA gene strongly supports the view that these fungi should be classified in the Urediniomycetes (Swann & Taylor, 1993). This classification is also supported by 5S rRNA sequence data (Blanz & Unseld, 1987), cellular carbohydrate spectra (Prillinger *et al.*, 1990, 1991a) and by the spindle pole body ultrastructure (McLaughin *et al.*, 1995). The phylogenetic analyses also allow these fungal taxa to be placed in a hierarchy above the family rank; these relationships were ambiguous when based solely on phenotypic data.

The genus *Mixia* has been placed in the ascomycetes since its first description in 1911 but SSU rRNA gene sequence data show that it belongs to the Urediniomycetes (Nishida *et al.*, 1995). *Mixia osmundae* shows its closest relationship to the Sporidiales group (Fig. 3), but it is impossible at present to conclude that it belongs to this group as it may form a monophyletic group distinct from the other species when more data are included in the tree. Assuming that rates of nucleotide substitutions are reasonably constant for all species being compared, long branches on phylogenetic trees may be due to the absence

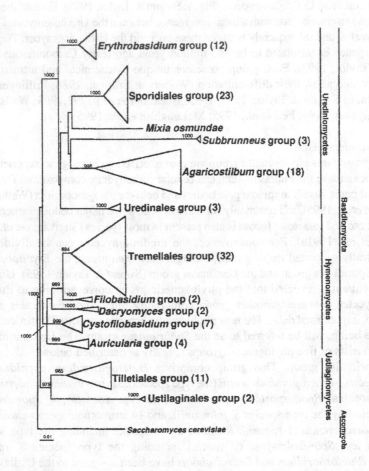

Figure 3. Phylogenetic tree showing the relative depth of one hundred and twenty basidiomycete species and *Saccharomyces cerevisiae* as an outgroup based on analysis of small subunit rRNA gene sequences. The unrooted tree was constructed from evolutionary distances (Kimura, 1980) by using the neighbour-joining method (Saitou & Nei, 1987). The numbers on the branches indicate the bootstrap confidence levels derived from 1,000 resamplings. The bar indicates the distance corresponding to one base change per 100 nucleotide positions. The numbers in parentheses indicate the number of species analysed.

of taxa which have yet to be studied for one reason or another, or to the fact that such taxa are now extinct (Kurtzman & Blanz, 1998). More studies are needed to resolve the evolutionary relationship of *M. osmundae*.

The *Erythrobasidium* group consists of *Erythrobasidium hasegawianum*, which is known to possess unique characteristics among the basidiomycetous yeasts, *Naohidea sebacea* (formerly *Platygloea sebacea*), two species from the genus *Rhodotorula*,

Rhodosporidium dacryoideum and eight species from the genus *Sporobolomyces*. *Erythrobasidium hasegawianum* and *Sporobolomyces elongatus* contain Q-10(H$_2$) as a major ubiquinone isoprenologue; a property which is unique among basidiomycetous yeasts. *Erythrobasidium hasegawianum* exhibits teliospore-forming yeast type chemotaxonomic characteristics such as are seen in the type strain of *Rhodosporidium* (*R. toruloides*) of the Urediniomycetes. However, the morphology of the sexual conidia and the life cycle resemble those of the Filobasidiaceae in the Hymenomycetes. In short, the species shows mixed phenotypic characteristics typical of separate fungal taxa. Twelve species are included in this group, although no common phenotypic characteristics have been found.

The *Agaricostilbum* group encompasses *Agaricostilbum hyphaenes*, *Mycogloea macrospora*, the ballistoconidium-forming anamorphic yeast genera *Bensingtonia* (8 species including the type species *B. ciliata*), *Sporobolomyces* (5 species), the stalked-conidium-forming yeasts *Kurtzmanomyces* and *Sterigmatomyces*, and the teliospore-forming yeast *Kondoa malvinella*. Nucleotide sequence diversity among the species within the *Agaricostilbum* group is high compared to that among the species belonging to the other four groups in the Urediniomycetes (Fig. 3).

The basidial morphology may be open to systematic questions, as both *A. hyphaenes* and *M. macrospora* have similar basidial morphologies, albeit under special conditions. Members of each of these species are capable of producing multiple gasterois basidiospores from the basidial compartment, this type of reproduction is most typically seen in *A. hyphaenes*. In contrast, *M. macrospora* possesses single, ballistic basidiospores.

The *Subbrunneus* group is a newly established group consists of three species from the genus *Sporobolomyces*. Common phenotypic characteristics for the three species of the subbrunneus group have yet to be found.

The Uredinales group. This group consists of the rust fungi, *Cronartium ribicola*, *Gymnoconia nitens* and *Nyssopsora echinata*. Members of all of the species representing this phylogenetic group are parasitic and show similarities in both the ultrastructure of the septum and the spindle pole body (Kahn & Kimbrough, 1982; McLaughlin *et al.*, 1995).

3.2.2 *Hymenomycetes*

The fungi belonging to this phylogenetic group possess complex septa with membranous pore caps of various configurations, in addition to "globular" spindle pole body morphology (Wells, 1994). These fungi commonly contain glucose as a major cellular carbohydrate together with mannose and xylose (Sugiyama *et al.*, 1985; Prillinger *et al.*, 1990, 1991a). Xylose is almost uniformly lacking in other basidiomycetes hence its presence in Hymenomycetes serves to distinguish them from members of the Urediniomycetes and Ustilaginomycetes.

Our phylogenetic tree differs from that of Swann & Taylor (1995), presumably due to the employment of different analytical methods. The tree presented here (Fig. 3) does not include the two major lineages suggested by Swann & Taylor (1995) which correspond to two subclasses of the Hymenomycetes; the Hymenomycetidae and Tremellomycetidae. The presence of five phylogenetic groups in the Hymenomycetes is strongly supported by high bootstrap values (Fig. 3). For convenience, these groups are called the *Auricularia* group; the *Cystofilobasidium* group; the *Dacryomyces* group;

the *Filobasidium* group and the *Tremellales* group. The *Tremellales* group consists of a wide variety of fungi; the ballistoconidium-forming yeasts *Bulleromyces* and *Bullera*, the stalked-conidium-forming yeasts *Fellomyces, Kockovaella, Sterigmatosporidium* and *Tsuchiyaea, Filobasidiella neoformans,* and *Tremella* species. The *Filobasidium* group encompasses *F. floriforme* and *Cryptococcus albidus;* the *Dacryomyces* group, *D. chrysospermus* and *Heterotextus alphinus*; the *Cystofilobasidium* group two species of the genus *Cystofilobasidium, Mrakia, Udeniomyces* and *Xanthophyllomyces*; and the *Auricularia* group, *Auricularia auricula, Athelia, Boletus* and *Coprinus.* At present, relatively few members of the Hymenomycetes have been the subject of SSU rRNA gene sequence analysis and so their precise phylogenetic classification is unclear.

3.2.3 *Ustilaginomycetes*
This phylogenetic group encompasses two major evolutionary branches typified by the Tilletiales and the Ustilaginales (Fig. 3). The former includes the monotypic *Sympodiomycopsis paphiopedili* and the sickle shape ballistoconidium-forming anamorphic fungi, *Tilletiopsis, Tilletia* and *Tilletiaria.* The Ustilaginales group includes *Ustilago hordei* and *Ustilago maydis.* The teliospore-forming yeast *Rhodosporidium* was once thought to be related to the Ustilaginales group due to the resemblance of its life cycle in the sexual generation of *Ustilago*; however, this resemblance is not reflected by its phylogeny. In contrast, the anamorphic yeast *Sympodiomycopsis paphiopedili* was placed in the basidiomycetous yeast group due to its simple septal pore (Suh *et al.*, 1993) until its placement in the Ustilaginomycetes group was demonstrated by SSU rRNA sequence data. The ultrastructure of the hyphal septal pores of *Tilletia caries* (Roberson & Luttrell, 1989) and *U. maydis* (Ramberg & McLaughlin, 1980; O'Donnell & McLaughlin, 1984) is of the dolipore type. In contrast to that of the Urediniomycetes and Hymenomycetes groups, the Ustilaginomycetes group is not homogeneous with respect to septal-pore ultrastructure.

4   Phylogenetic Relationships Among Lower Fungi

Lower fungi generally lack hyphal pores resulting in polykaryotic hyphae. Phylogenetic analyses based on SSU rRNA gene sequences suggest that the lower fungi arose earlier than higher fungi (Bruns *et al.*, 1992). The lower fungi encompass two phylogenetic groups; the chytridiomycetes and zygomycetes (Fig. 1). It is speculated that a loss of flagella occurred in many phylogenetic groups of lower fungi causing diversification (Fig. 4; Nagahama *et al.*, 1995; Sugiyama *et al.*, 1995). Nevertheless, it has been pointed out that phylogenetic relationships deduced from nucleotide sequence data do not match with classifications based on traditional morphological characteristics.

4.1   CHYTRIDIOMYCOTA

Zoospores which possess flagella are generally considered to be adapted to the aquatic environment whereas non-motile spores have evolved in the terrestrial environment. Most fungi form non-motile spores but certain chytridiomycetes form spores with flagella and consequently, are often regarded as the most primitive fungi.

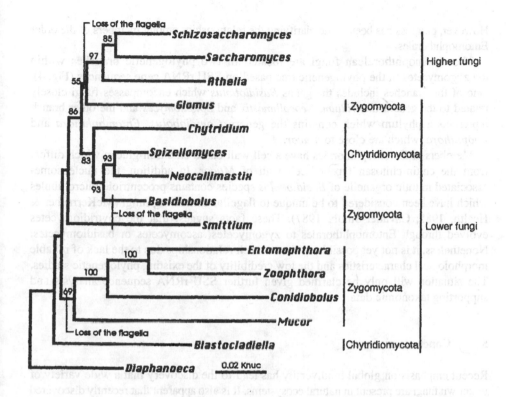

Figure 4. Phylogenetic tree of the lower fungi (adapted from Nagahama *et al.*, 1995 with modifications). The tree was constructed from evolutionary distances (Kimura, 1980) by using the neighbor joining method (Saitou & Nei, 1987). The bootstrap values were derived from 1000 resamplings; values greater than 50% are shown. The distance which corresponds to two base changes per 100 nucleotide positions is indicated by the bar. The lineages in which flagella have been lost are shown.

Chytridiomycetes form multi-branched hyphae and most reproduce sexually by gamete conjugation which suggests a relationship to the zygomycetes. Nucleotide analyses of SSU rRNA genes indicate that the chytridiomycetes are not monophyletic but can be divided into two major groups (Fig. 4; Nagahama *et al.*, 1995). It is now thought that the chytridiomycetes may have lost flagella during their diversification into a number of branches which lead to the evolution of the zygomycetes.

4.2 ZYGOMYCOTA

This group consists of the classes Trichomycetes and Zygomycetes. The Zygomycetes includes seven orders, namely the Dimargaritales, Endogonales, Entomophthorales, Glomales, Kickxellales, Mucorales and Zoopagales, which were mainly established on morphological criteria. Recent phylogenetic data suggest that the conventional seven orders may not necessarily reflect the relationships deduced from morphological studies.

However, progress has been made clarifying the relationships among members of the order Entomophthorales.

The entomophthoralean fungi are located in two phylogenetic branches within the zygomycetes in the phylogenetic tree based on SSU rRNA gene sequences (Fig. 4). One of the branches includes the genus *Basidiobolus* which encompasses fungi closely related to the genera *Chytridium, Neocallimastix* and *Spizellomyces* and the other branch represents a phylum which contains the genera *Conidiobolus, Entomophthora* and *Zoophthora*, which are close to *Mucor*.

Members of Entomophthorales have a cell wall based on chitin-glucan which differs from the chitin-chitosan type of cell wall of *Mucor*. In addition, the nucleosome-associated mitotic organelle of *Basidiobolus* species contains procentriole microtubules which have been considered to be unique to flagellar-containing fungi (McKerracher & Health, 1985; Cavalier-Smith, 1987). These facts suggest that the chytridiomycetes evolved through Entomophthorales to zygomycetes, ascomycetes or basidiomycetes. Nonetheless, it is not yet possible to discuss such relationships due to the lack of reliable morphological characteristics and the low credibility of the existing phylogenetic studies. The situation will only be clarified given further SSU rRNA sequence analysis and supporting taxonomic data.

5   Conclusions

Recent emphasis on global biodiversity has lead to the discovery that a wide variety of unknown fungi are present in natural ecosystems. It is also apparent that recently discovered fungi often include types which do not fit into conventional taxonomic groups based on phenotypic properties.

The application of rDNA sequence analysis is having an important impact on fungal systematics by introducing a phylogenetic component to classification but this development does raise the question: "Should fungal groups be defined on the basis of their phenotypic properties or on the basis of molecular data?" What, for instance, should be done with a group of fungi that have an identical phenotype but are located on completely unrelated phylogenetic branches in phylogenetic trees derived from SSU rRNA gene sequences? Alternatively, what stance should be taken when a group of fungi form a distinct evolutionary clade but have no common phenotypic characteristics?

If phylogenetic taxonomy were accepted as the natural taxonomic system it would still be unrealistic to establish a taxonomy without reference to phenotypic characteristics. There is therefore an urgent need to define a unified approach to the roles of phylogenetic and phenotypic data in fungal systematics as well as to study protein-coding genes in addition an integrated to rRNA genes. Such system might provide a new stable, natural classification of the fungi.

"Note addendum in proof: Recent findings in phylogenetic relationships of approximately 500 species of the hemiascomycetes (ascomycetous yeasts) from analysis of LSU rDNA partial sequences are described Kurtzman & Robnett (1998). In Fig. 3, *Cystofilobasidium* group has been proposed as Cystofilobasidiales (Fell *et al.*; 1999), therefore the *Cystofilobasidium* group is currently renamed as the Cystofilobasidiales."

6 References

Banno, I. (1967). Studies on the sexuality of *Rhodotorula*. *Journal of General and Applied Microbiology* **13**, 167-196.

Barr, M.E. (1987). *Prodromus to Class Loculoascomycetes*. Amherst, Massachusetts: Newell, Inc.

Barr, M.E. (1990). Prodromus to nonlichenized, pyrenomycetous members of class Hymenoascomycetes. *Mycotaxon* **39**, 43-184.

Berbee, M.L. (1996). Loculoascomycete origins and evolution of filamentous ascomycete morphology based on 18S rRNA gene sequence data. *Molecular Biology and Evolution* **13**, 462-470.

Berbee, M.L. & Taylor, J.W. (1993). Dating the evolutionary radiations of the true fungi. *Canadian Journal of Botany* **71**, 1114-1127.

Blanz, P.A. & Unseld, M. (1987). Ribosomal RNA as a taxonomic tool in mycology. In *The Expanding Realm of Yeast-like Fungi*, pp. 247-258. Edited by G.S. Hoog, M.T. Smith & A.C.M. Weijman. Amsterdam: Elsevier Science.

Boekhout, T., Fonseca, A., Sampaio, J.-P. & Golubev, W.I. (1993). Classification of heterobasidiomycetous yeasts: characteristics and affiliation of genera to higher taxa of heterobasidiomycetes. *Canadian Journal of Microbiology* **39**, 276-290.

Boekhout, T., Fell, J.W. & O'Donnell, K. (1995). Molecular systematics of some yeast-like anamorphs belonging to the Ustilaginales and Tilletiales. *Studies in Mycology* **38**, 175-183.

Bowen, A.R., Chen-Wu, J.L., Momany, M., Young, R., Szaniszlo, P.J. & Robbins, P.W. (1992). Classification of fungal chitin synthases. *Proceedings of National Academy of Sciences of the United States of America* **89**, 519-523.

Bruns, T.D., White, T.J. & Taylor, J.W. (1991). Fungal molecular systematics. *Annual Review of Ecological Systematics* **22**, 525-564.

Bruns, T.D., Vilgalis, R., Barns, S.M., Gonzalez, D., Hibbett, D.S., Lane, D.J., Simon, L., Stickel, S., Szaro, T.M. & Weinberg, W.G. (1992). Evolutionary relationships within the fungi: analyses of nuclear small subunit rRNA sequences. *Molecular Phylogenetics and Evolution* **1**, 231-241.

Cavalier-Smith, T. (1987). The origin of fungi and pseudofungi. In *Evolutionary Biology of the Fungi*, pp. 339-353. Edited by A.D.M. Rayner, C.M. Brasier & D. Moore. Cambridge: Cambridge University Press.

Fell, J.W., Statzell-Tallman, A. & Lutz, M.J. (1992). Partial rRNA sequences in marine yeasts: a model for identification of marine eukaryotes. *Molecular Marine Biology and Biotechnology* **1**, 175-186.

Fell, J.W., Boekhout, T. & Freshwater, D.W. (1995). The role of nucleotide sequence analysis in the systematics of the yeast genera *Cryptococcus* and *Rhodotorula*. *Studies in Mycology* **38**, 129-147.

Fell, J.W., Roeijmans, H. & Boekhout, T. (1999). Cystofilobasidiales, a new order of basidiomycetous yeasts. *International Journal of Systematic Bacteriology* **49**, 907-913.

Goto, S. & Sugiyama, J. (1970). Studies on Himalayan yeasts and molds (IV). Several asporogenous yeasts, including two new taxa of *Cryptococcus*. *Canadian Journal of Botany* **48**, 2097-2101.

Goto, S., Sugiyama, J., Hamamoto, M. & Komagata, K. (1987). *Saitoella*, a new anamorph genus in the Cryptococcaceae to accommodate two Himalayan yeast isolates formerly identified as *Rhodotorula glutinis*. *Journal of General and Applied Microbiology* **33**, 75-85.

Guého, E., Kurtzman, C.P. & Peterson, S.W. (1989). Evolutionary affinities of heterobasidiomycetous yeasts estimated from 18S and 25S ribosomal RNA sequence divergence. *Systematics and Applied Microbiology* **12**, 230-236.

Hori, H. & Osawa, S. (1987). Origin and evolution of organisms as deduced from 5S ribosomal RNA sequences. *Molecular Biology and Evolution* **4**, 445-472.

Kahn, S.R. & Kimbrough, J.W. (1982). A reevaluation of the basidiomycetes based upon septal and basidial structures. *Mycotaxon* **15**, 103-120.

Kimura, M. (1980). A simple method for estimating evolutionary rate of base substitutions through comparative studies of nucleotide sequences. *Journal of Molecular Evolution* **16**, 111-120.

Kramer, C.L. (1973). Protomycetales and Taphrinales. In *The Fungi, An Advanced Treatise*, Vol. 4A, pp. 33-41. Edited by G.C. Ainsworth, F.K. Sparrow & S.A. Sussman. London: Academic Press.

Kreger-van Rij, N.J.W. (Ed..) (1984). *The Yeasts, A Taxonomic Study*, 3rd Edn. Amsterdam: Elsevier Science.

Kurtzman, C.P. & Robnett, C. (1991). Phylogenetic relationships among species of *Saccharomyces, Schizosaccharomyces, Debaryomyces* and *Schwanniomyces* determined from partial ribosomal RNA sequences. *Yeast* **7**, 61-72.

Kurtzman, C.P. & Blanz, P.A. (1998). Ribosomal RNA/DNA sequence comparisons for assessing phylogenetic relationships. In *The Yeasts, A Taxonomic Study*, 4th Edn., pp. 69-74. Edited by C. P. Kurtzman & J. W. Fell. Amsterdam: Elsevier Science.

Kurtzman, C.P. & Fell J.W. (Eds.) (1998). *The Yeasts, A Taxonomic Study*, 4th Edn. Amsterdam: Elsevier Science.

Kurtzman, C.P. & Robnett, C.J. (1998). Identification and phylogeny of ascomycetous yeasts from analysis of nuclear large subunit (265) ribosomal DNA partial sequences. *Antonie van Leeuwenhoek* **73**, 331-371.

Luttrell, E.S. (1951). Taxonomy of the pyrenomycetes. *University of Missouri Studies* **3**, 1-20.

McKerracher, L.J. & Health, I.B. (1985). The structure and cycle of the nucleus-associated organelle in two species of *Basidiobolus*. *Mycologia* **77**, 412-417.

McLaughlin, D.J., Frieders, E.M. & Lü, H.S. (1995). A microscopist's view of heterobasidiomycete phylogeny. *Studies in Mycology* **38**, 91-109.

Nagahama, T., Sato, H., Shimazu, M. & Sugiyama, J. (1995). Phylogenetic divergence of the entomophthoralean fungi: evidence from nuclear 18S ribosomal RNA gene sequences. *Mycologia* **87**, 203-209.

Nishida, H. & Sugiyama, J. (1994). Archiascomycetes: Detection of a major new lineage within the Ascomycota. *Mycoscience* **35**, 361-366.

Nishida, H., Blanz, P.A. & Sugiyama, J. (1993). The higher fungus *Protomyces inouyei* has two group I introns in the 18S rRNA gene. *Journal of Molecular Evolution* **37**, 25-28.

Nishida, H., Ando, K., Ando, Y., Hirata, A. & Sugiyama, J. (1995). *Mixia osmundae*: Transfer from the Ascomycota to the Basidiomycota based on evidence from molecules and morphology. *Canadian Journal of Botany* **73** (Supplement 1), S660-S666.

O'Donnell, K.L. & McLaughlin, D.J. (1984). Ultrastructure of meiosis in *Ustilago maydis*. *Mycologia* **76**, 468-485.

Prillinger, H., Dörfler, C., Laaser, G. & Hauska, G. (1990). A contribution to the systematics and evolution of higher fungi: yeast-types in the basidiomycetes. Part III: *Ustilago*-type. *Zeitschrift für Mykologie* **56**, 251-278.

Prillinger, H., Deml, G., Dörfler, C., Laaser, G. & Lockau, W. (1991a). A contribution to the systematics and evolution of higher fungi: yeast-types in the basidiomycetes. Part II: *Microbotryum*-type. *Botanical Acta* **104**, 5-17.

Prillinger, H., Laaser, G., Dörfler, C. & Ziegler, K. (1991b). A contribution to the systematics and evolution of higher fungi: yeast-types in the basidiomycetes. Part IV: *Dacrymyces*-type, *Tremella*-type. *Sydowia* **43**, 170-218.

Ramberg, J.E. & McLaughlin, D.J. (1980). Ultrastructural study of promycelial development and basidiospore initiation in *Ustilago maydis*. *Canadian Journal of Botany* **58**, 1548-1561.

Reynolds, D.R. & Taylor, J.W. (Eds.) (1993). *The Fungal Holomorph, Mitotic, Meiotic and Pleomorphic Speciation in Fungal Systematics*. Wallingford, UK: CAB International.

Roberson, R.W. & Luttrell, E.S. (1989). Dolipore septa in *Tilletia*. *Mycologia* **81**, 650-653.

Saitou, N. & Nei, M. (1987). The neighbor-joining method: A new method for reconstructing phylogenetic trees.

Molecular Biology and Evolution 4, 406-425.

Spatafora, J.W. (1995). Ascomal evolution of filamentous ascomycetes: evidence from molecular data. *Canadian Journal of Botany* 73 (Supplement 1), S811-S815.

Sugiyama, J., Fukagawa, M., Chiu, S.-W. & Komagata, K. (1985). Cellular carbohydrate composition, DNA base composition, ubiquinone systems, and diazonium blue B color test in the genera *Rhodosporidium, Leucosporidium, Rhodotorula*, and related basidiomycetous yeasts. *Journal of General and Applied Microbiology* 31, 519-550.

Sugiyama, J., Nagahama, T. & Nishida, H. (1995). Fungal diversity and phylogeny with emphasis on 18S ribosomal DNA sequence divergence. In *Microbial Diversity in Time and Space*, pp. 41-51. Edited by R. Colwell, U. Simidu & K. Ohwada. New York: Plenum Press.

Suh, S.-O. & Sugiyama, J. (1993). Phylogeny among the basidiomycetous yeasts inferred from small subunit ribosomal DNA sequence. *Journal of General Microbiology* 139, 1595-1598.

Suh, S.-O. & Sugiyama, J. (1994). Phylogenetic placement of the basidiomycetous yeasts *Kondoa malvinella* and *Rhodosporidium dacryoidum* and the anamorphic yeast *Sympodiomycopsis paphiopedili* by means of 18S rRNA gene sequence analysis. *Mycoscience* 35, 367-375.

Suh, S.-O., Hirata, A., Sugiyama, J. & Komagata, K. (1993). Septal ultrastructure of basidiomycetous yeasts and their taxonomic implications with observations on the ultrastructure of *Erythrobasidium hasegawianum* and *Sympodiomycopsis paphiophedili*. *Mycologia* 85, 30-37.

Swann, E.C. & Taylor, J.W. (1993). Higher taxa of basidiomycetes: an 18S rRNA gene perspective. *Mycologia* 85, 923-936.

Swann, E.C. & Taylor, J.W. (1995). Phylogenetic perspectives on basidiomycete systematics: evidence from the 18S rRNA gene. *Canadian Journal of Botany* 73 (Suppl. 1), S862-S868.

Taylor, J.W., Swann, E.C. & Berbee, M.L. (1994). Molecular evolution of ascomycete fungi: Phylogeny and conflict. In *Ascomycete Systematics: Problems and Perspectives in the Nineties*, pp. 201-212. Edited by D.L. Hawksworth. New York: Plenum Press.

Wells, K.W. (1994). Jelly fungi then and now. *Mycologia* 86, 18-48.

Wolters, J. & Erdman, V.A. (1986). Cladistic analysis of 5S rRNA and 16S rRNA secondary and primary structure-The evolution of eukaryotes and their relation to archaebacteria. *Journal of Molecular Evolution* 24, 152-166.

Yamada, Y., Maeda, K., Nagahama, T. & Banno, I. (1991). The phylogenetic relationships of the Q9-equipped genus *Schwanniomyces* Klöcker (Saccharomycetaceae) based on the partial sequences of 18S and 26S ribosomal RNAs. *Journal of General and Applied Microbiology* 37, 523-528.

III   SOIL, PLANTS AND INSECTS

4   MOLECULAR ECOLOGY OF MYCORRHIZAL FUNGI

PETER JEFFRIES [1] and JOHN C. DODD [2]

[1] Department of Biosciences, University of Kent, Canterbury, Kent,
   CT2 6NJ, UK
[2] International Institute of Biotechnology, Biotechnology
   MIRCEN/Department of Biosciences, PO Box 228, Canterbury,
   Kent CT2 7YW, UK

1   Introduction

Mycorrhizal relationships are mutualistic symbioses between plant roots and certain soil fungi. Most species of plants are able to form them and in many ecosystems the fungus is essential to the plant for efficient mineral nutrient uptake and hence optimum plant health and development. The importance of mycorrhizas in a number of applied situations (e.g. agriculture, forestry, land revegetation) has been stressed elsewhere (see Jeffries & Dodd, 1991; Sieverding, 1991; Gianinazzi & Schüepp, 1994; Pfleger & Lindermann, 1994; Smith & Read, 1997) and it is now accepted that the belowground activity of mycorrhizal fungi has a profound effect on aboveground plant ecology.

Mycorrhizal fungi are important in the cycling of mineral nutrients that sustain soil fertility. It follows that information regarding the population ecology of mycorrhizal fungi is important for our understanding of ecosystem dynamics in general. Unfortunately the current status of the systematics of mycorrhizal fungi is inadequate to meet this demand. The arbuscular mycorrhizal fungi (AMF), characterized by the production of intracellular absorptive structures (the arbuscules), are amongst the most widespread of soil fungi, yet their taxonomy requires considerable revision. The second most important group are the ectomycorrhizal fungi (ECMF), characterized by extracellular structures such as the sheath and Hartig net. In this group there is a well-established taxonomy for the fungi involved, but there are considerable difficulties in matching mycorrhizal relationships in the field to the equivalent fungal taxa described from aboveground sporulating structures. Thus although considerable progress has been made in cataloguing the morphological appearance of ectomycorrhizal syntheses between tree hosts and specific fungal species, the goal of elucidating the structure and development of mycorrhizal communities is still an elusive target.

F. G. Priest and M. Goodfellow (eds.), Applied Microbial Systematics, 73-105
© 2000 Kluwer Academic Publishers. Printed in the Netherlands.

Information regarding mycorrhizal community dynamics is important for a number of reasons:
- for the resolution of species complexes and the unit of diversity
- to determine the degree of host and niche specificity
- to follow the dynamics of successions or replacement of mycorrhizal fungi
- to monitor gene flow and variability within populations of mycorrhizal fungi
- to provide biological predictability from a broader system of classification than currently used

This information is usually sought by determining the number and identity of individual mycorrhizal fungi within an ecosystem, an exercise which might seem simple but in practice is fraught with difficulties. One approach has been to identify and count the spores of AMF or the fruit bodies of ECMF within an ecosystem over several samplings and thus estimate where individual mycelia exist. However, sporulation or fruit body production can be sporadic (even soil-dependent for AMF) or may not reflect recent belowground events, particularly for AMF.

There are two essential yet quite different questions to be answered:
- which mycorrhizas correspond to which species or isolates of mycorrhizal fungi? i.e. an identification role to assess inter- and intraspecific genetic variation
- what is the extent of each individual mycorrhizal mycelium within the soil ecosystem? i.e. a mapping question involving ecological distribution of genotypes

The first question has now been tackled using molecular methodologies as an alternative way to recognise individual mycorrhizal types and to identify the mycobiont associated with plant roots. Again, the exercise is not simple as it raises questions such as "what is a species in terms of the AMF?" The determination or mapping of the extent of individual mycelia is more time-consuming. For the ECMF, it has traditionally been studied by obtaining large numbers of isolates from a given area and pairing cultures to determine whether or not they are vegetatively compatible (i.e. inferring that they are of the same clone). This is not possible for AMF which are unculturable in axenic conditions. Molecular methods again offer confirmatory evidence of relative similarities between different isolates from the same ecosystem and can provide a faster way of determining distributions of genetically-distinct mycelia. It must be stressed, however, that the study of population ecology of mycorrhizal fungi must begin with traditional approaches (morphology, vegetative compatibility analysis) in order to understand the relevance of the molecular data that are accumulated.

A final aspect to be considered is the determination of those fungi which are functionally significant in an ecosystem. Mere presence does not indicate whether a fungus plays an important role in an ecosystem. Molecular approaches are now being developed that will enable the *in situ* detection of the activity of enzymes that may have important functions in the sustainability of particular ecosystems. Correlating these activities with particular taxonomic entities is crucial and depends absolutely on understanding the taxonomic diversity within populations of mycorrhizal fungi in the soil.

2 Types of Mycorrhizal Fungi

2.1 MYCORRHIZAL RELATIONSHIPS

There are a variety of types of mycorrhizal relationships including arbuscular, arbutoid, ectomycorrhizal, ericoid, monotropoid and orchid (Smith & Read, 1997). In this chapter, only the two most ecologically important groups, AMF and ECMF will be considered. Information regarding the applications of PCR-based techniques to study the ericoid mycorrhizal fungi can be found in the recent review by Lanfranco *et al.* (1998). Although AMF and ECMF confer very similar benefits on their host plants, taxonomically they are very different fungi and must be dealt with separately in terms of their systematics.

2.2 ARBUSCULAR MYCORRHIZAL FUNGI

Arbuscular mycorrhizal fungi are associated with a wide range of plant families including both herbaceous and arboreal species. The fungal partners in this symbiosis are currently placed in the order Glomales of the Zygomycota. Little is known about their biodiversity, especially at the specific- and sub-specific levels, yet their importance in nutrient uptake by plants has been intensively studied (e.g., Sanders *et al.*, 1975; Gianinazzi-Pearson & Gianinazzi, 1986) and reviewed (Smith & Read, 1997).

2.3 ECTOMYCORRHIZAL FUNGI

In contrast to AMF, the ECMF form associations with a limited range of plant species, but among these species are most of the temperate forest trees as well as some important tropical tree species such as the Dipterocarpaceae. There are also taxonomic differences from AMF as the fungi which form ectomycorrhizal relationships comprise a range of taxa with representatives throughout the fungal kingdom. Most species belong to the Basidiomycota (the group that forms macroscopic fruit bodies such as mushrooms and puffballs), but others belong to the Ascomycota (including the gastronomically important truffle fungi). There are also a few representatives from the Zygomycota and these now appear to be taxonomically distinct from the fungi which form arbuscular mycorrhizas. All these fungi are classified on the basis of the morphology of their fruiting bodies and their systematics is well established. In practical terms, however, the problem is to match up distinct types of mycorrhizas found on tree roots with their respective taxa in the absence of fruit bodies in the field.

3 Traditional Approaches

3.1 ARBUSCULAR MYCORRHIZAL FUNGI

The study of diversity among AMF is hampered by the limited range of useful morphological characters of spores which are currently used for taxonomic ranking, and their inconsistent or incomplete use in species descriptions. Various molecular techniques have been applied, in recent years, in an effort to overcome this problem. Although isozymes and

DNA sequence data can distinguish between fungi at any taxonomic level, the species and genus concept in the Glomales is not well-defined, and further work is needed to establish the level of identification useful in studies of the biology and biodiversity of these ecologically-important fungi. These aspects will be considered below in the appropriate sections.

Thaxter (1922) provided the basis for the taxonomy of AMF and both mycorrhizal and non-mycorrhizal fungi were classified together in one genus, *Endogone*. In 1953, the first link was made between *'Endogone'* (= *Glomus mosseae*) spores and the mycorrhizal habit of the fungus (Mosse, 1953). This was followed by the informal description of several 'spore types' (Mosse & Bowen, 1968), some of which formed arbuscular mycorrhizas. The Linnean classification of Gerdemann & Trappe (1974) established 7 genera, *Acaulospora, Endogone, Gigaspora' Glaziella, Glomus, Modicella* and *Sclerocystis*, within the family *Endogonaceae*. Since then, *Glaziella* and *Modicella* have been removed and two additional genera, *Entrophospora* and *Scutellospora*, have been added.

The genus *Endogone* is not thought to contain members capable of forming arbuscular mycorrhizas, unlike members of the other six genera. *Endogone* was therefore separated, remaining in its own family and order, and those fungi known or supposed to form arbuscular mycorrhizas were placed in their own order, Glomales, with three families: Glomaceae *(Glomus* and *Sclerocystis),* Acaulosporaceae *(Acaulospora* and *Entrophospora)* and Gigasporaceae *(Gigaspora* and *Scutellospora)* as shown in Table 1 (Morton & Benny, 1990). The latter revision encompassed a phylogenetic perspective for the Glomales whereby it contains only those fungi acquiring carbon via arbuscules. This therefore assumes that the arbuscule is the key unifying structure and is the site of nutrient exchange. Whilst there are problems in this assumption, it does provide a basis for further study of the phylogeny of the group using molecular approaches.

TABLE 1.   The current taxonomy of the Glomales according to Morton & Benny (1990). The family
              Endogonaceae is placed in its own order and is not included

---

Order: Glomales Morton and Benny
  Suborder: *Glomineae* Morton and Benny
    Family: *Glomaceae* Pirozynski and Dalpé
      Genus: *Glomus* Tulasne and Tulasne
      Genus: *Sclerocystis* (Berkeley and Broome) Almeida and Schenck
    Family: *Acaulosporaceae* Morton and Benny
      Genus: *Acaulospora* (Gerdemann and Trappe) Berch
      Genus: *Entrophospora* Ames and Schneider
  Suborder: *Gigasporineae* Morton and Benny
    Family: *Gigasporaceae* Morton and Benny
      Genus: *Gigaspora* (Gerdemann and Trappe) Walker and Sanders
      Genus: *Scutellospora* Walker and Sanders

Many of the approximately 150 or more currently named species in the order Glomales (Walker & Trappe, 1993) were described from field-collected material, and are therefore of unknown mycorrhizal status, but it is assumed that they are all capable of forming arbuscules in roots. Many are synonymous and the number of true morphospecies may be significantly less than 150. Indeed, as Rosendahl *et al.* (1994) stated "the uncertainty about which structures are the most important as taxonomic characters is a problem in glomalean systematics, but more consistency is likely as species are redescribed".

Attempts have been made to differentiate taxa of AMF on the basis of the anatomy of the fungal colonization in roots using pure cultures (Abbott, 1982). This has met with limited success in field-based ecological surveys but with a trained eye useful data have been gained using this approach (Brundrett & Bougher, 1995). The use of the different stains (e.g. Acid Fuchsin, or Trypan Blue) has shown that at the current level of genus there are differences in how the stain reacts with the fungal mycelium *in planta*. *Acaulospora* spp. tend to stain poorly and produce lobed vesicles compared with most species of *Glomus* (Brundrett *et al.*, 1990; Brundrett & Bougher, 1995; Dodd, unpublished data) which produce more evenly-shaped oval vesicles in the roots. The genera *Gigaspora* and *Scutellospora*, in contrast, do not form vesicles at all but form auxiliary cell clusters in the soil which differ in morphology across the two genera. It is uncertain as to whether vesicles and auxiliary cells perform a similar temporary storage function for AMF. There also appears to be differences in the architecture of the extraradical mycelium between and within genera (Dodd, 1994). At a fine structural level, the mycelial characters can be important in systematics. For example, septum ultrastructure was important in the removal of two genera (*Complexipes* and *Glaziella*), previously thought to be related to *Glomus*, from the Zygomycota. Both of these genera had septa which are typical of the Ascomycota and were consequently transferred (Gibson, 1984; Thomas & Jackson, 1982). Recently Smith & Read (1997) highlighted the anatomical differences in two types of colonization:

- 'Arum-type': - essentially the fungus spreads within the root cortex via intercellular hyphae with short side-branches invaginating inner cortical cells where profuse dichotomous branching leads to arbuscule formation. This is the most frequently observed type of colonization.
- 'Paris-type': - where there is extensive development of intracellular coiled hyphae spreading from cell to cell within the cortex with arbuscules developing from these coils. There is little intercellular hyphal growth.

The importance of these differences on function and diversity of AMF is uncertain but the authors suggest that the Paris-type may reflect a non-mutualistic state between plant and fungus, with only the latter benefiting.

3.2 ECTOMYCORRHIZAL FUNGI

As stated earlier, there is a wide taxonomic spread of fungi included here, with over 5000 species being involved (Molina *et al.*, 1992). The Basidiomycota contains most ECMF and there are several orders containing putative ectomycorrhizal symbionts (Table 2). These fungi have traditionally been described and classified on the basis of the form of the basidiomata and the characteristics of the spores that are produced by them. A wide range of keys is available for identification on this basis (e.g. Sims *et al.*, 1995), but most are not restricted

TABLE 2.   Orders and representative genera of ectomycorrhizal fungi

Basidiomycota	
Agaricales	*Laccaria*
Boletales	*Suillus*
Cantharellales	
Cortinariales	
Gauteriales	
Gomphales	
Hymenogastrales	
Lycoperdales	
Melanogastrales	
Phallales	
Russulales	
Sclerodermatales	*Pisolithus*
Thelephorales	*Thelephora*
Ascomycota	
Elaphomycetales	
Pezizales	
Tuberales	*Tuber*
Zygomycota	
Endogonales	*Endogone*
	Sclerogone

to ectomycorrhizal species. In axenic culture, however, most of these fungi do not sporulate and hence only mycelial characters are available for identification. Nevertheless, this allows for the comparison of a range of cultural and biochemical characters. Dickinson & Hutchinson (1997) used 25 such characters to analyse the numerical taxonomy of 160 isolates of ecto-mycorrhizal agarics, boletes, and related gasteromycetes. They concluded that the level at which cultural characteristics prove most effective needs careful evaluation, as some could have evolved in parallel in separate clades. Thus ecological as well as taxonomic groups were also recognised.

Identification of the fungus from the ectomycorrhizal tissues is much more difficult and a variety of mycorrhizal types have been described, both from artificially synthesized dual symbioses and from field-collected material (Ingleby *et al.*, 1990; Agerer, 1996). In the latter case, it is difficult to ascertain the identity of the mycobiont in a given ectomycorrhiza unless a clear mycelial linkage is observed from a basidiome to the hyphae of the ecto-mycorrhizal sheath. This has lead to the use of artificial binomials being used for ectomycorrhizal morphotypes in which the tree species is known but the mycobiont has not been matched with a described fungal species. In exceptional cases, the production of specific metabolites can be diagnostic, as reported by Agerer *et al.* (1995) where ectomycorrhizas of *Picea abies*, previously described as *"Piceirhiza nigra"*, were shown to be formed by a member of the Thelephoraceae by virtue of the presence of thelephoric acid.

Like many other fungal groups, individual isolates from within particular species of ECMF can be classified into somatic incompatibility groups. Isolates from different incompatibility groups will not anastomose and thus remain as distinct mycelial individuals. It is commonly considered that somatically compatible secondary mycelia belong to the same genet, which is defined as a genetically distinct unit or assemblage within a species (Brasier & Rayner, 1987; Dahlberg & Stenlid, 1995). Genets can fragment into ramets by asexual or mycelial propagation, but the resulting ramets should all be somatically compatible and able to anastomose. The term clones includes both genets and ramets and emphasizes the clonal propensity of fungi (Dahlberg & Stenlid, 1995). The genet is usually taken as the basic unit of population structure. Traditionally genets have been determined by somatic incompatibility testing, but recently molecular approaches have brought the validity of this practice into question once Jacobson *et al.* (1993) found that somatically compatible isolates of *Suillus granulatus* were not necessarily genetically identical on the basis of RAPD analysis.

Traditional approaches to the mapping of basidiomata have a number of drawbacks. For example the work is often long term (may need observations over a period of 10 years or more to adequately document the sporadic fruiters) and the fruiting period is frequently seasonal. Non-sporulating fungi will not be detected and even when detected the above-ground structures may not correlate with ectomycorrhizal populations belowground. In addition, correct identification of the larger fungi requires experienced workers.

Morphological typing of ectomycorrhizal roots is subject to variation with environmental conditions or host, but their characteristics are usually well preserved and can be distinguished with experience (Agerer, 1996). There are some excellent descriptions of ectomycorrhizal morphology for specific host-fungus combinations (Ingleby *et al.*, 1990; Agerer, 1996) but, by their very nature, these databases are limited to those combinations currently included. It can also be difficult to successfully excavate roots such that their mycorrhizal structures remain intact. Cultural studies from fruitbodies or ectomycorrhizal roots on the other hand are labour intensive, and may be inadequate and miss fungi that cannot be grown on standard isolation media. Despite these drawbacks, there are many examples of studies of ECMF in particular ecosystems where the inherent ectomycorrhizal morphotypes have been described in detail and used to track changes in ECMF ecology (e.g. Bradbury *et al.*, 1998; Kranabetter & Wylie, 1998; Ursic & Peterson, 1997).

4   Molecular Approaches

4.1   TECHNIQUES USED

Many of the problems of identification of mycorrhizal fungi in the absence of sporulating structures can be addressed by molecular approaches, although it must be stressed that these tools have been mostly used to identify the fungus but not to assess their ecological significance. A prime objective of community analysis is to differentiate species, so molecular markers are chosen that:(1) are universally conserved among the organisms studied,
(2) show sufficient variability to discriminate species and
(3) retain within-species fidelity.

The nuclear-encoded ribosomal RNA (rRNA) can fulfil the criteria outlined above and within this area of the genome a number of discrete regions may be found which exhibit different degrees of variability, e.g., the internal transcribed spacer (ITS) region is moderately conserved, whilst the intergenic spacer (IGS) region is most variable and offers considerable scope for discrimination of genotypes. A large number of primers are now available which have been designed to amplify these regions in a wide variety of taxa, whilst others are available which amplify this region in fungi only, or in specific groups of fungi. PCR-RAPD analyses can also be powerful tools for detecting inter- and intraspecific differences, especially where the chosen regions coding for rRNA possess insufficient variation for adequate discrimination between isolates. A third method to design taxon-specific probes involves the cloning and sequencing of specific genes from ECMF using degenerate primers. Again the ribosomal RNA genes have been targetted, especially the mitochondrial large subunit rRNA gene. This approach takes more time but offers a rational approach and can also be used to study gene activity within the mycorrhizal symbiosis, providing appropriate genes are chosen. Two specific examples which have been reported in ECMF are chitin synthase and glyceraldehyde-3-phosphate dehydrogenase (Mehmann et al., 1994; Kreuzinger et al., 1996).

The use of isozyme profiling also offers a molecular tool to identify and discriminate mycorrhizal fungi. The detection of taxon-specific enzymes can be used to identify specific fungi from mycorrhizal roots collected in the field, or can be used to assess the degree of relatedness among fungal isolates. This technique can be developed to indicate the activity of certain enzymes within the mycorrhizal symbiosis, which can infer functional significance as well as presence.

4.2   ARBUSCULAR MYCORRHIZAL FUNGI

The ecological value of using a taxonomic structure based on morphology for investigating the communities of AMF in different biomes is questionable, as suggested by Sanders et al. (1996). The link between morphology and ecological function is not obvious, although there are some possible examples in Glomus coronatum and its ubiquity in mediterranean ecosystems, or Entrophospora colombiana which is only found in soils below pH 5 (Jeffries & Dodd, 1996; Rosendahl, 1996). These examples may prove erroneous as more extensive surveys are performed. The misidentification of AMF in past studies has not helped and persists even today, hence we support the plea for voucher specimens to be maintained for surveys and experimentation (Smith & Read, 1997). There are clear generic trends evident, however, with species of Gigaspora absent and Scutellospora rare in Northern European soils but frequent and diverse in the tropics and in dune sand ecosystems (Dodd pers. obs.; Koske, 1987). Molecular techniques can give information about the genetic diversity of AMF which could explain these phenomena but the fact that single spores contain hundreds or thousands of nuclei (Fig. 1) and the inference that they may be heterokaryotic complicates this analysis. The molecular study of the systematics of AMF and their application has been driven along following the first introduction of antibody protocols, polyacrylamide gel electrophoresis and ultimately PCR approaches into microbiological research. Any lagging behind other microbial groups has been the result of the inability to culture AMF in vitro.

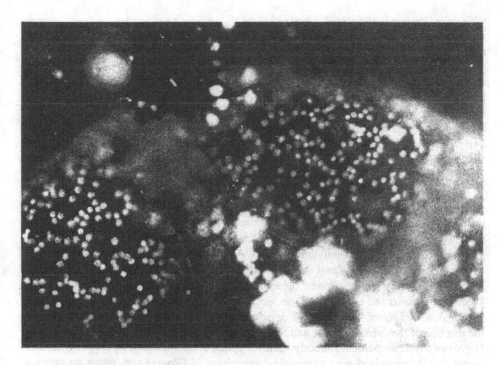

Figure 1. DAPI staining of nuclei in a crushed spore of *Gigaspora rosea* EC-3 (courtesy of Nadarajah, Newsam & Dodd).

4.2.1 *Immunological detection and identification*

Immunological methods have proved useful for the detection of many fungi in soil, roots and other plant material (Frankland *et al.*, 1981; Priestley & Dewey, 1993) and these techniques were some of the first employed when alternative molecular approaches to the identification of AMF were being tested. Kough *et al.* (1983) produced polyclonal antisera against extracted wall preparations from *Glomus versiforme* (formerly *G. epigaeum*). The antisera cross-reacted with the spore surfaces of two other species of *Glomus* but not with other glomalean fungi or non-mycorrhizal fungi in immunofluorescence tests.

Aldwell *et al.* (1983) used lyophilized extraradical mycelium to produce polyclonal antisera which was used in indirect enzyme-linked immunosorbent assay (ELISA) tests. These antisera showed some potential for discriminating between genera when tested against the soluble fraction of extraradical mycelium. This group (Aldwell *et al.*, 1985) also tested six antisera raised against extraradical mycelium of two species of *Glomus*, two species of *Gigaspora* (then classified separately as *Gigaspora* and *Scutellospora*), one *Acaulospora* and one *Sclerocystis* (now classified in *Glomus*). Fourteen isolates of different species of AMF were tested as above (Aldwell *et al.*, 1983) and *Gigaspora* spp. and *Acaulospora* were shown to be immunologically closer to each other than to *Glomus* and *Sclerocystis*. Sanders *et al.* (1992) raised polyclonal antisera against soluble spore extracts of *A. laevis* (LPA1) and *G. margarita* (LPA2) and found, using a dot immunobinding assay (DIBA), a similar strong cross-reactivity between *Acaulospora* and *Gigaspora* but not with another isolate of *A. laevis* from China.

Thingstrup *et al.* (1995) raised polyclonal antisera against the combined soluble and particulate fraction of spores of an isolate of *Scutellospora heterogama* (BEG35) and found that all antisera showed a high degree of specificity for the homologous fungus in Ouchterlony double diffusion tests. Immunostaining of Western blots was less specific but major cross reactions only occurred within the *Scutellospora* spp. tested. The use of similarly-raised polyclonal antisera against antigens on extraradical mycelium has been less successful in some cases (Dodd, 1994) but of potential use in microcosm studies in other cases (Aldwell & Hall, 1986).

A monoclonal antibody to crushed spores of *Glomus intraradices* (FL208) was used to detect young hyphae of AMF across all genera (Wright *et al.*, 1996). This immunofluorescent material ("glomalin") was also present on immature spores of *Glomus*. An immunoreactive protein extracted from hyphae of *Gigaspora* and *Glomus* species is probably a glycoprotein. This has potential use for identifying hyphae of AMF in soil and root sections and differentiating them from other soil fungi in microcosm studies. Moreover, monoclonal antibodies raised against an isolate of *Glomus occultum* distinguished this isolate from fifteen other species of *Glomus* and twenty-nine species of AMF using indirect ELISA (Wright *et al.*, 1987). A screening of isolates of *G. occultum* revealed differences in the intensity of the reaction of the antibody to the antigens (Oramas-Shirey & Morton, 1990). Hahn *et al.* (1996) raised a monoclonal antibody against hyphae of *Glomus monosporum* (Whs) and found that it recognized only extraradical mycelium of *Glomus* spp. using an indirect immunofluorescent labelling technique.

The application of these antisera to ecological studies has been limited but some data have proved interesting and others misleading. Wright & Morton (1989), for example, used their monoclonal antibody in a modified DIBA technique to stain fungal tissues of *Glomus occultum* on a nitro-cellulose squash of the crushed colonized root and were able to detect and quantify the mycelium in whole root systems. Friese & Allen (1991) used polyclonal antisera raised against spores of an isolate of *Gigaspora margarita* in a field study to monitor the fate and spread of an introduced isolate. They used fluorescent antisera to detect the extraradical mycelium of the exotic *Gigaspora* in a field site where no other members of that genus were purported to be present. Results indicated that the introduced fungus could survive for up to 2 years after inoculation in sagebrush steppe in Wyoming, USA. Later work, however, indicated that not only was the antiserum not isolate-specific but that it cross-reacted strongly with spores and hyphae of *Scutellospora calospora* (Weinbaum *et al.*, 1996). However, this antibody and that of another raised against *Acaulospora elegans* were used to monitor survival of the two AMF in cross-inoculation studies (introduced as exotics) with some apparent success (Weinbaum *et al.*, 1996). Thingstrup *et al.* (1995) showed that their polyclonal antiserum could be used to track the presence of *Scutellospora* species in colonized roots of several plants using immunostaining of Western blots. This approach has potential for monitoring the presence and function of this genus of AMF which has proved difficult to work with and culture (Dodd & Thomson, 1994).

It appears that the use of soluble spore components as immunogens has produced antisera with more potential widespread use. Gianinazzi & Gianinazzi-Pearson (1992) have raised and used polyclonal antisera to detect and localize mycelium of an isolate of *Gigaspora margarita* in roots by immunocytochemistry. Since the corresponding antigens were only present in metabolically-active hyphae they were able to monitor living colonization using ELISA (Ravolanirina & Gianinazzi-Pearson, 1992). A polyclonal antiserum produced against soluble proteins of spores of *Gigaspora rosea* (BEG9) showed a high specificity with homologous antigens from this same isolate in DIBA and ELISA tests (Cordier *et al*., 1996). The antibody did recognise other species of *Gigaspora* but not AMF outside this genus, a result similar to that found for the polyclonal antisera produced for *Scutellospora heterogama* (Thingstrup *et al*., 1995). Cordier *et al*. (1996), however, used the polyclonal antiserum to distinguish *Gigaspora rosea* in mixed colonization with *Glomus* spp. in leek roots. Other work by the same group (Gianinazzi-Pearson *et al*., 1994) used polyclonal and monoclonal antisera to detect β (1-3) glucans in inner spore and hyphal walls of AMF from the Glomaceae and Acaulosporaceae but not in species from the Gigasporineae. The systematic information to be gained from such studies is limited by the lack of complete screens of the raised antisera against AMF across the genera. The use of antisera may bear fruit, from a phylogenetic viewpoint, when specific molecules (proteins) can be studied as successfully shown by Gianinazzi-Pearson *et al*. (1994).

4.2.2 *Isozymes*

Isozyme characters are easily generated after electrophoretic separation of proteins (Sen & Hepper, 1986; Rosendahl & Sen, 1992) but their interpretation is more difficult. Putative loci and alleles should be identified and secondary bands ignored. Isozyme analyses exist for several species of *Glomus* (Hepper *et al*., 1988; Rosendahl, 1989; Rosendahl *et al*., 1994; Dodd *et al*., 1996). The major problem has been the extraction of sufficient protein from spores (between 20 300 spores are needed depending on the species) to obtain enzyme activity for sufficient stains.

Dodd *et al*. (1996), were able to screen many isolates of the morphologically similar *G. mosseae* and *G. coronatum*. Morphological and molecular (DNA fingerprints, isozymes and SDS profiles) characters were compared to assess the variation between and within the two groups of isolates of each species. The only morphological criterion for separating the taxa had been the colour of the spores but the use of malate dehydrogenase (MDH) and esterase loci clearly separated members of the two species. This approach has proved useful in other taxa including the genus *Gigaspora* (Bago *et al*., 1998). Interestingly, isolates of *Glomus clarum* from outside the tropics produced different MDH isozyme profiles compared with the same morphospecies (commonly referred to as *G. manihotis*) isolated from the tropics and this may represent an example of biogeographic separation (Rosendahl *et al*., 1994).

There have been several attempts to use the presence of specific isozymes in colonized root extracts to monitor the presence and spread of AMF in microcosm studies (Gianinazzi-Pearson & Gianinazzi, 1976; Hepper *et al*., 1988; Dodd *et al*., unpublished). Gianinazzi-Pearson & Gianinazzi (1976) first showed the presence of a mycorrhizal-specific alkaline phosphatase in onion roots colonized by *G. mosseae* (BEG12). Hepper *et al*. (1988), were the first to attempt to study how competition between different AMF for root cortex niches developed in microcosm studies. Three species of *Glomus* were used to colonize leeks as spatially separated inocula in a soil/sand substrate. Using esterase, glutamate oxaloacetate transaminase and peptidase enzyme stains as markers, these authors showed that *Glomus* sp. (E3) appeared unable to colonize roots if either *G. caledonium* or *G. mosseae* (BEG12) were present. Dodd *et al.* (unpublished) found similar results when *Acaulospora tuberculata* (BEG41), *Glomus geosporum* (BEG11) and *G. manihotis* (INDO-1) were compared on leeks. *Glomus geosporum* was detected using esterase and MDH isozymes as the sole fungus colonizing leek when the other two were co-inoculated. These data were supported by the parallel checking of the anatomy of colonization in plant roots; *A. tuberculata* stains poorly with Trypan Blue and has peculiar lobed vesicles whereas *G. manihotis* forms large vesicles and stains well. In further work, two AMF with similar anatomies of colonization, *Glomus geosporum* (BEG11) and *G. mosseae* (BEG25), were tested on leek to determine whether they could be distinguished *in planta* using diagnostic MDH and esterase isozymes (Fig. 2). In this case both AMF were detected inside the roots together but both isozyme stains were needed to confirm this.

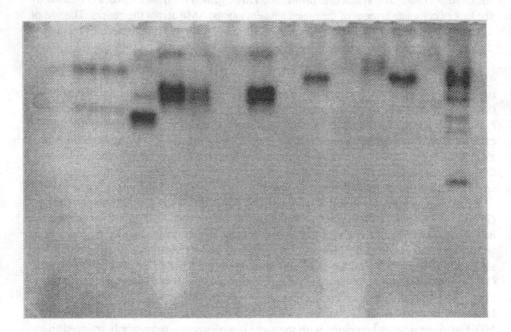

Figure 2. Extracts of spores of different AMF electrophoresed on non-denaturing polyacrylamide gel stained for malate dehydrogenase activity. The putative loci can be used to investigate the inter- and intra-specific variation in morphologically similar 'species'.

Recent work has been designed to identify diagnostic isozymes for species of AMF colonising the desert shrub *Anthyllis cytisoides* (Tisserant *et al.*, 1998). The high level induction of an alkaline phosphatase isozyme specific to mycorrhizas formed by *Glomus microaggregatum* (BEG56) was used to monitor the short-term fate of the fungus when outplanted into a desertified ecosystem in south-east Spain. The usefulness of this marker on its own is debatable, but in unison with nucleic acid-based techniques (described later) it could provide an assessment of function as well as presence of an AMF in plant roots. The potential for further applications of this in field studies is not clear but it could be useful where an exotic isolate is introduced and its long term fate monitored to assess its ecological role.

4.2.3 *Fatty acid methyl ester profiles*

Glomalean fungal spores contain abundant fatty acids (Jabaji-Hare, 1988) with between 130 and 500 spores required for extraction to give a definitive profile (Graham *et al.*, 1995). In a study of 53 isolates from 24 species of AMF, analysis of fatty acid data revealed clusters of groups at intraspecific, specific and generic levels (Graham *et al.*, 1995). There appeared to be a correlation between the total number of fatty acid methyl esters (FAMEs) per spore and spore size, with *Glomus occultum* showing the least and *Gigaspora gigantea* the most variation. Figure 3 shows the pair group comparison between the FAME profiles from non-mycorrhizal and mycorrhizal roots of Sudan grass and spores of isolates of *Glomus intraradices, Gigaspora margarita* and *Gigaspora rosea.* Whilst the spore profiles produced tight clusters, the FAME profiles for roots colonized by *Gigaspora* were similar to those in non-mycorrhizal roots. The FAME profiles in roots colonized by *G. intraradices* were distinct from those in non-mycorrhizal roots. The latter may be the result of the intense vesicle/spore formation by *G. intraradices* isolates in roots compared with the lack of vesicles in roots colonized by members of the Gigasporaceae. The technique was less robust when spores were parasitised, this could affect the data if careful selection of the propagules were not undertaken.

A distinct clustering of two *Glomus* species, which were clearly distinct from an isolate of *Scutellospora heterogama* and another of *Gigaspora rosea,* even when reproduced on different hosts, has similarly been recovered from FAME profiles (Bentivenga & Morton, 1994a). Interestingly, phospholipid fatty acid signatures have been used recently as biomass indicators for AMF in microcosm experiments (Olsson *et al.*, 1997). Grandmougin-Ferjani *et al.* (1997) recently surveyed the sterol and fatty acid contents of 16 species of the Glomales and found a total absence of ergosterol, unlike other Zygomycetes. They suggested that sterol profiles of AMF are closer to higher plants than fungi. Despite the relative success of this methodology for AMF, there are no equivalent examples of its application to ECMF.

4.2.4 *Nucleic acids*

The first report of DNA amplified from AMF using PCR and sequenced was from the 18S rRNA gene using isolates of *Glomus intraradices* and *Gigaspora margarita* (Simon *et al.*, 1992). The DNA was amplified from a few spores using universal primers. Subsequently the sequences of the 18S rRNA gene were found to differ between species of AMF and a Glomales-specific primer was designed (VANS1) which was used to amplify and sequence the gene from 12 different species representing all genera in the Glomales (Simon *et al.*, 1993a). These data were used to produce a phylogeny for the Glomales, which supported fossil (Pyrozynski & Dalpé, 1989) and other taxonomic evidence for the divergence of this lineage in Cambrian times (over 500 million years ago). Berbee & Taylor (1993) reported on the evolutionary radiations of the true fungi based on the 18S rRNA gene sequences calibrated using fossil evidence (see Chapter 2) which

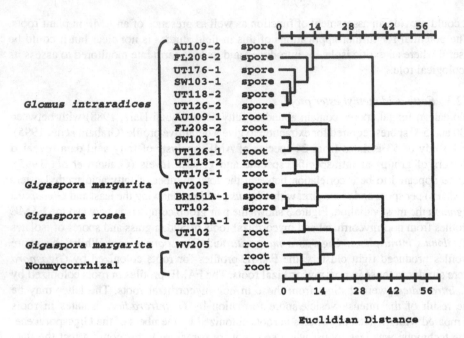

Figure 3. An example of the use of FAME profiles to detect AMF in roots. The dendogram shows the results of unweighted pair group analysis of similarities between FAME profiles of AMF spores alone, along with mycorrhizal and non-mycorrhizal Sudan grass roots (reproduced with permission of Graham *et al.*, 1995).

indicated that the Glomaceae diverged some 490 million years ago. Confirmation of this molecular clock was provided by the discovery of arbuscules in fossils of an early Devonian land plant (Remy *et al.*, 1994) proving unequivocally that AMF were established more than 400 Myr ago, and were already established as a symbiosis with root sytems prior to the evolution of land plants. The ECMF (or at least the ascomycete and basidiomycete lineages) evolved more recently, certainly after plants had invaded the land. This relative difference in evolution and origin may partially explain the lack of speciation in AMF compared with the ECMF which co-evolved with a rapidly developing terrestrial flora.

Recent work, which attempted to use the VANS1 primers in studies of populations of field-collected spores of *Scutellospora* in bluebell (*Hyacinthoides non-scripta*) woods in the north of England showed that the nucleotide sequences corresponding to this primer site varied (Clapp *et al.*, 1995; Sanders *et al.*, 1996). Since up to three base pair mismatches over a 21 base pair length might reduce the efficiency of annealing of the primer (explaining difficulties found by various groups in using this primer, Clapp *et al.*, 1995), Sanders *et al.* (1996) suggested that the putative phylogeny of Simon *et al.* (1993a) might be misleading and may only reflect the phylogeny of the 18S ribosomal gene sequences which were selected from the spores of *Scutellospora* themselves. Thus a high level of diversity in the 18S gene would then have been missed. Nevertheless, the fact remains that phylogeny based on the 18S gene sequence is a good reflection of that derived using alternative approaches.

Similar problems arose when nucleotide sequence data for the ITS region of an isolate of *Glomus mosseae* were gained. Fingerprints from ITS regions generated by PCR-RFLP from single spores from pot cultures themselves initiated from single spores sometimes revealed multiple banding patterns which did not add up to the original size of the ITS (Sanders *et al.*, 1995). Cloning and sequencing of the amplified ITS regions showed that different sequences of the ITS were present in single spores.

More recent data on inter- and intraspecific variation of ITS regions of *G. coronatum, G. mosseae* and other *Glomus* spp. showed that the intraspecific variation could be greater than the interspecific variation (Lloyd-MacGilp *et al.*, 1996). Franken and Gianinazzi-Pearson (1996) generated yet another different ITS sequence from *G. mosseae* (BEG12) compared with those of Sanders *et al.* (1995) and also found different rRNA clusters for *Scutellospora castanea* (BEG1) in line with observations by Clapp *et al.* (1995). Recent PCR-RFLP work using the ITS1-ITS4 region and various species of AMF from the Glomales provided further evidence for synonomizing several species e.g. *Glomus clarum* and *G. manihotis* (Redecker *et al.,* 1997) supporting isozyme and morphological data (Rosendahl *et al.*, 1994). Three species of *Gigaspora* (*G. albida, G. candida* & *G. rosea*) were shown to have identical RFLP patterns supporting morphological, isozyme and 18S rRNA sequence data (Bago *et al.*, 1998).

It has been estimated that spores of *Glomus* can contain up to 5000 nuclei (cited in Giovannetti & Gianinazzi-Pearson, 1994; Fig. 1) and it is normal that fungal nuclei contain multiple copies of the rRNA genes. It is not clear whether these nucleotide sequence differences are due to a heterozygous genome in a single nucleus or to a heterogeneous population of nuclei in a spore.

Zézé *et al.* (1997) have studied the genetic variation within spore populations of *Gigaspora margarita* (BEG34) using minisatellite-primed PCR and concluded that only a heterokaryotic status of the nuclear population could account for the results. Rosendahl and Taylor (1997) analysed intraspecific and interspecific genetic variation among AMF by AFLP (Amplified Fragment Length Polymorphisms) using DNA extracted from single spores. Their data provided clear evidence for substantial genetic variation between spores taken from the same isolate. They also suggested that these fungi reproduce clonally and concluded that the AFLP technique can be used in population genetic studies.

The degree of genetic diversity within natural populations of spores of an individual species in a community has also been highlighted by Sanders *et al.* (1995) who compared the ITS regions of 10 morphologically-identical spores of a *Glomus* sp. and found that none were identical. Another technique, RAPD-PCR, has been successfully used to detect polymorphisms in spores of different isolates of *Glomus mosseae* as well as other AMF (Wyss & Bonfante, 1993). Interestingly, whilst the similarity between fingerprints of an isolate of *G. mosseae* maintained in different laboratories in Europe over a 12-year period was high, there was slight variation indicating that genetic differences could occur over a short time (Sanders *et al.*, 1996). A non-polymorphic RAPD band was subsequently used to design specific primers (P0-M3) for *G. mosseae* and these were shown to specifically amplify the DNA from spores of this species and from isolates of G. mosseae colonising roots in greenhouse studies (Lanfranco et al., 1995; Dodd et al., 1996).

The same RAPD approach was adopted by another group to generate isolate-specific primers for isolates of Glomus mosseae and Gigaspora margarita which allowed detection of the AMF in mixtures of spores or when colonizing roots in greenhouse studies (Abbas et al., 1996). Recent studies using the SSU rRNA sequences of the intercytoplasmic bacterium-like organisms (BLOs) found in the cytoplasm of spores and germ tubes of G. margarita and a Scutellospora sp. indicated that they were specific to this family (Gigasporaceae) and absent in those species of the Glomaceae and Acaulosporaceae that were checked (Bianciotto et al., 1996). These BLO's may therefore offer a new character for studying the phylogeny of glomalean fungi.

Various techniques have been employed in attempts to study the ecology of AMF by detecting the fungus in its mycorrhizal state. Simon et al. (1992) used their VANS1 primer with the universal primer NS21 to amplify DNA directly from leek roots colonized by G. vesiculiferum. This approach has been successful in similar greenhouse studies using another species of Glomus colonizing three other plant species (Di Bonito et al., 1995). Simon et al. (1993b) designed family-specific primers and used fluorescent single-strand conformation polymorphism (SSCP)-PCR to detect AMF colonizing roots in pot studies. Edwards et al. (1997) used competitive PCR to quantify the presence of an isolate of Glomus mosseae colonising 6-week-old leek roots but did not cross test other plant/ fungal combinations. Van Tuinen et al. (1998) have used 25S rDNA-targeted PCR to detect four species of AMF in roots of leek and onion in experimental microcosms. They further showed that more than one fungus could exist in the majority of root fragments tested. They suggested that the advantage of their approach was in the potential use of relatively crude DNA preparation and even the use of previously stained root material. Data gainedby this technique indicate that the effects of abiotic factors on the ecological interactions between unique AMF in the root can be studied.

There have, however, been few in vivo studies. As mentioned earlier, Clapp et al. (1995) found difficulties with the VANS1 primer when used for the amplification of DNA of AMF from field-grown bluebell roots. These authors removed the host plant DNA using a subtractive hybridisation called SEAD (Selective Enrichment of Amplified DNA) and subsequent amplifications with VANS1 were successful. Interestingly, the use of the original procedure on onion roots in greenhouse trials had no need of the SEAD technique to obtain successful amplification of G. mo.seae DNA although the PO-M3 primers were more reliable (cited in Sanders et al., 1996).

Despite the obvious limitations, the use of ribosomal gene sequences has provided first indications of the distribution of AMF within the roots of one plant (bluebell) in a natural ecosystem (Clapp et al., 1995). Amplified DNA from spores at the bluebell site were compared with sequences emerging from the mycorrhizal roots of the adjacent bluebells. The 5' end of the 18S rRNA gene was used to characterise the distribution of AMF in time and space. Initial studies showed that three genera of AMF could be detected in a single root of bluebell using the SEAD technique, including sequences from Glomus which were unexpected as few Glomus spores had been isolated from the rhizosphere of the bluebells (Clapp et al., 1995). This supports the idea that spore data are of little use in the study of the active AMF in natural ecosystems but does not, however, preclude their usefulness in monoculture agricultural systems (Dodd *et al.*, 1990). The sequencing of the direct PCR products from the family-specific primers (Simon *et al.*, 1993b) and VANS1 were not successful as more than one sequence was present. It appears, therefore, that the use of sequence variation obtained from colonized roots in natural ecosystems and habitats, cannot at present be associated with the species of AMF. Extrapolation of these data to infer species diversity should not occur until the range of sequences for individual taxa has been determined. This is nowhere better indicated than in the work of Helgason *et al.* (1998) who showed that there was a reduction in sequence diversity from roots of various plants (of partial fungal small subunit RNA) in arable soils compared with woodland. To confirm that there was less **species diversity** in arable soils a parallel study (Koch's Postulates) would have been needed. This could be done by getting the AMF to sporulate and testing the sequences from the spores produced to see if similar sequence diversity was found. This dual approach to the ecology remains to done.

Attempts to detect AMF in soil extracts are also ongoing but few examples have been published. Claassen *et al.* (1996) have shown the potential to extract DNA from an alfisol mixed with sand which had a mycorrhizal grass plant grown in it. Subsequent amplification by PCR of the DNA in the extracts and primers (VANS1 & NS4) from the nuclear 17S rRNA were successful from the mycorrhizal treatments and not the non-mycorrhizal plants.

4.3   ECTOMYCORRHIZAL FUNGI

4.3.1 *Immunological detection and identification*
An early use of fluorescent polyclonal antibodies for identification of ECMF and their corresponding mycorrhizas was reported by Schmidt *et al.* (1974) for *Thelephora terrestris* and *Pisolithus tinctorius (= P. arhizus)*. Cross-reactivity, particularly for the antibody raised against the latter fungus, would have limited the applicability for identification purposes but the authors considered that the technique had considerable promise. Most of the heterologous cross-reactions could be eliminated by adsorption with the cross-reacting fungus without significantly affecting the activity of the homologous system. This potential has not yet been exploited although equivalent work was carried out on the non-mycorrhizal basidiomycete *Mycena galopus* (Chard & Gray, 1985 a,b) in which many of the problems of cross-reactivity were addressed.

Alternative approaches using monoclonal antibodies have not progressed because although antibodies which react to very specific epitopes can be obtained, the distribution of these epitopes can range across a number of taxonomically distinct fungi. This problem was encountered by Marshall *et al.* (1997) when attempting to produce species-specific antibodies against the ECMF *Paxillus involutus*.

In summary, despite the success now achieved for several non-mycorrhizal fungi (e.g. Schots *et al.*, 1994), it appears that the problems of cross-reactivity experienced using immunological approaches to the identification of ECMF have meant that DNA-based techniques are now favoured.

4.3.2 Isozymes

Genetic variation in natural populations of Suillus tomentosus provided the topic of one of the earliest applications of isozyme analysis to ECMF (Zhu et al., 1988). Considerable variation in eight enzymes was noted when 43 isolates from four geographic regions were analysed. Seven identical bands, presumably representing 7 monomorphic loci, were observed in all isolates. Fourteen bands, however, showed considerable polymorphism which indicated that this approach might be used to identify isolates at both the inter- and intraspecific level.

A similar approach was used to determine intraspecific variation in Suillus collinitus from under eleven different natural Pinus stands (El Karkouri et al., 1996). There was a high level of intraspecific variation but most of the forty-three isolates fell into two main geographically separate groups showing 35% dissimilarity in isozyme banding patterns. A specific and highly active acid phosphatase isoform was detected in only one group.

A correlation has been drawn between somatic compatibility groups and isozyme phenotypes in Suillus bovinus and S. variegatus (Sen, 1990), and this work has been extended to characterize fungal diversity in Scots pine ectomycorrhizas from natural humus microcosms using both isozyme and PCR-RFLP analyses (Timonen et al., 1997a). Ten morphotypes of ectomycorrhizas were formed on the tree roots and all ten displayed different esterase isozyme profiles. Two introduced species, Paxillus involutus and Suillus bovinus, grown in the forest humus gave identical isozyme patterns to those recorded earlier from pure-culture syntheses (Sen, 1990). A characteristic S. bovinus diagnostic band for esterase was detected in two white morphotypes suggesting these may be conspecific. In a complementary nucleic acid analysis, different RFLPs of the ITS placed the white morphotypes into two groups, one also corresponding to S. bovinus (Timonen et al., 1997a). Diagnostic, species-specific esterase bands were detected in intact external mycelium as well as in ectomycorrhizal roots. Esterase profiles were also used to type interspecific and intraspecific genets of Suillus spp. in order to determine the outcome of interactions between mixed inoculations (Timonen et al., 1997b).

Pisolithus species have also been subjected to protein profiling. Analysis of total protein profiles following 1D SDS-PAGE of polypeptides extracted from mycelia of Pisolithus isolates was used to group isolates on the basis of host species or geographic origin (Burgess et al., 1995). Polypeptide groups could also be correlated within groupings made on the basis of basidiospore morphology. The most important species in this genus, P. arhizus (= P. tinctorius) exhibits considerable morphological variation and the use of molecular approaches is aiding discrimination of different morphotypes within the species currently recognized.

Isozyme analysis using esterase, malate dehydrogenase and glucose-6-phosphate dehydrogenase (G6-PDH) has been used to distinguish groupings within a wordwide collection of 16 isolates of P. arhizus (Sims et al., 1999). In general, isozyme profiles could be correlated with geographic location, although some exceptions occurred where specific profiles seemed isolate-specific. For some enzymes (e.g. G-6-PDH), isozyme patterns for all isolates from the same plantation were identical, whereas for others (e.g. MDH) more polymorphisms occurred and very different profiles could be obtained from the same set of isolates (Fig. 4). Intraspecific variation has also been investigated within Laccaria bicolor by comparing isozyme profiles for leucine aminopeptidase to demonstrate the presence of

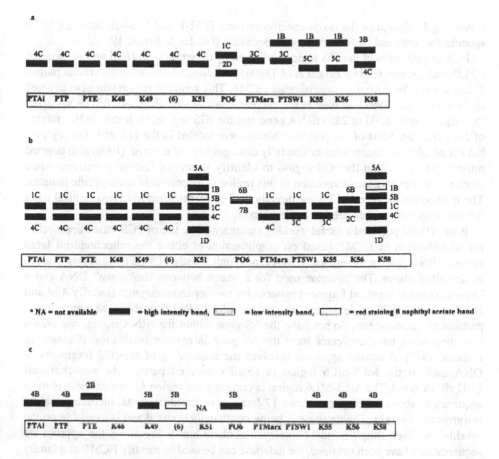

Figure 4. Depiction of isozyme profiles produced by a range of geographically-disparate isolates of *Pisolithus arhizus* stained for (a) malate dehydrogenase, (b) esterase and (c) glucose 6-P dehydrogenase activity (from Sims *et al.*, 1999).

more than one mycorrhizal, dikaryotic genotype on a single host (De La Bastide *et al.*, 1994). The usefulness of this approach as a genetic marker was generally confirmed by the results of RAPD analysis.

4.3.3 *Nucleic acids*

There has been rapid progress in the use of nucleic acid analysis for distinguishing ECMF since Armstrong *et al.* (1989) used RFLP techniques to demonstrate isolate-specific hybridization patterns with a DNA probe for basidiomycetous rRNA genes. Most techniques now use a PCR-based amplification of specific DNA sequences (Egger, 1995). For example, Gardes *et al.* (1991) demonstrated the use of primers ITS1 and ITS4 to amplify the ITS region of the nuclear-encoded rRNA gene repeat to distinguish isolates of *Laccaria* and *Thelephora* from pure cultures and ectomycorrhizal material. This approach was refined following the design of the taxon-specific primers ITS1-F and ITS4-B, intended to be specific for fungi and basidiomycetes respectively (Gardes & Bruns, 1993).

PCR amplification of the ITS region, followed by digestion with *Hinf*I and subsequent RFLP analysis, was used by Erland *et al.* (1994) to distinguish the ectomycorrhizal fungus *Tylospora fibrillosa* from a range of other ECMF. This genomic region was also targeted by Henrion *et al.* (1992), as well as the nuclear small subunit (SSU or 18S) rRNA gene, the large subunit (LSU or 28S) rRNA gene and the IGS region, to discriminate isolates of *Laccaria* spp. Most of the polymorphisms were located in the ITS and IGS regions, but not all selected isolates were successfully distinguished. Albee *et al.* (1996) also targeted polymorphisms within the IGS region to identify strains of *Laccaria proxima* since there was sufficient genetic variation in this region to discriminate intraspecific isolates. The method has the potential for use directly with DNA extracted from mycorrhizal roots but taxon-specific probes need to be developed.

Bruns (1996) provided a useful working summary of the use of PCR-based approaches for identification of ECMF based on amplification of either the mitochondrial large subunit rRNA gene or the nuclear rRNA repeat unit. Firstly ITS-RFLP analysis is used, as described above. The criterion used for a match between "unknown" DNA and a known control is identical fragment patterns for two separate enzymes (usually *Alu*I and *Hinf*I). Bruns (1996) is less keen on the use of IGS-RFLP analysis, as some fungi, particularly ascomycetes, do not have the 5S gene within the rRNA repeat, and others including some basidiomycetes have the 5S gene in reverse orientation (Cassidy & Pukkila, 1987). A second approach involves the sequencing of selected fragments of DNA such as the ML5-ML6 region (a small conserved part of the mitochdrional LSU rRNA gene). The ML5-ML6 region is currently the region for which there is most sequence database for 80 genera from 17 families of Basidiomycota, including mainly ectomycorrhizal taxa. The database is being continually updated and is available on the World-Wide Web (http://mendel.berkeleyedu/boletus.html). Providing the appropriate sequence data have been obtained, the database can be used to identify ECMF at a family or sub-family level. Higher levels of resolutioncan then be tackled using more stringent approaches such as PCR-RFLP matching. This approach is the only method useful for the matching of total unknowns with other RFLP patterns (Bruns, 1996) but its use can suffer from the limited resolution that is possible within this region. It may be superceded by sequence data comparisons within the more variable ITS region as the global database for this region continues to grow by default. Chambers *et al.* (1998) obtained sequence data for the ITS region from an ECMF isolated from *Pisonia grandis* and compared their data with those in the EMBL database. High homology with existing sequences suggested it belonged in the Thelephoraceae, and that it represented a *Tormentella* species or closely related fungus. The third method used by Bruns is the taxon-specific oligonucleotide probe designed to differentiate selected taxa. This approach was followed by Gardes *et al.* (1991) for *Laccaria,* and Bruns & Gardes (1993) for suilloid fungi again using part of the LSU rRNA gene as a target. Comparisons among partial sequences from this gene were used by the latter authors to determine empirically six probes that would be specific to particular taxa. Only partial success was reported as none of the probes exhibited their absolute intended specificity. However, all showed a high degree of specificity that was correlated with known or suspected phylogenetic affinities. For example, the S1 probe hybridized exclusively to *Suillus*, but failed to hybridize with one of the 17 species tested. The probes were also sucesssfully used to confirm the identity of field-collected material (Bruns & Gardes, 1993).

An alternative approach for the identification of non-sporulating fungal cultures is RAPD fingerprinting. This technique was evaluated for *Hydnangium* and *Laccaria* by Tommerup *et al.* (1995) who were able to demonstrate intraspecific differences in both these genera. The authors were concerned about the reported irreproducibility of RAPD analysis, but optimized their techniques such that they were able to satisfy themselves that particular RAPD fingerprints could be reproduced in any laboratory provided the same set of reaction and thermocycle conditions are used. Haudek *et al.* (1996) also used this approach to type 17 strains of ECMF and concluded that the technique was too sensitive to use at a generic level, but valuable to identify individual strains within a species. A major drawback of RAPD analysis is that it cannot be directly applied to DNA extracted from mycorrhizal roots because of interference by the excess of plant DNA present (Haudek *et al.* 1996).

RAPD fingerprinting has revealed a high degree of interspecific variability within *Tuber*, an ectomycorrhizal genus within the Ascomycota which is gastronomically significant in forming truffles as the reproductive phase. In contrast to the situation described for the basidiomycete fungi, there is a much lower degree of intraspecific variation (Lanfranco *et al.*, 1993). Because of its economic importance, this genus has also been subjected to ITS and IGS analysis for the identification of fruitbodies, mycelia at the species level and for identification of the mycobiont from mycorrhizal roots (Henrion *et al.*, 1994: Paolocci *et al.*, 1995; Amicucci *et al.*, 1996: Mello *et al.*, 1996). More recently, Longato & Bonfante (1997) have demonstrated the use of direct amplification of microsatellite sequences to reveal species-specific fingerprints for 11 species of *Tuber*, and also to detect intraspecific differences.

The examples above illustrate the use of nuclear probes for identification, but there are fewer examples where nucleic acid typing has been applied to the field situation to discriminate and map individual genotypes. An excellent example of this technology is exemplified in the mapping of a colony of *Armillaria bulbosa* (Smith *et al.*, 1992). Although this fungus is not an ECMF, this study used a combination of RAPD analysis and vegetative incompatibility studies to show that a single colony of this fungus occupied a minimum of 15 ha. Extrapolation led the authors to suggest the colony weighed in excess of 10000 kg and was at least 1500 years old ("the oldest and largest living organism in the world").

The same approach has been used for ECMF, and RAPD analysis was used to show that a single genetic individual of *Laccaria bicolor,* an early-stage ECMF, could remain associated with a host root system for at least three years in the field, and would vary significantly in spatial distribution over this period (De La Bastide *et al.*, 1994). The study was begun in a 4-year old plantation of *Picea abies* where trees had not been inoculated but possessed mycorrhizas of *Thelephora terrestris* at outplanting. The fact that a single genotype dominated the plantation inferred that this individual was able to replace and exclude other genotypes within the fungal population, an observation which has great significance for ECMF inoculation programmes. It is currently not known if a chosen isolate will remain mycorrhizal on an inoculated seedling after transplanting and thus confer the expected benefits to survival and growth. If another genotype dominates the outplanting site, then replacement may occur thus negating the benefits of prior inoculation. In a related study, the symbiotic efficiency of seven isolates of *Laccaria bicolor* obtained from a single basidiome was studied, and large differences in seedling growth stimulation were observed in the nursery (Di Battista *et al.*, 1996) yet no differences in restriction patterns of mitochondrial small rDNA or ITS and IGS1 regions were noted.

Suillus bovinus is another ECMF that has been subjected to nucleic acid analysis. On the basis of somatic incompatibility testing, Dahlberg & Stenlid (1994) studied the spatial distribution of genets of *S. bovinus* in five Scots Pine stands differing in history and age. With increasing forest size, the size of genets increased while the number of genets decreased. There was an estimated 700-5700 genets ha^{-1} (maximum size 1.7-5.3 m) in younger forests and 30-120 ha^{-1} (maximum size 17.5 m) in older ones. Gryta *et al.* (1997) used nucleic acid polymorphisms in the mitochondrial genome to study diversity within populations of *Hebeloma cylindrosporum* associated with pine trees growing in coastal sand dune ecosystems. Four molecular methods were used to demonstrate the presence of several different genets at each site, or even on the same root. When sites were sampled again after a three-year interval, none of the genotypes identified in the first year of sampling were detected suggesting a rapid turnover of individual mycelia. The largest genet of an ECMF yet reported was identified as *Suillus pungens* by Bonello *et al.* (1998) in a *Pinus muricata* forest. It covered an area of over 300 m^2 with a maximum measured dimension of 40 m by 14 m. The mapping was accomplished by typing collections of basidiomata using sequence-based markers and SSCP analysis.

Pisolithus arhizus is an important ECMF in warm climates where it is an early stage fungus which aids establishment of tree seedlings, particularly in poor soils (see section 4.4.2.). Current taxonomy places almost all isolates in a single species but recent studies have shown that there is considerable physiological diversity within the taxon (e.g. colonial morphology, growth rate differences, mycelial pigmentation, spore ornamentation). It might be expected that relative effectiveness in the mycorrhizal symbiosis would also show variation. European and North American isolates show considerable similarity, whilst Asian and Australian isolates show more intraspecific diversity. Molecular analyses (both isozyme profiling and PCR-RFLP) have confirmed intraspecific variation and provide more rapid tools to type new isolates. In a study of 62 isolates of *Pisolithus* collected from a 90 km^2 area of New South Wales (NSW), Anderson *et al.* (1998a) used RAPD and ITS-RFLP analysis to demonstrate that at least two of the isolates represented a different species than the common *P. arhizus*. This was confirmed by ITS sequence analysis which showed that these two isolates had 98.2% sequence homology with one another but only 83.2 to 85.6% homology with the other isolates. A similar approach was used to estimate the size and distribution of mycelial individuals of *P. arhizus* within another much smaller field site in NSW (Anderson *et al.*, 1998b). All basidiomata were collected from within 200 m of one another and 33 individuals were recognized, comprising one large (>30 m diam.) and several smaller (<3 m diam) ones. In a wider study, Junghans *et al.* (1998) used PCR-RAPD analysis of 20 isolates of *P. arhizus* from different geographic locations (Brazil, France, USA) and hosts (*Eucalyptus, Pinus*) to reveal a high degree of genetic diversity, although two main groups were evident, broadly correlated with geographic origin and host. The alternative approach of sequence analysis within the ITS and IGS1 regions was used by Martin *et al.* (1998) to compare 52 basidiomata of *P. arhizus* collected from under *Afzelia*, *Eucalyptus* and *Pinus* hosts in Kenya. Three separate host-associated morphotypes could be distuished and these were confirmed as genetically different. The authors considered that the three types probably represented separate biological species, and there was close sequence homology between some of the Kenyan types and those from NSW reported by Anderson *et al.* (1998a). In a wider study of PCR-RFLP variation of the ITS region of a diverse collection of ECMF, Farmer & Sylvia (1998) also noted that intraspecific variation in *P. arhizus* (and also *Cenococcum geophilum*) was marked, providing further evidence that these taxa represent species complexes.

Scleroderma species occupy a similar ecological niche to *Pisolithus*. Members of this genus are also widespread and important early colonisers of nursery beds and young seedlings of tropical forest trees (Sims *et al.*, 1995, 1997). Morphological studies combined with molecular analysis support the traditional view that several species exist and could be rapidly identified from mycorrhizal root samples providing the appropriate PCR-RFLP analyses are carried out.

As stated earlier, ITS-RFLP analysis of *Tylospora fibrillosa* following *Hin*fI digestion has proven useful in distinguishing this fungus from a large number of other basidiomycetes (Erland *et al.* 1994). This methodology was extended into field studies and used to screen single ectomycorrhizas sampled from a Norway spruce forest in south Sweden (Erland, 1995). In this forest 98% of randomly collected mycorrhizas had the general macroscopic features of those formed by *T. fibrillosa*, and fungal ITS was successfully amplified from 93% of these. Five distinct fungal RFLP patterns were recovered. The *T. fibrillosa* pattern was found in 21% of the analyses, demonstrating that this fungus is likely to be of significant ecological importance within the stand. The other four RFLP patterns belong to unknown mycobionts and represented 27%, 20%, 11% and 3% of the ectomycorrhizas analysed. Two different fungi were detected in 22% of the mycorrhizas. Obviously on a morphological basis alone, these data could not be easily obtained as all five mycobionts produced similar ectomycorrhizal structures, thus the molecular approach is useful as a rapid alternative for determining specific identifications of mycorrhizal symbionts from field-collected material.

PCR-RFLP analysis of the ITS region was used by Pritsch *et al.* (1997) to determine the identity of ECMF from 16 morphotypes of ectomycorrhizas formed within a 60 yr old black alder stand. Eight of the mycorrhizal types were identified when compared to restriction patterns obtained from basidiomata found in the locality, but the eight others remained unidentified. Four of the identified ECMF types had identical profiles to all the corresponding basidiomata collected from the site, despite the geographical distance of about 300 km between the sampling locations. In contrast, intraspecific variation between basidiomata from the different locations was noted for *Paxillus rubicundulus*. Dahlberg *et al.* (1997) reported the use of ITS-RFLP analysis to be a substantial improvement over the sorting of ectomycorrhizae into distinct morphotypes during the study of species diversity in an old-growth spruce forest in Sweden. Their study emphasised the fact that ascomycetes probably accounted for about 20% of the mycorrhizal abundance and yet no specific primers are available to target this broad taxon. Using root samples collected from a mixed forest of *Pseudotsuga menziesii* and *Pinus muricata*,

Horton & Bruns (1998) employed a combination of morphotyping and PCR-RFLP methods to show that the most abundant ECMF were associated with both hosts. Seventy-four per cent of the biomass of ectomycorrhizal root tips were colonized by members of the Thelephoraceae and Russulaceae.

5   Conclusions

Bentivenga & Morton (1994b) clearly stated the current state of the art in the classification of AMF: 'we are at a critical juncture in the transition from a classification based on alpha taxonomy to one derived from systematic studies' and further '..when classifications are based on phylogenies they provide an invaluable tool in testing relationships using other character data sets (biochemical, biogeographic, ecological genetic, physiological, ultrastructural)'. This situation is also true of ECMF, and other fungi in general. Traditionally these organisms have been classified on the basis of morphology alone and the existing taxonomy does not necessarily reflect phylogenetic relationships. The synthesis of modern molecular and biochemical approaches to systematics with the existing knowledge offers an excellent opportunity to clarify relationships on an evolutionary basis. In this respect, the fungal systematist will be able to take a more holistic approach to identification as is much more common in bacteriology (see other chapters in this volume). The description of ectomycorrhizae of *Lactarius lignyotus* on the host *Picea abies* (Kraigher *et al.*, 1995), which included both morphological and molecular data, is an example of this polyphasic approach. Equally valuable is the application of modern methods to clarify relationships between and within species 'groups' of *Glomus* described using traditional approaches (Dodd *et al.*, 1996).

It is more difficult to take an ecological view. Distribution patterns of species diversity are unknown for AMF and limited for ECMF (Gardes & Dahlberg, 1996) and ericoid mycorrhizal fungi (Perotto *et al.*, 1996) but the modern approaches to identification of field-collected material that have been described in this chapter can help plug this gap. By using molecular methods based on PCR, Gardes & Bruns (1996) were able to identify the mycobionts directly from mycorrhizas in nearly all of the ectomycorrhizal morphotypes they found within a natural pine forest community. This type of study lays a foundation for determining intraspecific variation across ecosystems.

Rosendahl (1996) set out the current debate on the species concept and its usefulness in an ecological approach to classification. The ecological species concept works in plant pathology for classifying groups of fungal pathogens according to their host range e.g. downy mildews (Hall, 1996). It is not relevant for AMF yet as they have similar adaptive zones but further detailed surveys of where they occur, using some or all of the methodologies above, could allow safer identification of AMF *in situ*.

The relationship between genetic divergence of AMF and spore morphology is clearly complicated and more work is needed on a greater numbers of species and isolates of species and different genes. The use of the two major glomalean collections, INVAM and BEG, will allow more thorough testing of *bona fide* material and this should be considered as a priority for future multidisciplinary studies (e.g. Dodd *et al.*, 1996; Bago *et al.* 1998). In the case of ectomycorrhizal fungi, the culture collections are much more fragmented and systematic studies are confined to individual research groups. Identification and validation of cultures has a particularly important role where the commercial use of mycorrhizal inoculum is envisaged. Accurate identification and consistency of product is vital for the success of inoculation programmes, and this demands effective taxonomic back-up.

Although this chapter has been primarily concerned with new developments, it must not be assumed that new techniques will supercede the old. In contrast, there is a need for integration of traditional and modern approaches in future studies. An interesting point is raised by Bruns (1996) who notes that the people in his laboratory spend so much time sorting ectomycorrhizae by hand from a particular region or ecosystem in order to prepare DNA, that they become familar enough with the morphological types that they are able to identify the common ones visually under a dissecting microscope without the need to resort to molecular analysis! This could equally be the case with AMF but an apparent aversion to quality time on the light microscope by many scientists in the area has led to a belief that molecular diagnostics are the only option. At the end of the day it is paramount that the organism is not forgotten.

Acknowledgements

We thank Dr. Justin Clapp for helpful comments on part of the manuscript.

6 References

Abbas, J.D., Hetrick, B.A.D. & Jurgenson, J.E. (1996). Isolate specific detection of mycorrhizal fungi using genome specific primer pairs. *Mycologia* **88**, 939-946.

Abbott, L.K. (1982). Comparative anatomy of vesicular-arbuscular mycorrhizas formed on subterranean clover. *Australian Journal of Botany* **30**, 485-499.

Agerer, R. (1996). *Colour Atlas of Ectomycorrhizae*. 10th delivery. Einhorn-Verlag Eduard Dietenberger GmbH, Schwäbisch Gmünd.

Agerer, R., Klostermeyer, D. & Steglich, W. (1995). *Piceirhiza nigra*, an ectomycorrhiza on *Picea abies* formed by a species of Thelephoraceae. *New Phytologist* **131**, 377-380.

Albee, S.R., Mueller, G.M. & Kropp, B.R. (1996). Polymorphisms in the large intergenic spacer of the nuclear ribosomal repeat identify *Laccaria proxima* strains. *Mycologia* **88**, 970-976.

Aldwell, F.E.B. & Hall I.R. (1986). Monitoring the spread of *Glomus mosseae* through soil infested with *Acaulospora laevis* using serological and morphological techniques. *Transactions of the British Mycological Society* **87**, 131-134.

Aldwell, F.E.B., Hall, I.R. & Smith, J.M.B. (1983). Enzyme-linked immunosorbent assay (ELISA) to identify endomycorrhizal fungi. *Soil Biology and Biochemistry* **15**, 377-378.

Aldwell, F.E.B., Hall, I.R. & Smith, J.M.B. (1985). Enzyme-linked immunosorbent assay as an aid to taxonomy of the Endogonaceae. *Transactions of the British Mycological Society* **84**, 399-402.

Amicucci, A., Rossi, I., Potenza, L., Zambonelli, A., Agostini, D., Palma, F. & Stocchi, V. (1996). Identification of ectomycorrhizae from *Tuber* species by RFLP analysis of the ITS region. *Biotechnology Letters* **18**, 821-826.

Anderson, I.C., Chambers, S.M. & Cairney, J.W.G. (1998a). Molecular determination of genetic variation in *Pisolithus* isolates from a defined region in New South Wales, Australia. *New Phytologist* **138**, 151-162.

Anderson, I.C., Chambers, S.M. & Cairney, J.W.G. (1998b). Use of molecular methods to estimate the size and distribution of mycelial individuals of the ectomycorrhizal basidiomycete *Pisolithus tinctorius*. *Mycological Research* **102**, 295-300.

Armstrong, J.L., Fowles, N.L. & Rygiewicz, P.T. (1989). Restriction fragment length polymorphisms distinguish ectomycorrhizal fungi. *Plant and Soil* **116**, 1-7.

Bago, B., Bentivenga, S.P., Brenac, V., Dodd, J.C., Piché, Y. & Simon, L. (1998). Molecular analysis of *Gigaspora*. *New Phytologist* **139**, 581-588.

Bentivenga, S.P. & Morton, J.B. (1994a). Stability and heritability of fatty acid methyl ester profiles of glomalean endomycorrhizal fungi. *Mycological Research* **98**, 1419-1426.

Bentivenga, S.P. & Morton, J.B. (1994b). Systematics of glomalean endomycorrhizal fungi: current views and future directions. In *Mycorrhizae and Plant Health*, pp. 283-308. Edited by F.L. Pfleger, & R.G. Linderman. St. Paul, Minnesota: American Phytopathological Society Press.

Berbee, M.L. & Taylor, J.W. (1993). Dating the evolutionary radiations of the true fungi. *Canadian Journal of Botany* **71**, 1114-1127.

Bianciotto, V., Bandi, C., Minerdi, D., Sironi, M., Tichy, H.V. & Bonfante, P. (1996). An obligately endosymbiotic mycorrhizal fungus itself harbours obligately intracellular bacteria. *Applied and Environmental Microbiology* **62**, 3005-3010.

Bonello, P., Bruns, T.D. & Gardes, M. (1998). Genetic structure of a natural population of the ectomycorrhizal fungus *Suillus pungens*. *New Phytologist* **138**, 533-542.

Bradbury, S.M., Danielson, R.M. & Visser, S. (1998). Ectomycorrhizas of regenerating stands of lodgepole pine (*Pinus contorta*). *Canadian Journal of Botany* **76**, 218-227.

Brasier, C.M. & Rayner, A.D.M. (1987). Whither terminology below the species level in fungi? In *Evolutionary Biology of the Fungi*, pp. 379-388. Edited by A.D.M. Rayner, C.M. Brasier & D. Moore. Cambridge: Cambridge University Press.

Brundrett, M.C., Murase, G. & Kendrick, B. (1990). Comparative anatomy of roots and mycorrhizae of common Ontario tress. *Canadian Journal of Botany* **62**, 2128-2134.

Brundrett, M.C. & Bougher, N. (1995). Mycorrhizal associations in the Alligator Rivers Region. Part 1 - Experimental Methods. Open file record 116. Supervising Scientist for the Alligator Rivers Region, Canberra.

Bruns, T.D. (1996). Identification of ectomycorrhizal fungi using a combination of PCR-based approaches. In: *Fungal Identification Techniques*, pp. 116-123. Biotechnology (1992-94). Brussels: EUR 16510.

Bruns, T.D. & Gardes M. (1993). Molecular tools for the identification of ectomycorrhizal fungi - taxon-specific oligonucleotide probes for suilloid fungi. *Molecular Ecology* **2**, 233-242.

Bruns, T.D., Szaro, T.M., Gardes, M., Cullings, K.W., Pan, J.J., Taylor, D.L., Horton, T.R., Kretzer, A., Garbelotto, M. & Li, Y. (1998). A sequence database for the identification of ectomycorrhizal basidiomycetes by phylogenetic analysis. *Molecular Ecology* **7**, 257-272.

Burgess, T., Malajczuk, N. & Dell, B. (1995). Variation in *Pisolithus* based on basidiome and basidiospore morphology, culture characteristics and analysis of polypeptides using 1D SDS-PAGE. *Mycological Research* **99**, 1-13.

Cassidy, J.R. & Pukkila, P. (1987). Inversion of 5S ribosomal RNA genes within the genus *Coprinus*. *Current Genetics* **12**, 33-36.

Chambers, S.M., Sharples, J.M. & Cairney, J.W.G. (1998). Towards a molecular identification of the *Pisonia* mycobiont. *Mycorrhiza*, **7** 319-321.

Chard, J.M. & Gray, T.R.G. (1985a). Use of an anti-*Mycena galopus* serum as an immunofluorescence reagent. *Transactions of the British Mycological Society* **84**, 243-249.

Chard, J.M. & Gray, T.R.G. (1985b). Purification of an antigen characteristic for *Mycena galopus.* *Transactions of the British Mycological Society* 84, 235-241.

Claassen, V.P., Zasoski, R.J. & Tyler, B.M. (1996). A method for direct soil extraction and PCR amplification of endomycorrhizal fungal DNA. *Mycorrhiza* 6, 447-450.

Clapp, J.P., Young, J.P.W., Merryweather, J. & Fitter, A.H. (1995). Diversity of fungal symbionts in arbuscular mycorrhizas from a natural community. *New Phytologist* 130, 259-265.

Cordier, C., Gianinazzi-Pearson, V. & Gianinazzi, S. (1996). An immunological approach for the study of spatial relationships between arbuscular mycorrhizal fungi *in planta.* In: *Mycorrhizas in Integrated Systems from Genes to Plant Development,* pp. 25-30. Proceedings of 4th European Symposium on Mycorrhizas, Granada, Spain 11-14 July 1994.

Dahlberg, A. & Stenlid, J. (1994). Size, distribution and biomass of genets in populations of *Suillus bovinus* (L.:Fr.) Roussel revealed by somatic incompatibility. *New Phytologist* 128, 225-234.

Dahlberg, A. & Stenlid, J. (1995). Spatiotemporal patterns in ectomycorrhizal populations. *Canadian Journal of Botany* 73, S1222-S1230.

Dahlberg, A., Jonsson, L. & Nylund, J.-E. (1997). Species diversity and distribution of biomass above and below ground among ectomycorrhizal fungi in an old-growth Norway spruce forest in south Sweden. *Canadian Journal of Botany* 75, 1323-1335.

De La Bastide, P.Y., Kropp, B.R. & Piché, Y. (1994). Spatial distribution and temporal persistence of discrete genotypes of the ectomycorrhizal fungus *Laccaria bicolor* (Maire) Orton. *New Phytologist* 127, 547-556.

Di Battista, C., Selosse, M.-C., Bouchard, D., Stenström, E. & Le Tacon, F. (1996). Variations in symbiotic efficiency, phenotypic characters and ploidy level among different isolates of the ectomycorrhizal basidiomycetes *Laccaria bicolor* strain S 238. *Mycological Research* 100, 1315-1324.

Di Bonito, R., Elliot, M.L. & Des Jardin, E.A. (1995). Detection of an arbuscular mycorrhizal fungus in roots of different plant species with PCR. *Applied and Environmental Microbiology* 61, 2809-2810.

Dickinson, T.A. & Hutchison, L.J. (1997). Numerical taxonomic methods, cultural characters, and the systematics of ectomycorrhizal agarics, boletes and gasteromycetes. *Mycological Research* 101, 477-492.

Dodd, J.C. (1994). Approaches to the study of the extraradical mycelium of arbuscular mycorrhizal fungi. In *Impact of Arbuscular Mycorrhizas on Sustainable Agriculture and Natural Ecosystems,* pp. 147-166. Edited by S. Gianinazzi & H. Schüepp, Basel, Switzerland: Birkhauser Press.

Dodd, J.C. & Thomson, B. (1994). The screening and selection of inoculant arbuscular-mycorrhizal and ectomycorrhizal fungi. *Plant and Soil* 159, 149-158.

Dodd, J.C., Arias, I., Koomen, I. & Hayman, D.S. (1990). The management of vesicular-arbuscular mycorrhizal populations in acid-infertile soils of a savanna ecosystem. II.- The effects of inoculation and pre-crops on the native VAMF spore populations. *Plant and Soil* 122, 241-247.

Dodd, J. C., Rosendahl, S., Giovannetti, M., Broome, A., Lanfranco, L. & Walker, C. (1996). Inter- and intraspecific variation within the morphologically similar arbuscular mycorrhizal fungi *Glomus mosseae* and *Glomus coronatum. New Phytologist* 133, 113-122.

Edwards, S.G., Fitter, A.H. & Young, J.P.W. (1997). Identification of an arbuscular mycorrhizal fungus, *Glomus mosseae,* within plant roots by competitive polymerase chain reaction. *Mycological Research* 101, 1440-1444.

Egger, K.N. (1995). Molecular analysis of ectomycorrhizal fungal communities. *Canadian Journal of Botany* 73, S1415-S1422.

El Karkouri, K., Cleyet-Marel, J.-C. & Mousain, D. (1996). Isozyme variation and somatic incompatibility in populations of the ectomycorrhizal fungus *Suillus collinitus. New Phytologist* 134, 143-153.

Erland, S. (1995). Abundance of *Tylospora fibrillosa* ectomycorrhizas in a South Swedish spruce forest measured by RFLP analysis of the PCR-amplified rDNA ITS region. *Mycological Research* **99**, 1425-1428.

Erland, S., Henrion, B., Martin, F., Glover, L.A. & Alexander, I.J. (1994). Identification of the ectomycorrhizal basidiomycete *Tylospora fibrillosa* Donk by RFLP analysis of the PCR-amplified ITS and IGS regions of ribosomal DNA. *New Phytologist* **126**, 525-532.

Farmer, D.J. & Sylvia, D.M. (1998). Variation in the ribosomal DNA internal transcribed spacer of a diverse collection of ectomycorrhizal fungi. *Mycological Research* **102**, 859-865.

Franken, P. & Gianinazzi-Pearson, V. (1996). Construction of genomic phage libraries of the arbuscular mycorrhizal fungi *Glomus mosseae* and *Scutellospora castanea* and isolation of ribosomal RNA genes. *Mycorrhiza* **6**, 167-173.

Frankland, J.C., Bailey, A.D., Gray, T.R.G. & Holland, A.A. (1981). Development of an immunological technique for estimating mycelial biomass of *Mycena galopus* in leaf litter. *Soil Biology and Biochemistry* **13**, 87-92.

Friese, C.F. & Allen, M.F. (1991). Tracking the fates of exotic and local VA mycorrhizal fungi: methods and patterns. *Agriculture Ecosystems and Environment* **34**, 87-96.

Gardes, M. & Bruns, T.D. (1993). ITS primers with enhanced specificity for basidiomycetes - application to the identification of mycorrhizae and rusts. *Molecular Ecology* **2**, 113-118.

Gardes, M. & Bruns, T.D. (1996). Community structure of ectomycorrhizal fungi in a *Pinus muricata* forest: above- and below-ground views. *Canadian Journal of Botany* **96**, 1572-1583.

Gardes, M. & Dahlberg, A. (1996). Mycorrhizal diversity in artic and alpine tundra: an open question. *New Phytologist* **133**, 147-157.

Gardes, M., White, T.J., Fortin, J.A., Bruns, T.D. & Taylor, J.W. (1991). Identification of indigenous and introduced symbiotic fungi in ectomycorrhizae by amplification of nuclear and mitochondrial ribosomal DNA. *Canadian Journal of Botany* **69**, 180-190.

Gerdemann, J.W. & Trappe, J.M. (1974). The Endogonaceae in the Pacific Northwest. *Mycological Memoirs* **5**, 1-76.

Gianinazzi, S. & Schüepp, H. (Eds.) (1994). *Impact of Arbuscular Mycorrhizas on Sustainable Agriculture and Natural Ecosystems*, Basel, Switzerland: Birkhauser Press.

Gianinazzi-Pearson, V. & Gianinazzi, S. (1976). Enzymatic studies on the metabolism of vesicular-arbuscular mycorrhiza. II. Soluble alkaline phosphatase specific to mycorrhizal infection in onion roots. *Physiological Plant Pathology* **12**, 45-53.

Gianinazzi-Pearson, V. & Gianinazzi, S. (1986). *Physiological and Genetical Aspects of Mycorrhizae. Proceedings of the 1st European Conference on Mycorrhizae.* INRA, Paris.

Gianinazzi-Pearson, V., Lemoine, M.-C., Arnould, C., Gollotte, A. & Morton, J.B. (1994). Localisation of β(1-3) glucans in spore and hyphal walls of fungi in the Glomales. *Mycologia* **86**, 477-484.

Gianinazzi, S. & Gianinazzi-Pearson, V. (1992). Cytology, histochemistry and immunocytochemistry as tools for studying structure and function in endomycorrhiza. *Methods in Microbiology* **24**, 109-139.

Gibson, J.L. (1984). *Glaziella aurantiaca* (Endogonaceae): zygomycete or ascomycete? *Mycotaxon* **20**, 325-328.

Giovannetti, M. & Gianinazzi-Pearson, V. (1994). Biodiversity in arbuscular mycorrhizal fungi. *Mycological Research* **98**, 703-715.

Graham, J.H., Hodge, N.C. & Morton, J.B. (1995). Fatty acid methyl ester profiles for characterisation of Glomalean fungi and their endomycorrhizae. *Applied and Environmental Microbiology* **61**, 58-64.

Grandmougin-Ferjani, A., Dalpé, Y, Hartmann, M-A., Laruelle, F., Couturier, D. & Sancholle, M. (1997). Taxonomic aspects of the sterol and Δ11-hexadecenoic acid (C16:D11) distribution in arbuscular mycorrhizal spores. In *Physiology, Biochemistry and Molecular Biology of Plant Lipids*, pp. 195-197. Edited by J.P. Williams, M.U. Khan & N.W. Lem. Dordrecht: Kluwer Academic Publishers.

Gryta, H., Debaud, J-C., Effose, A., Gay, G. & Marmeisse, R. (1997). Fine-scale structure of populations of the ectomycorrhizal fungus *Hebeloma cylindrosporum* in coastal sand dune forest ecosystems. *Molecular Ecology* 6, 353-364.

Hahn, A., Göbel, C. & Hock, B. (1996). Monoclonal antibodies against hyphal surface antigens of arbuscular mycorrhizal fungi. In *Mycorrhizas in Integrated Systems from Genes to Plant Development*, pp. 39-42. Proceedings of 4th European Symposium on Mycorrhizas, Granada, Spain 11-14 July 1994.

Hall, G.S. (1996). Modern approaches to species concepts in downy mildews. *Plant Pathology* 45, 1009-1026.

Haudek, S.B., Gruber, F., Kreuzinger, N., Göbl, F. & Kubicek, C.P. (1996). Strain typing of ectomycorrhizal basidiomycetes from subalpine Tyrolean forest areas by random amplified polymorphic DNA analysis. *Mycorrhiza* 6, 35-41.

Helgason, T., Daniell, T.J., Husband, R., Fitter, A.H. & Young, J. P. W. (1998). Ploughing up the wood-wide web? *Nature* 394, 431.

Henrion, B., Le Tacon, F. & Martin, F. (1992). Rapid identification of genetic variation of ectomycorrhizal fungi by amplification of ribosomal RNA genes. *New Phytologist* 122, 289-298.

Henrion, B., Chevalier, G. & Martin, F. (1994). Typing truffle species by PCR amplification of the ribosomal DNA spacers. *Mycological Research* 98, 37-43.

Hepper, C.M., Sen, R., Azcon-Aguilar, C. & Grace, C. (1988). Variation in certain isozymes amongst different geographical isolates of the vesicular-arbuscular mycorrhizal fungi *Glomus clarum*, *Glomus monosporum* and *Glomus mosseae*. *Soil Biology and Biochemistry* 20, 51-59.

Horton, T.R. & Bruns, T.D. (1998). Multiple-host fungi are the most frequent and abundant ectomycorrhizal types in a mixed stand of Douglas fir (*Pseudotsuga menziesii*) and bishop pine (*Pinus muricata*). *New Phytologist* 139, 331-339.

Ingleby, K., Mason, P. A., Last, F. T. & Fleming, L. V. (1990). *Identification of Ectomycorrhizas*. *ITE Research Publication no. 5*. London: HMSO.

Jahaji-Hare, S. (1988). Lipid and fatty acid profiles of some vesicular-arbuscular mycorrhizal fungi: contribution to taxonomy. *Mycologia* 80, 622-629.

Jacobson, K.M., Miller, O.K. Jr. & Turner, B.J. (1993). Randomly amplified polymorphic DNA markers are superior to somatic incompatibility tests for discriminating genotypes in natural populations of the ectomycorrhizal fungus *Suillus granulatus*. *Proceedings of the National Academy of Sciences of the United States of America* 90, 9159-9163.

Jeffries, P. & Dodd, J.C. (1991). The use of mycorrhizal inoculants in forestry and agriculture. In *Handbook of Applied Mycology vol. 1, Soil & Plants*, pp.155-185. Edited by D.K. Arora, B. Rai, K.G. Mukerji, & G. Knudsen, NewYork: Dekker.

Jeffries, P. & Dodd, J.C. (1996). Functional ecology of mycorrhizal fungi in sustainable soil-plant systems. In *Mycorrhizas in Integrated Systems from Genes to Plant Development*, pp. 497-501. Proceedings of 4th European Symposium on Mycorrhizas, Granada, Spain 11-14 July 1994.

Junghans, D.T., Gomes, E.A., Guimares, W.V., Barros, E.G. & Araœjo, E.F. (1998). Genetic diversity of the ectomycorrhizal fungus *Pisolithus tinctorius* based on RAPD-PCR analysis. *Mycorrhiza* 7, 243-248.

Koske, R.E. (1987). Distribution of VA mycorrhizal fungi along a latitudinal temperature gradient. *Mycologia* 79, 55-68.

Kough, J., Malajczuk, N. & Linderman, R.G. (1983). The use of the indirect immunofluorescent technique to study the vesicular-arbuscular fungus *Glomus epigaeum* and other *Glomus* species. *New Phytologist* 94, 57-62.

Kraigher, H., Agerer, R. & Javornik, B. (1995). Ectomycorrhizae of *Lactarius lignyotus* on Norway spruce, characterised by anatomical and molecular tools. *Mycorrhiza* 5, 175-180.

Kranabetter, J.M. & Wylie, T. (1998). Ectomycorrhizal community structure across forest openings on naturally regenerated western hemlock seedlings. *Canadian Journal of Botany* 76, 189-196.

Kreuzinger, N., Podeu, R., Gruber, F., Göbl, F. & Kubicek, C.P. (1996). Identification of some ectomycorrhizal basidiomycetes by PCR amplification of their *gpd* (glyceraldehyde-3-phosphate dehydrogenase) genes. *Applied and Environmental Microbiology* **62**, 3432-3438.

Lanfranco, L., Perotto, S. & Bonfante, P. (1998). Applications of PCR for studying the biodiversity of mycorrhizal fungi. In *Applications of PCR in Mycology*, pp.107-124. Edited by P.D. Bridge, D.K. Arora, C.A. Reddy & R.P. Elander. Wallingford, UK: CAB International.

Lanfranco, L., Wyss, P., Marzachi, C. & Bonfante, P. (1993). DNA probes for identification of the ectomycorrhizal fungus *Tuber magnatum* Pico. *FEMS Microbiology Letters* **114**, 245-252.

Lanfranco, L., Wyss, P., Marzachi, C. & Bonfante, P. (1995). Generation of RAPD-PCR primers for the identification of isolates of *Glomus mosseae*, an arbuscular mycorrhizal fungus. *Molecular Ecology* **4**, 61-68.

Lloyd-Macgilp, S.A., Chambers, S.M., Dodd, J.C., Fitter, A.H., Walker, C. & Young, J.P.W. (1996). Diversity of the internal transcribed spacers within and among isolates of *Glomus mosseae* and related arbuscular mycorrhizal fungi. *New Phytologist* **133**, 103-112.

Longato, S. & Bonfante, P. (1997). Molecular identification of mycorrhizal fungi by direct amplification of microsatellite regions. *Mycological Research* **101**, 425-432.

Marshall, M., Gull, K. & Jeffries, P. (1997). Monoclonal antibodies as probes for fungal wall structure during morphogenesis. *Microbiology* **143**, 2255-2265.

Martin, F., Delaruelle, C. & Ivory, M. (1998). Genetic variability in intergenic spacers of ribosomal DNA in *Pisolithus* isolates associated with pine, eucalyptus and Afzelia in lowland Kenyan forests. *New Phytologist* **139**, 341-352.

Mehmann, B., Brunner, I., & Braus, G.H. (1994). Nucleotide sequence variation of chitin synthase genes among ectomycorrhizal fungi and its potential use in taxonomy. *Applied and Environmental Microbiology* **60**, 3105-3111.

Mello, A., Nosenzo, C., Meotto, F. & Bonfante, P. (1996). Rapid typing of truffle mycorrhizal roots by PCR amplification of the ribosomal DNA spacers. *Mycorrhiza* **6**, 417-421.

Molina, R., Massicotte, H. & Trappe, J.M. (1992). Specificity phenomena in mycorrhizal symbiosis: Community-ecological consequences and practical implications. In *Mycorrhizal Functioning- an Integrated Plant-Fungal Process*, pp. 357-423. Edited by M.F. Allen. New York: Chapman & Hall.

Morton, J.B., & Benny, G.L. (1990). Revised classification of arbuscular mycorrhizal fungi (Zygomycetes): A new order, Glomales, two new suborders, Glominae and Gigasporinae, and two new families, Acaulosporaceae and Gigasporaceae, with an emendation of Glomaceae. *Mycotaxon* **37**, 471-491.

Mosse, B. (1953). Fructifications associated with mycorrhizal strawberry roots. *Nature* **171**, 974.

Mosse, B. & Bowen, G.D. (1968). A key to the recognition of some Endogone spore types. *Transactions of the British Mycological Society* **51**, 469-483.

Olsson, P.A., Baath, E. & Jakobsen, I. (1997). Phosphorus effects on the mycelium and storage structures of an arbuscular mycorrhizal fungus as studied in the soil and roots by analysis of fatty acid signatures. *Applied and Environmental Microbiology* **63**, 3531-3538.

Oramas-Shirey, M. & Morton, J.B. (1990). Immunological stability among different geographic isolates of the arbuscular mycorrhizal fungus *Glomus occultum*. *Abstracts of the 90th Annual Meeting of the American Society of Microbiology*, p. 311.

Paolocci, F., Angelini, P., Cristofari, E., Granetti, B. & Arcioni, S. (1995). Identification of *Tuber* spp. and corresponding ectomycorrhizae through molecular markers. *Journal of Science Food & Agriculture* **69**, 511-517.

Perotto, S., Actis-Perino, E., Perugini, J. & Bonfante, P. (1996). Molecular diversity of fungi from ericoid mycorrhizal roots. *Molecular Ecology* **5**, 123-131.

Pfleger, F.L. & Lindermann, R.G. (Eds.) **(1994).** In *Mycorrhizae and Plant Health*. Minnesota, USA: APS Press.

Priestley, R.A. & Dewey, F.M. (1993). Development of a monoclonal antibody immunoassay for the eyespot pathogen *Pseudocercosporella herpotrichoides*. *Plant Pathology* **42**, 403-412.

Pritsch, K., Boyle, H., Munch, J.C. & Buscot, F. (1997). Characterisation and identification of black alder ectomycorrhizas by PCR/RFLP analyses of the rDNA internal transcribed spacer (ITS). *New Phytologist* **137**, 357-369.

Pyrozynski, K.A. & Dalpé, Y. (1989). Geological history of the Glomaceae with particular reference to mycorrhizal symbiosis. *Symbiosis* **7**, 1-36.

Ravolanirina, F. & Gianinazzi-Pearson, V. (1992). VA endomycorrhization of microplants of grapevine: an immunological approach. In *Interactions between Plants and Microorganisms*, pp. 186-198. Edited by J.N. Wolf. Stockholm: International Foundation for Science,

Redecker, D., Thierfelder, H., Walker, C. & Werner, D. (1997). Restriction analysis of PCR-amplified internal transcribed spacers of ribosomal DNA as a tool for species identification in different genera of the order Glomales. *Applied and Environmental Microbiology*, **63**, 1756-1761.

Remy, W., Taylor, T., Haas, H. & Kerp, H. (1994). Four-hundred-million-year-old vesicular arbuscular mycorrhizae. *Proceedings of the National Academy of Sciences of the United States of America* **91**, 11841-11843.

Rosendahl, S. (1989). Comparisons of spore-cluster forming *Glomus* species (Endogonaceae) based on morphological characteristics and isoenzyme banding patterns. *Opera Botanica* **100**, 215-223.

Rosendahl, S. (1996). A practical approach to the species concept in the Glomales. In *Mycorrhizas in Integrated Systems from Genes to Plant Development*, pp. 15-18. Proceedings of 4th European Symposium on Mycorrhizas, Granada, Spain 11-14 July 1994.

Rosendahl, S. & Sen, R. (1992). Isozyme analysis of mycorrhizal fungi and their mycorrhiza. *Methods in Microbiology* **24**, 169-194.

Rosendahl, S. & Taylor, J.W. (1997). Development of multiple genetic markers for studies of genetic variation in arbuscular mycorrhizal fungi using AFLP. *Molecular Ecology* **6**, 821-829.

Rosendahl, S., Dodd, J.C. & Walker, C. (1994). Taxonomy and phylogeny of the Glomales. In *Impact of Arbuscular Mycorrhizas on Sustainable Agriculture and Natural Ecosystems*, pp. 1-12. Edited by S. Gianinazzi & H. Schüepp. Basel, Switzerland: Birkhauser Press

Sanders, F.E., Mosse, B. & Tinker, P.B. (1975). *Endomycorrhizas*. London: Academic Press.

Sanders, I.R., Ravolanirina, F., Gianinazzi-Pearson, V., Gianinazzi, S. & Lemoine, M. C. (1992). Detection of specific antigens in the vesicular-arbuscular mycorrhizal fungi *Gigaspora margarita* and *Acaulospora laevis* using polyclonal antibodies to soluble spore fractions. *Mycological Research* **96**, 477-480.

Sanders, I.R., Alt, M., Groppe, K., Boller, T. & Wiemken, A. (1995). Identification of ribosomal DNA polymorphisms among and within spores of the Glomales: application to studies on the genetic diversity of arbuscular mycorrhizal fungal communities. *New Phytologist* **130**, 419-427.

Sanders, I.R., Clapp, J.P. & Wiemken, A. (1996). The genetic diversity of arbuscular mycorrhizal fungi in natural ecosystems - a key to understanding the ecology and functioning of the mycorrhizal symbiosis. *New Phytologist* **133**, 123-134.

Schmidt, E.L., Biesbrock, J.A. & Bohlool, B.B. (1974). Study of mycorrhizae by means of fluorescent antibody. *Canadian Journal of Microbiology* **20**, 137-139.

Schots, A., Dewey, F.M. & Oliver, R. (1994). *Modern Assays for Plant Pathogenic Fungi: Identification, Detection and Quantification*. Wallingford: CAB International.

Sen, R. (1990). Isozymic identification of individual ectomycorrhizas synthesized between Scots pine (*Pinus sylvestris L.*) and isolates of two species of *Suillus*. *New Phytologist* **114**, 617-626.

Sen, R. & Hepper, C.M. (1986). Characterisation of vesicular-arbuscular mycorrhizal fungi (*Glomus* spp.) by selective enzyme staining following polyacrylamide gel electrophoresis. *Soil Biology and Biochemistry* 18, 29-34.

Sieverding, E. (1991). *Vesicular-Arbuscular Mycorrhizal Management in Tropical Agrosystems*. Germany: GTZ Publishers.

Simon, L., Lalonde, M. & Bruns, T. (1992). Specific amplification of 18S fungal ribosomal genes from vesicular-arbuscular endomycorrhizal fungi colonizing roots. *Applied and Environmental Microbiology* 58, 291-295.

Simon, L., Bousquet, J., L., Vesque, R. & Lalonde, M. (1993a). Origin and diversification of endomycorrhizal fungi and coincidence with vascular land plants. *Nature* 363, 67-69.

Simon, L., Vesque, R. & Lalonde, M. (1993b). Identification of endomycorrhizal fungi colonizing roots by fluorescent single-strand conformation polymorphism-polymerase chain reaction. *Applied and Environmental Microbiology* 59, 4211-4215.

Sims, K., Watling, R. & Jeffries, P. (1995). A revised key to the genus *Scleroderma. Mycotaxon* 56, 403-420.

Sims, K., Watling, R., De La Cruz, R. & Jeffries, P. (1997). Ectomycorrhizal fungi of the Philippines: a preliminary survey and notes on the geographic biodiversity of the Sclerodermatales. *Biodiversity and Conservation* 6, 45-58.

Sims, K.P., Sen, R., Watling, R. & Jeffries, P. (1999). Species and population structures of *Pisolithus* and *Scleroderma* identified by combined phenotypic and genomic marker analysis. *Mycological Research* 103, 449-458.

Smith, M.L., Bruhn, J.N. & Anderson, J.B. (1992). The fungus *Armillaria bulbosa* is among the largest and oldest living organisms. *Nature* 356, 428-431.

Smith, S.E. & Read, D.J. (1997). In *Mycorrhizal Symbiosis*, 2nd Edn. Edited by J.L. Harley, & S.E. Smith. London: Academic Press.

Thaxter, R. (1922). A revision of the Endogonaceae. *Proceedings of the American Academy of Arts and Sciences* 57, 291-351.

Thingstrup, I., Rozycka, M., Jeffries, P., Rosendahl, S. & Dodd, J.C. (1995). Detection of the arbuscular mycorrhizal fungus *Scutellospora heterogama* within roots using polyclonal antisera. *Mycological Research* 99, 1225-1232.

Thomas, G.W. & Jackson, R.M. (1982). *Complexipes moniliformis* - ascomycete or zygomycete? *Transactions of the British Mycological Society* 79, 149-151.

Timonen, S., Tammi, H. & Sen, R. (1997a). Characterization of the host genotype and fungal diversity in Scots pine ectomycorrhiza from natural humus microcosms using isozyme and PCR-RFLP analyses. *New Phytologist* 135, 313-323.

Timonen, S., Tammi, H. & Sen, R. (1997b). Outcome of interactions between genets of two *Suillus* spp. and different *Pinus silvestris* genotype combinations: identity and distribution of ectomycorrhizas and effects on early seedling growth in N-limited nursery soil. *New Phytologist* 137, 691-702.

Tisserant, B., Brenac, V., Requena, N., Jeffries, P. & Dodd, J. C. (1998). The detection of *Glomus* spp. (arbuscular mycorrhizal fungi) forming mycorrhizas in three plants, at different stages of seedling development, using mycorrhiza-specific isozymes. *New Phytologist* 138, 225-239.

Tommerup, I. C., Barton, J. E., & O'Brien, P. A. (1995). Reliability of RAPD fingerprinting of three basidiomycete fungi, *Laccaria, Hydnangium* and *Rhizoctonia. Mycological Research* 99, 179-186.

Ursic, M. & Peterson, R.L. (1997). Morphological and anatomical characterization of ectomycorrhizas and ectendomycorrhizas on Pinus strobus seedlings in a southern Ontario nursery. *Canadian Journal of Botany* 75, 2057-2072.

Van Tuinen, D., Jacquot, E., Zhao, B., Gollotte, A. & Gianinazzi-Pearson, V. (1998). Characterisation of root colonisation profiles by a microcosm community of arbuscular mycorrhizal fungi using 25S rDNA-targeted nested PCR. *Molecular Ecology* 7, 879-887.

Walker, C. & Trappe, J.M. (1993). Names and epithets in the Glomales and Endogonales. *Mycological Research* **97**, 339-344.

Weinbaum, B.S., Allen, M.F. & Allen, E.B. (1996). Survival of arbuscular mycorrhizal fungi following reciprocal transplanting across the Great Basin, USA. *Ecological Applications* **6**, 1365-1372.

Wright, S.F., Franke-Snyder, M., Morton, J.B. & Upadhyaya, A. (1996). Time-course study and partial characterization of a protein on hyphae of arbuscular mycorrhizal fungi during active colonization of roots. *Plant and Soil* **181**, 193-203.

Wright, S.F. & Morton, J.B. (1989). Detection of vesicular-arbuscular mycorrhizal fungus colonization of roots by using a dot-immunoblot assay. *Applied and Environmental Microbiology* **55**, 761-763.

Wright, S.F., Morton, J.B., Sworobuk, J.E. (1987). Identification of a vesicular-arbuscular mycorrhizal fungus by using monoclonal antibodies in an enzyme-linked immunosorbent assay. *Applied and Environmental Microbiology* **53**, 2222-2225.

Wyss, P. & Bonfante, P. (1993). Amplification of genomic DNA of arbuscular-mycorrhizal (AM) fungi by PCR using short arbitrary primers. *Mycological Research* **97**, 1351-1357.

Zézé, A., Sulistyowati, E., Ophel-Keller, K., Barker, S. & Smith, S. (1997). Intersporal genetic variation of *Gigaspora margarita*, a vesicular arbuscular mycorrhizal fungus, revealed by m13 minisatellite-primed PCR. *Applied and Environmental Microbiology*, **63**, 676-678.

Zhu, H., Higginbotham, K.O. & Dancik, B.P. (1988). Intraspecific genetic variability of isozymes in the ectomycorrhizal fungus *Suillus* tomentosus. *Canadian Journal of Botany* **66**, 588-594.

Note added in proof

These papers published in 1999 represent further important advances in the area understanding the genetics and hence systematics of the Glomales.

Lanfranco, L., Delpero, M. & Bonfante, P. (1999). Intrasporal variability of ribosomal sequences in the endomycorrhizal fungus *Gigaspora margarita*. *Molecular Ecology* **8**, 37-45.

Hijri, M., Hosny, M., van Tuinen, D. & Dulieu, H. (1999). Intraspecific ITS polymorphism in *Scutellospora castanea* (Glomales, Zygomycota) is structured within multinucleate spores. *Fungal Genetics and Biology* **26**, 141-151.

Hosny, M., Hijri, M., Passerieux, E. & Dulieu, H. (1999). rDNA units are highly polymorphic in *Scutellospora castanea* (Glomales, Zygomycota). *Gene* **226**, 61-71.

5 SYSTEMATICS OF LEGUME NODULE NITROGEN FIXING BACTERIA

AGRONOMIC AND ECOLOGICAL APPLICATIONS

HEITOR L. C. COUTINHO [1], VALÉRIA M. DE OLIVEIRA [2]
and FÁTIMA M. S. MOREIRA [3]

[1] *Embrapa Solos, Rua Jardim Botânico 1024, Rio de Janeiro -
RJ, 22460-000, Brazil;*
[2] *Fundação Tropical de Pesquisas e Tecnologia André Tosello,
R. Latino Coelho 1301, Campinas - SP, 13087-010, Brazil;*
[3] *Departamento de Ciências do Solo, Universidade Federal de Lavras,
Caixa Postal 37, Lavras - MG, 37200-000, Brazil.*

1   Introduction

Biological nitrogen fixation (BNF) is easily the most studied microbial process applied
to agriculture. It consists of the reduction of atmospheric dinitrogen (N_2), unavailable to
higher plants, into ammonium (NH_4^+), an assimilable form of this nutrient. The BNF
process is performed solely by microorganisms, the majority inhabiting the soil ecosystem,
and is considered to be the most relevant component of the global nitrogen cycle
(Ishizuka, 1992). The best known diazotrophs (dinitrogen fixing bacteria) are rhizobia,
which are able to establish a symbiotic relationship with plants of the family Leguminosae,
hereby called legumes. This symbiosis is characterized by a highly specific association
between plant and bacteria. Particular varieties of legume species recognize specific strains
of rhizobia, which are able to infect the legume roots. This process triggers the expression
of certain plant genes, resulting in the development of nodules around the site of invasion.
In the interior of the nodules the bacteria undergo morphological and physiological
transformations, becoming nitrogen fixing bacteroides and supplying the plant with nutrient
in exchange for carbon-rich material derived from plant photosynthesis.

BNF contributions to the economics of legume crop production are very significant,
since the need for fertilizer applications can be greatly reduced. The inoculation of legume
seeds with diazotrophs is a common practice in several countries. In Brazil, soybean inoculation with strains of *Bradyrhizobium elkanii* and *Bradyrhizobium japonicum* is a great success,
and is responsible for a reduction in production costs of approximately 1.3 billion US$ per
year (Siqueira & Franco, 1988). Apart from the economic benefits, reductions in fertilizer
applications may enhance the environmental quality of agricultural production, since there
are lower rates of leaching of nitrates to both surface and ground waters (Boddey *et al.*, 1984).

F. G. Priest and M. Goodfellow (eds.), Applied Microbial Systematics, 107-134
© 2000 *Kluwer Academic Publishers. Printed in the Netherlands.*

The last decade has witnessed a redirection of global policies of development. This is characterized by assimilation of the concept of sustainability into projects funded by governmental and non-governmental agencies. This has led to the establishment of a common goal to all of those involved in the agricultural business, namely the achievement of sustainable production systems with minimal environmental impact and absence of the degradation of natural resources. Sustainable agricultural systems can benefit from BNF by using legumes at any stage of the production line. Agroforestry systems may contain tree legume species, which can supply fodder, fuelwood or shade, in consortia with annual crops, such as maize or sorghum. Tree legumes may also be used as wind barriers or live fences with high efficiency and low cost. In crop rotation systems, grasses like maize, wheat, sugarcane, or pastures, are followed by legume crops such as soybeans, common beans or peanuts, in order to fertilize the soil with the nitrogen fixed by rhizobia as well as producing an extra income. Legumes may also be used as green manure whereby fixed nitrogen transported to the leaves enriches the soil after decomposition of the plant residues and mineralization of the organic nitrogen.

In order to maximize the benefits rendered by biological nitrogen fixation, research efforts originally focused on the isolation and screening of diazotrophic bacteria that could be used as inoculants in agricultural systems. This initial stage was very successful with the development of commercial inoculants for legumes worldwide. Later, the introduction of more efficient nitrogen fixing strains to soil containing native or naturalized populations of rhizobia was hampered by low competitiveness or lack of adaptation of the inoculant strains to the soil environment (Neves & Rumjanek, 1997). This led to an increased interest in the ecology of nitrogen fixing microorganisms. More recently, global concerns regarding the potential loss of biodiversity due to human activities have boosted initiatives to assess, catalogue, preserve and monitor the diversity of nitrogen fixing microorganisms in soils of diverse ecosystems.

Microbial systematic tools are required for the development of research projects in all of the areas mentioned in the previous paragraph. Therefore, advances in rhizobial research have followed closely the development of methodologies to classify, identify, differentiate, and detect specific microorganisms in the environment. In this chapter we review the application of microbial systematics to elucidate agronomic and ecological aspects of legume nodule, nitrogen fixing bacteria and the symbiotic process. Initially, the taxonomy of rhizobia will be placed in a historical perspective. This will be followed by a consideration of the application of DNA probe technology to the study of rhizobia and, finally, we conclude with remarks on future prospects for solving agronomic and ecological problems with the aid of modern techniques used in microbial systematics.

2   Current Knowledge: Classification

2.1   HISTORICAL PERSPECTIVE

A comprehensive review of the early history of *Rhizobium* taxonomy, which is summarized here, was brilliantly presented by Nutman (1987). The occurrence of nitrogen fixation in legumes was first hypothesized by Boussingault in 1837 (cited by Nutman, 1987), who was followed by others, who failed to consider nodules as the actual sites of the nitrogen fixation process. They were criticized by Liebig, who stated that only manure could provide mineral elements found in plant ashes (Nutman, 1987). The strong personality of Liebig discouraged others from trying to clarify something that had been evident since ancient times, namely, why had legumes more nitrogen content than other plants? Finally, Hellriegel and Wilfarth (1888) demonstrated the occurrence of nitrogen fixation in legumes. Their only opponent was Frank, who argued against nitrogen fixation at scientific meetings. Ironically, Frank (1889) was the one who proposed *Rhizobium leguminosarum*, the bacterium which had been isolated from legume nodules by Beijerinck (1888), who had named it "Bacillus radicicola".

For many years, the study of legume-*Rhizobium* symbiosis was mainly restricted to a few legume species of the sub-family Papilionoideae, most of them agricultural crops from temperate regions. The taxonomy of rhizobia was based on the cross-inoculation concept, which assumed that members of each *Rhizobium* species were able to nodulate a specific group of host legumes (Jordan & Allen, 1974; see Table 1). There are many exceptions to this "rule" and, as a result, it failed as a criterion for species classification (Graham, 1976). The cross-inoculation criterion has not been considered since the first edition of *Bergey's Manual of Systematic Bacteriology* (Jordan, 1984).

TABLE 1.   Classification of *Rhizobium* species according to cross-inoculation groups (Jordan & Allen, 1974)

Group	*Rhizobium* species	Hosts
I	R. leguminosarum	Lens, Pisum, Vicia
I	R. trifolii	Trifolium
I	R. phaseoli	Phaseolus angustifolius, P. multiflorus P. vulgaris,
I	R. meliloti	Medicago, Melilotus, Trigonella
II	R. lupini	Lupinus, Ornithopus
II	R. japonicum	Glycine
II	Unclassified	Cowpea group

2.2   THE VALUE OF PHENOTYPIC TESTS

Molecular characterization tools, particularly nucleic sequence analysis, are essential for the classification of microorganisms (see next section). However, many microbiology laboratories do not yet have access to the infrastructure required for routine molecular biological practices and some methods are expensive and time consuming when applied to large numbers of strains. Thus, a first approach in *Rhizobium* classification and identification requires the analysis of characteristics that are easily assessed. Rhizobial cells are Gram-negative, aerobic rods lacking endospores, usually containing granules of poly-β-hydroxybutyrate, which are refractile under phase-contrast microscopy.

Graham (1976) pointed out that three characteristics were useful for rhizobial classification: growth rate, pH reaction on yeast mannitol agar (YMA) medium, and the type of flagella.

Generation times vary a great deal among *Rhizobium* strains, namely from 1.4 to 44.1 hours (Martinez-Dretz & Arias, 1972; Tan & Broughton, 1981; Kennedy & Greenwood, 1982; Hernandez & Focht, 1984). The same is true with regards to the growth rates of rhizobia (Lim & Ng, 1977; Gross *et al.*, 1979; Moreira, 1991; Moreira *et al.*, 1993). Recently, growth rates and generation times proved to be useful characteristics for the description of new rhizobia taxa. Members of the genus *Bradyrhizobium*, the species *Bradyrhizobium liaoningense*, and the genus *Mesorhizobium* are slow, very slow, and intermediary growers, respectively. SDS-PAGE analysis of total proteins grouped the majority of the very slow growing strains isolated from Brazilian native forest species with known *Bradyrhizobium* species except for three strains which did not cluster with any other rhizobial species (Moreira *et al.*, 1993). Further characterization is necessary to verify if these very slow growing strains belong to *B. liaoningense* or to other species of this genus.

The type of flagella is a conserved characteristic within each genus (Graham, 1976). It is important that all strains used in comparative analysis of flagella types are grown on the same growth media (Table 2). The type of flagella is consistent with the phylogenetic divisions of rhizobia (see Fig. 1), which makes this characteristic an important tool for preliminary analyses of collections of strains. The pH reaction resulting from growth on YMA is also usually consistent within genera: *Mesorhizobium*, *Rhizobium*, and *Sinorhizobium* species have an acid reaction or in some cases the pH is not modified, whereas *Azorhizobium* and *Bradyrhizobium* species produce alkaline reactions. Other cultural characteristics of members of each of these genera are shown in Table 3.

TABLE 2.   Types of flagella found in members of the different genera of rhizobia (Graham, 1976)

Type of flagella	*Azorhizobium*	*Bradyrhizobium*	*Mesorhizobium*	*Rhizobium*	*Sinorhizobium*
One lateral	+	-	-	-	-
One subpolar	-	+	+	-	+
Peritrichous	+	-	+	+	+

Patterns of utilization of carbon sources do not seem to distinguish between members of rhizobial taxa, not even at the genus level, except for *Azorhizobium* (Table 4). The ability to metabolize sugars is not a conserved trait among rhizobia. A significant proportion of the carbon sources tested can give variable results at the intrageneric level (Dreyfus *et al.*, 1988; de Lajudie *et al.*, 1994; Jarvis *et al.*, 1997). It is not possible, therefore, to classify rhizobia on the basis of carbon utilization patterns. *Bradyrhizobium* and *Mesorhizobium* strains are able to utilize the greatest number of carbon sources, while *Azorhizobium* strains utilize only 52% of the sources tested in the studies summarized in Table 4. Utilization of mannitol and sucrose, except for *Azorhizobium* strains, is widespread among rhizobial genera. However, it must be considered that these are the common components of the usual isolation media and can act as a selective pressure for strains utilizing these sources. The same is true for pH, since common isolation media have a pH around 7.0. It is important to make slight modifications in pH, carbohydrate source or other medium components in order to isolate strains that do not grow on conventional YMA medium.

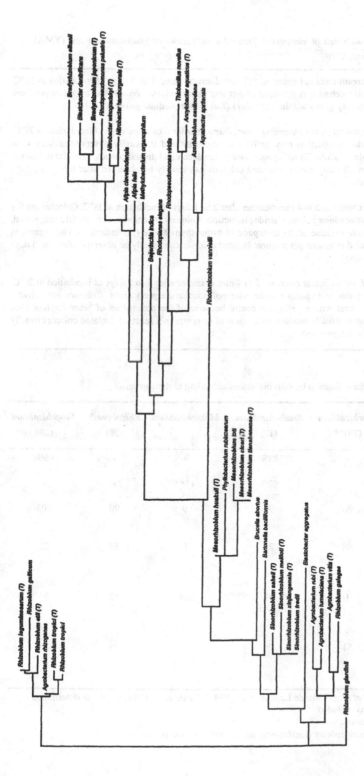

Figure 1. Phylogenetic tree of rhizobia species and their closest relatives derived by maximum likelihood analysis of small-subunit rRNA sequences. The tree is abstracted from that provided by the Ribosomal Database Project (Maidak *et al.*, 1997) by submitting the 16S rRNA sequences of *R. giardinii* and *R. gallicum* (GenBank accession numbers U86344 and AF008130, respectively) against the following RDP sequences: Afp.cleve, Ntb.winog2, Ntb.hambu2, Rps.palus5, Bdr.japon4, Bdr.jpn_76, Blb.denitr, Afp.felis, Mlb.organ2, Bei.indica, Rm.vanniel, Rps.viridi, Thb.novell, A.nc.aquati, Azr.cauli2, Aqb.spirit, Rhp.elegan, Rhb.etli42, Rhb.legu11, Ag.rhizog2, Rhb.tropi8, Rhb.tropi6, Srh.saheli, Srh.meli17, Srh.fredi2, Blb.aggreg, Ag.vitis3, Rhb.galega, Ag.rubi3, Ag.tumefa6, Bar.bacill, Bru.aborts, Rhb.loti, Rhb.tiansh, Rhb.ciceri, Plb.rubiac, Rhb.huaku4. The scale indicates the evolutionary distance as the number of changes expected per sequence position (Olsen *et al.*, 1994). The species *Sinorhizobium medicae*, *S. teranga*, *Mesorhizobium mediterraneum*, and *Bradyrhizobium liaoningense* are not represented in the tree.

TABLE 3.   Cultural characteristics of members of rhizobial genera grown on yeast-mannitol-agar (YMA) media (Vincent, 1970)

Azorhizobium - circular, cream coloured colonies, 0.5 mm diameter after 2 to 5 days of incubation at 28°C. Very little extracellular polysaccharide is produced (much less than *Bradyrhizobium*). The organism produces an alkaline reaction and usually grows within 3 to 5 days (fast to intermediate growers).

Bradyrhizobium - circular colonies, not exceeding 1 mm diameter after 6 or more days of incubation at 28°C. Extracellular polysaccharide production may be little to abundant and is usually observed in very slow growing strains (after 10 days). Colonies are opaque, rarely translucent, white and convex, granular in texture. The organism produces an alkaline reaction. Isolated colonies can usually be observed after 6 or more days (slow to very slow growers).

Mesorhizobium - circular colonies, 0.5 to 4 mm diameter after 2 to 7 days of incubation at 28°C. Colonies usually coalesce due to copious extracellular polysaccharide production. Colonies are convex, semi-translucent, raised, most with a yellowish centre because of the absorption of bromothymol blue (pH indicator). The organisms usually produce either an acid or neutral pH reaction. Isolated colonies can usually be observed after 4 to 5 days (fast to intermediate growers)

Rhizobium and Sinorhizobium - circular colonies, 2 to 4 mm diameter after 2 to 5 days of incubation at 28°C. Colonies usually coalesce due to copious extracellular polysaccharide production. Colonies are convex, semi-translucent, raised, most with a yellowish centre because of the absorption of bromothymol blue (pH indicator). The organisms usually produce either an acid or neutral pH reaction. Isolated colonies usually appear within 2 to 3 days (fast growers).

TABLE 4.   Carbon source utilization by rhizobia strains according to their genera

	Azorhizobium (20)**	*Bradyrhizobium* (17)	*Mesorhizobium*	*Rhizobium** (29)	*Sinorhizobium* (29)
Mannitol	-	88%	+	+	96%
Sucrose	-	82%	+***	+	+
Number of other C sources tested	90	90	9	90	103
% of C sources on which no strain is able to grow	48	22	0	33	23
% of C sources on which at least one strain is able to grow	1	52	n.d.	42	56
% of C sources on which all strains are able to grow	51	26	n.d	25	21

Data taken from Dreyfus *et al.* (1988); de Lajudie *et al.* (1994); Jarvis *et al.* (1997); n.d. not determined
*   *Sinorhizobium* strains included.
**   Number of strains tested.
*** Some strains of *Mesorhizobium tianshanense* are not able to grow on sucrose.

2.3   PHYLOGENETICS

The development of molecular genetic tools, represented by the polymerase chain reaction (PCR) (Mullis & Faloona, 1987), along with the concept of evolutionary clocks (Kimura, 1983; Woese, 1987), provided the means to base microbial taxonomy on phylogenetic relationships. *Rhizobium* phylogeny, based on sequences of genes coding for the small subunit ribosomal RNA (16S rRNA), confirmed the close relationship between the genera *Agrobacterium* and *Rhizobium* suggested by Fred *et al.* (1932), and showed them to be distant relatives of *Azorhizobium* and *Bradyrhizobium*, which are closely related to *Afipia* (responsible for human cat-scratch infections), *Blastobacter* (a denitrifier), and *Rhodopseudomonas* (a purple phototroph), all in the alpha subdivision of the class *Proteobacteria* (Fig. 1). Molecular systematics also promoted a breakthrough in the taxonomy of rhizobia. The results of successive reclassifications and introduction of new taxa led to the classification of all known rhizobia into 6 genera and 26 species (Fig. 1; Table 5), contrasting with the 4 species previously described in *Bergey's Manual of Systematic Bacteriology* (Jordan, 1984). The evidence provided by the phylogenetic analyses points to the need for a review of the current suprageneric classification of rhizobia.

Only a few of the recently-described rhizobial species, for example *Rhizobium etli, R. tropici, Sinorhizobium saheli,* and *S. teranga* were based on the characterization of isolates from nodules of tropical legume species (Martinez-Romero *et al*, 1991; Segovia *et al.*, 1993; de Lajudie *et al.*, 1994), thereby underscoring the point that taxonomic knowledge of the micro-symbionts of tropical forest species is scarce, as is that of their hosts. It is probable that tropical rhizobia represent an untapped, and largely unexplored, source of microbial diversity with clear, and measurable, economic and ecological value.

2.4   TROPICAL RHIZOBIAL DIVERSITY

Almost a century after the first *Rhizobium* was isolated, Graham (1976) and Allen & Allen (1981) noted that only about 15% of Leguminosae species - the third largest family of flowering plants with about 19,000 species - were known to have the ability to form nodules. This information deficit was mainly attributed to species of tropical origin.

Extensive surveys of the nodulation capacity of tropical Leguminosae species were made recently in the Brazilian Amazonian and Atlantic forests (Faria *et al.*, 1989; Moreira *et al.*, 1992). This work led to the isolation of a large collection of Brazilian rhizobial strains, which have been studied by many workers (Moreira, 1991; Moreira *et al.*, 1993, 1995; Loureiro *et al.*, 1994, 1995; Coutinho *et al.*, 1995; Hollanda *et al.* 1996; Moreira *et al.*, 1998). Comparative analyses of some of these Brazilian strains with others from Africa have revealed similarity among certain groups, as well as potentially new species (de Lajudie *et al.*, 1994; Dupuy *et al.*, 1994). Other geographical regions have also been explored, and the results show a high degree of diversity among rhizobia isolated from nodules of both woody and non-woody legumes (Rinaudo *et al.*, 1991; Zhang *et al.*,1991; Oyaizu *et al.*, 1992; Haukka & Lindström, 1994; Novikova *et al.*, 1994; van Rossum *et al.*, 1995). Additional new genera and species of rhizobia can be expected as more strains are isolated from legume species in other regions of the world and examined by microbial taxonomists. Indeed, advances in molecular ecology and systematics may allow the discovery of new taxa in natural habitats without the need to cultivate the microorganisms (Stephen *et al.*, 1996).

TABLE 5.   Current classification of validly described bacteria able to fix nitrogen in association with leguminous plants

Genera	Species
Rhizobium (Frank, 1889)	**R. leguminosarum* (Frank, 1889) with biovars *phaseoli*, *trifolii* and *viceae* (Jordan, 1984)
	R. tropici (Martinez-Romero *et al.*, 1991)
	R. etli (Segovia *et al.*, 1993)
	R. gallicum (Amarger *et al.*, 1997)
	R. giardinii (Amarger *et al.*, 1997)
	R. hainanense (Chen *et al.*, 1997)
	R. huautlense (Wang *et al.*, 1998)
	R. mongolense (van Berkum *et al.*, 1998)
Sinorhizobium (de Lajudie *et al.*, 1994)	**S. meliloti* (Dangeard, 1926; de Lajudie *et al.*, 1994)
	S. fredii (Scholla & Elkan, 1984; de Lajudie *et al.*, 1994)
	S. saheli (de Lajudie *et al.*, 1994)
	S. teranga (de Lajudie *et al.*, 1994)
	S. medicae (Rome *et al.*, 1996b)
	M. amorphae (Wang *et al.*, 1999)
	M. plurifarium (de Lajudie *et al.*, 1998b)
Mesorhizobium (Lindström *et al.*, 1995; Jarvis *et al.*, 1997)	**M. loti* (Jarvis *et al.*, 1982)
	M. huakuii (Chen *et al.*, 1991
	M. ciceri (Nour *et al.*, 1994a)
	M. tianshanense (Chen *et al.*, 1995)
	M. mediterraneum (Nour *et al.*, 1995)
Undefined	*Rhizobium galegae* (Lindström, 1989)
Bradyrhizobium (Jordan, 1982)	**B. japonicum* (Jordan, 1982)
	B. elkanii (Kuykendall *et al.*, 1992)
	B. liaoningense (Xu *et al.*, 1995)
Azorhizobium (Dreyfus *et al.*, 1988)	**A. caulinodans* (Dreyfus *et al.*, 1988)
Allorhizobium (de Lajudie *et al.*, 1998a)	**A. undicola* (de Lajudie *et al.*, 1998a)

* type species.

2.5   THE NEED FOR MINIMAL STANDARDS

Rhizobial diversity is still a "black box" due to the scarce knowledge available on the vast majority of extant leguminous species and their microsymbionts. The ability of Leguminosae species to form effective symbiotic relationships with rhizobia is estimated to be unknown in about 80% of that plant family. Consequently, the characteristics of *Rhizobium* species nodulating hosts belonging to this group is not known. However, the last decade has been extremely important as studies revealed a huge diversity of previously unknown rhizobia, which are now well documented and preserved in appropriate microbial culture collections.

In order to avoid the introduction of poorly described new taxa, minimal standards for the description of new rhizobial genera and species were proposed by Graham *et al.* (1991). It was recommended that phylogenetic as well as phenotypic (symbiotic, cultural, morphological) traits are needed for the valid description of rhizobial species. The authors also recommended that relatively large numbers of strains should be represented in studies designed to describe new taxa. In addition, it is important that collaborative studies are undertaken so as to reach adequate standards for the description of new rhizobial taxa. One way to achieve this is by integrating the efforts of strain isolation and phenetic characterization carried out by workers in less well equipped laboratories with the phylogenetic analyses possible in institutes which have more developed microbiological facilities. This kind of collaboratiom can be very effective provided all stages of the process are planned jointly by the partners involved, and training of students and personnel, as well as technological transfer is accounted for.

3   The Application of Molecular Probe Technology

The need to study the genetics and regulation of important biochemical cycles mediated by microorganisms initiated the development of numerous molecular techniques, some of which turned out to be of great value in microbial systematics. Nucleic acid probes may be designed for the detection of bacteria that carry particular genes, such as *nif* (nitrogen fixation) or *nod* (nodulation) genes, or unique, taxon signature sequences. The latter can be transposons, insertion sequences, or oligonucleotide sequences of rRNA genes (Barkay *et al.*, 1985; Fredrickson *et al.*, 1988). The application of DNA probes to study natural microbial populations became even more practical with the development of PCR technology (Mullis & Faloona, 1987). This technique promotes a significant increase in the probability of the detection of rare sequences in a mixture of heterologous DNA. This is achieved through the exponential amplification of target sequences.

Initially, the genes involved in the symbiotic process were seen as the best candidates for the design of probes or PCR primers (Hahn & Hennecke, 1987). Technological advances enabled the development of rapid DNA sequencing procedures as well as facilitating phylogenetic studies. The ribosomal gene regions turned out to be sites of great interest for probe designers aiming to detect specific rhizobial taxa in the environment (Oliveira *et al.*, 1997).

3.1   SYMBIOTIC GENES

The obvious need to understand the regulatory mechanisms of the nitrogen fixation process of the legume-*Rhizobium* symbiosis, which involves bacteria-plant recognition events, bacterial invasion, nodule formation, and nitrogen fixation *per se*, led to the mapping and sequencing of the major genes involved in the symbiotic process, namely the *nod* (nodulation) and *nif* (nitrogen fixation) genes. The conservation, number of copies, and occurrence of these genes have been evaluated in different strains, species, and genera of rhizobia. The symbiotic genes were later found to be useful for research into the systematics of rhizobia (Masterson *et al.*, 1985; Cadahia *et al.*, 1986; Harrison *et al.*, 1988; Young, 1993; Dobert *et al.*, 1994; Lindström *et al.*, 1995).

Analyses of *nod* and *nif* gene regions pointed to the need for reclassifications of rhizobia taxa. The number of copies of the *nif* genes was the basis for the assignment of *Rhizobium leguminosarum* biovar *phaseoli* group II to the species *Rhizobium etli* (Segovia *et al.*, 1993). Data from restriction fragment length polymorphism (RFLP) using *nod* and *nif* genes as probes provided supporting evidence that *Bradyrhizobium japonicum* should be split into two groups (Stanley *et al.*, 1985; Hahn & Hennecke, 1987; Minamisawa *et al.*, 1992). Further corroborating molecular data resulted in strains belonging to DNA homology group II being reclassified as *Bradyrhizobium elkanii* (Kuykendall *et al.*, 1992). Similarly, most of the fast growing rhizobial strains that nodulate soybeans were found to contain their *nod* and *nif* genes on plasmids; these organisms differ from the slow growing strains, which carry their symbiotic genes in the chromosome (Masterson *et al.*, 1985). The fast growing soybean rhizobia are now classified as a separate species in a different genus, namely *Sinorhizobium fredii* (de Lajudie *et al.*, 1994).

Specific taxonomic problems, some of which were rather difficult to elucidate, have been resolved by sequencing symbiosis genes. Strain Or191, classified as *Rhizobium* sp., nodulates both alfalfa (*Medicago sativa*) and bean plants (*Phaseolus vulgaris*). In an attempt to assign this strain to a species, Eardly *et al.* (1992) sequenced its *nifH* gene and compared it with those of reference rhizobia species. The *nifH* sequence of strain Or191 differed significantly from those of the other test strains. However, its 16S rDNA sequence resembled those of a group of strains of *R. leguminosarum* bv. *phaseoli*, although differing from them by possessing only a single copy of the gene. This example clearly illustrates the need to characterize strains using a variety of genotypic and phenotypic tools in order to achieve a polyphasic taxonomic approach of the rhizobia.

The degree of conservation of *nif* and *nod* genes varies according to the species and genus that carries them, reflecting the evolutionary status of the different taxa. The divergence of *nod* genes among rhizobia has been studied (Young & Johnston, 1989; Lindström *et al.*, 1995). The degree of sequence conservation may determine the usefulness of symbiotic genes as sites for annealing oligonucleotide probes or PCR primers specific to different taxonomic levels from strain to genus. The more conserved a gene is within a species the more likely it is that a species-specific probe can be generated. However, it is more difficult to find a probe site specific to the strain level. It is important to determine to what extent genes are conserved among members of different taxa before spending time and effort in the pursuit of highly specific probing sites.

Analyses of RFLP using gene fragments as probes is a simpler and more affordable way to assess the distribution and conservation of a particular gene within a taxon than sequencing the whole gene of several strains or isolates. This was the method used to demonstrate the high conservation of *nif* genes in chickpea (*Cicer arietinum* L.) rhizobia, both fast and slow growers, and their absence in the isolated plasmids from the 27 strains analysed (Cadahia *et al.*, 1986). The *nif* genes are highly conserved among diazotrophs, as confirmed by sequence alignment and cluster analysis of *nifH* genes from divergent species, including ancient archaeal and clostridial groups (Manjula & Rakesh, 1990). These workers found two related but discrete clusters of *nifH* sequences from rhizobia, which were grouped according to their location in the genome, whether chromosomal or plasmidial.

Probes specific to *nif* or *nod* genes can be used to detect and quantify symbiotic plasmids in *Rhizobium* strains. This was done for *R. leguminosarum* biovar *trifolii* by using a repeated sequence incorporating the *nifH* promoter as a symbiotic plasmid specific probe (Harrison *et al.*, 1988). The investigators found a negative correlation between the number and size of Sym plasmids and the nitrogen fixation effectiveness of the strains carrying them.

Molecular probe technology is also helping in the development of methods for the selective isolation of rhizobia from soil. Laguerre *et al.* (1993 a, b) used a chromosomal probe to identify soil isolates of *R. leguminosarum*, and *nod* probes to assign isolates to biovars *viciae* or *trifolii*. A more refined and specific medium has also been developed for isolating *R. meliloti* from soil (Bromfield *et al.*, 1994). The authors used a *nodH* species-specific probe to evaluate the usefulness of the selective medium.

3.2   RIBOSOMAL GENES

Ribosomal nucleic acids are considered to be the most useful biopolymers for comparative studies in microbial molecular ecology. The rRNA genes (rDNA) are universally distributed, have a high degree of conservation, and can accumulate more or less variability in different sections of the molecule (Lane *et al.*, 1985). Bacterial rRNAs are classified, according to their centrifugation sedimentation rates, as 5S, 16S, or 23S. The 16S rRNA is associated with the smaller subunit of the ribosome, and is the most well studied molecule. One of the greatest advantages of rRNA sequence information is its cumulative nature, i.e., once obtained, it can be deposited in specialized databases, eg. RDP, GenBank, EMBL databases, thereby becoming available for future studies.

The automation of nucleic acid sequencing procedures resulted in a large number of microbial rRNA sequences becoming available in databases. These data can be used to build phylogenetic trees, depicting relationships between test organisms. The alignment of sequences from groups of microorganisms of any particular interest may allow one to find regions of the molecule that are valuable for designing oligonucleotide probes specific to different taxonomic levels, from universal probes, which can detect the domains Archaea, Bacteria, or Eucarya, to family, species or strain-specific probes. A great variety of ribosomal probes of different levels of specificity have been developed and used in microbial ecology (Pace *et al.*, 1986; Laguerre *et al.*, 1993 a, b; Louvrier *et al.*, 1995; Oliveira *et al.*, 1997).

Finding adequate sites in the 16S rRNA molecule that can be used to design oligonucleotide probes or PCR primers specific to rhizobial taxa is not an easy task, primarily because of the high similarity shown between rhizobial species and their phylogenetic close relatives. Species of *Azorhizobium caulinodans*, *Bradyrhizobium japonicum*, *Mesorhizobium* (*ciceri*, *huakuii*, and *loti*), and *Rhizobium* (*etli*, *leguminosarum*, and *tropici*) are closely related to *Aquabacter spiritensis*, *Rhodopseudomonas palustris*, *Phyllobacterium rubiacearum*, and *Agrobacterium rhizogenes*, respectively (Young, 1996). Members of the different taxa of a phylogenetic group may differ only in a few 16S rDNA nucleotides, and this makes it difficult to design an effective probe or primer specific to the species or genus level as such probes must be at least 10 base pairs long. In the case of PCR primers, the nucleotide bases that confer specificity to the primer should be concentrated at the 3'-end of the oligonucleotide, as this is the binding site of the enzyme DNA polymerase. These difficulties probably explain the low number of ribosomal probes or primers that are specific to rhizobial species or genera compared to the wealth of data on *nod* and *nif* probes.

An alignment of 16S rRNA sequences provided only one potential probing site specific for *R. tropici* and *R. leguminosarum*, but this site was unable to exclude *Agrobacterium rhizogenes* (Oliveira *et al.*, unpublished results). For the same reason, a 16S rDNA-directed probe specific to *Bradyrhizobium* spp. could not be designed due to the high similarity of this species with *Rhodopseudomonas palustris*. However, a potential probing site specific to *Bradyrhizobium elkanii* was detected and the probe designed was shown to be specific to that species and to several other *Bradyrhizobium* spp. isolated from tropical tree and shrub legumes (Oliveira *et al.*, 1997).

The large subunit of ribosomal RNA, that is 23S rRNA, contains a higher degree of sequence variation than the corresponding 16S rRNA subunit and, hence, can be a potential target for probes. A probe specific to the 23S rDNA of *Bradyrhizobium japonicum* was designed and found to be effective (Springer *et al.*, 1993). These authors not only sequenced this molecule from *Bradyrhizobium japonicum* but also from representatives of closely related species. A highly variable region at the 5'-end of the 23S rRNA gene of rhizobia and agrobacterial strains has been identified as a potential target for the design of specific probes and PCR primers (Evguenieva-Hackenberg & Selenska-Pobell, 1995).

Similarly, a phylogenetic analysis using 16S rDNA sequences of five strains of bacteriochlorophyll-synthesizing rhizobia isolated from stem nodules of members of *Aeschynomene* species showed them to be homogeneous and members of the same line of descent as *Bradyrhizobium japonicum* (Wong *et al.*, 1994). These workers determined potential target sites for the design of probes specific to the photosynthetic rhizobia.

A promising genomic region for probing and priming sites is the spacer region between the 16S and 23S rRNA genes. This region sometimes contains tRNAs, as well as DNA segments with variable nucleotide sequences. Considerable variation can occur in the size and sequence of this region between strains or species (Honeycutt *et al.*, 1995; Gürtler & Stanisich, 1996). It seems likely that PCR amplifications with primers directed to conserved sequences flanking the intergenic spacer region (IGS) may reveal size variations. If these are not resolved, the amplified fragment can be digested by endonucleases generating polymorphisms among analysed strains to reflect sequence variations. The technique of amplifying a part of the genome and digesting the product with restriction enzymes is called PCR-RFLP. This technique has been used to identify species or strains of bacteria (Gürtler & Stanisich, 1996).

4 Agronomic and Ecological Applications

The development of molecular systematics opened up an array of possibilities for the resolution of current problems found in the application of *Rhizobium* technology in agriculture, as well as in understanding the ecology of these microorganisms and the role that different rhizobial taxa play in nitrogen cycling. Not all of these aspects are currently being investigated, since they require further development or simplification of techniques for *in situ* detection and identification of members of specific taxa. However, a great deal of ecological work has been carried out, particularly diversity studies in different biomes. There has been some progress in the application of systematics tools for the quality control of rhizobial inoculants and on the selection of improved inoculant strains. A few of the most recent advances on the application of modern molecular systematic techniques to improve inoculant quality and to evaluate rhizobia diversity in different environments will now be considered.

4.1 IMPROVING THE QUALITY OF INOCULANTS

One of the major challenges of *Rhizobium* technology is the development of improved inoculants, that is, ones containing strains able to fix nitrogen with greater efficiency, to compete well with native rhizobia for nodule occupancy, and to survive and persist under a wide range of environmental conditions.

The first step in the development of a good inoculant involves screening good quality material. This necessitates access to a suitable culture collection. A good culture collection requires personnel and a suitable infra-structure to promote sound and reliable systematics research. This involves thorough characterization of strains by both conventional techniques and advanced methods, including chemosystematic and molecular tools.

Examples of the application of advanced systematic methods to investigate the performance and composition of rhizobial inoculants can be found in the literature. For example, two of the strains that are used in the formulation of the Brazilian commercial soybean inoculant were identified as members of the species *Bradyrhizobium elkanii* through analysis of their 16S rDNA sequences (Rumjanek *et al.*, 1993). DNA fingerprinting (RAPD) coupled with pyrolysis mass spectrometry (Py-MS) enabled the identification of Italian soybean nodule isolates as strains of different European inoculants (Kay *et al.*, 1994). The same approach was used to identify soybean nodule isolates from the Brazilian cerrados. Some of these isolates showed genetic drift, which was detected by Py-MS and correlated with their superior adaptation to the cerrado soils (Coutinho, 1993; Coutinho *et al.* 1999a).

A major problem in the use of inoculants containing strains with greater nitrogen fixation efficiency is their generally low competitiveness compared to native or naturalized rhizobia. In this case, inoculations yield no response, since the majority of the nodules will be occupied by rhizobial strains already present in the soil. In some areas of the USA, strains of *Bradyrhizobium japonicum* serogroup USDA 123 that were introduced in the past, prevail in soybean nodules. They are so competitive that the attempts to use more efficient strains were unsuccessful (Kramicker & Brill, 1986).

Competition experiments with varying numbers of indigenous, ineffective rhizobia in the soil and different seed inoculum levels were performed by Tas *et al.* (1996).

These workers developed *R. galegae* species-specific primers that were used in competition experiments along with strain-specific primers developed earlier by Tas *et al.* (1994). Crushed nodules were used as DNA templates for PCR reactions. It was demonstrated that the ineffective strains present in the soil were very competitive and impaired growth and nodulation of the plants inoculated with the effective strain. Tas and his colleagues suggested that the use of PCR with specific primers is the most powerful technique available for ecological studies, since it may discriminate between closely related strains of the same species and hence can be used to amplify specific fragments from soil or nodule samples containing mixed populations. However, although analysis of PCR-amplified 16S rRNA genes is effective for identifying species, it is not so useful for typing at the infra-specific level, that is, at the level usually required for competition studies.

The success of rhizobial inoculation procedures also depends on how the inoculant is produced, stored, and distributed. Quality standards need to be followed, such as the minimum number of cells of the recommended strain per gram or millilitre of the carrier, and the degree of purity with an established limit for contaminant cells. Furthermore, the growing concern over the quality of rhizobial inoculants led to a restriction in the degree of tolerance regarding contaminant cells and standards of minimum rhizobial cell numbers in commercial inoculants (Olsen *et al.*, 1994).

It is clear that methods are needed for the detection of specific strains of rhizobia not only for ecological studies but also for the quality control of inoculants. Methods that enable the typing of a rhizobial strain without affecting its physiology in nature can be used to detect and help to enumerate specific strains in an inoculant. DNA fingerprinting techniques, such as RAPD, REP-PCR, or RFLP, can be employed for this purpose, provided the generated fingerprints are unique for each strain. For example, primers homologous to the extremities of a repeated sequence (RSα) of *Bradyrhizobium japonicum* were designed and shown to amplify fragments of the expected size exclusively from DNA of *Bradyrhizobium elkanii* and *B. japonicum* strains (Hartmann *et al.*, 1996). These workers recommend the use of PCR with these primers as a tool to assess the quality of soybean inoculants. Their results also showed that the amplified RSα fragment is largely conserved among *B. japonicum* strains, and hence may be used as a screening method to identify them among field isolates. They adapted the PCR method in order to quantify the cell numbers of *B. japonicum* in liquid inoculants as well as in growth media. By increasing the number of PCR cycles employed, the detection limit was of the order of 16 cells per gram, which is quite promising.

4.2   EXPLORING RHIZOBIAL DIVERSITY

Perhaps the most relevant consequence of the development of modern molecular systematics has been the possibility to investigate rhizobial diversity in soils of different biomes of the world. The use of DNA probes and fingerprinting techniques allow a better understanding of the taxonomic diversity of nodule isolates from different legume species growing in distinct environments. Strains that were once indistinguishable can now be discriminated and typed, and this allows their fate, either when introduced or following environmental perturbation, to be followed. Nodule occupancy can now be easily and unequivocally established, and competition studies facilitated.

Some of the advances in research on rhizobial diversity are reviewed below. The discovery of new species and genera can be considered to be a feedback from applied sciences to pure taxonomy.

4.2.1 *Indirect evaluation of rhizobial diversity*

Indirect evaluations of microbial diversity are defined here as those that rely on characterization data of isolated and cultured microorganisms.

Genotypic differences and relatedness among members of a taxon can be assessed by sequence polymorphisms of their common genes. Hence, RFLP analyses of *nod* and *nif* genes have been widely used in studies of the diversity of different rhizobial species. Different probes directed towards the *nod* gene region have been applied to study the diversity of *Rhizobium leguminosarum* bv. *phaseoli*, bv *trifolii*, and bv *viciae* (Laguerre *et al.*, 1993a). It was possible to group the restriction patterns of the *nod* probe according to the main nodulation host, and this enabled the assignment of strains to the biovar level. Biovars *trifolii* and *viciae*, with legume hosts belonging to the genera *Trifolium* and *Vicia*, respectively, showed greater diversity of their *nod* regions than biovar *phaseoli*, which nodulates legumes belonging to the genus *Phaseolus*. The diversity of *Rhizobium meliloti* strains isolated from Italian soils was also evaluated by RFLP analysis of a 25-kb *nod* gene region located in the symbiotic plasmid: these data were complemented by the results of RAPD and RFLPs analyses of the ribosomal intergenic spacer region (Paffetti *et al.*, 1996).

Rhizobial sequence diversity and the distribution of symbiotic genes may vary according to the soil environment of the rhizobia inhabitants. The restriction patterns of *nodDABC* genes of *Bradyrhizobium japonicum* strains isolated from nodules of soybeans grown in tropical soils of Thailand were found to be different from those isolated in Japan or the USA (Yokoyama *et al.*, 1996). A *nifH* probe was used to discriminate between rhizobia isolated from nodules of bean plants (*Phaseolus vulgaris*) grown in Kenyan soils showing different bulk pH values. The majority of rhizobia from a neutral soil (pH 6.8) had multiple copies of the *nifH* gene and were specific nodulants of *P. vulgaris* whereas most of the strains from an acid soil (pH 4.5) possessed a single copy of *nifH* and were characterized by forming effective nodules on a broad range of host legumes (Anyango *et al.*, 1995).

The use of probes for the characterization of strains can be greatly facilitated and improved by the application of PCR rather than by the use of conventional Southern hybridizations methods. The evolution of probe technology applied to rhizobial research was discussed by Laguerre *et al.* (1996). Instead of using probes and a conventional RFLP procedure, these investigators designed oligonucleotide primers directed towards the extremities of the *nifDK* and *nodD* gene regions and digested the resultant PCR products with restriction endonucleases. Primers were designed that were specific to the *nif* genes of all rhizobia, as well as *nod* primers specific to *Rhizobium etli* and the different biovars of *R. leguminosarum*. These investigators advocate the use of PCR-based methods rather than Southern hybridizations which are somewhat more laborious and time-consuming. A primer specific to the *nif* gene has been used in single primer PCR reactions (similar to RAPD) to obtain reproducible, unique amplification profiles for each rhizobial strain; this procedure is very useful for rhizobial diversity and for inoculant quality control studies (Richardson *et al.*, 1995).

The diversity of rhizobia isolated from root and stem nodules of legume species growing in the Philippines was analysed by sequencing of their 16S rRNA and six new species were proposed (Gamo et al., 1991). The genetic diversity of Bradyrhizobium strains isolated from soybean nodules in Thailand was also evaluated by their nod genes, and compared with those of Bradyrhizobium elkanii and B. japonicum strains isolated in Japan and the USA (Yokoyama et al., 1996). The workers were of the view that the Thai isolates belonged to new species of Bradyrhizobium.

Early classifications of rhizobia were dependent on host range studies. One of the advantages of using molecular techniques is that they can be used to unravel the taxonomic status of strains which nodulate diverse legume species. It has been shown that eight Rhizobium strains isolated from nodules of lucerne (Medicago sativa) also nodulate bean plants (Phaseolus vulgaris) (Eardly et al., 1992). When a 260-bp segment of the 16S rDNA of these strains was sequenced, it was shown that they were different from the Rhizobium leguminosarum, R. meliloti, and R. tropici strains which commonly nodulate lucerne and beans. It seems likely that these eight strains are members of a new Rhizobium species.

A study of the diversity and ecology of bradyrhizobia isolated from surface and deep soil under Acacia albida trees in Senegal was carried out using SDS-PAGE analyses (Dupuy et al., 1994). The strains belonged to the Bradyrhizobium-Rhodopseudomonas complex. This assignment was confirmed by hybridization of their genomic DNA using total rRNA from the Bradyrhizobium japonicum type strain as a probe. However, some of the isolates belonged to a separate lineage as far removed from B. japonicum as the latter is from the genus Afipia. Consequently, these isolates also seem to belong to a new species.

The 5'-half of the 23S rDNA gene is highly variable. Restriction analysis of this region enabled the clustering of strains currently referred to as Rhizobium "hedysari". These organisms were well discriminated from other species of the family Rhizobiaceae (Agrobacterium tumefaciens, Rhizobium galegae, R. leguminosarum, and R. meliloti). However, these strains, when PCR fingerprinting methods were applied, gave unique patterns (Selenska-Pobell et al., 1996). Consequently, it can be concluded that analysis of rhizobial 23S rDNA provides useful taxonomic information.

Some results demonstrate the enormous unknown diversity of rhizobia. An extensive biodiversity survey was performed by Oyaizu et al. (1992) on 117 rhizobial strains isolated from 91 legume species and 28 reference strains of Rhizobium and Bradyrhizobium obtained from bacterial culture collections. These workers sequenced a 157-bp segment of the 16S rDNA of the test strains and assigned them to sixteen discrete groups, which correlated with DNA-DNA reassociation data thereby suggesting the existence of at least 16 species in this sample of microorganisms. DNA-DNA reassociation was also used to demonstrate that three genomic groups are present among rhizobia isolated from bean plants grown in French soils (Laguerre et al., 1993b). One group was equated with Rhizobium leguminosarum whereas the others were considered to be new genomic species encompassing strains that nodulate both Leucaena leucocephala and Phaseolus vulgaris. Similarly, Amarger et al. (1994) examined 287 bean-nodulating rhizobia from four different geographical locations in France and concluded that they belonged to either Rhizobium leguminosarum or to R. tropici. The latter contains strains that nodulate both Leucaena leucocephala and Phaseolus vulgaris. These studies suggest that Rhizobium tropici is widely distributed in

French soils. Phylogenetic analysis of 16S rRNA gene sequences of *Rhizobium* strains that nodulate bean plants highlighted four clusters which corresponded to *Rhizobium etli*, *R. leguminosarum*, *R. tropici*, and to a single membered cluster (van Berkum *et al.*, 1996).

The highly variable intergenic spacer regions (IGS) between ribosomal genes have also been considered as adequate DNA regions to be analysed by those willing to investigate the diversity of bacteria (Barry *et al.*, 1991). Consensus primers directed to IGS-located tRNA genes have been used to analyse the genetic variability and phylogenetic relationships among isolates of *Pseudomonas solanacearum* (Seal *et al.*, 1992), *Streptococcus* spp. (McClelland *et al.*, 1992), and *Xanthomonas albilineans* (Honeycutt *et al.*, 1995). Variation has also been observed in the number and size of the IGS regions in diverse microorganisms, leading some workers to advocate the use of IGS analysis as a general and quick method for typing bacteria (Barry *et al.*, 1991; Jensen *et al.*, 1993).

Restriction analysis of the ribosomal IGS regions have been applied recently to characterize rhizobia, and rDNA intergenic spacer primers universal to all rhizobia designed (Laguerre *et al.*, 1996). Several polymorphisms have been observed at both the intra-generic and the intra-specific levels (Nour *et al.*, 1994b; Laguerre *et al.*, 1996; Selenska-Pobell *et al.*, 1996). This method was compared with other DNA fingerprinting techniques, such as REP-PCR and RAPD, in the characterization of 43 strains of *Rhizobium leguminosarum*. Intraspecific polymorphisms, independent of the biovar status of the strains, were found with all of the methods (Laguerre *et al.*, 1996). However, IGS PCR-RFLP data are easier to analyse than complex banding patterns obtained using other fingerprinting techniques; the method is also useful for the analysis of a large number of strains. The diversity of chickpea rhizobia was also studied using this method in association with phenotypic tests; these studies showed that two distantly related groups of strains were present amongst the isolates (Nour *et al.*, 1994b).

Analyses which include the 16S rRNA gene along with the spacer region increase the number of restriction sites and, consequently, the discriminative power of the method. Seventy-three strains of rhizobia isolated from *Medicago trunculata* were analysed by this procedure and this led to the discovery of two genotypic groups, which were confirmed by DNA-DNA hybridization data suggesting that the strains represent different genomic species (Rome *et al.*, 1996a).

The use of symbiotic plasmid-located *nod* probes allowed the assignment of *Rhizobium leguminosarum* strains to the biovar level, with biovars *trifolii* and *viciae* showing greater diversity in their *nod* regions than biovar *phaseoli* (Laguerre *et al.*, 1993a). Identification of this species to the biovar level is not possible when the target of the analyses are chromosomal regions or when DNA fingerprinting methods are employed (Laguerre *et al.*, 1996). Molecular assessments of rhizobial diversity should encompass analyses on both the chromosome and the symbiotic plasmids.

4.2.2 *Direct evaluation of rhizobial diversity*

The indirect diversity assessments reported above were based on rhizobia isolated from legume nodules. The obvious limitation of this approach is that, even for the so called "promiscuous" legume species, the plant will select the *Rhizobium* strains that will occupy its nodules. Consequently, biodiversity analyses based on data from isolate characterizations probably underestimate the extent of rhizobial diversity. Similarly, there is no guarantee that representative strains of rhizobia will be isolated directly from soil on selective or semi-selective media (Bromfield *et al.*, 1994; Louvrier *et al.*, 1995), which means that this approach is also likely to underestimate rhizobial diversity.

The limitations outlined above are particularly serious in studies designed to assess the impact of anthropic activities (agricultural, industrial, and mining) on the extent of rhizobial diversity. In such studies it is essential to monitor all rhizobial types in soils under different intensities of management. Assessments of the extent to which different human activities cause impact on biodiversity are top priorities to nations that have signed the Convention on Biological Diversity (UNEP, 1992). It is, therefore necessary to develop methods which will overcome the limitations presently imposed upon indirect evaluations of the diversity of soil microorganisms.

The development and application of nucleic acid probes offer a fast and reliable method to study environmental samples, such as soil samples. Probes can be used to study culturable and non-culturable organisms, and those which do not grow on isolation media (Pickup, 1991). Direct evaluations of microbial diversity are defined here as those which do not require the isolation and culture of the microorganisms under study.

Soil is a complex and difficult environment. Microbial ecologists have been applying molecular systematics techniques intensively to detect bacteria in soil samples in an attempt to overcome the difficulties associated with the isolation and growth of microorganisms. Detection methods which do not depend on *in vitro* microbial growth are usually based on the amplification of target DNA fragments followed by nucleic acid hybridization. The first step is to obtain the microbial DNA from the environmental sample. Strategies commonly involve the separation of microbial cells from soil particles, followed by lysis and recovery of their DNA (indirect lysis method) or the disruption of cells directly in the soil samples prior to alkaline extraction of DNA (Pickup, 1991). A rapid method for the direct extraction of DNA from soil involves mechanical lysis by bead beating aided by sodium-dodecyl-sulphate (SDS), lysozyme and low temperature treatment, followed by cold phenol extraction, precipitation of the DNA, and purification with caesium chloride and potassium acetate, spermidine-HCl or glass milk. The resulting DNA is usually amplifiable by PCR (Smalla *et al.*, 1993).

A molecular ecological study on Australian soil samples, using direct lysis of the microbial cells present in the soil matrix, 16S rRNA amplification, sequencing, and hybridization with taxon specific probes, revealed the occurrence of phylogenetically distinct groups associated with new, unknown lineages within the *Eubacteria* domain (Liesack & Stackebrandt, 1992). This approach was also used to investigate the genetic diversity of streptomycetes in soil samples taken from a subtropical forest in Australia. These results also revealed extensive microbial diversity with the majority of the 113 analysed 16S rRNA sequences belonging to new, unknown taxa, which were phylogenetically distinct from known, culturable bacteria (Stackebrandt *et al.*, 1993).

The molecular methodology for direct, specific microbial detection, which is qualitative in nature, can also be adapted to enumerate bacteria in soil. Rosado *et al.* (1996) developed a quantitative PCR method to detect and enumerate *Paenibacillus azotofixans* in soil and rhizosphere of wheat (*Triticum aestivum* L.). Following DNA extraction from soil, the 16S rRNA genes were amplified by using specific primers, and hybridized with an internal probe to the amplified fragment. The method was used to evaluate the impact of water stress on the population of *P. azotofixans* in the environment, and to overcome the need to culture the bacteria.

The ribosomal IGS may also provide probing sites for rhizobial species. Oliveira *et al.* (1999) sequenced and aligned the IGS of five rhizobial strains belonging to two different species. The resultant data allowed the design of PCR primers specific to *Rhizobium leguminosarum* and/or *R. tropici*. Direct detection of rhizobia has been accomplished using these primers to amplify their IGS from DNA extracted from agricultural soils under different management practices. The PCR products were separated using denaturing gradient gel electrophoresis (DGGE), which has been used to reveal sequence variability among rhizobial strains (Vallaeys *et al.*, 1997). This procedure allows the separation of DNA fragments that are of the same size but different in their nucleotide sequences. The separation is achieved because differences in sequence composition alter the electophoretic mobility of a DNA molecule when subjected to a denaturing gradient (Fischer & Lerman, 1983). The resultant sequence heterogeneity is directly related to the genetic diversity of the populations of *Rhizobium leguminosarum* and *R. tropici* in the soil. This procedure is being used to evaluate the applicability of rhizobial diversity as an indicator of the environmental impact of different agricultural management schemes (Coutinho *et al.* 1999b).

This procedure was also used in a molecular ecology study designed to analyse the genetic diversity of microbial populations. DNA fragments derived from the amplification of a variable region of the 16S rDNA of complex microbial populations were analysed, and the occurrence of up to 10 different fragments of rDNA detected in microbial communities of different origins (Muyzer *et al.*, 1993). Thus, the use of DGGE to analyse a mixture of PCR products enables the direct detection of different genomic species in environmental samples.

A methodological framework for the molecular evaluation of rhizobial diversity in soil is presented in Figure 2. The starting point of the methodology is soil-extracted total DNA. The initial approach will depend on the availability of oligonucleotide primers to amplify rhizobial ribosomal gene sequences from soil DNA. If one is interested in the diversity of a particular rhizobial species for which specific primers are available, their rDNA can be amplified and the PCR products separated by DGGE (or TGGE, i.e., temperature gradient gel electrophoresis). This will produce fingerprints that will reflect the diversity of genotypes or strains that are present in sufficient numbers to be detected in the soil samples analysed. A refinement of the analyses may be accomplished by sequence analysis of DNA fragments purified from bands excised from the DGGE gels. A different approach would be to use universal primers to amplify eubacterial ribosomal gene sequences from soil DNA. This procedure would require an extra step in order to identify among the PCR products those that derive from rhizobia. DGGE (or TGGE) analyses followed by hybridization with rhizobial specific probes would enable the evaluation of rhizobial diversity. Again, DGGE bands which hybridise to rhizobial-specific probes could have their DNA sequences analysed. If access to DGGE technology is a handicap, one could construct an eubacterial rDNA clone library. Detection of rhizobial rDNA containing clones could be achieved by hybridization with specific probes, while analysis of their DNA sequences would enable diversity evaluation. Methods for sequence analysis may vary from full or partial sequencing to more simple evaluations such as analysis of rDNA restricted fragments polymorphisms.

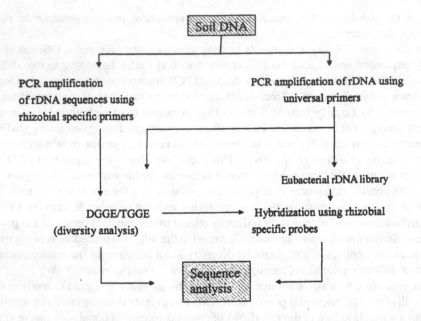

Figure 2. Methodological framework for analysing the molecular diversity of rhizobia in soil without the need for isolation or growth of bacteria in the laboratory.

5   Concluding Remarks

The systematics of rhizobia has undergone revolutionary changes in the past fifteen years, that is, from the moment the definition of rhizobial species was no longer based solely on host range. Legume nodule, nitrogen fixing bacteria, can now be classified in up to six different genera, belonging to at least three divergent lineages in the α subgroup of the *Proteobacteria* (Young, 1996). The molecular evidence also shows that changes are needed at the suprageneric level. The family *Rhizobiaceae*, for instance, contains phylogenetically diverse rhizobial genera such as *Rhizobium* and *Bradyrhizobium*, and excludes *Rhodopseudomonas*, a genus closely related to *Bradyrhizobium*.

Rhizobial strains previously indistinguishable may now be recognized and assigned to different species. The methods responsible for this taxonomic advance open up an array of possibilities in the fields of ecology, environmental impact assessment, and development of inoculants (Coutinho *et al.* 1999b). More innovations are expected in the near future, with automation of *in situ* hybridization techniques coupled with advanced electron microscopy. This will allow the investigation of subjects such as the chemical signalling mechanisms between symbiotic partners, which may shed more light on the processes and environmental factors involved in competition for nodulation and survival, and for the persistence as saprophytes of particular strains, species, or even genera of rhizobia. Succession and dynamics of rhizobial populations in soils undergoing environmental stress are also a field of interest aimed at better management and optimizatiom of the ecological and agronomic benefits of biological nitrogen fixation. Last, but definitely not least, advances in rhizobial systematics will unravel even further the extent of the diversity of diazotrophic microorganisms and how much of this variatiom is related to the actual process of nitrogen fixation of soils in ecosystems under different environmental pressures.

6   References

Allen, O.N. & Allen, E.K. (1981). *The Leguminosae. A Source Book of Characteristics, Uses and Nodulation.* Madison, University of Wisconsin Press.

Amarger, N., Bours, M., Revoy, F., Allard, M.R. & Laguerre, G. (1994). *Rhizobium tropici* nodulates field-grown *Phaseolus vulgaris* in France. *Plant and Soil* 161, 147-156.

Amarger, N., Macheret, V. & Laguerre, G. (1997). *Rhizobium gallicum* sp. nov. and *Rhizobium giardinii* sp. nov., from *Phaseolus vulgaris* nodules. *International Journal of Systematic Bacteriology* 47, 996-1006.

Anyango, B., Wilson, K.J., Beynon, J.L. & Giller, K.E. (1995). Diversity of rhizobia nodulating *Phaseolus vulgaris* L. in two Kenyan soils with contrasting pHs. *Applied and Environmental Microbiology* 61, 4016-4021.

Barkay, T., Fouts, D.L. & Olson, B.H. (1985). Preparation of a DNA gene probe for detection of mercury resistance genes in gram-negative bacterial communities. *Applied and Environmental Microbiology* 49, 686-692.

Barry, T., Colleran, G., Glennon, M., Dunican, L.K. & Gannon, F. (1991). The 16S/23S ribosomal spacer region as a target for DNA probes to identify eubacteria. *PCR Methods and Applications* 1, 51-56.

Beijerinck, M.W. (1888). Die Bacterien der Papilionaceeknollchen. Botanische Zeitung 46, 726-735, 741-750, 757-771, 781-790, 797-804.

Boddey, R.M., Chalk, P.M., Victoria, R.L. & Matsui, E. (1984). Nitrogen fixation by nodulated soybean under tropical field conditions estimated by the 15_N isotope dilution technique. *Soil Biology and Biochemistry* 16, 583-589.

Bromfield, E.S.P., Wheatcroft, R. & Barran, L.R. (1994). Medium for direct isolation of *Rhizobium meliloti* from soils. *Soil Biology and Biochemistry* 26, 423-428.

Cadahia, E., Leyva, A., Ruiz-Argueso, T. (1986). Indigenous plasmids and cultural characteristics of rhizobia nodulating chickpeas (*Cicer arietinum* L.). *Archives of Microbiology* 146, 239-244.

Chen, W.X., Li, G.S., Qi, Y.L., Wang, E.T., Yuan, H.L. & Li, J.L. (1991). *Rhizobium huakuii* sp. nov. isolated from the root nodules of *Astragalus sinicus. International Journal of Systematic Bacteriology* 41, 275-280.

Chen, W.X., Tan, Z.Y., Gao, J.L., Li, Y., Wang, E.T. (1997). *Rhizobium hainanense* sp. nov., isolated from tropical legumes. *International Journal of Systematic Bacteriology* 47, 870-873.

Chen, W., Wang, E., Wang, S., Li, Y., Chen, X. & Li, Y. (1995). Characteristics of *Rhizobium tianshanense* sp. nov., a moderately and slowly growing root nodule bacterium isolated from an arid saline environment in Xinjiang, People's Republic of China. *International Journal of Systematic Bacteriology* 45, 153-159.

Coutinho, H.L.C. (1993). *Studies of Bradyrhizobia from the Brazilian Cerrados.* Ph.D. thesis, University of Bristol.

Coutinho, H.L.C., Oliveira, V.M., Hollanda, L.M., Moreira, F.M.S. & Franco, A.A. (1995). Diversity of rhizobia isolated from nodules of legume species occurring in the Atlantic and Amazonian rainforests p. 162.

In 7th *International Symposium of Microbial Ecology, Abstracts*, Santos-SP: Brazil.

Coutinho, H.L.C., Kay, H.E., Manfio, G.P., Neves, M.C.P., Ribeiro, J.R.A., Rumjanek, N. & Beringen J.E. (1999a). Molecular evidence for shifts in polysaccharide composition associated with adaptation of soybean *Bradyrhizobium* strains to the Brazilian cerrado soils. *Environmental Microbiology*, in press.

Coutinho H.L.C., Oliveira, V.M., Manfio, G.P., Rosado, A.S. (1999b). Evaluating the microbial diversity of soils samples: Methodological innovations. *Anais da Academia Brasileira de Ciências* 71, 3-11.

Dangeard, P.A. (1926). Recherches sur les tubercules radicaux des Légumineuses. *Botaniste* 13, 1-275.

de Lajudie, P., Willems, A., Pot, B., Dewettinck, D., Maestrojuan, G., Neyra, M., Collins, M.D., Dreyfus, B., Kersters, K. & Gillis, M. (1994). Polyphasic taxonomy of rhizobia: emendation of the genus *Sinorhizobium* and description of *Sinorhizobium meliloti* comb. nov., *Sinorhizobium saheli* sp. nov., and *Sinorhizobium teranga*, sp. nov. *International Journal of Systematic Bacteriology* 44, 715-733.

de Lajudie, P., Laurent-Fulele, E, Willems, A, Torck, U., Coopman, R., Collins, M.D., Kersters, K., Dreyfus, B. & Gillis, M. (1998a). *Allorhizobium undicola* gen. nov., sp. nov., nitrogen-fixing bacteria that efficiently nodulate *Neptunia natans* in Senegal. *International Journal of Systematic Bacteriology* 48, 1277-1290.

de Lajudie, P., Willems, A, Nick, G., Moreira, F., Molouba, F., Hoste, B., Torck, U., eNeyra, M., Collins, M.D., Lindstrom, K., Dreyfus, B. & Gillis, M. (1998b). Characterization of tropical tree rhizobia and description of *Mesorhizobium plurifarium* sp. nov. *International Journal of Systematic Bacteriology* 48, 369-382.

Dobert, R.C., Breil, B.T. & Triplett, E.W. (1994). DNA sequence of the common nodulation genes of *Bradyrhizobium elkanii* and their phylogenetic relationship to those of other nodulating bacteria. *Molecular Plant Microbe Interactions* 7, 564-572.

Dreyfus, B., Garcia, J.L. & Gillis, M. (1988). Characterization of *Azorhizobium caulinodans* gen. nov., sp. nov., a stem-nodulating nitrogen-fixing bacterium isolated from *Sesbania rostrata*. *International Journal of Systematic Bacteriology* 38, 89-98.

Dupuy, N., Willems, A., Pot, B., Dewettinck, D., Vandenbruaene, I., Maestrojuan, G., Dreyfus, B., Kersters, K., Collins, M. D. & Gillis, M. (1994). Phenotypic and genotypic characterization of bradyrhizobia nodulating the leguminous tree *Acacia albida*. *International Journal of Systematic Bacteriology* 44, 461-473.

Eardly, B.D., Young, J.P.W. & Selander, R.K. (1992). Phylogenetic position of *Rhizobium* sp. strain Or 191, a symbiont of both *Medicago sativa* and *Phaseolus vulgaris*, based on partial sequences of the 16S rRNA and *nif*H genes. *Applied and Environmental Microbiology* 58, 1809-1815.

Evguenieva-Hackenberg, E. & Selenska-Pobell, S. (1995). Variability of the 5'-end of the large subunit rDNA and presence of a new short class of RNA in *Rhizobiaceae*. *Letters in Applied Microbiology* 21, 402-405.

Faria, S.M., Lewis, G.P., Sprent, J.I. & Sutherland, JM (1989). Occurrence of nodulation in the Leguminosae. *New Phytologist* 111, 607-619.

Fischer, S.G. & Lerman, L.S. (1983). DNA fragments differing by single basepair substitutions are separated in denaturing gradient gels: correspondence with melting theory. *Proceedings of the National Academy of Sciences of the USA* 80, 1579-1583.

Frank, B. (1889). Ueber die Pilzsymbiose der Leguminosen. *Berichte der Deutschen Botanischen Gesellschaft* 7, 332-346.

Fred, E.B., Baldwin, I.L. & McCoy, E. (1932). *Root nodule bacteria and leguminous plants*. Madison: University of Wisconsin Press.

Fredrickson, J.K., Bezdicek, D.F., Brickman, F.J. & Li, S.W. (1988). Enumeration of Tn5 mutant bacteria in soil by using a most-probable-number DNA hybridization technique and antibiotic resistance. *Applied and*

Environmental Microbiology **54**, 446-453.

Gamo, T., Itoh, A., Sawazaki, A., Manguiat, I.J. & Mendoza, D.M. (1991). Rhizobia collected from leguminous plants in the Philippines. *Bulletin of the National Institute of Agrobiological Resources* **6**, 111-129.

Graham, P.H. (1976). Identification and classification of root nodule bacteria. In *Symbiotic Nitrogen Fixation in Plants.* Edited by P.S. Nutman, pp. 99-112. Cambridge, Cambridge University Press.

Graham, P.H., Sadowsky, M.J., Keyser, H.H., Barnet, Y.M., Bradley, R.S., Cooper, J.E., De Ley, D.J., Jarvis, B.D.W., Roslycky, E.B., Strijdom, B.W. & Young, J.P.W. (1991). Proposed minimal standards for the description of new genera and species of root- and stem-nodulating bacteria. *International Journal of Systematic Bacteriology* **41**, 582-587.

Gross, D.C., Vidaver, A.K. & Klucas, R.U. (1979). Plasmids, biological properties and efficacy of nitrogen fixation in *Rhizobium japonicum* strains indigenous to alkaline soils. *Journal of General Microbiology* **114**, 257-266.

Gürtler, V. & Stanisich, V.A. (1996). New approaches to typing and identification of bacteria using the 16S-23S rDNA spacer region. *Microbiology* **142**, 3-16.

Hahn, M. & Hennecke, H. (1987). Conservation of a symbiotic DNA region in soybean root nodule bacteria. *Applied and Environmental Microbiology* **53**, 2253-2255.

Harrison, S.P., Jones, D.G., Schunmann, P.H.D., Forster, J.W. & Young, J.P.W. (1988). Variation in *Rhizobium leguminosarum* biovar *trifolii* Sym plasmids and the association with effectiveness of nitrogen fixation. *Journal of General Microbiology* **134**, 2721-2730.

Hartmann, A., Gomez, M., Giraud, J.J. & Revellin, C. (1996). Repeated sequence RSα is diagnostic for *Bradyrhizobium japonicum* and *Bradyrhizobium elkanii. Biology and Fertility of Soils* **23**, 15-19.

Haukka, K. & Lindström, K. (1994). Pulse-field gel electrophoresis for genotypic comparison of *Rhizobium* bacteria that nodulate leguminous trees. *FEMS Microbiology Letters* **119**, 215-220.

Hellriegel, H. & Wilfarth, H. (1888). Untersuchungen uber die Stickstoffernahrung der Gramineen und Leguminosen. *Beitraege zur Vereinigung Deutschen Zuckerindustrie.* pp. 234.

Hernandez, B.S. & Focht, D.D. (1984). Invalidity of the concept of slow growth and alkali production in cowpea rhizobia. *Applied and Environmental Microbiology* **48**, 206-210.

Hollanda, L.M., Coutinho H.L.C. & Manfio G.P. (1996). Caracterização de rizóbios através de perfis de crescimento em carboidratos. In *Anais do XIII Congresso Latino Americano de Ciência do Solo, CD-ROM.*

Honeycutt, R.J., Sobral, B.W.S. & McClelland, M. (1995). tRNA intergenic spacers reveal polymorphisms diagnostic for *Xanthomonas albilineans. Microbiology* **141**, 3229-3239.

Ishizuka, J. (1992). Trends in biological nitrogen fixation research and application. *Plant and Soil* **141**, 197-209.

Jarvis, B.D.W., Pankhurst, C.E. & Patel, J.J. (1982). *Rhizobium loti,* a new species of legume root nodule bacteria. *International Journal of Systematic Bacteriology* **32**, 378-380.

Jarvis, B.D.W., van Berkum, P., Chen, W.X., Nour, S.M., Fernandez, M.P., Cleyet-Marel, J.C. & Gillis, M. (1997). Transfer of *Rhizobium loti, Rhizobium huakuii, Rhizobium ciceri, Rhizobium mediterraneum* and *Rhizobium tianshanense* to *Mesorhizobium* gen. nov. *International Journal of Systematic Bacteriology* **47**, 895-898.

Jensen, M.A., Webster, J.A. & Straus, N. (1993). Rapid identification of bacteria on the basis of polymerase chain reaction-amplified ribosomal DNA spacer polymorphisms. *Applied and Environmental Microbiology* **59**, 945-952.

Jordan, D.C. (1982). Transfer of *Rhizobium japonicum* Buchanan 1980 to *Bradyrhizobium* gen. nov., a genus of slow growing root-nodule bacteria from leguminous plants. *International Journal of Systematic Bacteriology* **32**, 136-139.

Jordan, D.C. (1984). *Rhizobiaceae.* In *Bergey's Manual of Systematic Bacteriology,* vol. 1, pp. 234-256.

Edited by N.R. Kreig & J.G. Holt. Baltimore: Williams and Wilkins.

Jordan, D.C. & Allen, O.N. (1974). *Rhizobiaceae.* In *Bergey's Manual of Determinative Bacteriology,* *8th* edition, pp. 261-264. Edited by R.E. Buchanan & N.E. Gibbons. Baltimore: Williams & Wilkins.

Kay, H.E., Coutinho, H.L.C., Fattori, M., Manfio, G.P., Goodacre, R., Nuti, M.P., Basaglia, M. & Beringer, J.E. (1994). The identification of *Bradyrhizobium japonicum* strains isolated from Italian soils. *Microbiology* **140**, 2333-2339.

Kennedy, L.D. & Greenwood, R.M. (1982). 6-phosphogluconate and glucose-6-phosphate dehydrogenase activities, growth rate and acid production as taxonomic criteria for *Rhizobium. New Zealand Journal of Science* **25**, 361-366.

Kimura, M. (1983). *The Neutral Theory of Molecular Evolution.* New York, Cambridge University Press.

Kramicker, B.J. & Brill, W.J. (1986). Identification of *Bradyrhizobium japonicum* nodule isolates from Wisconsin soybean farms. *Applied and Environmental Microbiology* **51**, 487-492.

Kuykendall, L.D., Saxena, B., Devine, T.E. & Udell, S.E. (1992). Genetic diversity in *Bradyrhizobium japonicum* Jordan 1982 and a proposal for *Bradyrhizobium elkanii* sp. nov. *Canadian Journal of Microbiology* **38**, 501-505.

Laguerre, G., Bardin, M. & Amarger, N. (1993a). Isolation from soil of symbiotic and nonsymbiotic *Rhizobium leguminosarum* by DNA hybridization. *Canadian Journal of Microbiology* **39**, 1142-1149.

Laguerre, G., Geniaux, E., Mazurier, S., Rodriguez-Casartelli, R. & Amarger, N. (1993b). Conformity and diversity among field isolates of *Rhizobium leguminosarum* bv. *viciae,* bv. *trifolii* and bv. *phaseoli* revealed by DNA hybridization using chromosome and plasmid probes. *Canadian Journal of Microbiology* **39**, 412-419.

Laguerre, G., Mavingui, P., Allard, M.R., Charnay, M.P., Louvrier, P., Mazurier, S.I., Rigottier-Gois, L. & Amarger, N. (1996). Typing of rhizobia by PCR DNA fingerprinting and PCR-restriction fragment length polymorphism analysis of chromosomal and symbiotic gene regions: Application to *Rhizobium leguminosarum* and its different biovars. *Applied and Environmental Microbiology* **62**, 2029-2036.

Lane, D.L., Pace, B., Olsen, G.J., Stahl, D.A., Sogin, M.L. & Pace, N.R. (1985). Rapid determination of 16S ribosomal RNA sequences for phylogenetic analyses. *Proceedings of the National Academy of Sciences, USA* **82**, 6955-6959.

Liesack, W. & Stackebrandt, E. (1992). Occurrence of novel groups of the domain *Bacteria* as revealed by analysis of genetic material from an Australian terrestrial environment. *Journal of Bacteriology* **174**, 5072-5078.

Lim, G.& Ng, H.L. (1977). Root nodules of some tropical legumes in Singapore. *Plant and Soil* **46**, 317-327.

Lindström, K. (1989). *Rhizobium galegae,* a new species of legume root nodule bacteria. *International Journal of Systematic Bacteriology* **39**, 365-367.

Lindström, K., Paulin, L., Roos, C. & Suominen, L. (1995). Nodulation genes of *Rhizobium galegae.* In *Nitrogen Fixation: Fundamentals and Applications*, pp. 365-370. Edited by I.A. Tikhonovich, N.A. Provorov, V.I. Romanov & W.E. Newton. Dordrecht, The Netherlands, Kluwer Academic Publishers.

Lindström, K., van Berkum, P., Gillis, M., Martinez, E., Novikova, N. & Jarvis, B. (1995). Report of the roundtable on *Rhizobium* taxonomy. In *Nitrogen Fixation: Fundamentals and Applications,* pp.# 807-81. Edited by I.A. Tikhonovich, N.A. Provorov, V.I. Romanov & W.E. Newton. Dordrecht, The Netherlands: Kluwer Academic Publishers.

Loureiro, M.F., de Faria, S.M., James, E.K., Pott, A. & Franco, A.A. (1994). Nitrogen-fixing stem nodules of the legume *Discolobium pulchellum* Benth. *New Phytologist* **128**, 283-295.

Loureiro, M.F., James, E.K., Sprent, J.I. & Franco, A.A. (1995). Stem and root nodules on the tropical legume *Aeschynomene fluminensis. New Phytologist* **130**, 531-544.

Louvrier, P., Laguerre, G. & Amarger, N. (1995). Semiselective medium for isolation of *Rhizobium leguminosarum* from soils. *Soil Biology and Biochemistry* **27**, 919-924.

Maidak, B.L., Olsen, G.J., Larsen, N., Overbeek, R., McCaughey, M.J. & Woese, C.R. (1997). The RDP (Ribosomal Database Project). *Nucleic Acids Research* **25**, 109-111.

Manjula, M. & Rakesh, T. (1990). Cluster analysis of genes for nitrogen fixation from several diazotrophs. *Journal of Genetics* **69**, 67-78.

Martinez-Dretz, G. & Arias, A. (1972). Enzymatic basis for differentiation of *Rhizobium* into fast and slow-growing groups. *Journal of Bacteriology* **109**, 467-470.

Martinez-Romero, E., Segovia, L., Mercante, F.M., Franco, A.A., Graham, P. & Pardo, M.A. (1991). *Rhizobium tropici*, a novel species nodulating *Phaseolus vulgaris* L. beans and *Leucaena* sp. trees. *International Journal of Systematic Bacteriology* **41**, 417-426.

Masterson, R.V., Prakash, R.K. & Atherly, A.G. (1985). Conservation of symbiotic nitrogen fixation gene sequences in *Rhizobium japonicum* and *Bradyrhizobium japonicum*. *Journal of Bacteriology* **163**, 21-26.

McClelland, M., Petersen, C. & Welsh, J. (1992). Length polymorphisms in tRNA intergenic spacers detected by using the polymerase chain reaction can distinguish streptococcal strains and species. *Journal of Clinical Microbiology* **30**, 1499-1504.

Minamisawa, K., Seki, T., Onodera, S., Kubota, M. & Asami, T. (1992). Genetic relatedness of *Bradyrhizobium japonicum* field isolates as revealed by repeated sequences and various other characteristics. *Applied and Environmental Microbiology* **58**, 2832-2839.

Moreira, F.M.S. (1991). *Caracterização de Estirpes de Rizóbio Isoladas de Espécies Florestais Pertencentes a Diversos Grupos de Divergência de Leguminosae Introduzidas ou Nativas da Amazônia e Mata Atlântica.* Tese de Doutorado em Ciências do Solo. Seropédica, Universidade Federal Rural do Rio de Janeiro.

Moreira, F.M.S., Silva, M.F. & Faria, S.M. (1992). Occurrence of nodulation in legume species in the Amazon region of Brazil. *New Phytologist* **121**, 563-570.

Moreira, F.M.S., Gillis, M., Pot, B., Kersters, K. & Franco, A.A. (1993). Characterization of rhizobia isolated from different divergence groups of tropical Leguminosae by comparative polyacrylamide gel electrophoresis of their total proteins. *Systematic and Applied Microbiology* **16**, 135-146.

Moreira, F.M.S., Martinez-Romero, E., Segovia, L. & Franco, A.A. (1995). Genetic diversity of rhizobia and bradyrhizobia from native tropical species characterized by multilocus enzyme electrophoresis, p. 88. In *7th International Symposium of Microbial Ecology, Abstracts*, Santos-SP, Brazil.

Moreira, F.M.S., Haukka, K. & Young, J.P.W. (1998). Biodiversity of rhizobia isolated from a wide range of forest legumes in Brazil. *Molecular Ecology* **7**, 889-895.

Mullis, K.B. & Faloona, F.A. (1987). Specific synthesis of DNA *in vitro* via a polymerase-catalyzed chain reaction. *Methods of Enzymology* **155**, 335-351.

Muyzer, G., De Waal, E.C. & Uitterlinden, A.G. (1993). Profiling of complex microbial populations by denaturing gradient gel electrophoresis analysis of polymerase chain reaction-amplified genes coding for 16S rRNA. *Applied and Environmental Microbiology* **59**, 695-700.

Neves, M.C.P. & Rumjanek, N.G. (1997). Diversity and adaptability of soybean and cowpea rhizobia in tropical soils. *Soil Biology and Biochemistry* **29**, 889-895.

Nour, S.M., Fernandez, M.P., Normand, P. & Cleyet-Marel, J.C. (1994a). *Rhizobium ciceri* sp. nov., consisting of strains that nodulate chickpeas (*Cicer arietinum* L.). *International Journal of Systematic Bacteriology* **44**, 511-522.

Nour, S.M., Cleyet-Marel, J.C., Beck, D., Effosse, A. & Fernandes, M.P. (1994b). Genotypic and phenotypic diversity of *Rhizobium* isolated from chickpea (*Cicer arietinum* L.). *Canadian Journal of Microbiology* **40**, 345-354.

Nour, S.M., Cleyet-Marel, J.C., Normand, P. & Fernandes, M.P. (1995). Genomic heterogeneity of strains nodulating chickpea (*Cicer arietinum* L.) and description of *Rhizobium mediterraneum* sp. nov.

International Journal of Systematic Bacteriology **45**, 640-648.

Novikova, N.I., Pavlova, E.A., Vorobjev, N.I. & Limenshchenko, E.V. (1994). Numerical taxonomy of *Rhizobium* strains from legumes of the temperate zone. *International Journal of Systematic Bacteriology* **44**, 734-742.

Nutman, P.S. (1987). Centenary lecture. *Philosophical Transactions of the Royal Society of London* **317**, 69-106.

Oliveira, V.M., Rosato, Y.B., Coutinho, H.L.C. & Manfio, G.P. (1997). Design of a 16S rRNA-directed oligonucleotide probe for *Bradyrhizobium* tropical strains. In *11th International Congress on Nitrogen Fixation, Proceedings*, p.581. Edited by C. Elmerich, A. Kondorosi & W.E. Newton. Dordrecht, The Netherlands: Kluwer Academic Publishers.

Oliveira, V.M., Coutinho, H.L.C., Sobral, B.W.S., Guimarães, C.T., van Elsas, J.D. & Manfio, G.P. (1999). Discrimination of *Rhizobium tropici* and *R. leguminosarum* strains by PCR-specific amplification of 16S-23S rDNA spacer region fragments and denaturing gradient gel electrophoresis (DGGE). *Letters in Applied Microbiology* **28**, 137-141.

Olsen, G.J., Matsuda, H., Hagstrom, R. & Overbeek, R. (1994). fastDNAml: A tool for construction of phylogenetic trees of DNA sequences using maximum likelihood. *Computer Applications in Biosciences* **10**, 41-48.

Olsen, P.E., Rice, W.A., Bordeleau, L.M. & Biederbeck, V.O. (1994). Analysis and regulation of legume inoculants in Canada: the need for an increase in standards. *Plant and Soil* **161**, 127-134.

Oyaizu,H., Naruhashi, N. & Gamou, T. (1992). Molecular methods of analysing bacterial diversity: the case of rhizobia. *Biodiversity and Conservation* **1**, 237-249.

Pace, N.R., Stahl, D.A., Lane, D.J. & Olsen, G.J. (1986). The analysis of natural microbial populations by ribosomal RNA sequences. *Advances in Microbial Ecology* **9**, 1-55.

Paffetti, D., Scotti, C., Gnocchi, S., Fancelli, S. & Bazzicalupo, M. (1996). Genetic diversity of an Italian *Rhizobium meliloti* population from different *Medicago sativa* varieties. *Applied and Environmental Microbiology* **62**, 2279-2285.

Pickup, R.W. (1991). Development of molecular methods for the detection of specific bacteria in the environment. *Journal of General Microbiology* **137**, 1009-1019.

Richardson, A.E., Viccars, L.A., Watson, J.M. & Gibson, A.H. (1995). Differentiation of *Rhizobium* strains using the polymerase chain reaction with random and directed primers. *Soil Biology and Biochemistry* **27**, 515-524.

Rinaudo, G., Orenga, S., Fernandez, M.P., Meugnier, H. & Bardin, R. (1991). DNA homologies among members of the genus *Azorhizobium* and other stem- and root- nodulating bacteria isolated from the tropical legume *Sesbania rostrata*. *International Journal of Systematic Bacteriology* **41**, 114-120.

Rome, S., Brunel, B., Normand, P., Fernandez, M. & Cleyet-Marel, J.C. (1996a). Evidence that two genomic species of *Rhizobium* are associated with *Medicago truncatula*. *Archives of Microbiology* **165**, 285-288.

Rome, S., Fernandez, M.P., Brunel, B., Normand, P. & Cleyet-Marel, J.C. (1996b). *Sinorhizobium medicae* sp. nov., isolated from annual *Medicago* spp. *International Journal of Systematic Bacteriology* **46**, 972-980.

Rosado, A.S., Seldin, L., Wolters, A.C. & van Elsas, J.D. (1996). Quantitative 16S rDNA-targeted polymerase chain reaction and oligonucleotide hybridization for the detection of *Paenibacillus azotofixans* in soil and the wheat rhizosphere. *FEMS Microbiology and Ecology* **19**, 153-164.

Rumjanek, N.G., Dobert, R.C., van Berkum, P. & Triplett, E.W. (1993). Common soybean inoculant strains in Brazil are members of *Bradyrhizobium elkanii*. *Applied and Environmental Microbiology* **59**, 4371-4373.

Scholla, M.H. & Elkan, G.H. (1984). *Rhizobium fredii* sp. nov., a fast-growing species effectively nodulates soybeans. *International Journal of Systematic Bacteriology* **34**, 484-486.

Seal, S.E., Jackson, L.A. & Daniels, M.J. (1992). Use of tRNA consensus primers to indicate subgroups of *Pseudomonas solanacearum* by polymerase chain reaction amplification. *Applied and Environmental Microbiology* **58**, 3759-3761.

Segovia, L., Young, J.P.W. & Martinez-Romero, E. (1993). Reclassification of American *Rhizobium leguminosarum* biovar *phaseoli* type I strains as *Rhizobium etli* sp. nov. *International Journal of Systematic Bacteriology* **43**, 374-377.

Selenska-Pobell, S., Evguenieva-Hackenberg, E., Radeva, G. & Squartini, A. (1996). Characterization of *Rhizobium* 'hedysari' by RFLP analysis of PCR amplified rDNA and by genomic PCR fingerprinting. *Journal of Applied Bacteriology* **80**, 517-528.

Siqueira, J.O. & Franco, A.A. (1988). Biotecnologia do Solo: Fundamentos e Perspectivas. pp. 236. Lavras, FAEPE/ABEAS/MEC/ESAL.

Smalla, K., Cresswell, N., Mendonãa-Hagler, L.C., Wolters, A.C. & van Elsas, J.D. (1993). Rapid DNA extraction protocol from soil for polymerase chain reaction-mediated amplification. *Journal of Applied Bacteriology* **74**, 78-85.

Springer, N., Ludwig, W. & Hardarson, G. (1993). A 23S rRNA targeted specific hybridization probe for *Bradyrhizobium japonicum*. *Systematic and Applied Microbiology* **16**, 468-470.

Stackebrandt, E., Liesack, W. & Goebel, B.M. (1993). Bacterial diversity in a soil sample from a subtropical Australian environment as determined by 16S rDNA analysis. *FASEB Journal* **7**, 232-236.

Stanley, J., Brown, G.G. & Verma, P.S. (1985). Slow-growing *Rhizobium japonicum* comprises two highly divergent symbiotic types. *Journal of Bacteriology* **163**, 148-154.

Stephen, J.R., McCaig, A.E., Smith, Z., Prosser, J.I. & Embley, T.M. (1996). Molecular diversity of soil and marine 16S rRNA gene sequences related to β-subgroup ammonia-oxidizing bacteria. *Applied and Environmental Microbiology* **62**, 4147-4154.

Tan, I.K.P. & Broughton, W.J. (1981). Rhizobia in tropical legumes. XIII. Biochemical basis of acid and alkali reactions. *Soil Biology and Biochemistry* **13**, 389-393.

Tas, E., Kaijalainen, S., Saano, A. & Lindström, K. (1994). Isolation of a *Rhizobium galegae* strain-specific DNA probe. *Microbiological Releases* **2**, 231-237.

Tas, E., Leinonen, P., Saano, A., Räsänen, L.A., Kaijalainen, S., Piippola, S., Hakola, S. & Lindström, K. (1996). Assessment of competitiveness of rhizobia infecting *Galega orientalis* on the basis of plant yield, nodulation, and strain identification by antibiotic resistance and PCR. *Applied and Environmental Microbiology* **62**, 529-535.

UNEP (United Nations Environment Programme). (1992). *Convention on Biological Diversity*. Nairobi, UNEP.

Vallaeys, T., Topp, E., Muyzer, G., Macheret, V., Laguerre, G., Rigaud, A. & Soulas, G. (1997). Evaluation of denaturing gradient gel electrophoresis in the detection of 16S rDNA sequence variation in rhizobia and methanotrophs. *FEMS Microbiology and Ecology* **24**, 279-285.

van Berkum, P., Beyene, D. & van Berkum, B.D. (1996). Phylogenetic relationships among *Rhizobium* species nodulating the common bean (*Phaseolus vulgaris* L.). *International Journal of Systematic Bacteriology* **46**, 240-244.

van Berkum, P., Beyene, D., Bao, G., Campbell, T.A. & Eardlly, B.D. (1998). *Rhizobium mongolense* sp. nov. is one of three rhizobial genotypes identified which nodulate and form nitrogen-fixing symbioses with *Medicago ruthenica* [(L.) Ledebour]. *International Journal of Systematic Bacteriology* **48**, 13-22.

van Rossum, D., Schuurmans, F.P., Gillis, M., Muotcha, A., van Verseveld, H.W., Stouthamer, A.H. & Boogerd, F.C. (1995). Genetic and phenetic analyses of *Bradyrhizobium* strains nodulating peanut (*Arachis hypogaea* L.) roots. *Applied and Environmental Microbiology* **61**, 1599-1609.

Vincent, J.M. (1970). *A Manual for the Practical Study of Root-Nodule Bacteria*. Oxford and Edinburgh:

Blackwell Scientific Publications.

Wang, E.T., van Berkum, P., Beyene, D., Sui, X.H., Dorado, O., Chen, W.X. & Martínez-Romero, E. (1998). *Rhizobium huautlense* sp. nov., a symbiont of *Sesbania herbacea* that has a close phylogenetic relationship with *Rhizobium galegae*. *International Journal of Systematic Bacteriology* **48**, 687-699.

Wang, E.T., van Berkum, P., Sui, X.H., Beyene, D., Chen, W.X. & Martínez- Romero, E. (1999). Diversity of rhizobia associated with *Amorpha fruticosa* isolated from Chinese soils and description of *Mesorhizobium amorphae* sp. nov. *International Journal of Systematic Bacteriology* **49**, 51-65.

Woese, C.R. (1987). Bacterial evolution. *Microbiological Reviews* **51**, 221-271.

Wong, F.Y.K., Stackebrandt, E., Ladha, J.K., Fleischman, D.E., Date, R.A. & Fuerst, J.A. (1994). Phylogenetic analysis of *Bradyrhizobium japonicum* and photosynthetic stem-nodulating bacteria from *Aeschynomene* species grown in separated geographical regions. *Applied and Environmental Microbiology* **60**, 940-946.

Xu, L.M., Ge, C., Cui, Z., Li, J. & Fan, H. (1995). *Bradyrhizobium liaonigensis* sp. nov. isolated from the root nodules of soybeans. *International Journal of Systematic Bacteriology* **45**, 706-711.

Yokoyama, T., Ando, S., Murakami, T. & Imai, H. (1996). Genetic variability of the common *nod* gene in soybean bradyrhizobia isolated in Thailand and Japan. *Canadian Journal of Microbiology* **42**, 1209-1218.

Young, J.P.W. (1993). Molecular phylogeny of rhizobia and their relatives. In *New Horizons in Nitrogen Fixation* p. 587-592. Edited by R. Palacios, J. Mora & W.E. Newton. Proceedings of the 9th International Congress on Nitrogen Fixation, Cancun, Mexico, December 6-12, 1992. Dordrecht, The Netherlands: Kluwer Academic Publishers.

Young, J.P.W. (1996). Phylogeny and taxonomy of rhizobia. *Plant and Soil* **186**, 45-52.

Young, J.P.W. & Johnston, A.W.B. (1989). The evolution of specificity in the legume-*Rhizobium* symbiosis. *Trends in Ecology and Evolution* **4**, 341-349.

Zhang, X., Harper, R., Karsisto, M. & Lindström, K. (1991). Diversity of *Rhizobium* bacteria isolated from the root nodules of leguminous trees. *International Journal of Systematic Bacteriology* **41**, 104-113.

6 RECENT DEVELOPMENTS IN SYSTEMATICS AND THEIR IMPLICATIONS FOR PLANT PATHOGENIC BACTERIA

JOHN M. YOUNG

Landcare Research, Private Bag 92170, Auckland, New Zealand

1 Introduction

The systematics of bacteria is at a fascinating stage; the past fifteen years have seen a revolution in the concepts guiding bacterial classification and identification. As bacteria have been investigated using an ever wider range of methods, they are increasingly shown to be diverse in character, making them less amenable to grouping in taxonomic hierarchies. New methods allow the recognition of distinct populations, but they do not necessarily permit easy identification of isolates or their allocation to appropriate taxa. A historical perspective was given in Young *et al.* (1992) and a recent review (Vandamme *et al.*, 1996) gives a survey of current methods used in systematics. The present account takes as its points of departure the Approved Lists of Bacterial Names (Skerman *et al.*, 1980) and the review of Young *et al.* (1992). It focuses on the major trends in systematics with special reference to plant pathogenic bacteria, and attempts to indicate their major consequences for plant pathologists.

2 The Users

The systematics and nomenclature of plant pathogenic bacteria serve the purposes of several interest groups. Those most centrally and practically involved are researchers of disease cause and control, those concerned with the ecology of pathogens, diagnosticians, and those involved in advisory and quarantine services. These practitioners form a group with relatively common interests although there is some divergence in their motivations. Another important interest group is the systematists themselves.

In the past, the systematics of plant pathogenic bacteria was undertaken by plant pathologists who sought to clarify the relationships of the organisms with which they worked. For example, Burkholder, Dowson, Dye, Lelliott, and Starr set out to establish coherent taxonomic groups for the pathogens in which they were interested. Their foresight led to the creation of several collections of organisms - the International Collection of Phytopathogenic Bacteria (ICPB), the International Collection of Micro-organisms from Plants (ICMP) and the National Collection of Plant Pathogenic Bacteria (NCPPB) - which have become international resources for ecological and systematics studies.

F. G. Priest and M. Goodfellow (eds.), Applied Microbial Systematics, 135-163

Since then, drawing upon these collections, specialist systematists have made major taxonomic revisions which have been generally informative in indicating taxonomic relationships of greater precision. However, as will be discussed, some of their discoveries about the relationships of bacteria have created practical problems connected with the use of nomenclature, and may have revealed an inherent limitation in formal bacterial systematics.

Plant pathologists identify bacterial isolates in two general situations. For quarantine and diagnostics, the object is commonly to establish whether or not particular isolates are members of selected taxa; as pathogens and as members of particular species or pathovars. For this purpose, recent studies have expanded upon past successes with the result that there are many powerful probe methods to interrogate isolates as to their identity. These users have been well-served by the advances made in identification methods. By contrast, in ecological studies, the objective is commonly to establish the identity of pathogenic and non-pathogenic isolates without specific bacterial taxa in question. For these studies, modern revisions in classification have for the most part established more natural bacterial groupings but they commonly have not been accompanied by keys or methods by which isolates can be allocated to their taxonomic groups (Young et al., 1989).

3   Early Expectations

Before 1980, bacterial systematics was based largely on the comparison of descriptions of phenotypes. Initially these were of single-character biochemical and physiological tests (staining reactions, cell and flagellar morphology, pigment production, and biochemical and nutritional tests) which represented a small component of the total bacterial phenotype. The advent of the computer in the analysis of taxanomic data and the development of numerical taxonomy (Sneath & Sokal, 1973) for the study of extensive bacterial collections paved the way for modern bacterial systematics. These, as well as preliminary genomic studies of plant pathogenic *Pseudomonas* (Palleroni et al., 1973; Pecknold & Grogan, 1973) and *Xanthomonas* (Murata & Starr, 1973) strains indicated that the bacterial taxa as then described would be further refined by amalgamation or by subdivision. At that time the objective of systematics was to establish generic and species circumscriptions, descriptions of taxa, and where appropriate, the identity of plant pathogens as pathovars. The idea that bacteria might be formally allocated to an extensive hierarchy had been abandoned. There was little to contradict the assertion of Buchanan & Gibbons (1974) that ".... a complete and meaningful hierarchy [of bacterial taxa] is impossible". The state of bacterial systematics in 1980 was such that it was possible reliably to key a given isolate to genus and then determine species identity by phenotypic characters. Some phenotypic species and genera as taxonomic entities were crude by today's standards but they served well many general needs of practitioners, especially in the diagnosis of groups of bacteria with simple, commonly expressed activities.

4   The Approved Lists of Bacterial Names

A signal step in bacterial systematics was made in 1980 with the publication of the Approved Lists of Bacterial Names (Skerman *et al.*, 1980), which recognized as valid only those names of bacterial species for which there was a modern description and at least one extant strain which could be accepted as the type, or name-bearing, strain. Descriptions were made primarily in terms of biochemical and nutritional tests, and identification by determinative keys was relatively easy. The ~28,000 previously reported species names, most of which were uninformative, were reduced to ~2000 valid species and subspecies names. In 1980 then, there was a close correspondence between valid names, classification as it was best understood at the time, and methods of identification of taxa.

Since 1980 there has been an extensive accumulation of taxonomic data gathered by many methods and especially by molecular methods based on genomic analysis. Perhaps in these 20 years as much useful data has been gathered for analysis as in the preceding century. In 1980, there were about 270 valid genus names for bacteria (Skerman *et al.*, 1980). In mid-1999, including synonyms, there were 980 validly published genera and 5450 species. These quantitative measures do not indicate the scale of the qualitative changes that have also occurred. The impact on the nomenclature of plant pathogenic bacteria, as on other groups of bacteria, has been considerable. In 1980 there were 75 species and subspecies of plant pathogenic bacteria classified in 10 genera (Young *et al.*, 1978; Skerman *et al.*, 1980). In 1996 there were about 118 species and subspecies assigned to 25 genera (Young *et al.*, 1996). With the exception of four new genera, and 17 previously unreported species, most of the activity in bacterial systematics has generally resulted from revision by subdivision of known taxa.

5   Modern Classifications

Until 1980, modern bacterial systematics was guided almost entirely by phenetic concepts of classification in which ecological relationships were considered paramount. After 1980, with the development of molecular methods, phylogenetic relationships have become overwhelmingly important. Phylogenetic relationships are considered to reflect the ancestral relationships of bacteria

5.1   PHENETIC CLASSIFICATION

The goal of *phenetic classification* is to create clustered groups of strains, established as a hierarchy of species and genera on the basis of their overall similarities (Cowan, 1978; Goodfellow & O'Donnell, 1993). Thus members of a species share high levels of similarity and are more similar to species in their own genus than to members of other genera. Members of a genus also share common characters and can be identified accordingly. Phenetic classification has the advantage that a nomenclature based on it permits users to make predictions about the characters of related taxa (Lelliott, 1972; Goodfellow & O'Donnell, 1993). For instance, among plant pathogens, the ecological interactions of one taxon can be taken as the basis for behaviour of related taxa and the establishment of control measures against one bacterial species can form the basis for control of related organisms.

Early phenetic classifications relied on data provided by single-character tests. This approach produced over-simple and not very accurate classifications because only a limited expression was given to a relatively small proportion of the total bacterial phenotype. Multiple-character tests (also called polyphasic tests; Young *et al.*, 1992) such as cell wall composition, fatty acid and protein profiling, isoprenoid quinone and polyamine comparisons, as well as comparisons of DNA and RNA composition and sequences as part of the bacterial phenotype, (and now including comparison by such methods as pyrolysis mass spectrometry (Magee, 1993)) have the advantage that large components of a phenotype could be compared directly. Such methods have seemed to offer a way of moving towards the phenetic ideal of comparing total bacterial phenotypes, and have played an important role in circumscribing specific bacterial groups including plant pathogenic bacteria (Willems *et al.*,1992; Urakami *et al.*, 1994; Gillis *et al.*, 1995).

5.2   PHYLOGENETIC CLASSIFICATION

The alternative goal of systematics, which has been paramount in recent years, is the provision of a *phylogenetic classification* of bacteria based on ancestral relationships (Swofford *et al.*, 1996). Surprisingly, the term "phylogeny" is rarely precisely defined. If it is to have any precise meaning (to distinguish it from with "phenetic" or "polyphasic" classification) then it can be taken to mean that taxa sharing more recent common ancestry *in time* can be considered to be more closely related to one another than they are to other taxa. If "phylogeny" does not have this meaning then it is not clear what precise meaning it does have. Such vagaries as "representing evolutionary relationships" are not informative (Young *et al.*, 1992). The assumption that has fuelled phylogenetic classification is that the analysis of conserved sequence data can permit the inference of historical relationships of bacteria. This assumption, together with the ready ability to amplify conserved DNA using the polymerase chain reaction (PCR), and the ability to sequence this DNA, has seen an explosion of phylogenetic studies. Commonly these studies have been based on 16S rDNA or 23S rDNA, which are universally found in bacteria. Of course, the ubiquity of 16S rDNA does not, of itself, mean that it ensures accurate phylogenies. However the 16S rDNA molecule has seemed to provide a key to the systematization of relationships at the generic and higher taxonomic levels (Stackebrandt & Woese, 1984). It was this phylogenetic philosophy and these molecular methods which were largely responsible for recently reported family and genus revisions.

5.3   CLASSIFICATION BASED ON DNA

An important shift in emphasis since 1980, associated with the move to phylogenetic classification, has been away from classification based on phenotype expressed in cell structure and metabolic processes, to classification based on the genotype. Thus rDNA sequencing is seen as providing a framework for taxa at generic levels and above, DNA-rRNA hybridization and rDNA sequencing provides a framework for generic comparison,

DNA-DNA hybridization is the standard for determining species, and ribotyping, amplified fragment length polymorphism (AFLP), and restriction fragment length polymorphism (RFLP) methods permit discrimination of strains (Swings & Hayward, 1990). This tendency has developed to see genomic data as central to systematic interpretation (Wayne *et al.*, 1987), other data either being considered to confirm genomic analyses or to be set aside. However, recently it has been proposed that, when taxa (species) are characterized solely or primarily on the basis of genomic data, they be discriminated as "genomovars" (Ursing *et al.*, 1995). As will become evident the application of this concept could be of considerable utility if a consensus can be reached as to when it should be applied in lieu of a species proposal.

5.4   POLYPHASIC CLASSIFICATION

Colwell (1970) introduced the term *polyphasic taxonomy* to refer to classifications based on a consensus of all available methods; single-character tests, as well as multiple-character tests, including phenotypic and genomic data. She reserved *phenetic classification* for classifications based on numerical analyses. There appears to be no difference in principle between "polyphasic classification" (Colwell, 1970) and "phenetic classification" as defineed by Cowan (1978). It is in Cowan's sense that "phenetic classification" has been widely used since then (Young *et al.*, 1992; Goodfellow & O'Donnell, 1993). Since then, Goodfellow *et al.* (1997) has used "polyphasic taxonomy" as a synonym for "phenetic classification" in Cowan's sense. Vandamme *et al.*, (1996) also adopted Colwell's term "polyphasic taxonomy", by which they proposed that classifications be based on a consensus of data gathered by all available methods, i.e. phenetic classification. However, Vandamme *et al.*, (1996) also proposed that polyphasic classifications will be consistent with phylogenetic classification, which they take to be the relationships established using 16S or 23S rRNA sequence data. These writers need to make clear whether they interpret such sequence data as another component of information contributing to a wholly phenetic classification, or whether they take the very radical view that their polyphasic interpretation is an amalgamation of phenetic classification (based on overall similarities) and of phylogenetic classification (based on relationships in time). If they take this latter view, then they need to specify the assumptions involved that would make such a classification coherent.

5.5   INTEGRATION OF CLASSIFICATION

One expectation of modern systematics has been that the increasing numbers of multiple-character methods would form congruent circumscriptions of taxa; that in future, genera and species would be more precisely circumscribed and methods would be available for allocation of isolates to those groups (Goodfellow & O'Donnell, 1993, Vandamme *et al.*, 1996; Goodfellow *et al.*, 1997). As greater attention was turned to genomic and multiple-character methods, so Wayne *et al.* (1987) attempted to give guidance to the practices associated with the proposal of new species. They suggested a quantitative definition of bacterial species as population, whose strains share more than 70% DNA-DNA hybridization and have ΔT_m of less than 5° C. This definition has come to form the standard for species circumscriptions in the past decade. The implications and problems with this species standard were discussed earlier (Young *et al.*, 1992) and have been elaborated by Palys *et al.* (1997) who note the failure of species established on this arbitrary basis to correspond to ecologically distinct populations. Wayne *et al.* (1987) also urged that hybridization data be supported by phenotypic data. In proposing this, they anticipated that phenotypic data would support the genomic framework; that there would be a congruence of systematics using different methods and that this would give expression to phylogenetic relationships. There is no equivalent definition for the circumscription of higher taxa; genus, family and so on. The implications of these lacunae are developed below.

There is no assumption in phylogenetic inference that closely related organisms will necessarily be similar in phenotype. Certainly phylogenetic classifications have often led to radical proposals for the reclassification of taxa and subsequently more detailed phenetic studies have supported the proposal in general terms, leading to broadly coherent classifications. The revision of *Pseudomonas,* summarized in Kersters *et al.* (1996), is a good example. Where phenetic support is lacking it is assumed that future studies will provide justification for the classification. However, where congruence is not established, it follows that robust descriptions of taxa will not be derived. In practice, establishing genera and species on the basis of genomic data for phylogenetic classification has weakened the emphasis on phenotypic descriptions and hence our ability to allocate isolates (Young *et al.*, 1992). For many plant pathogens, it is practically impossible to identify isolates to generic levels without resort to sequence analysis or DNA reassociation investigation, unless the origin of the pathogen, its host plant, is known (see below). Instances of this problem are to be found in the genera *Acidovorax, Burkholderia* and *Ralstonia*. This should not be a surprise, but its consequences must be considered.

A major consequence of modern systematics has been to bring in to relief the contrasting poles of our assumptions concerning bacterial evolution. Phylogenetic systematics and classification assume the monophyletic origins of present day taxa and our ability to trace their ancestry, in the way supposed to be the case for higher organisms. For this to be possible, evolutionary change is assumed to occur through the natural selection of mutations occurring in a relatively stable genome, and environmental forces function primarily to induce change. According to this view, stability is a function of genome structure, and evolutionary change is the result of environmental pressures. An alternative view is that the bacterial genome is highly mutable, both because mutation rates are high and because horizontal gene transfer is a common activity involving all sites of the genome. According to this view, natural selection plays a primary role both in maintaining the stability of observed taxa, sometimes as unrecognized polyphyletic entities, as well as causing adaptational changes leading to the evolution of new taxa. Interpretations based on assumptions of genome stability (Woese, 1987) are may be giving way to those predicated on greater plasticity of genotype and phenotype.

Both for phenetic and phylogenetic classifications, it is assumed that the phenotype and genotype are relatively stable to environmental selection pressures. If variability is high, then the natural hierarchies expressed in both kinds of classification will not prove to be as coherent as is assumed in the present development of nomenclature. If this proved to be the case, then it would be helpful if the pace at which new nomenclature was generated was slowed so that the fundamental nature of bacterial relationships could be explored. This point may not have been reached, but the issue needs to be considered.

5.6 RECENT COMPREHENSIVE MULTIPLE-CHARACTER STUDIES

If multiple-character methods are to contribute to the circumscription of species (Wayne *et al.,* 1987, Goodfellow & O'Donnell,1993; Vandamme *et al.,* 1996; Goodfellow *et al.,* 1997) then they should be expected to give coherent, congruent classifications consistent with relationships expressed by genomic data. In a specific example (Janse *et al.,* 1996), *Pseudomonas avellanae* was established as a new species distinct from *Pseudomonas syringae.* This new species corresponds to one of the minor genomic groups of *P. syringae sensu lato,* and includes *P. syringae* pathovar (pv.) *theae* (L. Gardan, pers. comm.). DNA-DNA hybridization data and phenotypic reactions (LOPAT tests - Lelliott *et al.,* (1966) - for levan production, oxidase activity, potato rot, arginine dihydrolase activity and tobacco hypersensitivity reaction) show that *P. avellanae* and *P. syringae* are closely related, with *P. aeruginosa* and *P. fluorescens* as outlying species (in agreement with Shafik, 1994, L. Gardan, pers. comm.). By contrast, whole protein and fatty acid data show *P. avellanae* to be a species more distantly related to *P. syringae,* outlying *P. corrugata* and *P. fluorescens* (Janse *et al.,* 1996). How then are these incongruent data to be interpreted? What are the relationships between these species?

The interrelationships of multiple-character analyses have recently been comprehensively illuminated by a multi-laboratory investigation of the genus *Pseudomonas,* reported in *Systematic and Applied Microbiology* 19 (4) 1996. In this issue of the journal, papers are presented which consider the relationships of many *Pseudomonas* species, including plant pathogenic species, comprehensively examined by 16S RNA sequence analysis (Moore *et al.,* 1996), ribotyping (Brosch *et al.* 1996), fatty acid content of whole-cell hydrolysates and phospholipid fractions (Vancanneyt *et al.,* 1996a), SDS-PAGE of whole-cell protein (Vancanneyt *et al.,* 1996b), westprinting (Tesar *et al.,* 1996), and Biolog and BioMerieux API Biotype-100 systems (Grimont *et al.,* 1996).

The relationships among fluorescent *Pseudomonas* species, indicated by the different methods described above are shown in Table 1. In most cases, the allocation of isolates to particular species is supported by most or all methods. The treatment of *P. fluorescens* using Biotype 100 and Biolog microtitre plates (Grimont *et al.,* 1996) and by SDS-PAGE of whole cell proteins (Vancanneyt *et al.,* 1996b) shows considerable heterogeneity and may indicate a need for further species subdivisions. However, from the standpoint of the internal consistency expected in the data, and using known DNA-DNA reassociations as the basis for congruence, the notable feature of the remaining accumulated data is the different relationships indicated between species by the different multiple-character methods (Table 1). Some incongruence may indicate further species subdivisions as noted; other incongruence does not. How are these data to be reconciled? For example, should a proximate congruence of relationships of DNA-DNA reassociation data and whole cell protein data be expected? There may be explanations as to how bacteria with relatively homologous DNA overall produce whole cell proteins which are relatively heterogeneous, but it is not clear which implied classification is to be preferred or how the different classifications are to be reconciled. Of note is the presence of strains of *Burkholderia* and

TABLE 1. Comparison of plant pathogenic *Pseudomonas* species from 16S rRNA sequence data (Moore et al., 1996), ribotyping (Brosch et al., 1996), SDS-PAGE of whole-cell protein (Vancanneyt et al., 1996b), fatty acid content of whole cell hydrolysates (Vancanneyt et al., 1996a), Westprinting (Tesar et al., 1996,) and Biolog and BioMérieux Biotype-100 systems (Grimont et al., 1996). Species are shown in the groups generated by each method as blocks in each column. The relative positions of species groups in the columns is not necessarily significant. Taxa that are considered to be closely related (Young et al., 1992) are indicated by either [ ] or [[ ]]. Species placements which are anomalous in groups are underlined. Authentic *Pseudomonas* species are members of the Gamma sub-class of the *Proteobacteria*. Species marked with an asterix* are members of the Beta-sub-class.

16S rRNA	Ribotyping	SDS-PAGE	FAME (whole cell)	Westprinting	Biolog	Biotype 100
[[marginalis]]	corrugata	asplenii	agarici	[[marginalis]]	asplenii	corrugata
[[tolaasii]]	[syringae]	fuscovaginae	corrugata	[[tolaasii]]	fuscovaginae	[[marginalis]]
[viridiflava]		[[tolaasii]]	[[marginalis]]			fuscovaginae/asplenii
	[caricapapayae]	[[marginalis]]	[[tolaasii]]	cichorii	[[tolaasii]]	
[syringae]	cepacia *			[viridiflava]	corrugata	[caricapapayae]
[amygdali]	meliae	corrugata	[caricapapayae]	[caricapapayae]	[[marginalis]]	[coronafaciens]
[coronafaciens]	[ficuserectae]	[viridiflava]	[coronafaciens]	[ficuserectae]		cichorii
[ficuserectae]	[amygdali]	[coronafaciens]	[syringae]	[coronafaciens]	[caricapapayae]	[viridiflava]
		cichorii	[viridiflava]	[syringae]	[coronafaciens]	
cichorii	[[marginalis]]				[viridiflava]	[[marginalis]]
	[coronafaciens]	[ficuserectae]	asplenii	asplenii	cichorii	[[tolaasii]]
asplenii		[syringae]	cichorii			
	cichorii	[amygdali]	ficuserectae	agarici	cepacia *	cepacia *
agarici	[[tolaasii]]		fuscovaginae			
		agarici			[[marginalis]]	agarici
	fuscovaginae	cepacia *	cepacia *		agarici	
	asplenii	solanacearum	solanacearum *		solanacearum *	[syringae/amygdali]
	agarici	(Burkholderia)				[ficuserectae]
	[viridiflava]	[caricapapayae]			[amygdali]	
	solanacearum *				[syringae]	solanacearum *
	[[marginalis]]				[ficuserectae]	
	cichorii					

Ralstonia spp. within the *Pseudomonas* cluster using SDS-PAGE, FAME, and Biolog. Incongruities of this kind pose serious questions either about the actual relationships of these taxa or of the validity and capacity of multiple-character methods to establish coherent relationships. Brosch *et al.* (1996) note the limitation which point mutation sets on ribotyping as a method for determining relationships, but if similar caveats are placed on other methods then what contribution do multiple character methods offer to systematics except as identification methods? Are there different species relationships depending on whether DNA-DNA hybridization, protein analysis, fatty acid profiles, polyamine analyses, or RFLP data are compared? How are the species to be circumscribed? These data raise potentially important questions about the natural relationships of bacteria and our ability to understand them. If these different outcomes of multiple-character studies using different methods are a general phenomenon, then it is difficult to see how the expectations either of phenetic or of phylogenetic systematics above species levels can be realized. The immediate impact is on nomenclature. What significance do names have if there is doubt about the establishment of a coherent base of systematics data?

6   Generic Revisions

6.1   THE CORYNEFORM GENERA

The Gram-positive coryneform plant pathogenic bacteria were divided into three genera (*Clavibacter, Curtobacterium,* and *Rhodococcus*) on the basis of genotypic studies, with support from cell wall chemistry. Recently, the establishment of the genus *Rathayibacter* by the division of *Clavibacter* was proposed primarily on DNA-DNA reassociation studies and supporting phenotypic data (Zgurskaya *et al.,* 1993). In studies of 16S rRNA sequence and RFLP, Lee *et al.* (1997b) have suggested that the *Clavibacter* and *Rathayibacter* species can be divided into three genera, comprising (1) *C. michiganensis* and *C. xyli,* (2) *R. iranicus* and *R. tritici,* and (3) *R. rathayi.* There are no phenetic data to support these taxonomic subdivisions.

6.2   PSEUDOMONAS

The genus *Pseudomonas* was circumscribed on the basis of rod-like cell morphology, polar insertion of flagella, and oxidative metabolism by Palleroni (1984). Its heterogeneous nature and the need for formal subdivision was signalled by the pioneering work of Palleroni *et al.* (1973), who showed that *Pseudomonas* comprised three groups of species which could be distinguished by DNA-rRNA hybridization. This work was confirmed by further comprehensive DNA-rRNA hybridizations (De Vos & De Ley, 1983; De Vos *et al.,* 1985) and subsequently by 16S rDNA sequence analysis (Kersters *et al.,* 1996). Thus, plant pathogenic *Pseudomonas* species are now allocated to five genera: *Acidovorax, Burkholderia, Herbaspirillum, Pseudomonas* and *Ralstonia* (Table 2). Unfortunately, new determinative methods are still not available by which the non-fluorescent genera can be discriminated one from another. This problem is discussed below.

TABLE 2   Plant pathogenic genera and species which have been validly published and which share general
genomic and phenotypic congruence. Species names are from Young *et al.* (1996) unless published
more recently. Sources of new names are indicated in footnotes. Supra-generic groups are based on
DNA-rRNA homology and 16S rRNA sequence data. Numbers in parenthesis represent pathovars of the preceding taxon

---

Division: Firmicutes (Gram-positive genera)
   Family: no family classification in Firmicutes

               Arthrobacter   *ilicis*[1]

               Bacillus   *megaterium (1)*

               Clavibacter   *iranicus*[2], *michiganensis* subsp. *michiganensis*, *michiganensis*
                                        subsp. *insidiosus*, *michiganensis* subsp. *nebraskensis*, *michiga-*
                                        nensis subsp. *sepedonicus*, *michiganensis* subsp. *tessellarius*,
                                        rathayi[2], *tritici*[2], *xyli* subsp. *xyli*, *xyli* subsp. *cynodontis*

               Curtobacterium   *flaccumfaciens (4)*

               Nocardia   *vaccinii*

               Rathayibacter[2]

               Rhodococcus   *fascians*

Division: Gracilicutes (Gram-negative genera)
   Class: Proteobacteria
     Alpha sub-class
       Family: Acetobacteriaceae

               Acetobacter   *aceti, pasteurianus*

   Family: Rhizobiaceae
   Agrobacterium   *rhizogenes, rubi, tumefaciens, vitis*

   Family: not classified
   Rhizomonas   *suberifaciens*

Beta sub-class
   Family: Comamonadaceae

     Acidovorax   *avenae* subsp. *avenae*, *avenae* subsp. *cattleyae*, *avenae* subsp.
                                       citrulli, konjaci

   Xylophilus   *ampelinus*

   Family: not named
   Burkholderia[a]   *andropogonis, caryophylli, cepacia, gladioli (3), glumae, plan-*
                                  tarii, solanacearum

   Herbaspirillum   *rubrisubalbicans*[b]

   Ralstonia[3]

TABLE 2   continued

Gamma sub-class

    Family: Enterobacteriaceae

        Enterobacter   *cancerogenus, dissolvens, nimipressuralis*

        Erwinia   (a) *alni[c], amylovora, mallotivora, nigrifluens, psidii, quercina, rubri-*

                                      faciens, salicis, stewartii[4], tracheiphila

                    (b) *ananatis[4], herbicola* (2)[4]*, uredovora[4]*

                    (c) *cacticida, carotovora* subsp. *carotovora, carotovora* subsp.

                                      atroseptica, carotovora subsp. *betavasculorum, carotovora*

                                      subsp. *wasabiae, chrysanthemi* (7)*, cypripedii, rhapontici*

    Pantoea[4]

    Family: Pseudomonadaceae (requires emendation)

        Pseudomonas[5]   (a) *agarici, asplenii, avellanae[d], fluorescens* (1)*, fuscovaginae,*

                                      marginalis (3)*, tolaasii*

                    (b) *amygdali, caricapapayae, ficuserectae, meliae, syringae* (45)*,*

                                      viridiflava

                    (c) *cichorii*

                    (d) *corrugata*

                    (e) *syzygii[6]*

    Family: not named

        Xanthomonas[7]   *albilineans, axonopodis, campestris* (145)*, fragariae, oryzae* (2)

        Xanthomonas[8]   *albilineans, arboricola* (5)*, axonopodis* (43)*, bromi, campestris* (7)*,*

                                  codiaei, cucurbitae, fragariae, hortorum (4)*, hyacinthi, melonis,*

                                    oryzae (2)*, pisi, populi, sacchari, theicola, translucens* (9)*, vasicola,*

                                  vesicatoria

    Family: Not classified

        Xylella   *fastidiosa*

Division: Tenericutes

    Class: Mollicutes

        Family: Spiroplasmataceae

             Spiroplasma   *citri, kunkelii, phoeniceum*

Genera of uncertain affinity

    Rhizobacter   *dauci*

    Streptomyces   *acidiscabies, albidoflavus, candidus, caviscabies[e], collinus, inter-*

                                medius, ipomoeae, scabies, setonii, wedmorensis

TABLE 2   continued

Species which are misclassified	
Pseudomonas beteli (= *betle*)	(*Xanthomonas*?)
Pseudomonas cissicola	*Xanthomonas* sp.[f]
Pseudomonas flectens	(generically misclassified)
Pseudomonas hibiscicola	(*Xanthomonas*?)

[1] The pathogen is probably a member of *Clavibacter* (Young *et al.*, 1992).

[2] The genus *Rathayibacter* is very similar to *Clavibacter*, showing slight differences in DNA-DNA reassociation values and in unsaturated menaquinones in cell wall components.

[3] The genus *Ralstonia* has been established principally on the basis of 16S rRNA sequence differences. *Ralstonia solanacearum* has been allocated here to *Burkholderia*.

[4] Species of *Pantoea* have been classified in the genus *Erwinia*. See Young *et al.*(1992) for discussion.

[5] The genus *Pseudomonas* is discussed in detail in the text and in Young *et al.* (1992).

[6] *Pseudomonas syzygii* is properly classified with *Burkholderia/Ralstonia* (Seal *et al.* 1993).

[7] *Xanthomonas* sensu Bradbury (1984) is discussed in the text.

[8] *Xanthomonas* sensu Vauterin *et al.* (1995) is discussed in the text

[a] Yabuuchi *et al.* (1992); [b] Baldani *et al.* (1996); [c] Surico *et al.* (1996); [d] Janse *et al.* (1996); [e] Goyer *et al.* (1996); [f] Hu *et al.* (1997)

[*] (Note added in proof) Hauben *et al.* (1999) have revised the genus *Erwinia* by the creation of two genera, *Brenneria* and emended *Pectobacterium*, based on a comparative sequence analysis of 16S rDNA data.

The allocation of species to these five genera is based on differences in their 16S rDNA sequences. These generic allocations are often supported by multiple-character methods including DNA-rRNA hybridization, fatty acid profiling, PAGE protein profiles and polyamine analysis. Without question the basic differentiation of the non-fluorescent, PHB-accumulating bacteria from *Pseudomonas* is sound. However, the sub-division of genera justified by phylogenetic inferences based solely on 16S analyses will have important consequences if uncritically pursued. For instance, our unpublished data indicate that *Acidovorax, Comamonas* or *Variovorax* may not merit discrimination as genera based on differences in 16S rRNA sequences. However, if they are differentiated as separate genera, then *Burkholderia andropogonis* and *B. caryophylli* could equally be considered as members of two separate genera, apart from *B. cepacia* and *B. gladioli*. These taxa are phenotypically distinct, although there are no obvious characters which would stand as generic determinants if they were divided.

Pseudomonas sensu stricto is represented by organisms that are characteristically fluorescent, have one or more polar flagella, are strictly oxidative, and do not accumulate poly-β-hydroxybutyrate (PHB). Species now contained in the genus *Pseudomonas* are recorded in Moore *et al.* (1996.). *Pseudomonas corrugata,* as a non-fluorescent, PHB-positive member (Scarlett *et al.*, 1978) of the *Pseudomonas* rRNA group (De Vos *et al.*, 1985), is anomalous. Most species are usually well established taxa (but see *P. syringae* below).

6.3 AGROBACTERIUM

The family *Rhizobiaceae* (Jordan, 1984) is in the process of revision based on the phylogenetic interpretation of 16S rRNA sequence data (Young & Huakka, 1996). So far, three closely related nitrogen-fixing genera have been recognized; *Mesorhizobium, Rhizobium,* and *Sinorhizobium* (as well as *Azorhizobium* and *Bradyrhizobium*). If these genera are accepted, then a consistent interpretation of 16S rRNA data (Sawada *et al.* 1993; Willems & Collins 1993) will require that *Agrobacterium rhizogenes* be allocated to *Rhizobium sensu stricto,* and that *Rhizobium galegae* be removed from this genus and perhaps included in *Agrobacterium.* An alternative classification might recognize a single genus, *Rhizobium,* which excluded *Azorhizobium* and *Bradyrhizobium* but comprised the closely related nitrogen-fixing symbionts at present in *Mesorhizobium, Rhizobium,* and *Sinorhizobium,* as well as the agrobacterial species with oncogenic potential at present in *Agrobacterium.*

6.4 PANTOEA

Analysing 16S rRNA data, Kwon *et al.* (1997) showed that *Erwinia* species were distributed in four clusters; cluster 1: *Erwinia [Pantoea] ananatis, E. herbicola, E. millettiae, E. stewartii,* and *E. uredovora*; cluster 2: *E. amylovora, E. cypripedii, E. persicinus* and *E. rhapontici*; cluster 3: *E. carotovora* and; cluster 4: *E. nigrifluens, E. rubrifaciens,* and *E. salicis.* Cluster one contains representatives of all known *Pantoea* spp. Clusters one and two are outlying groups, both separated from the adjacent clusters three and four by members of other genera. Acceptance of *Pantoea* implies subdivision of *Erwinia* into four genera. The creation of *Pantoea* must be considered in the wider context of generic classification within the family *Enterobacteriaceae.* This is discussed in Young *et al.* (1992) and elsewhere in this chapter.

With these examples it should be remembered that, by contrast with species (Goodfellow *et al.,* 1997), few attempts have been made to provide objective criteria for the circumscription of higher taxonomic ranks. The decision that any particular node in a cladogram represents the determining point for the division of genera, and that all members below share a common name, or that any selection of characters represents the circumscription of a taxon, is entirely a subjective one. Because the original imprecise generic divisions form the basis of subsequent revisions, there has been a strong tendency to create new genera by the division. This has had the effect of producing many genera with smaller numbers of species. We should perhaps ask what interest is served by the creation of ever finer taxonomic subdivisions, sometimes unsupported by phenetic data, or only supported by small numbers of characters.

7 Species Revisions

7.1 PSEUDOMONAS SYRINGAE

Until recently, *Pseudomonas syringae* represented practically all bacteria circumscribed as group I in the LOPAT determinative scheme. Gardan *et al.* (1992) re-established

P. savastanoi (with the pathovars *savastanoi, glycinea* and *phaseolicola*) on the basis of DNA-DNA hybridization data. Gardan *et al.* (1997) have reported a further three major genomic groups containing most known pathovars, and four minor genomic groups, all of which can be proposed as species if supporting phenotypic data becomes available. However, it is unlikely that determinative tests based on such methods as API BioMerieux will be available for these species. So far, results of other phenotypic or multiple-character tests have not been systematically established as part of the circumscriptions of these genomic species.

7.2   XANTHOMONAS SPP.

The genus *Xanthomonas* was revised recently and the five existing species have been reallocated in 19 newly circumscribed species (Vauterin *et al.*, 1995). Of the 150 known pathovars, more than 90 have been allocated to these new species. Several of the proposed species determinations were based solely on DNA-DNA reassociation studies although the authors elsewhere have expressed reservations concerning the arbitrariness of this method (Vandamme *et al.*, 1996). Species circumscriptions were supported by SDS-PAGE of proteins (Vauterin *et al.*, 1995) but some species exhibited internal protein heterogeneity. Thus, for genomic species representing unique pathogens - *X. albilineans, X. bromi, X. cassavae, X. cucurbitae, X. hyacinthi, X. oryzae, X. populi, X. sacchari, X. theicola* - SDS-PAGE patterns were homogeneous. As well, some species comprising several pathovars - *X. campestris, X. translucens, X. vasicola* - exhibited uniform patterns. In *X. hortorum*, pvs *hederae* and *vitians* were similar by SDS-PAGE, whereas pv. *pelargonii* showed a different pattern. In *X. arboricola*, pvs *juglandis* and *pruni* and a part of pv. *corylina* were homogeneous, whereas other pv. *corylina* strains were distinct. *Xanthomonas arboricola, X. axonopodis, X. campestris* and *X. hortorum* gave heterogeneous patterns. Strains of *X. axonopodis*, representing 43 pathovars, were more heterogeneous than were those of any other *Xanthomonas* species. This was reflected in the loose genomic relatedness of most of the strains it contained. Furthermore, SDS-PAGE subgroups within *X. axonopodis* did not always correspond with pathovars (examples: *alfalfae, citri, phaseoli*).

Phenotypic support for the species proposals was based on fingerprinting by Biolog using nutritional tests. However, Biolog data in Vauterin *et al.* (1995) gave only a measure of the probability of the reaction of a strain to tests but did not give a method of isolate identification. If the standard was applied that a determinative test was considered positive when more than 90% of strains gave a positive reaction, negative when fewer than 10% of strains gave a positive reaction, and otherwise indeterminate, then the tests will not reliably discriminate strains of *X. axonopodis* from *X. campestris* and X. melonis, *or strains of* X. arboricola, *X. cassavae and X. hortorum, or strains of "X. translucens"*, *X. vasicola* and *X. vesicatoria*. *Xanthomonas* species were also supported by fatty acid profile comparisons on a statistical basis, but not with sufficient discrimination to allow the reliable allocation of isolates or the recognition of new species (Vauterin *et al.*, 1996).

Subsequently, summarizing methods to discriminate *Xanthomonas* spp., Vandamme *et al.* (1996) reported the capacity for available multiple-character methods (whole cell proteins, fatty acids and biochemical tests) to discriminate only *X. albilineans, X. axonopodis,*

X. fragariae, X. populi, and (using fatty acids and biochemical tests) *X. oryzae.* It is hard to know to what extent this is a retreat from the original proposal (Vauterin *et al.,* 1995). Furthermore, in this case it is not obvious that phenetic data have produced a coherent classification of the genus. On this analysis, it is hard to share these authors' enthusiasm for the classification and hence the nomenclature so generated.

Viewed overall, it seems clear that *Xanthomonas* is indeed composed of a number of groups, some as well-established species (Vauterin *et al.,* 1995, and related papers). However, it is in the nature of some of the other groups within *Xanthomonas* that they cannot be circumscribed by methods that permit strain allocation, nor even ready discrimination as species overall, except perhaps by DNA-DNA hybridization. If bacterial systematics is pursued in the same way, then further species divisions in other genera can be expected, with the attendant practical difficulties.

8   Statistical Analysis of Sequence Data

The theoretical basis of sequence analyses in systematics is complex (Swofford *et al.* 1996). The uncritical application of simple algorithms can lead to proposals of classifications which may be little better than those which they replace. Sneath (1989), Swofford *et al.* (1996), and Goodfellow *et al.* (1997) have drawn attention to the need for improved and careful statistical analyses of sequence data in order to establish the validity of branches. Moore *et al.* (1996) have proposed nine rRNA lineages in *Pseudomonas,* but the relatively low bootstrap values suggest that many branches may not be reliable and Sneath (pers. comm.) concludes that the lineage differences of only *P. mendocina, P. putida* and *P. resinovorans* are statistically significant. Reliability of sequence analyses should be confirmed by other statistical methods for *Clavibacter* and *Rathayibacter* (Lee *et al.,* 1997b), for *Burkholderia* and *Ralstonia* (Yabuuchi *et al.,* 1995), and for *Xanthomonas* (Hauben *et al.,* 1997).

9   The Pace of Nomenclature Revision

More than many other interest groups, those involved in plant pathology have valued insights of contemporary systematics reported as changes in nomenclature. However, since the revisions are based on ever finer subdivisions of taxa without a concomitant improvement in phenotypic circumscription, the time must come when the price of a constantly changing nomenclature may be considered to be too high. Diagnostic, advisory, and quarantine services are best served by stable nomenclature which reflects a simple connection between a disease and its pathogen (Hayward, 1972). Plant pathogenic bacteria are the subject of extensive legislation connected with plant quarantine. Particular named pathogens are the subject of exclusions for many countries. The progressive revisions of genera based on phylogenetic refinements impose a significant burden on practitioners, without obvious benefit. The recent allocation of *Pseudomonas solanacearum* to *Burkholderia* (Yabuuchi *et al.,* 1992) and then to *Ralstonia* (Yabuuchi *et al.,* 1995) is a case in point. Furthermore, current interpretations of the Bacteriological Code of Nomenclature

(Lapage *et al.*, 1990) has been taken to foreclose on the use of earlier valid nomenclature for the publication of new or revised names. For instance, the creation of genomic species in *Xanthomonas* in the sense of Vauterin *et al.* (1995) has been interpreted as forbidding the allocation and application of pathovars in *X. campestris* in the sense of Bradbury (1984). The effect of this interpretation is to inhibit the formal reporting of pathogens. Pathologists should feel confident that the use of any and all validly published nomenclature of their choice is permissible in reporting.

Most users of nomenclature, as it is applied to plant pathogenic bacteria, make the assumption that nomenclature reflects evolution of natural groups of bacteria in relation to particular environments; they assume the benefits of a phenetic taxonomy. In this respect the plant pathology community differs from the public health community for whom nomenclatural stability and the need for routine identification seems to be dominant. For instance although the systematics of the family *Enterobacteriaceae* is in need of substantial revision with the possible amalgamation of some of the closely related genera; *Citrobacter, Enterobacter, Escherichia, Klebsiella, Salmonella*, and *Shigella* (Brenner, 1984), such revisions are impeded by the understandable concern at possible confusion it could cause.

10   Identification

The dynamic relationships between classification, nomenclature, and identification have been noted (Young *et al.*, 1992); without adequate phenotypic descriptions it is impossible to allocate isolates to taxa; to make strain identifications; without identified strains it is impossible to perform comparative taxonomy, and without comparative taxonomy there are no phenotypic descriptions. Stackebrandt & Goebel (1994) considered the rapid and reliable identification of isolates to be the most important task in taxonomy. The search now is for some character or small set of characters found universally in bacteria, revealed in a single analytic step or by a small number of steps, which expresses such variation as to reflect those species and genus relationships indicated by consensus analyses. The desirable features of the favoured methods of analysis are that they should be quick, simple, reliable, and reproducible, should not depend on high levels of theoretical skill, particularly for trouble-shooting, and should be cheap. API BioMerieux and Biolog systems seem to fulfil many of the criteria. However, they seem to have reached their limit insofar as they can be used to discriminate some but not all Gram-negative genera and species recently established using multiple-character methods. Their utility in reliably identifying Gram-positive plant pathogenic bacteria seems to be more limited (Harris-Baldwin & Gudmestad, 1996). Future alternative methods may involve the analysis of data gathered using multiple character methods, such as DNA restriction profiles or SDS-PAGE of proteins, comparing the pattern of unknown isolates with computer-stored databases (Priest & Williams 1993; Vauterin *et al.* 1993). As yet, the reproducibility of patterns poses practical limits to the application of such methods.

Of course, strains of taxa (species and genera) which were described using ancient technologies and for which determinative keys were developed can still be readily keyed out. The problem arises for strains of taxa based on genomic and multiple-character methods alone. How are isolates to be identified unless source ecological information is also available? There are now many named species existing as epiphytes which, if isolated from a plant other than their host, would be almost impossible to identify directly according to the most recent nomenclature; examples are the discrimination of *Pantoea* spp. from *Enterobacter agglomerans* and from the *Erwinia herbicola* group, the differentiation of *Acidovorax* spp. from *Burkholderia* spp. and from *Ralstonia* spp. (except by the comparative analysis of sequence data), species in the recent revision of *Xanthomonas* spp., and differentiation of *Clavibacter* spp. from *Rathayibacter* spp.

10.1   DETERMINATIVE KEYS DOWN OR DETERMINATIVE KEYS UP?

Keys down. In the past, determinative keys were an expected component of the descriptions of taxa. Determinative keys were available for genera, and for species within genera, i.e., it was possible to key down through the taxonomic hierarchy to identify isolates with increasing precision. This is no longer always the case: the focus on polyphasic, multiple-character, and phylogenetic classification has been at the expense of simple identification methods. It is now common that strains cannot readily be allocated to their genus. For example, there are few if any generic determinative tests by which an isolate can be distinguished as being a member of *Acidovorax* (in the family *Comamonaduceae*), or *Burkholderia* (in its unnamed family), or *Ralstonia solanacearum*, or *P. corrugata* (a non-fluorescent species in *Pseudomonas sensu stricto*). There are few tests by which strains of *Clavibacter* can be distinguished uniquely from *Rathayibacter*.

Keys up. For some isolates, species characterization is possible by direct reference to the species determinative tests. For instance, non-fluorescent pathogenic isolates from known host plants can be keyed to species and hence to genus in *Acidovorax* or *Burkholderia* spp. using species determinative keys (Hu *et al.*, 1991; JMY unpublished data). The same may be true for *Clavibacter* and *Rathayibacter* (Zgurskaya *et al.*, 1993). Probabilistic methods (Firrao & Locci, 1989; Priest & Williams, 1993), developed in Biolog and API systems, can be used in a similar way. However, the essential first step is to discover which selection of determinative keys is relevant. For this, it is usually necessary to have prior knowledge of the source host plant which restricts the scope of the probable identity the organism.

10.2   PROBES

Molecular techniques have made an unambiguously positive contribution to probe methods for the detection and identification of specified organisms. The methods are available in the form of research tools and as commercial kits. There is now a choice of many probes with different levels of sensitivity that can be used to establish the presence of specified bacteria, as clonal groups, infrasubspecies, and higher taxonomic groups. Probes have been developed as specific antigens, as specific DNA oligonucleotides, and as DNA primers which amplify specific sequences in target organisms. That they target a particular organism and do not permit general identification of isolates may be seen as a limitation for general identification. A promising possibility for general identification may be REP-PCR (Louws *et al.*, 1994). Positive identification still depends on the inclusion of authenticating reference strains.

A further development of probe methods has been the ability to identify pathogenic populations *in situ*, whether in plant tissue and plant-associated environments, in seed, or as epiphytes. Just one example is the use of nested PCR for the detection of *Clavibacter michiganensis* subsp. *sepedonicus* (Lee *et al.*, 1997a). As identification tools these can be expected to be valuable for ecological studies of specific pathogens and applied in quarantine eradication programmes. For some of the more elegant methods, a limitation on their routine use is cost.

Of perhaps the greatest importance, molecular probes are now available to identify fastidious pathogens which can not be cultivated by conventional means or can be cultivated only with difficulty. Thus, Saillard *et al.* (1996) report on the identification of *Spiroplasma citri* from symptom-free plant tissue by PCR-amplification of the spiralin gene of the immuno-captured bacterium. Methods of this kind permit an increasing number of pathogens - such as the mycoplasmas, mycoplasma-like organisms, and the pathogens that are difficult to culture, in *Spiroplasma* and *Xylella*, to be studied in detail.

11   Systematics and Economics

11.1   CLASSIFICATION

Throughout the world economic revisions have had a significant impact on research in systematics and on consequential activities. Unless systematics has been shown to contribute to functional activities for which there are social or economic returns, its support has been increasingly restricted. Only a small number of institutes are protected from such economic pressures. For this reason phylogenetic research may become increasingly vulnerable where the driving interest is in practical applications; divorced from phenotypic, ecologically based characterizations of bacteria, it is difficult to justify in practical market terms. International collaboration will support research in systematics where there is a common interest. A problem will be to establish confident relationships between those who need identifications and laboratories where the systematics and procedural skills are available. This is especially true for diagnosticians and researchers on the African, Asian, and South American continents where facilities are most rare and the potential for sources of new plant pathogens is greatest.

11.2   IDENTIFICATION

A consequential effect of phylogenetic systematics is that it has raised the cost and increased the difficulty of making identifications for practical purposes according to modern nomenclature. The costs of multiple-character methods are not thought of as great, yet there are usually substantial capital costs, a need for dedicated skilled technical operators, and the consideration of overheads and depreciation of equipment. Furthermore, the proliferation of methods (fatty acid, SDS-PAGE and polyamine profiling, and recently, ribotyping, and westprinting) effective for various taxonomic groups and at different taxonomic ranks, means that any one laboratory can develop expertise only in a few methods. Suggestions that identification be centralized in one country must take account of the fact that the countries most in need of technical support have restrictions on overseas payments in hard currencies.

The difficulties of identifying pathogenic strains to species impact on the naming of pathovars. To name a new pathovar now entails proving a unique host range for the strains (Young *et al.*, 1992) - a not inconsiderable task - and then allocating them to a genus and species. Given the unreliability of reported methods, this demands resort to DNA-DNA reassociation. Thus, accurate circumscription of the pathovars in *Xanthomonas* involves up to 20 DNA-DNA hybridizations. The situation described above will also apply if *Pseudomonas syringae* is subdivided into species based on genomic data (Gardan *et al.*, 1997), if supporting phenotypic circumscriptions of the new species are not developed.

12   Pathovars

Before 1980, the nomenclature of plant pathogenic bacteria was in a chaotic state, as bad as or worse than for other bacterial groups. Besides a plethora of synonyms and illegal names, the systematics of plant pathogens was beset by the "new pathogen - new species" concept (Stolp *et al.*, 1965), by which a new species name was proposed for each new host/pathogen combination. The decision to recognize species as valid only if they were represented by a legitimate name, a modern description, and a type strain required pathologists to recognize that many plant pathogenic species differed only in host range, a character not considered to form part of a species description in terms of the Bacteriological Code of Nomenclature. With these restrictions, many pathogens were considered to be members of the same species (Sands *et al.*, 1970; Dye & Lelliott, 1974, Dye & Kemp, 1977). The problem of naming pathogens was resolved by the proposal to recognize the *infra-subspecific* term "pathovar" for populations of pathogens within species (Young *et al.*, 1978, 1992) and the creation of the Standards for Naming Pathovars (Dye *et al.*, 1980). These standards were intended to obviate a special problem for plant pathogenic bacteria by providing a nomenclature complementary to that for higher taxa (Young *et al.*, 1992). The driving force for the pathovar scheme as a special purpose classification was a wish to meet the formal nomenclatural needs of pathologists while at the same time preserving a rational natural classification for bacteria at higher levels. Since 1980, little has occurred to reduce the utility of pathovar usage.

The pathovar scheme provides a useful nomenclature for the distinct pathogens in *Curtobacterium flaccumfaciens, Burkholderia gladioli, Pseudomonas* spp. and *Xanthomonas* spp., which form distinct sub-populations of strains sharing common pathogenicity genes. The value of the nomenclature is its focus on the ecologically significant genes of these environmentally specialized organisms. This has been the source of fruitful research into pathogenic subdivisions, with attention to the particular genes for pathogenicity (Stall, 1995). Within pathovars there exist subdivisions (as races) by which host specificity may be further subdivided. Like all attempts to pigeonhole nature into convenient groups, the pathovar scheme is not universally successful. Thus, pathogenic groups with wide and overlapping host ranges, as in *Erwinia chrysan-themi, Burkholderia/Ralstonia solanacearum*, and the pathogens of *Agrobacterium* (which have a significant component of their pathogenicity genes borne on plasmids), are not readily susceptible to formal classification in terms of names based on pathogenicity (Young *et al.*, 1992). Interestingly, there is an increasing number of cases where similar, perhaps identical, pathogenicity genes are to be found in different genomic species. Further revisions can be expected to lead to the creation of additional new species which have similar pathogenicity characteristics (as in the examples of "*X. axonopodis* pv. *vesicatoria*" and *X. vesicatoria* in Vauterin *et al.*, 1995). The nomenclatural problem created is for species for which identification in practice depends on distinct pathogenic characters. This problem does not apply to pathovar classification since the same pathogenicity characters cannot, by definition, occur in the same species.

At first sight, it may not be obvious why plant pathogenic bacteria were considered to warrant a formal nomenclature outside the Bacteriological Code. The pathovar scheme was introduced because so many plant pathogens are specifically named in plant quarantine legislation. Without formality of nomenclature for plant pathogenic bacteria international quarantine could become chaotic. For human and animal pathogens, the claims for a simple unvarying and unconfusing nomenclature take precedence over the claims for a natural classification, whether phenetic or phylogenetic. Thus, the generic differentiation of *Escherichia* and *Shigella* within the family *Enterobacteriaceae* is not supportable solely on the basis of taxonomic data, but these taxa are maintained because of their conventional use in human and animal diagnostics. This has frozen the nomenclature of the *Enterobacteriaceae* and inhibited rational revision of the family into a smaller number of genera. Similarly, the species names *Yersinia pestis* and *Y. pseudotuberculosis* would be revised were it not for a proper fear of confusion over the names associated with a dangerous human pathogen. Even the innocuous *Bacillus thuringiensis*, a species with strains sometimes carrying toxin-producing plasmids, though perhaps a synonym of *B. cereus*, has been retained by entomologists studying this bacterium as a biocontrol agent (Chapter 7). This nomenclatural protection has never been claimed for names of plant pathogenic species, hence the need for stability at the level of infrasubspecies.

13   Change to the Standards for Pathovars?

As taxonomic revisions show that bacterial taxa are not readily circumscribed, and hence that strains are difficult to allocate to them, so it may be necessary to employ alternative measures for naming bacterial groups of interest. Special purpose nomenclature may offer an interim step when new organisms are isolated. For instance, plant pathologists with concerns connected with quarantine regulations and control may find it efficacious to establish the pathogenic characteristics of a new pathovar and to give it a name, before the organism has been characterized and allocated to any species. Thus a new pathogen to (say) aubergine (*Solanum melongena* var. *esculentum*) could be named *Xanthomonas* sp. pv. *lelliottiae* (see Young *et al.* (1992) for a discussion of the use of non-descriptive names) following a demonstration of its pathogenic character. This approach would have the merit of permitting publication of names for pathogens with descriptions and pathotype strains for quarantine and diagnostic purposes by laboratories lacking the highly specialized techniques now demanded for modern systematics. As the pathogen became better understood through more intensive study, so its systematic position could be more precisely established. This proposal would require a revision of the Standards for Naming Pathovars to permit application of formal names of pathovars to genera without a species epithet. Because pathovars are sometimes designated without reference to the Standards, there would need to be formal provision for the regular audit of pathovar names for which formal recognition was given. Alternatively, the Bacteriological Code of Nomenclature can be interpreted, as it was in the past, to allow pathogens to be allocated to any validly published species name, such as *X. campestris* in the sense of Bradbury (1984).

14   Horizontal Gene transfer

The transfer of genes between unrelated bacteria has been recognized since 1928. Antibiotic and heavy metal resistance, and the transfer of nitrogen-fixing genes have long been recognized as commonly occurring events (Veal *et al.*, 1992). The mechanisms by which exchange occurred in nature, i.e.; the direct exchange of genes between bacteria, or exchange through the mediation of infectious viruses, have been the subject of intensive study. Indeed, the application of these mechanisms in the laboratory have themselves formed the foundation of modern molecular biology. In general, gene transfer has been considered to be confined to the exchange of small adaptive functional genes (Goodfellow *et al.*, 1997). For some, the denial of gene transfer as significant in bacterial evolution could be seen to take the form of a dogma (Woese, 1987). However, increasing numbers of reports suggest that genetic interchange may be more common than has been appreciated.

Recently, detailed evidence of gene homologies in unrelated bacteria has raised questions as to the relative significance in bacterial evolution of mutation lines as compared with horizontal gene transfer. Cohan (1994, 1996) and Jaenecke *et al.* (1996) suggest that genetic interchange may be more common than has been appreciated. For example, genes associated with symbiosis in nitrogen-fixing bacteria (Young & Huakka, 1996), and homologous *hrp* genes implicated in the pathogenicity and resistance mechanisms regulating the interactions between bacterial pathogens and their plant and animal hosts (Gough *et al.*, 1992), are found in separate sub-classes of the *Proteobacteria*. These observations are more readily explained by horizontal gene transfer than as having a common ancestral origin. Sequence discontinuities in *Escherichia coli* also suggest gene transfer (Milkman & Bridges, 1993), and recently Haubold & Rainey (1996), studying *Pseudomonas*, concluded that there was frequent large-scale genetic transfer between strains. The presence of related genes in unrelated bacteria is now widely documented. The presence of foreign sequences in bacterial species, discovered by eliminating shared DNA (Lan & Reeves 1996; Sagerson *et al.* 1997), is strong support for exchange. An integron gene system with the specific function of facilitating gene exchange has been reported as widespread in bacteria (Hall & Collis 1995; Mazel *et al.*, 1998), together with evidence of the presence of foreign functional genes. It is not impossible that the incongruence noted earlier between multiple-character methods with *Pseudomonas* is in part a product of gene transfer between species. Perhaps most striking, the transfer of parts of 16S rRNA sequences has been demonstrated between genomic species in *Aeromonas* (Sneath, 1993) and in *Rhizobium* (Eardly *et al.*, 1996). A recent report that the highly conserved state of 16S rRNA makes it a prime candidate as a vector for gene transfer (Strätz *et al.*, 1996) is especially thought-provoking. The exchange of house-keeping genes, or of extended chromosomal sequences containing large numbers of genes that are central to bacterial metabolic activity have not been well-documented though there are tantalizing indications of this possibility (Syvanen, 1994). Genetic exchange representing up to 10% of the bacterial chromosome and involving most of the genes associated with symbiosis in nitrogen-fixing bacteria have been demonstrated in nature (Sullivan & Ronson, 1998).

There is nothing in evolutionary theory to contradict the exchange of conserved house-keeping genes. In fact, theory and observation supports the promiscuous exchange of any and all genes (though not between any and all bacterial taxa, and not inevitably leading to expression), providing a potential for adaptability to novel environments by the recipient bacteria. If horizontal gene transfer is a significant factor in bacterial genetics, then bacterial genera and species might better be viewed as groups of organisms sharing common core chromosome structures while individually receiving genes for all possible metabolic processes from unrelated bacteria. If so, then the current view of bacterial evolution and diversity, and the assumptions supporting bacterial classifications may well need re-evaluation (Syvanen, 1994). As yet there is insufficient evidence from which to determine the significance for bacterial systematics of horizontal gene transfer, although it is regarded by molecular ecologists as a natural explanation of commonly observed, systematics-related phenomena involving gene variation. Plant pathogenic bacteria are usually metabolically active only in associations with their host in conditions that could almost be considered as unmixed cultures. Therefore their populations may be relatively less prone to gene exchange with other bacteria than are those bacteria present as mixed populations in natural environments. Gene transfer may therefore have a smaller, possibly a significantly smaller, impact on plant pathogen classification. If confirmed as a widespread occurrence, gene transfer would have considerable importance in the interpretation of all relationships among bacteria and particular relevance to those bacterial taxa of significance in medicine, industry, and the environmental and plant sciences, including plant pathogens.

15   What Is To Be Done?

Perhaps it is unavoidable that a formal nomenclatural hierarchy tends to impose a mis-leading pattern on our perception of bacterial relationships. We expect that individual strains can be grouped within species which are grouped within genera, which are grouped within families, and so on. Bacterial classification is assumed to have a similar basis to that observed in higher organisms: that phenotypic and phylogenetic relationships are approximately congruent and that phenotype stability is based primarily on internal genetic mechanisms. Unquestionably, the greatest contribution of recent studies in systematics has been to make clear the extent of the complexity of bacterial relationships. Recent molecular evidence suggests that bacterial phenotypes may not have a comparable level of structural stability and that mutation and perhaps horizontal gene transfer occur at rates such that the bacteria are not well conceived in terms of these assumptions. While it is likely that many bacterial taxa, as families, genera and species, will be susceptible to description in unambiguous terms, other taxonomic groups may reflect characteristics of a heterogeneous nature, which defy simple definition. The extent to which these alternatives reflect reality will shape concepts in systematics in future.

Schleifer and Stackebrandt (1983) noted that "modern systematists should not be content to work with two types of classification: an artificial one for practical purposes - and a natural one with no practical application." However, other users of systematics, dependent on it for practical purposes (they should not be overlooked), may be driven to special purpose nomenclature if systematists do not deliver satisfactory usable outputs. Nomenclature may have to be viewed more flexibly, and the application of special-purpose classifications such as pathovars for plant pathogens may need to be given more support. Systematists may need to give more consideration to the practical impact of nomenclatural revisions based on partial information. For practitioners who depend on nomenclature generated by specialists in systematics, a proper caution, even scepticism in accepting at face value, new proposals in nomenclature is not out of place.

As it becomes possible to characterize individual bacterial strains with greater and greater precision, so it becomes more difficult - and perhaps less appropriate - to allocate them to particular taxa. It would be helpful if Minimal Standards, recommended in the Bacteriological Code of Nomenclature, were developed for all well established genera to ensure that only well described taxa were reported and validated in the literature. Minimal Standards could guard against inadequate descriptions, but if excessively prescriptive they might exclude individual strains from a taxon because of a small number of aberrant characters. An alternative to species descriptions for taxa established using genomic data, but for which comprehensive supporting phenotypic descriptions are lacking, is the application of the term "genomovar" (Ursing *et al.*, 1995). However, these are only partial solutions which do not solve the fundamental problem. It may be that both species and genus concepts need to be more clearly, if flexibly, defined. These concepts have always been difficult to give satisfactory operational definitions. As the extent of population variation becomes clear, it may be that, rather than circumscribe species as smaller groups, it may serve our conceptual purposes to establish genera and species in wider terms.

Acknowledgement

Dr S.R. Pennycook for his thoughtful reading of the manuscript. This study was supported by the New Zealand Foundation for Research Science and Technology under contract no. CO9309.

16 References

Baldani, J.I., Pot, B., Kirchhof, G., Falsen, E., Baldani, V.L.D., Olivares, F.L., Hoste, B., Kersters, K., Hartmann, A., Gillis, M. & Döbereiner, J. (1996). Emended description of *Herbaspirillum*; inclusion of [*Pseudomonas*] *rubrisubalbicans*, a mild plant pathogen, as *Herbaspirillum rubrisubalbicans* comb. nov.; and classification of a group of clinical isolates (EA. Group 1) as *Herbaspirillum* species 3. *International Journal of Systematic Bacteriology* 46, 802-810.

Bradbury, J.F. (1984). Genus II. *Xanthomonas* Dowson 1939. In *Bergey's Manual of Systematic Bacteriology*,

Vol. 1, pp.199-210. Edited by N.R. Krieg & J.G. Holt. Baltimore, USA: The Williams & Wilkins Co.

Brenner, D.J. (1984). *Enterobacteriaceae* Rahn 1937. In *Bergey's Manual of Systematic Bacteriology*, Vol. 1, pp. 408-420. Edited by N.R. Krieg & J.G. Holt. Baltimore: The Williams & Wilkins Co.

Brosch, R., Lefèvre, M., Grimont, F. & Grimont, P.A.D. (1996). Taxonomic diversity of pseudomonads revealed by computer interpretation of ribotyping data. *Systematic and Applied Microbiology* **19**, 541-555.

Buchanan, R.E. & Gibbons, N.E. (1974). Introduction - on using the manual. In *Bergey's Manual of Determinative Bacteriology*, 8th edition, pp. 1-3. Edited by R.E. Buchanan & N.E. Gibbons. Baltimore: The Williams & Wilkins Co.

Cohan, F.M. (1994). Genetic exchange and evolutionary divergence in prokaryotes. *Trends in Ecology and Evolution* **9**, 175-180.

Cohan, F.M. (1996). The role of genetic exchange in bacterial evolution. *American Society for Microbiology Newsletter* **62**, 631-636.

Colwell, R.R. (1970). Polyphasic taxonomy and the genus *Vibrio*: Numerical taxonomy of *Vibrio cholerae*, *Vibrio parahaemolyticus*, and related *Vibrio* species. *Journal of Bacteriology* **104**, 410-433.

Cowan, S.T. (1978). A *Dictionary of Microbial Taxonomy*. Edited by L.R. Hill. London: Cambridge University Press.

De Vos, P. & De Ley, J. (1983). Intra- and intergeneric similarities of *Pseudomonas* and *Xanthomonas* ribosomal ribonucleic acid cistrons. *International Journal of Systematic Bacteriology* **33**, 487-509.

De Vos, P., Goor, M. Gillis, M. & De Ley, J. (1985). Ribosomal ribonucleic acid cistron similarities of phytopathogenic *Pseudomonas* species. *International Journal of Systematic Bacteriology* **35**, 169-184.

Dye, D.W. & Kemp, W.J. (1977). A taxonomic study of the plant pathogenic *Corynebacterium* species. *New Zealand Journal of Agricultural Research* **20**, 563-582.

Dye, D.W. & Lelliott, R.A. (1974). Genus *Xanthomonas* Dowson. In *Bergey's Manual of Determinative Bacteriology*, 8th edition, pp. 243-24. Edited by R.E. Buchanan & N.E. Gibbons. Baltimore: The Williams & Wilkins Co.

Dye, D.W., Bradbury, J.F., Goto, M., Hayward, A.C., Lelliott R.A. & Schroth, M.N. (1980). International standards for naming pathovars of phytopathogenic bacteria and a list of pathovar names and pathotype strains. *Review of Plant Pathology* **59**, 153-168.

Eardly, B.D., Wang, F.-S. Berkum & P. van. (1996). Corresponding 16S rRNA gene segments in *Rhizobiaceae* and *Aeromonas* yield discordant phylogenies. *Plant and Soil* **186**, 69-74.

Firrao, G. & Locci, R. (1989). Identification by probabilistic methods of plant pathogenic bacteria. *Annali di Microbiologia e Enzimologia* **39**, 81-91.

Gardan, L., Bollet, C., Abu Ghorrah, M., Grimont, F. & Grimont, P.A.D. (1992). DNA relatedness among the pathovar strains of *Pseudomonas syringae* subsp. *savastanoi* Janse (1982) and proposal of *Pseudomonas savastanoi* sp. nov. *International Journal of Systematic Bacteriology* **42**, 606-612.

Gardan, L., Shafik, H.L. & Grimont, P.A.D. (1997). DNA relatedness among pathovars of *P. syringae* and related bacteria. In Pseudomonas syringae *Pathovars and Related Pathogens*. Proceedings of the 5th International Working Group on *Pseudomonas syringae* pathovars and related pathogens, September 3-8, 1995, Berlin, pp. 445-448. Edited by K. Rudolph, T.J. Burr, J.W. Mansfield, D. Stead, A. Vivian & J. Von Kietzell. Dordrecht, Kluwer Academic Publications.

Gillis, M., Van Van, T., Bardi, R., Goor, M., Hebbar, P., Willems, A., Segers, P., Kersters, K., Heulin, T. & Fernanadez, M.P. (1995). Polyphasic taxonomy in the genus *Burkholderia* leading to an emended description of the genus and proposition of *Burkholderia vietnamiensis* sp. nov. for N$_2$-fixing isolates from rice in Vietnam. *International Journal of Systematic Bacteriology* **45**, 274-289.

Goodfellow, M. & O'Donnell, A.G. (1993). The roots of bacterial systematics. In *Handbook of New Bacterial Systematics*, pp. 3-54. Edited by M. Goodfellow & A.G. O'Donnell. London: Academic Press.

Goodfellow, M. Manfio, G.P. & Chun, J. (1997). Towards a practical species concept for cultivable bacteria. In *Species: The Units of Biodiversity*, pp. 25-59. Edited by M.F. Claridge, H.A. Dawah & M.R. Wilson. London: Chapman Hall.

Gough, C.L., Genin, S., Zischek, C. & Boucher, C.A. (1992). hrp genes of *Pseudomonas solanacearum* are homologous to pathogenicity determinants of animal pathogenic bacteria and are conserved among plant pathogenic bacteria. *Molecular Plant-Microbe Interactions* 5, 384-389.

Goyer, C., Faucher, E. & Beaulieu, C. (1996). *Streptomyces caviscabies* sp. nov., from deep pitted lesions in potatoes in Québec, Canada. *International Journal of Systematic Bacteriology* 46, 635-639.

Grimont, P.A.D., Vancanneyt, M., Lefèvre, M., Vandenmeulebroecke, K., Vauterin, L., Brosch, R., Kersters, K. & Grimont, F. (1996). Ability of Biolog and Biotype 100 systems to reveal the taxonomic diversity of pseudomonads. *Systematic and Applied Microbiology* 19, 510-527.

Hall, R.M. & Collis, C.M. (1995). Mobile gene cassettes and integrons: capture and spread of genes by site-specific recombination. *Molecular Microbiology* 15, 593-600.

Harris-Baldwin, A. & Gudmestad, N.C. (1996). Identification of phytopathogenic coryneform bacteria using Biolog automated microbial identification system. *Plant Disease* 80, 874-878.

Hauben, L., Vauterin, L., Swings, J. & Moore, E.R.B. (1997). Comparison of 16S ribosomal DNA sequences of all *Xanthomonas* spp. *International Journal of Systematic Bacteriology* 47, 328-335.

Hauben, L., Moore, E.R.B., Vanterin, L. Steenackers, M., Mergaert, J., Verdonck, L. & Swings, J. (1998). Phylogenetic position of phytopathogens within the Enterobacteriaceae. *Systematic and Applied Microbiology* 21, 384-397.

Haubold, B. & Rainey, P.B. (1996). Genetic and ecotypic structure of a fluorescent *Pseudomonas* population. *Molecular Ecology* 5, 747-761.

Hayward, A. C. (1972). The impact of changes in nomenclature on quarantine measures. In *Plant Pathogenic Bacteria 1971*. Proceedings of the 3rd International Conference on Plant Pathogenic Bacteria, Wageningen, 14-21 April, 1971, pp. 293-294. Edited by HP Maas Geesteranus. Wageningen: Pudoc.

Hu, F. P., Young, J.M. & Triggs, C.M. (1991). Numerical analysis and determinative tests for non-fluorescent plant-pathogenic *Pseudomonas* spp. and genomic analysis and reclassification of species related to *Pseudomonas avenae* Manns 1909. *International Journal of Systematic Bacteriology* 41, 516-525.

Hu, F.-P., Young, J.M., Stead, D.E. & Goto, M. (1997). Transfer of *Pseudomonas cissicola* (Takimoto 1939) Burkholder 1948 to the genus *Xanthomonas*. *International Journal of Systematic Bacteriology* 47, 228-230.

Jaenecke, S., de Lorenzo, V., Timmis, K.N. & D'az, E. (1996). A stringently controlled expression system for analysing lateral gene transfer between bacteria. *Molecular Microbiology* 21, 293-300.

Janse, J.D., Rossi, P., Angelucci, L., Scortichini, M., Derks, J.H.J., Akkermans, A.D.L., De Vrijer, R. & Psallidas, P.G. (1996). Reclassification of *Pseudomonas syringae* pv. *avellanae* as *Pseudomonas avellanae* (spec. nov.), the bacterium causing canker of hazelnut (*Corylus avellana* L.). *Systematic and Applied Microbiology* 19, 589-595.

Jordan, D.C. (1984). Family III. *Rhizobiaceae* Conn 1938. In *Bergey's Manual of Systematic Bacteriology*, Vol. 1, pp. 234-235. Edited by N.R. Krieg & J.G. Holt. Baltimore: The Williams & Wilkins Co.

Kersters, K., Ludwig, W., Vancanneyt, M., De Vos, P., Gillis, M. & Schleifer, K.H. (1996). Recent changes in the classification of pseudomonads: an overview. *Systematic and Applied Microbiology* 19, 465-477.

Kwon, S.-W., Go, S.-J., Kang, H.-W., Ryu, J.-C. & Jo, J.-K. (1997). Phylogenetic analysis of *Erwinia* species based on 16S rRNA gene sequences. *International Journal of Systematic Bacteriology* 47, 1061-1067.

Lan, R.T. & Reeves, P.R. (1996). Gene transfer is a major factor in bacterial evolution. *Molecular Biology and Evolution* 13, 47-55.

Lapage, S.P., Sneath, P.H.A., Lessel, E.F., Skerman, V.B.D., Seeliger, H.P.R. & Clark, W.A. (Sneath P.H.A.,

editor for this edition) (1990). *International Code of Nomenclature of Bacteria.* Washington: American Society of Microbiology. 188 pp.

Lee, I.-M., Bartoszyk, I.M., Gundersen, D.E., Mogen, B. & Davis, R.E. (1997a). Nested PCR for ultrasensitive detection of the potato ring rot bacterium, *Clavibacter michiganensis* subsp. *sepedonicus. Applied and Environmental Microbiology* **63**, 2625-2630.

Lee, I.-M., Bartoszyk, I.M., Gundersen-Rindal, D.E. & Davis, R.E. (1997b). Phylogeny and classification of bacteria in the genera *Clavibacter* and *Rathayibacter* on the basis of 16S rRNA gene sequence analysis. *Applied and Environmental Microbiology* **63**, 2631-2636.

Lelliott, R.A. (1972). The genus *Xanthomonas.* In *Plant Pathogenic Bacteria 1971.* Proceedings of the 3rd International Conference on Plant Pathogenic Bacteria, Wageningen, 14-21 April, 1971, pp. 269-272. Edited by HP Maas Geesteranus. Wageningen: Pudoc.

Lelliott, R.A., Billing, E. & Hayward, A.C. (1966). A determinative scheme for the fluorescent plant pathogenic pseudomonads. *Journal of Applied Bacteriology* **29**, 470-989.

Louws, F., Fulbright, D.W., Taylor Stephens, C. & De Bruin, F. (1994). Specific genomic fingerprints of phytopathogenic *Xanthomonas* and *Pseudomonas* pathovars and strains generated with repetitive sequences and PCR. *Applied and Environmental Microbiology* **60**, 2286-2295.

Magee, J. (1993). Whole-organism fingerprinting. In: *Handbook of New Bacterial Systematics*, pp. 383-427. Edited by M. Goodfellow and A.G. O'Donnell. London: Academic Press.

Mazel, D., Dychinco, B., Webb, V.A. & Davies, J. (1998). A distinctive class of integron in the *Vibrio cholerae* genome. *Science* **280**, 605-608.

Milkman, R. & Bridges, M.M. (1993). Molecular evolution of the *Escherichia coli* chromosome. IV. Sequence comparisons. *Genetics* **133**, 455-468.

Moore, E.R.B., Mau, M., Arnscheidt, A., Böttger, E.C., Hutson, R.A., Collins, M.D., Van Der Peer, Y., De Wachter, R. & Timmis, K.N. (1996). The determination and comparison of the 16S rRNA gene sequences of species of the genus *Pseudomonas* (*sensu stricto*) and estimation of the natural intrageneric relationships. *Systematic and Applied Microbiology* **19**, 478-492.

Murata, N. & Starr, M.P. (1973). A concept of the genus *Xanthomonas* and its species in the light of segmental homology of deoxyribonucleic acids. *Phytopathologische Zeitschrift* **77**, 285-323.

Palleroni, N.J. (1984). Genus I. *Pseudomonas* Migula 1894. In *Bergey's Manual of Systematic Bacteriology*, Vol. 1, pp. 141-199. Edited by N.R. Krieg & J.G. Holt. Baltimore: The Williams & Wilkins Co.

Palleroni, N.J., Kunisawa, R., Contopoulou, R. & Doudoroff, M. (1973). Nucleic acid homologies in the genus *Pseudomonas. International Journal of Systematic Bacteriology* **23**, 333-339.

Palys, T., Nakamura, L.K. & Cohan, F.M. (1997). Discovery and classification of ecological diversity in the bacterial world: the role of DNA sequence data. *International Journal of Systematic Bacteriology* **47**, 1145-1156.

Pecknold, P.C. & Grogan, R.G. (1973). Deoxyribonucleic acid homology groups among phytopathogenic *Pseudomonas* species. *International Journal of Systematic Bacteriology* **23**, 111-121.

Priest, F.G & Williams, S.T. (1993). Computer-assisted identification. In *Handbook of New Bacterial Systematics*, pp. 361-381. Edited by M. Goodfellow & A.G. O'Donnell. London: Academic Press.

Sagerson, C.G., Sun, B.I. & Sive, H.L. (1997). Subtractive cloning: past, present, and future. *Annual Review of Biochemistry* **66**, 751-783.

Saillard, C., Barthe, C., Bové, J.M. & Whitcomb, R.F. (1996). Diagnosis of *Spiroplasma* infections in plants and insects. In: *Molecular and Diagnostic Procedures in Mycoplasmology*, vol. 2, *Diagnostic Procedures*, pp. 466. Edited by J.G. Tully and S. Razin. San Diego: Academic Press.

Sands, D.C., Schroth, M.N. & Hildebrand, D.C. (1970). Taxonomy of phytopathogenic pseudomonads. *Journal of Bacteriology* **101**, 9-23.

Sawada, H., Ieki, H., Oyaizu, H. & Matsumoto, S. (1993). Proposal for rejection of *Agrobacterium tumefaciens* and revised descriptions for the genus *Agrobacterium* and for *Agrobacterium radiobacter* and for *Agrobacterium rhizogenes*. *International Journal of Systematic Bacteriology* **43**, 694-702.

Scarlett, C.M., Fletcher, J.T., Roberts, P. & Lelliott, R.A. (1978). Tomato pith necrosis caused by *Pseudomonas corrugata* n. sp. *Annals of Applied Biology* **88**, 105-114.

Schleifer, K.H. & Stackebrandt, E. (1983). Molecular systematics of prokaryotes. *Annual Review of Microbiology* **37**, 143-187.

Seal, S.E., Jackson, L.A., Young, J.P.W. & Daniels, M.J. (1993). Differentiation of *Pseudomonas solanacearum*, *Pseudomonas syzygii*, *Pseudomonas picketti*, and blood disease bacterium by partial 16S rRNA sequencing: construction of oligonucleotide primers for sensitive detection by polymerase chain reaction. *Journal of General Microbiology* **139**, 1587-1594.

Shafik, H.L. (1994). Taxonomie des *Pseudomonas* Phytopathogènes du Groupe de *Pseudomonas syringae*: étude Phénotypique et Génotypique. PhD Thesis, University of Angers, Angers, France.

Skerman, V.B.D., McGowan, V. & Sneath, P.H.A., (eds.) (1980). Approved lists of bacterial names. *International Journal of Systematic Bacteriology* **30**, 225-420.

Sneath, P.H.A. (1989). Analysis and interpretation of sequence data for bacterial systematists: the view of a numerical taxonomist. *Systematic and Applied Microbiology* **12**, 15-31.

Sneath, P.H.A. (1993). Evidence from *Aeromonas* for genetic crossing-over in ribosomal sequences. *International Journal of Systematic Bacteriology* **43**, 626-629.

Sneath, P.H.A. & Sokal, R.R. (1973). *Numerical Taxonomy: The Principles and Practice of Numerical Classification*. San Francisco: Freeman.

Stackebrandt, E. & Goebel, B.M. (1994). Taxonomic note: a place for DNA-DNA-reassociation and 16S rRNA sequence analysis in the present species definition in bacteriology. *International Journal of Systematic Bacteriology* **44**, 846-849.

Stackebrandt, E. & Woese, C.R. (1984). The phylogeny of the prokaryotes. *Microbiological Science* **1**, 117-122.

Stall, R.E. (1995). *Xanthomonas campestris* pv. *vesicatoria*. In *Pathogenesis and Host Specificity in Plant Diseases*, Vol. 1, pp.167-184. Edited by U.S. Singh, R.P. Singh & K. Kohmoto. Kidlington, Oxfordshire: Elsevier Science.

Stolp, H., Starr, M.P. & Baigent, N.L. (1965). Problems in speciation of phytopathogenic pseudomonads and xanthomonads. *Annual Review of Phytopathology* **3**, 231-264.

Strätz, M., Mau, M. & Timmis, K.N. (1996). System to study horizontal gene exchange among microorganisms without cultivation of recipients. *Molecular Microbiology* **22**, 207-215.

Surico, G., Mugnai, L., Pastorelli, R., Giovannetti, L. & Stead, D.E. (1996). *Erwinia alni*, a new species causing bark cankers of alder (*Alnus*) species. *International Journal of Systematic Bacteriology* **46**, 720-726.

Sullivan, J.T. & Ronson, C.W. (1998). Evolution of rhizobia by acquisition of a 500-Kb symbiosis island that integrates into a phe-rDNA gene. *Proceedings of the National Academy of Sciences, USA* **95**, 5145-5149.

Swings, J. & Hayward, A.C. (1990). Taxonomy. In *Methods in Phytopathology*, pp. 125-131. Edited by Z. Klement, K. Rudolph & D.C. Sands. Budapest: Akadémiai Kiadó.

Swofford, D.L., Olsen, G.J., Waddell P.J. & Hillis, D.M. (1996). Phylogenetic inference. In: *Molecular Systematics*, pp. 407-514. Edited by D.M. Hillis, C. Moritz & B.M. Mable. Massachusetts: Sinauer Associates.

Syvanen M. (1994). Horizontal gene transfer: evidence and possible consequences. *Annual Review of Genetics* **28**, 237-261.

Tesar, M., Hoch, C., Moore, E.R.B. & Timmis, K.N. (1996). Westprinting: development of a rapid immunochemical identification for species within the genus *Pseudomonas sensu stricto*. *Systematic and Applied Microbiology* **19**, 577-588.

Urakami, T., Ito-Yoshida, C., Araki, H., Kijima, T., Suzuki, K.-I. & Komagata, K. (1994). Transfer of

Pseudomonas plantarii and *Pseudomonas glumae* to *Burkholderia* as *Burkholderia* spp. and description of *Burkholderia vandii* sp. nov. *International Journal of Systematic Bacteriology* **44**, 235-245.

Ursing, J.B., Rosselló-Mora, R.A., García-Valdés, E. & Lalucat, J. (1995). Taxonomic note: a pragmatic approach to the nomenclature of phenotypically similar genomic groups. *International Journal of Systematic Bacteriology* **45**, 604.

Vancanneyt, M., Witt, S., Abraham, W.-R., Kersters, K. & Fredrickson, H.L. (1996a). Fatty acid content in whole-cell hydrolysates and phospholipid fractions of pseudomonads: a taxonomic evaluation. *Systematic and Applied Microbiology* **19**, 528-540.

Vancanneyt, M., Torck, U., Dewettinck, D., Vaerewijck, M. & Kersters, K. (1996b). Grouping of pseudomonads by SDS-PAGE of whole cell proteins. *Systematic and Applied Microbiology* **19**, 556-568.

Vandamme, P., Pot, B., Gillis, M., De Vos, P., Kersters, K. & Swings, J. (1996). Polyphasic taxonomy, a consensus approach to bacterial systematics. *Microbiological Reviews* **60**, 407-438.

Vauterin, L., Kersters, K. & Swings, J. (1993). Protein electrophoresis and classification. In *Handbook of New Bacterial Systematics*, pp. 251-280. Edited by M. Goodfellow & A.G. O'Donnell. London: Academic Press.

Vauterin, L., Hoste, B., Kersters, K. & Swings, J. (1995). Reclassification of *Xanthomonas*. *International Journal of Systematic Bacteriology* **45**, 472-489.

Vauterin, L., Yang, P. & Swings, J. (1996). Utilization of fatty acid methyl esters for the differentiation of new *Xanthomonas* species. *International Journal of Systematic Bacteriology* **46**, 298-304.

Veal, D.A., Stokes, H.W. & Daggard, G. (1992). Genetic exchange in natural microbial communities. *Advances in Microbial Ecology*, Vol. 12, pp. 383-430. Edited by K.C. Marshall. New York: Plenum Press.

Wayne, L.G., Brenner, D.J., Colwell, R.R., Grimont, P.A.D., Kandler, O., Krichevsky, M.I., Moore, L.H., Moore, W.E.C., Murray, R.G.E., Stackebrandt, E., Starr, M.P. & Trüper, H.G. (1987). Report of the *ad hoc* committee on the reconciliation of approaches to bacterial systematics. *International Journal of Systematic Bacteriology* **37**, 463-464.

Willems, A. & Collins, M.D. (1993). Phylogenetic analysis of rhizobia and agrobacteria based on 16S rRNA gene sequences. *International Journal of Systematic Bacteriology* **43**, 305-313.

Willems, A., Goor, M., Thielemans, S., Gillis, M., Kersters, K. & De Ley, J. (1992). Transfer of several phytopathogenic *Pseudomonas* species to *Acidovorax* as *Acidovorax avenae* subsp. *avenae* subsp. nov., comb. nov., *Acidovorax avenae* subsp. *citrulli*, *Acidovorax avenae* subsp. *cattleyae*, and *Acidovorax konjaci*. *International Journal of Systematic Bacteriology* **42**, 107-119.

Woese, C.R. (1987). Bacterial evolution. *Microbiological Reviews* **51**, 221-271.

Yabuuchi, E., Kosako,Y., Oyaizu, H., Yano, I., Hotta, H., Hashimoto, Y., Ezaki, T. & Arakawa, M. (1992). Proposal of *Burkholderia* gen.nov. and transfer of seven species of the genus *Pseudomonas* homology group II to the new genus, with the type species *Burkholderia cepacia* (Palleroni & Holmes 1981) comb. nov. *Microbiology and Immunology* **36**, 1251-1275.

Yabuuchi, E., Kosako, Y., Yano, I., Hotta, H. & Nishiuchi, Y. (1995). Transfer of two *Burkholderia* and an *Alcaligenes* species to *Ralstonia* gen. nov.: proposal of *Ralstonia pickettii* (Ralston, Palleroni and Doudoroff 1973) comb. nov., *Ralstonia solanacearum* (Smith 1896) comb. nov. and *Ralstonia eutropha* (Davis 1969) comb. nov. *Microbiology and Immunology* **39**, 897-904.

Young, J.M., Dye, D.W., Bradbury, J.F., Panagopoulos C.G. & Robbs, C.F. (1978). A proposed nomenclature and classification for plant pathogenic bacteria. *New Zealand Journal of Agricultural Research* **21**, 153-177.

Young, J.M., Bradbury, J.F. & Vidaver, A.K. (1989). The impact of molecular biological studies on the nomenclature of plant pathogenic bacteria. In *Plant Pathogenic Bacteria*. Proceedings of the 7th

International Conference on Plant Pathogenic Bacteria, Budapest, 11-16 June, 1989, pp. 659-661. Edited by Z. Klement. Budapest: Akadémiai Kiadó.

Young, J.M., Takikawa, Y., Gardan, L. & Stead, D.E. (1992). Changing concepts in the taxonomy of plant pathogenic bacteria. *Annual Review of Phytopathology* **30**, 67-105.

Young, J.M., Saddler, G.S., Takikawa, Y., De Boer, S.H., Vauterin, L., Gardan, L., Gvozdyak, R.I. & Stead, D.E. (1996). Names of plant pathogenic bacteria 1864-1995. *Review of Plant Pathology* **75**, 721-763.

Young, J.P.W. & Huakka, K.E. (1996). Diversity and phylogeny in the rhizobia. *New Phytologist* **133**, 87-94.

Zgurskaya, H.I., Evtushenko, L.I., Akimov, V.N. & Kalakoutskii, L.V. (1993). *Rathayibacter* gen. nov., including the species *Rathayibacter rathayi* comb. nov., *Rathayibacter iranicus* comb. nov., and six strains from annual grasses. *International Journal of Systematic Bacteriology* **43**, 143-149.

7   BACTERIA AND INSECTS

FERGUS G. PRIEST *and* SUSAN J. DEWAR

Department of Biological Sciences,
Heriot-Watt University, Edinburgh EH14 4AS, Scotland, UK.

1.   Introduction

Bacteria and insects have had the opportunity to evolve together in both mutualistic and parasitic relationships for some 250 million years. This has led to numerous interesting and often fascinating examples of insect/microbe interactions. The most complex relationships are those which involve endosymbiotic bacteria and their insect hosts (O'Neill, 1995). Bacteria such as *Buchnera* and *Wolbachia* are strict endosymbionts of numerous insects and have evolved sophisticated dependencies with their hosts.

Endosymbiotic bacteria of the genus *Buchnera* provide nutrients to their aphid hosts thereby allowing the insects to survive on carbohydrate-rich plant sap which is deficient in several essential amino acids. In return, the bacteria, which are unable to grow outside their hosts, are provided with protection and effective distribution. *Wolbachia* endosymbionts, which are widespread amongst insects and other arthropods, have a more bizarre contribution to their host's physiology which involves various aspects of sexual reproduction (Hart, 1995). In particular, *Wolbachia* "infection" of the reproductive tissues results in cytoplasmic incompatibility whereby uninfected females and infected males are unable to generate viable offspring whereas crosses between infected females and males, either infected or not, result in infected offspring. The bacterium is transmitted maternally (as are mitochondria), and this cytoplasmic incompatibility ensures vertical transmission of the bacterium through providing infected females with a reproductive advantage (see section 6.2).

Obligate pathogens of insects are less intimately involved with their hosts than the endosymbionts but nevertheless are dependent on their hosts for growth or completion of their life cycle. *Paenibacillus lentimorbus* and *Paenibacillus popilliae*, for example, can grow outside the body of Japanese beetle larvae (a favourite host) but in these circumstances cannot form the resistant endospores necessary for survival in the environment and reinfection of new beetle larvae. Something vital to the sporulation process is lacking *in vitro* and despite numerous attempts to produce spores *in vitro* the vital element(s) remains elusive.

The most common relationships between microorganism and insects with regard to the diversity of organisms involved are mutualism and opportunistic pathogenicity; that is, pathogenicity which is not essential for the completion of the life cycle of

F. G. Priest and M. Goodfellow (eds.), Applied Microbial Systematics, 165-202
© 2000 *Kluwer Academic Publishers. Printed in the Netherlands.*

the microorganism (Table 1). Mutualism, for example in the context of the normal gut microflora of insects, has been studied with regard to environmental health perspectives (possible transmission of human pathogens by insects (Greenberg *et al.*, 1970)), nutritional contributions of the bacterium to its host, for example in bees (Gilliam *et al.*, 1988a), and enhanced resistance of the insect to pathogens (Gilliam *et al.*, 1988b; Dillon & Charnley, 1995), has received far less attention than the opportunistic bacterial pathogens with their considerable potential for biological control of insect pests and vectors of disease. Several bacteria including *Bacillus thuringiensis*, the most successful of all biological insecticides, are in the category of opportunistic pathogens.

TABLE 1.   Some illustrative examples of bacterium/insect relationships

Bacterium	Taxonomic position	Typical insect host	Relationship with insect/disease
Bacillus sphaericus	Endospore-forming bacterium, rRNA group 3	*Anopheles* and *Culex* mosquito larvae	Opportune pathogen, death of host by toxicity and bacterial growth
Bacillus thuringiensis	Endospore-forming bacterium, rRNA group 1	Various coleopteran, dipteran and lepidopteran larvae	Opportune pathogen, death of host by ingestion and toxicity
Buchnera	γ- subdivision of class *Proteobacteria*	Aphids	Endosymbionts of bacteriocyte cells of the body cavity. Provides nutrition
Paenibacillus larvae	Fastidious, endospore-forming bacterium	Honey bee larvae	Obligate[2] pathogen. Causative agent of American foulbrood
Paenibacillus lentimorbus and *P. popilliae*	Fastidious, endospore-forming bacteria	Scarabaeid beetle larvae	Obligate pathogens. Causative agents of "milky disease"
Photorhabdus and *Xenorhabdus*	γ-subdivision of class *Proteobacteria*	Various insects	Symbionts of nematodes; pathogens of insects
Wolbachia	Rickettsial-like bacteria of the α-subdivision of class *Proteobacteria*	Various arthropods, mosquitoes and other insects	Endosymbionts of reproductive cells causing alterations to reproductive behaviour of host

[1] rRNA group 2 of the genus *Bacillus* has been established as a separate genus, *Paenibacillus* (see text for details).
[2] Obligate refers to the necessity for *in vivo* growth in the insect for completion of endospore formation.

Bacteria of the genera *Photorhabdus* and *Xenorhabdus* provide an interesting hybrid mutualistic/pathogenic relationship with their hosts. These bacteria are symbionts of the guts of certain nematode worms. When injected from the worm into an insect host during feeding of the worm, the bacteria grow in the insect and kill it. The insect cadaver then provides nutrients for the growth and development of the nematode which subsequently carries the bacteria off to the next unsuspecting insect target (Hurlburt, 1994). A wonderfully exotic way to deliver a pathogen into a target insect!

Despite the enormous diversity of the insect world and the fascinating relationships that insects display with microorganisms, there has been relatively little study of the normal microbial flora associated with insects. I suspect that there are at least two reasons for this. First, until recently the taxonomic tools necessary for the detailed study of the microorganisms were lacking. Simple attempts at cataloguing the range of bacteria encountered in insect guts, for example, were severely hampered by the inability to classify and identify accurately most of the organisms involved. Molecular systematics is now addressing this problem by providing classification and identification procedures based on 16S rRNA gene sequences and high definition approaches to strain characterization and typing (Priest & Austin, 1993). These approaches are essential for understanding so many other aspects of microbe/insect interactions, notably for unravelling the evolution of endosymbiotic bacteria, associating virulence with certain lineages of pathogen or analysing the genetic population structures of bacteria and their hosts. Second, there is an enormous gulf between entomologists and microbiologists which makes it difficult to adopt a multidisciplinary approach for studying insect/microbe interactions.

One of the major aims of this chapter is to bridge the entomology/microbiology divide by providing an introduction for the microbiologist to the fascination of microbes that associate with insects. We also hope that it will benefit entomologists wishing to learn more about the microbes that are vital to an understanding of their discipline. These objectives are realised by taking well-studied, illustrative examples of insect/microbe associations rather than supplying an exhaustive inventory of microbes that have been reported to be involved with insect life. In this way we intend to stress the systematic aspects of the organisms and highlight important biological issues connected with the evolution of mutualistic and parasitic relationships.

2   Opportunistic Pathogens

The microbial flora of the healthy insect is largely limited to the gut. The indigenous gut flora consists of a balance of bacteria which enter with food intake with a prevalence of those able to proliferate under the prevailing conditions exemplified by pH, availability of oxygen, carbon and nitrogen. Common bacteria include members of the *Enterobacteriaceae*, Gram-positive cocci including staphylococci and streptococci, and actinobacteria such as *Brevibacterium* (Bucher, 1981). The sick insect, on the other hand, usually succumbs to infection from without through the cuticle, in the case of fungi, or from within through the gut wall with bacteria. In both cases, septicemia (growth of the invading microorganism in the hemocoel of the insect) is the normal outcome followed by death of the insect. Typical examples of bacterial infections which follow this route are those caused by the obligate or primary pathogens *P. lentimorbus* and *P. popilliae* (see section.4.1).

Two interesting exceptions to the septicemic pathogens have proven to be the most important biological agents so far developed for the control of insect pests and vectors of disease. These are both endospore-forming bacteria of the genus *Bacillus*; namely *B. sphaericus* and *B. thuringiensis*. Both produce protein toxins which crystallize within the cell during sporulation in the form of parasporal bodies or inclusions (Fig. 1). When ingested by susceptible insect larvae these toxins dissolve at the high pH of the larval midgut and are lethal (Davidson, 1995; Gill, 1995; Charles *et al.*, 1996; Kumar *et al.*, 1996).

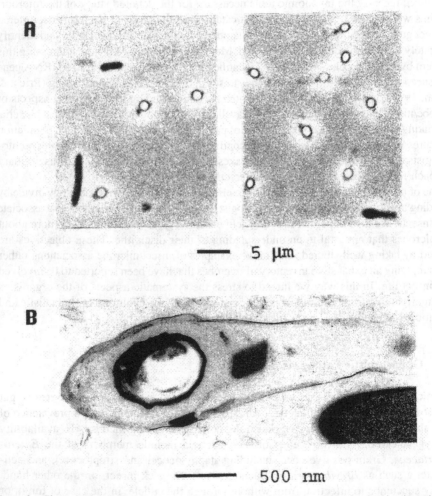

Figure 1.   Phase contrast (A) and electron (B) micrographs of *Bacillus sphaericus* showing insecticidal crystal proteins. Top left, strain 9002 (serotype 1a) and top right strain IAB 872 (serotype 48) with phase bright spores and small crystals in the form of dark spots associated with the spores. Bottom, thin section of strain 2362 (serotype 5a5b) with a spore-associated crystal and indications of the exosporium which surrounds the crystal/spore combination.

But there the similarity ends. *Bacillus sphaericus* and *B. thuringiensis* are phlyogenetically distinct and this realization helps explain other distinctive attributes of the organisms. They differ markedly in their physiology, their toxins bear little resemblance beyond the formation of crystalline inclusions, and the relationships between bacterium and insect are subtly different. Moreover, the systematics of the two bacteria provide their own problems which, in the case of *B. thuringiensis* are complex and far from solved. To place the complexities of the taxonomy of these pathogens in context, it is necessary to consider the classification of the genus *Bacillus* and current ideas on its phylogentic structure.

3   Aerobic Endospore-forming Bacteria

The genus *Bacillus* is a large group of Gram-positive bacteria unified by the instantly-recognizable traits of rod-shaped cells which produce endospores under aerobic growth conditions (Gordon *et al.*, 1973). This simple phenotypic description still provides an invaluable aid to the bench microbiologist requiring a generic assignment for new isolates, but it fails to encompass the extensive phylogenetic depth and diversity of aerobic, endospore-forming bacteria (AEFB).

Over the past 20 years it has become increasingly apparent that the genus *Bacillus* is the equivalent of many families of bacteria and actually encompasses several genera. This conclusion was best demonstrated by 16S rRNA sequence analysis of numerous strains which initially indicated five distinct lineages (Ash *et al.*, 1991a), each of which could be considered a genus. Interestingly, similar conclusions had previously been reached by numerical analysis of phenotypic characters (Priest *et al.*, 1988) and it was reassuring that both approaches gave largely congruent results indicating that the phylogenetic differences were largely reflected in phenetic variation. Since this comprehensive rRNA study, a group of highly acidophilic and thermophilic AEFB have been christened *Alicyclobacillus*, the thermophilic bacilli have been deemed distinct (Rainey *et al.*, 1994) and the alkaliphilic species have been assigned to at least one additional rRNA group (Nielsen *et al.*, 1994) resulting in at least eight generic-ranked phylogenetic lineages.

A phylogeny of AEFB based on representative species and those species which have some involvement with insects is shown in Figure 2. Some of these groups have now been given genus names, *Paenibacillus* for rRNA group 2 which includes such common organisms as *P. macerans* and *P. polymyxa* (Ash *et al.*, 1993), *Brevibacillus* for rRNA group 4 which includes *Brevibacillus laterosporus* (Shida *et al.*, 1996) and *Aneurinibacillus* for *A. anerinolyticus* (Shida *et al.*, 1996).

Insect bacterial relations have presumably evolved several times in AEFB since insect pathogenicity occurs in at least four lineages. *Bacillus thuringiensis* is a member of rRNA group 1 (which includes *B. subtilis*) although it is a peripheral member of this group (Fig. 2) and may in due course be separated from it. *Bacillus sphaericus* differentiates into a spherical spore and these round-spore forming bacteria are all gathered in rRNA group 2 (Ash *et al.*, 1991a). *"Bacillus larvae"* which has recently been reclassified in rRNA group 3 as *Paenibacillus larvae* (Ash *et al.*, 1993), causes American foulbrood in honey bee larvae. Similarly, it is evident from Figure 2 that the obligate insect pathogens *"B. lentimorbus"* and *"B. popilliae"* should be reclassified in the genus *Paenibacillus* (see section 4.1).

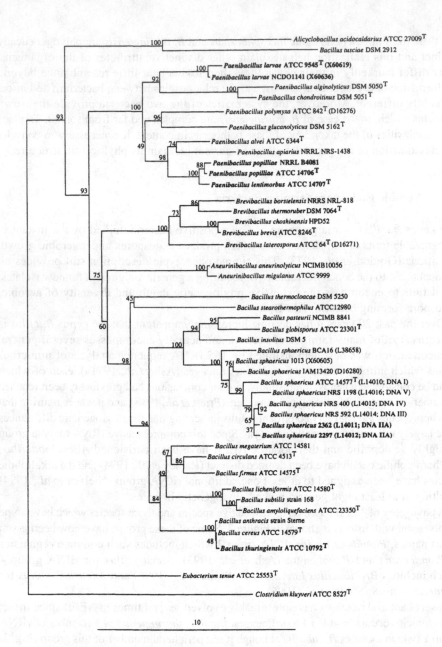

Figure 2. An evolutionary tree of insect pathogenic, endospore-forming bacteria derived from distance data calculated from a non-edited alignment of 16S rRNA gene sequences from insect pathogens (names in bold) and reference strains within the *Bacillaceae* family by using the neighbor-joining method. *Eubacterium tenue* and *Clostridium kluyveri* served as outgroups. The bootstrap values are given as the percentage times out of 500 replicates that a species or a strain to the right of the actual node occurred. The scale bar indicates nucleotide substitutions per site and the distance between two taxa are obtained by adding the horizontal lines connecting the two taxa (vertical lines have no phylogenetic meaning).

Finally, some strains of *Brevibacillus laterosporus* form crystals which are toxic to mosquitoes and blackfly larvae (Favret & Yousten, 1985) and certain *Br. brevis* strains have been noted for their mosquitocidal activities (Singer, 1996). All four forms of insect pathogenicity (*B. sphaericus, B. thuringiensis*, the obligate paenibacilli and *Br. laterosporus*) are believed to be distinct and so it appears that the close association of soil borne AEFB with insects has led to the evolution of insect pathogenicity on several occasions. The taxonomy of these bacteria will be discussed in the context of their interactions with insects in order to show how an appreciation of the taxonomy of the bacteria is essential to understanding their evolution and pathogenicity.

3.1   BACILLUS THURINGIENSIS

Bacillus thuringiensis was first isolated in Japan in 1901 from diseased silkworms (named "*Bacillus sotto*" at that time) and subsequently, and apparently independently in Germany from the diseased larvae of the Mediterranean mealmoth by Berliner [reviewed by (Stahly *et al.*, 1992; Kumar *et al.*, 1996)].

   Bacillus thuringiensis is a member of a group of four closely related *Bacillus* species, the status of which seems to invoke greater controversy the more the bacteria are studied. *Bacillus anthracis*, the causative agent of anthrax, *B. cereus*, a common saprophytic bacterium often responsible for food poisoning in man, *B. mycoides*, a soil bacterium apparently of little consequence, and *B. thuringiensis*, a pathogen of numerous insects are virtually indistinguishable except for their pathogenicity properties, and in the case of *B. mycoides* by an ability to form distinctive, rhizoidal colony morphology. *Bacillus mycoides* seems to be the most distinct and can be separated from the other species by DNA reassociation, fatty acid analysis and by an oligonucleotide probe based on a variable region of the 16S rRNA gene (Nakamura & Jackson, 1995; Wintzingerode *et al.*, 1997). The other three species are more homogeneous. The rRNA genes are virtually identical in all four taxa: indeed the 16S rRNA sequences from *B. anthracis* and one strain of *B. cereus* are identical and differed from a second strain of *B. cereus* by only one nucleotide. Moreover, *B. thuringiensis* shows only four differences from *B. cereus* (Ash *et al.*, 1991b) and DNA reassociation studies have failed to separate *B. cereus* and *B. thuringiensis* convincingly (Nakamura, 1994). Are these four taxa members of the same species or should they be retained as separate species status beyond the obvious value and utility of adopting distinct names for bacteria with distinct pathogenicity? This difficult question is addressed below.

3.2   SOME COMMENTS ON BACTERIAL POPULATIONS

A key concept in appreciating the complexity of the bacterial species is that of clonal populations (Maynard Smith, 1995). Members of some bacterial species seldom exchange DNA. They are poorly if at all transformable in nature and have few genetic elements capable of mobilizing the chromosome. These bacterial species develop as assemblies of clones in which each clone is genetically distinct and represents the direct offspring of a parent. Each member of the clone (isolate) is essentially identical. Members of a clone are widely distributed geographically and in time and can be recognized by identical patterns in fine typing techniques of high resolution such as multilocus enzyme electrophoresis (MLEE) in which cell contents are electrophoresed in a starch or polyacrylamide gel. Enzyme location within the gel is then visualized by histochemical staining using a colorimetric or fluorescent substrate. A battery of some 20 enzymes is compared between strains for differences in migration rates, given that these differences represent mutational changes in the genes responsible for the enzymes (Selander *et al.*, 1986). In this way, MLEE provides high resolution comparison of genetic variation between strains based on the 20 enzymes. Strains with the same or very similar patterns of isozymes are considered to be genetically identical and therefore members of the same clone.

Alternative approaches to population analysis are based on some form of chromosomal DNA analyses such as restriction enzyme digested DNA (Figs. 3 and 6) or random amplified polymorphic DNA analysis (RAPDS) (Fig. 5). Each provides a detailed comparison of strains based on their chromosomal composition and therefore identifies strains of identical or highly similar genetic background; namely members of the same clone (Eisenstein, 1990). Increasingly, comparison of gene sequences is being used to analyse population structures (Selander *et al.*, 1994).

In the environment, a positive mutation which allows a bacterium to out-compete other members of the species will result in the spread of that bacterium. In the absence of genetic exchange, the entire genome associated with that mutation remains intact as the bacterium replaces its brethren. Unsuccessful (non-competitive) clones become extinct and the successful one becomes widely disseminated. This purging of diversity is a process known as "periodic selection" and results in a distinctive clonal population structure. *Bacillus sphaericus*, and probably *B. thuringiensis*, have such structures to their populations. In species in which DNA exchange is frequent, for example naturally transformable bacteria such as *B. subtilis*, it is rare to find two isolates with identical chromosomes because strains are continually subject to incoming DNA. In these populations the chromosome becomes a mosaic of DNA sequences and clonal populations do not have the opportunity to develop (Istock *et al.*, 1992). *Bacillus sphaericus* and *B. subtilis* represent virtual extremes of clonal and non-clonal populations respectively, but many bacterial species lie between these model populations.

3.3   BACILLUS THURINGIENSIS REVISITED

Returning to the *B. thuringiensis* situation, toxicity in *B. anthracis* and *B. thuringiensis* is associated with plasmids. In the case of *B. anthracis*, two large plasmids, pXO1 and pXO2, code for exotoxin and capsule respectively. Natural avirulent strains lacking plasmids and genes for toxin and/or capsule production confuse this situation (Turnbull *et al.*, 1992) and such organisms can superficially be mistaken for *B. cereus* because they are not pathogenic. However, chromosomal typing based on restriction fragment length polymorphisms and PCR-based methods (RAPDS) show that *B. anthracis* strains comprise a homogeneous clonal population of bacteria which is similar to, but distinct from *B. cereus* (Henderson *et al.*, 1994). These bacteria are genetically different. Although the plasmid pXO1 and pXO2 can be transferred into *B. cereus* hosts in the laboratory, only bacteria with the *B. anthracis* genotype are able to accommodate the plasmids naturally. There is presumably some affinity between host and plasmid which confers a competitive advantage on the combination. The current concept is therefore of *B. anthracis* as a separate lineage of bacteria establishing itself as a new species. It is diverging from *B. cereus* given the competitive advantage offered by harboring the two plasmids i.e. death of the host animal and subsequent growth in the carcass. Growth of *B. anthracis* in the soil environment is very limited.

A similar situation may exist with *B. thuringiensis*. These bacteria are recognised by virtue of a parasporal "crystal" protein similar to that shown in Figure 1. This is the toxic element of the organism and defines the bacterium known as *B. thuringiensis*. Without the crystal *B. thuringiensis* is phenotypically indistinguishable from *B. cereus*, indeed it is more like *B. cereus* than is *B. anthracis* since the latter can be distinguished from *B. cereus* by a few phenotypic characters (Henderson *et al.*, 1994). The insect toxic crystal proteins of *B. thuringiensis* may be active against lepidopteran (butterfly and moth) larvae, dipteran (mosquito and blackfly) larvae or coleopteran (beetle) larvae although many crystals are of unknown (or perhaps no) toxicity. The products of these *cry* genes, which are generally located on large plasmids, are arranged into numerous classes according to sequence relationships and target range (Table 2).These classes have recently been modified and the older system (CryI, lepidopteran-; CryII, lepidopteran and dipteran-, CryIII, coleopteran- and CryIV, dipteran-active) has been abandoned in favour of a more objective classification using roman numerals to assign proteins to classes based on amino acid sequence homology [see (Kumar *et al.*, 1996; Bravo, 1997)]. In effect, Cry proteins which share less than 45% homology are assigned to different classes denoted by an arabic numeral and the second rank, denoted by an upper case letter is defined at the 75% homology level. The tertiary rank (indicated by a lower case letter) is defined at the 95% homology level.

The confusion between *B. cereus* and *B. thuringiensis* is actually becoming greater the more we learn of the molecular structures of these organisms. Several genome maps, which have been generated by pulsed field gel electrophoresis (PFGE) of chromosomal digests prepared with rare cutting enzymes, are now available for *B. cereus* and *B. thuringiensis* strains. These maps show considerable overlap with no clear distinctions between strains of the two species (Carlson *et al.*, 1996 a,b). The two organisms cannot be readily and consistently distinguished using phenotypic or molecular characteristics and from the systematists' viewpoint comprise the same species (Gordon *et al.*, 1973; Priest *et al.*, 1988). However, some insight into the evolution of *B. cereus* and *B. thuringiensis* can be gleaned by inquiring at the subspecies level; by comparing populations.

Bacillus thuringiensis strains are traditionally typed, using flagellar (H) antigens, into numerous serovars each of which is given a name (de Barjac & Frachon, 1990). There is some correlation between serovar and toxicity (as Cry protein) but serovar is not a consistently valid indicator of toxicity [reviewed by (Kumar *et al.*, 1996)]. For example, in the case of serotype 14 strains (serovar *israelensis* or Bti), these bacteria are essentially identical and have identical plasmid compositions (Kaji *et al.*, 1994; Priest *et al.*, 1994). These strains are toxic to mosquito larvae and form the basis of mosquito and blackfly control programmes in tropical countries (Walsh, 1986). *Bacillus thuringiensis* serovar *israelensis* strains are a good example of a clone in which plasmid composition, toxin types and host bacterium are consistent from wherever in the world the isolate originates. This is a very successful bacterium in nature.

However, in most other serotypes there is little correlation with toxicity (see Table 2). Strains of several serovars, for example *galleriae*, *kurstaki*, *morrisoni* and *sotto* may produce the same Cry protein such as Cry1Aa and strains of a single serotype, e.g. 7 (*aizawai*) may

TABLE 2. Some crystal proteins of *Bacillus thuringiensis*, associated serovars and target insects

Toxin	Old name	Serovar	Serotype	Target insects
Cry1Aa	CryIA(a)	*galleriae*	5a5b	Lepidoptera:
		kurstaki	3a3b3c	*Bombyx mori, Pieris*
		morrisoni	8a8b	*brassicae, Plutella*
		sotto	4a4b	*xylostella*
Cry1Ab	CryIA(b)	*aizawai*	7	Lepidoptera:
		alesti	3a3c	*Heliothis virescens,*
		kurstaki	3a3b3c	*Manduca sexta,*
		thuringiensis	1	*P. brassicae*
Cry1Ac	CryIA(c)	*kenyae*	4a4c	Lepidoptera:
		kurstaki	3a3b3c	*H. virescens,*
				M. sexta, P. brassicae
				Trichoplusa ni
Cry1Ba	CryIB	*aizawai*	7	Lepidoptera:
		thuringiensis	1	*P. brassicae*
Cry1Bb	ET5	*thuringiensis*	1	Lepidoptera
Cry1Bc	PEG5	*morrisoni*	8a8b	Lepidoptera
Cry1Ca	CryIC	*aizawai*	7	Lepidoptera:
		entomocidus	6	*Spodoptera exigua,*
		kenyae	4a4c	*Spodoptera littoralis*
Cry1Da	CryID	*aizawai*	7	Lepidoptera: *M. sexta*
Cry1Ea	CryIE	*kenyae*	4a4c	Lepidoptera:
		tolworthi	9	*M. sexta, S. exigua*
Cry1Fa	CryIF	*aizawai*	7	Lepidoptera:
		galleriae	5a5b	*Ostrinia nubilalis*
				S. exigua
Cry2A	CryIIA	*kurstaki*	3a3b3c	Lepidoptera: *H. virescens,*
				and Diptera: *Aedes aegypti*
Cry2Ab	CryIIB	*kurstaki* (cryptic)	3a3b3c	Lepidoptera: *M. sexta*
Cry2Ac	CryIIC	*shanghai*		Lepidoptera: *M. sexta, T. ni*

TABLE 2.   continued

Toxin	Old name	Serovar	Serotype	Target insects
Cry3Aa	CryIIIA	*tenebrionis*	8a8b	Coleoptera:*Leptinotarsa decem-lineata, Phaedon cochleariae*
Cry3Ba	CryIIIB	*tolworthi*	9	Coleoptera: *L. decemlineata*
Cry4Aa	CryIVA	*israelensis*	14	Diptera: *A. aegypti, Anopheles stephensi*
Cry4Ba	CryIVB	*israelensis*	14	Diptera: *A. aegypti, An. stephensi*
Cry5Aa	CryVA(a)	*darmstadiensis*	10a10b	Lepidoptera: *O. nubilalis* and Coleoptera: *L. decemlineata*
Cry9Ca	CryIH	*tolworthi*	9	Lepidoptera: *Agrotis* spp., *S. exigua, S. littoralis*
Cry10Aa	CryIVC	*israelensis*	14	Diptera: *A. aegypti*
Cry11Aa	CryIVD	*israelensis*	14	Diptera: *A. aegypti, Culex quinquefasciatus*
Cry11Ba	Jeg80	*jegathesan*	28a28c	Diptera: *Aedes togoi, C. quinquefasciatus*

Data taken from the following references: Aronson (1993), Lereclus *et al.* (1993), Delécluse *et al.* (1995) and Kuo & Chak (1996). For a full list of toxin genes and the new nomenclature see; http://epunix.biols.susx.ac.uk/home/Neil_Crickmore/Bt/index.html

produce numerous Cry proteins such as Cry1Aa, Cry1Ba, Cry1Ca and Cry1Fa. This does not mean that all *aizawai* strains will necessarily produce all these Cry proteins but in some cases such as the HD-1 strain of *kurstaki* several proteins are produced in a single crystal (Aronson, 1993). The mobility of the *cry* genes on plasmids obviously makes it impossible to correlate serotype with toxicity but nevertheless the serotyping scheme has provided a useful basis for *B. thuringiensis* typing.

Unfortunately there are no extensive studies of population structures of *B. cereus* and *B. thuringiensis* but in the few strains which have been examined for clonality some agreement does exist. Ribotyping, a form of chromosomal fingerprinting based on the distribution of restriction enzyme sites in and around the rRNA operons of the bacterium, revealed that strains of *B. thuringiensis* within serotypes generally had consistent ribotype patterns that were distinct from those of other serotypes/ribotypes. In other words, chromosomal structure largely matched serotype (Priest et al., 1994). In other studies, chromosomal PFGE for two strains of serotype *kurstaki* were identical while single representatives of other serotypes could be distinguished; i.e. no members of different serotypes showed the same *Not*I PFGE patterns (Carlson et al., 1994). Moreover, RAPDS analysis distinguished strains of individual serotypes but unfortunately in this study members of the same serotype were not compared (Brousseau et al., 1993). However, correlation between serotype and clonal populations based on some other typing procedure do not invariably match and the correlation is weak when assessed by MLEE. Members of the same serotype were distributed in several clones by this procedure and some MLEE clones contained representatives of several serotypes (Zahner et al., 1989). Nevertheless, despite the few strains included in the MLEE study, evidence of a clonal population structure was still present.

Whatever the typing procedure, one conclusion so far is that *B. cereus* and *B. thuringiensis* clones are intermingled but rarely, if ever are *B. cereus* and *B. thuringiensis* strains included in the same clone (Zahner et al., 1989; Carlson et al., 1994; Priest et al., 1994). Indeed, where MLEE has placed *B. cereus* and *B. thuringiensis* strains in the same clone, differences have been noted in PFGE patterns which show the clone to be heterogeneous and comprising more than one lineage (Carlson et al., 1994) . It is therefore tempting to consider *B. cereus* and *B. thuringensis* as a collection of intermingled clones which is diverging into species. It is possible that plasmid maintenance is a driving force in this process in the same way as it might be for *B. anthracis*. This assumes that plasmid transfer between bacterial cells occurs, but that there is a sufficiently strong requirement for a specific interaction between host and plasmid that only certain combinations of plasmid/host can compete and survive in the environment. It might be possible to force a plasmid into a foreign host in the laboratory, for example from *B. thuringiensis* to *B. cereus*, and this may also happen in the field (Jarrett & Stephenson, 1990), but maintenance of the plasmid in that host in the field requires specific host/plasmid factors.

In this scenario, a *B. thuringiensis* strain receives a plasmid and through recombination with a resident plasmid a recombinant *cry* gene might be produced with a slightly different toxicity range. The new Cry protein recognizes a new receptor borne by a different insect. This allows the bacterium to kill the new insect larva which happens to provide a conducive growth environment. The new plasmid/clone combination is therefore successful within this population of insects and becomes increasingly associated with that host. The new combination becomes fixed in the population and distributed widely. *Bacillus thuringiensis* serovar *israelensis*, the serotype 14 clone which is toxic towards mosquitoes and blackflies, is an excellent example of such an occurrence; strains from all around the world are identical in ribotype, whole cell protein patterns, plasmid profiles and toxin complement (Kaji et al., 1994; Priest et al., 1994). This bacterium is well on the way to becoming a new species and shows limited DNA reassociation with other strains of *B. cereus* and *B. thuringiensis* (Nakamura, 1994). Others serotypes such as *kurstaki* may also be evolving in a similar manner. This process would extend to include *B. cereus*. If a *B. cereus* cell were to receive a Cry-encoding plasmid from a strain of *B. thuringiensis* of a type such that it conferred a significant advantage on that cell, i.e. enabled it to make effective use of the insect toxicity, a stable association could ensue and the *B. cereus* cell would become a clonal member of *B. thuringiensis*.

In this context *B. cereus* and *B. thuringiensis* are different and identifiable populations within the same species. As distinct clonal populations they do not undergo significant chromosomal exchange, although plasmid transfer, and with it *cry* gene transfer, is extant but not rampant since the necessity of a suitable host for the plasmid excludes the successful establishment of most plasmid /host combinations. Thus only certain plasmid host combinations survive to become successful clones.

The species status of these taxa is not significant here beyond the purely pragmatic problem of whether *B. thuringiensis* strains are safe to use as biocontrol agents. *Bacillus cereus* strains may carry several food poisoning toxins; including three distinct enterotoxins of which one is also a hemolysin. Enterotoxin genes have also been found in strains of *B. thuringiensis* (Damgaard, 1995; te Giffel *et al.*, 1997) as might be expected for organisms so closely related to *B. cereus*. Biocontrol strains should therefore be tested for such genes and, if found, deletants produced or alternative strains used for field application.

Much of the above assumes a role for the crystal protein in natural selection; it assumes that the crystal confers a competitive advantage on *B. thuringiensis* populations. As we shall see in section 3.5 this is not so clear-cut and the role of the crystal in the environment is not necessarily obvious. However, before examining the role of the crystal in the environment, we shall consider the systematics and biology of a second crystaliferous bacterium, *B. sphaericus*.

3.4 A CLONAL POPULATION STRUCTURE TO GROUP IIA BACILLUS SPHAERICUS

Some strains of *B. sphaericus* are toxic to mosquito larvae of the genera *Anopheles* and *Culex* which are vectors of two tropical diseases; malaria and filariasis caused by a nematode worm. Mosquitocidal strains of *B. sphaericus* synthesize a crystal parasporal protein (Fig. 1) which comprises equimolar quantities of two proteins, one of about 42 kDa and the other about 51 kDa. When the crystal is ingested by susceptible mosquito larvae it dissolves in the alkaline midgut and insect proteases remove peptides from the amino and carboxy termini of both proteins to release highly active core toxins (Baumann *et al.*, 1991). The *B. sphaericus* toxin is a true binary toxin; both proteins are needed for the toxic effect (Charles *et al.*, 1996). For this reason these have been referred to as Bin (binary) proteins. The larger protein (BinA) binds to receptors in the insect midgut and promotes internalization of the smaller (BinB) protein which exerts its toxic activity in an unknown fashion (Charles, *et al.*, 1996). The result is that the mitochondria of the midgut epithelial cells swell the first hour after feeding and large vacuoles appear. The midgut swells and disintgrates and spores subsequently germinate and grow so that the cadaver becomes a bag of bacterial growth (Davidson, 1995). Other mosquitocidal toxins have been described in these bacteria (reviewed by Charles *et al.*, 1996), but will not be considered here.

Bacillus sphaericus provides a good example of the power of modern systematics for unravelling previously difficult problems. The basic question was; are the mosquitocidal strains different from non-pathogenic strains of *B. sphaericus*? There is now an answer to this question but a simple, complete solution to the identification of the mosquito-pathogenic strains has still to be found. *Bacillus sphaericus*, and most of its relatives in rRNA group 2 of the aerobic, endospore-forming bacteria, do not use carbohydrates as sources of carbon and energy but instead channel acetate and amino acids such as glutamate into the TCA cycle in a highly aerobic lifestyle (Russell *et al.*, 1989; Alexander & Priest, 1990). This results in these bacteria being negative for many of the traditional taxonomic tests such as acid production from sugars, hydrolysis of starch and the Voges Proskauer reaction. Thus all aerobic, round-spored bacilli were classified as *B. sphaericus* because of a lack of diagnostic features (Gordon *et al.*, 1973), which was tantamount to calling all oval-endospore forming bacilli *B. subtilis*! Obviously there could be great diversity in this group hidden by a lack of effective discriminatory phenotypic tests

The framework for *B. sphaericus* classification was provided by Krych *et al.* (1980) who divided strains into six DNA homology groups on the basis of DNA reassociation studies. All pathogenic strains were recovered in group IIA which was closely (about 70% DNA hybridization) related to non-pathogenic group IIB. The low levels of DNA reassociation between all six groups supported the establishment of new species for each group but names were not introduced because the groups could not be distinguished phenotypically. Numerical analysis of phenotypic features specifically chosen with regard to the physiology of the bacteria, recovered the same six groups lending weight to the idea of six species (Alexander & Priest, 1990). These same six groups have also been defined using 16S rRNA sequencing (Aquino de Muro & Priest, 1993), MLEE (Zahner *et al.*, 1989), ribotyping (Aquino de Muro *et al.*, 1992), and PFGE of restriction enzyme-digested chromosomal DNA (Zahner *et al.*, 1998). There is no doubt that they represent six taxa of species rank but, despite several attempts to find distinctive phenotypic features, reliable diagnostic characters have yet to be found.

The physiology of the round-spore forming bacilli is impressively and frustratingly uniform and although growth on almost 100 substrates, resistance to numerous antibiotics and other compounds, and countless other types of phenotypic tests have been studied, no features consistently distinctive for DNA homology groups have been obtained. This lack of phenotypic differentiation has precluded the introduction of individual species names and the confusing taxonomic situation of referring to mosquito-pathogenic types as *B. sphaericus* when they not; they are genotypically distinct from the non-pathogenic *B. sphaericus* *sensu stricto*, is retained. In this chapter the entomopathogens will be referred to as group IIA *B. sphaericus* in deference to their unique DNA homology group and recognition that they constitute a separate taxon.

Bacillus sphaericus strains have been allocated to H-serotypes in the same way as *B. thuringiensis* strains have been serotyped (de Barjac *et al.*, 1985). Group IIA *B. sphaericus* strains have been allocated to 9 serotypes which, because of early confusion over *B. sphaericus* classification do not run consecutively (Priest *et al.*, 1994). Ribotyping of group IIA strains gave a very different picture of the genetic structure of these bacteria from *B. thuringiensis*. Ribotype patterns do not correlate with serotype, instead all group IIA strains seem to have an identical ribotype structure suggesting a more homogeneous taxon than *B. thuringiensis* (Jahnz *et al.*, 1996). However, when strains are examined by PFGE of chromosomal DNA fragments generated by a rare cutting enzymes such as *SmaI* (see Fig. 3), banding patterns are consistent with a clonal population structure with members of the clones having identical chromosomal structure (Zahner *et al.*, 1998). Thus whereas clonality is revealed in some taxa by the relatively insensitive process of ribotyping, a more refined typing procedure is required in other taxa such as *B. sphaericus*.

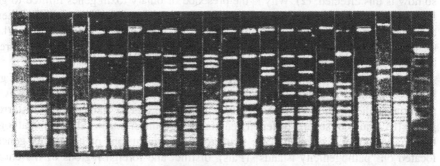

Figure 3. Pulsed field gel electrophoresis of *Sma*I-digested chromosomal DNA from some strains of *Bacillus sphaericus* representing various serotypes and countries of isolation. Lanes: 1, 9002 (India); 2, 4b1 (Nicaragua); 3, BDG2 (France); 4, SL 42 (USA); 5, LP7-A (Singapore); 6, LP14-8 (Singapore); 7, LP12-AS (Singapore); 8, LP1-G (Singapore); 9, 2362 (Nigeria); 10, IAB 460 (Ghana); 11, IAB 59 (Ghana); 12, 11 (Iraq); 13, B55 (India); 14, COK 31 (Turkey); 15, 2297 (Sri Lanka); 16, 2627 (Israel); 17, 2173 (India); 18, 2115 (Philippines); 19, IAB 872 (Ghana); 20, IMR 66.1S (Malaysia); 21, Pr-1 (Scotland), * indicates that the serotype of this strain has been inferred from the PFGE pattern; 22 Molecular weight markers (BioRad lambda ladder with approximately 50-kb increase in sizes). The samples were prepared as agarose plugs, digested with *Sma*I for 18 h at 25°C and electrophoresed in a 1% agarose gel (see Zahner *et al.*, 1998 for details).

Bacillus sphaericus clones have specific pathogenicity properties. For example, strains assigned to the clone associated with serotype 5a5b all show the same chromosomal structure and in the four cases which have been studied, have identical crystal protein structural genes (BinA2; BinB2) (Humphreys & Berry, 1998; Zahner *et al.*, 1998). This bacterium is the most common of all pathogenic strains of group IIA and represents a widely disseminated clone with strains isolated from all parts of the world. Other clones based on PFGE are less common, so certainly there is something about the combination of host strain and those crystal protein genes in serotype 5a5b strains which combine to make for a very successful bacterium. One of the other interesting features to emerge from these studies was that members of different clones may have identical *bin* genes (Zahner *et al.*, 1998). The occurrence of identical *bin* genes in bacteria of separate origin is strong, if not conclusive evidence for lateral transfer of *bin* genes.

Although all mosquito pathogenic strains are invariably members of DNA homology group IIA, this does not mean that all group IIA strains are pathogenic. Most strains were originally isolated and characterized on the basis of toxicity to mosquito larvae and consequently only toxic strains were collected. This resulted in the extensive collection of entomopathogenic strains in the "Collection of *Bacillus thuringiensis* and *Bacillus sphaericus*" housed in the Institut Pasteur, Paris. However, a recent ecological survey of group IIA strains isolated, not according to pathogenicity but on the basis of hybridization to a 16S rRNA gene probe for group IIA, failed to find toxic strains among 40 group IIA isolates from soil samples from Brazil and Scotland (Jahnz *et al.*, 1996). Therefore toxicity is not the norm for group IIA *B. sphaericus*, apparently many more non-toxic strains exist than toxic strains. This prompts at least three interesting questions with a taxonomic slant; (1). Can toxicity be lost and gained through gene transfer and, if so how is this effected, (2). Why is the presence of binary toxin genes limited to group IIA strains, if the genes are transmissible between strains why do we not find them in the closely related groups I, IIB and other DNA homology groups and (3). What is the ecological advantage to having a binary toxin. At present question 1 can be addressed but answers to the other two questions are more elusive.

bin genes are located on defined fragments of the chromosome of toxic strains of group IIA *B. sphaericus* rather than plasmids as is the norm for *B. thuringiensis*. It is evident from the studies mentioned above, that *bin* genes can be lost, gained and transferred to group IIA strains of several different genetic backgrounds. In several human pathogenic bacteria including *Escherichia coli,* virulence determinants and pathogenicity genes are located on "pathogenicity islands" (Pais), distinct and foreign regions of the chromosome (Hacker *et al.*, 1997). In addition to carrying virulence genes, the G+C content of Pais is often different from that of the remaining chromosome thus supporting the notion of an external origin. Pais of Gram-negative bacteria are often bounded by direct repeat sequences or insertion sequence elements to effect transmission. However, in Gram-positive bacteria such as *Clostridium difficile* or *Listeria monocytogenes* Pais are not flanked by direct repeat sequences and they lack mobility genes (Hacker *et al.*, 1997) This makes them more stable than their Gram-negative counterparts. Pais are also found on phages and plasmids, thus enabling their transmission between hosts. It seems quite likely that *bin* genes are located on a Pais, but this has yet to be demonstrated.

Following gene transfer two genetic processes must ensue; stabilization of the new genetic element and optimal expression. Expression of the *bin* toxin genes does not seem to be a problem. The promoter is active in a variety of *Bacillus* backgrounds including *B. subtilis* and *B. sphaericus sensu stricto* (Baumann & Baumann, 1989). However, it seems likely that *bin* genes, whether or not located on a conventional Gram-positive Pais, can only become established in a group IIA host since they have never been found outside this taxon. This may be because only the group IIA bacterium can receive the incoming DNA, or has the appropriate genetic constitution to incorporate the incoming DNA. Whichever, the ability to stabilize and maintain *bin* genes is obviously taxon specific, further underlining the species identity of DNA homology group IIA strains.

3.5   ECOLOGICAL ASPECTS OF BACILLUS THURINGIENSIS AND BACILLUS SPHAERICUS AND THE ROLE OF THE CRYSTAL

The hypotheses presented here for *B. thuringiensis* speciation and for the lateral distribution of toxin genes in *B. sphaericus* require that crystal formation offers some competitive advantage to its host thus providing selection of Cry$^+$ strains. The synthesis of so much protein is a tremendous drain on the bacterium and it must provide a distinct environmental advantage if the gene is to be mantained in a population. In *B. thuringiensis,* where the *cry* genes are generally plasmid borne, the plasmids are remarkably stable and crystal protein synthesis is not lost on laboratory subculture in the same way that pathogenic bacteria sometimes lose virulence attributes such as capsule synthesis. Although in *B. sphaericus,* the *bin* genes are chromosomal, their possession also appears to be a stable attribute.

The role of crystal protein synthesis is more easily explained in group IIA *B. sphaericus* because in some ways it is more intimately involved with its insect host than *B. thuringiensis*. Of the various types of group IIA strains, those of flagellar serotypes 5a5b and 26a26b provide some insight into the role of the crystal. The former strains invariably harbour the *binA2;binB2* gene combination. On the other hand, serotype 26a26b strains which contain *bin* genes have never been isolated. This suggests that the serotype 5a5b strains retain *bin* genes because they confer a competitive advantage which is not so effective for 26a26b strains. The latter cannot take advantage of the crystal protein for some reason, or perhaps it cannot accept the *bin* genes in the first instance.

An intriguing possibility was raised by Correa & Yousten (1995) who showed that spores from strains possessing a crystal protein were able to germinate, grow, and sporulate in larval cadavers, and in this way recycle in the environment. This suggests that the bacterium takes advantage of the dead larva to provide nutrients that promote extensive growth of the bacterium and its ultimate return to the environment in large numbers. Perhaps strains of serotype 26a26b are unable to germinate so effectively under these conditions, even if they were to possess a crystal protein, and thus the energy requirements of maintaining and expressing *bin* genes are not compensated by increased reproduction. This association of bacterium with the dead insect is reinforced by the fact that the endospore and crystal protein are both encapsulated by a membranous exosporium (Nicolas *et al.*, 1994) (Fig. 1), thus the insect consumes a deadly package of toxin and spore and the continued multiplication of the bacterium is encouraged.

The situation in *B. thuringiensis* is not so obvious because in most varieties the spore and crystal are not encapsulated within an exosporium (subsp. *finitimus* is an exception). Nevertheless, Heimpel and Angus recognized three types of *B. thuringiensis*/insect interaction which have stood the test of time (Heimpel & Angus, 1960). Type I insects (e.g. mosquitoes, the silkworm and tobacco hornworm) are affected by the crystal only, this is a straightforward toxicosis and the bacterium is irrelevant to the death of the insect. The most common situation is Type II in which the insect (e.g. cabbage looper or gypsy moth) is killed by the crystal but the spore is relevant to maximal toxicity. There is a spore/crystal synergy. Finally in the rare case of type III insects, such as the greater wax moth, the spores are toxic without crystal although the reasons for this are obscure. So, in most cases the synergy between spore and crystal are evidence of more than a simple toxicosis and suggest that the crystal is providing the spore with access to nutritive hemolymph and the spore is then growing and vegetative cells are contributing to the death of the larva. Indeed, toxins associated with vegetative cell growth are found in both *B. thuringiensis* (Estruch *et al.*, 1996) and group IIA *B. sphaericus* (Thanabalu *et al.*, 1993) which presumably aid pathgenicity during growth in the dying larva. So, when the larva dies it generally contains large numbers of spores and crystals and these will be deposited in the environment as a package ready for ingestion by another larva. This may be particularly important for *B. thuringiensis* which does not grow readily in the soil environment (Petras & Casida, 1985; West *et al.*, 1985) unless additional nutrients are present such as those provided by larval cadavers.

Given a role for the crystal in the natural environment, it might be expected that *B. thuringiensis* and group IIA *B. sphaericus* are unevenly distributed and predominate in habitats associated with the insect host. *Bacillus thuringiensis* serovar *israelensis* should perhaps predominate in countries in which *Aedes* mosquitoes (the most sensitive to its toxin) are abundant or serovar *tenebrionis* in areas where cottonwood leaf beetle or Colorado potato beetle (two highly sensitive beetle larvae) live. This is certainly so for the obligate pathogen *P. popilliae* which is found in soils associated with its host, the Japanese beetle. However, this does not seem to be the case with the opportunistic pathogens.

Ecological studies of *B. thuringiensis* were pioneered by Phyllis Martin in the 1980s who developed a selective isolation procedure for the bacterium based on inhibition of spore germination (Travers *et al.*, 1987). She and her colleagues discovered that 0.5 M sodium acetate inhibited the germination of *B. thuringiensis* spores while permitting the germination of many other *Bacillus* spores. In brief, pasteurized (to kill Gram-negative bacteria etc.) soil samples are suspended in acetate buffer and incubated for several hours to allow endospores to germinate. The suspension is then heat-treated again to kill the germinated vegetative cells but *B. thuringiensis* spores remain due to the acetate inhibition. The suspension is then plated onto agar to allow *B. thuringiensis* spores to germinate and grow. Since the selection is not completely effective and some other *Bacillus* species (in particular *B. cereus*) survive and grow, presumptive *B. thuringiensis* were confirmed by microscopy for the presence of crystals. Strains were further characterized using a "microdot" biochemical testing system in which agar droplets containing the media necessary for different biochemical tests were used as a miniaturized system rather like an API tray (Martin *et al.*, 1985). Such biochemical profiles based on eight physiological tests do not match with the serotyping scheme or enable prediction of toxicity, but allowed isolates from the same locality to be distinguished.

Armed with this simple selection procedure and biochemical characterization, Martin and her colleagues set about isolating *B. thuringiensis* from the environment in a world-wide study. The findings were illuminating; *B. thuringiensis* could be isolated from virtually all soil samples tested (Martin & Travers, 1989) and has since been found commonly on the leaves of plants (Smith & Couche, 1991). There was no correlation between insect habitat and the prevalence of *B. thuringiensis* although the bacterium is particularly common in SE Asia. Less extensive studies of the distribution of group IIA *B. sphaericus* showed that it was no more common in Recife, Brazil, where *Culex* mosquitoes are rife, than in Scotland where they are effectively absent (Jahnz *et al.*, 1996).

It is however important not to be confused by the ubiquitous distribution of *B. thuringiensis* and *B. sphaericus* group IIA spores in the environment. The ecology of endospore-forming bacteria is notoriously difficult to unravel because the recovery of a spore from a habitat does not necessarily mean that the bacterium was actively growing in that situation (Priest & Grigorova, 1990). For example, thermophilic bacilli are ubiquitous and can be recovered in large numbers from the icy depths of the sea bed where they could never grow, alkaliphilic bacilli that are able to grow only at pH levels higher than 9 are prevalent in normal soils as are acidophilic *Alicyclobacillus* species. The list of such physiological anomalies is extensive. Thus, the recovery of toxic *B. sphaericus* and *B. thuringiensis* strains from locations which lack target insects is to be expected.

There are three possible explanations for the widespread occurrence of *B. thuringiensis* and group IIA *B. sphaericus* which are not mutually exclusive.

(1). The spores have simply survived in a location and may not be growing to any extent. They have been disseminated by the wind or water and the soil is acting as a reservoir. This is presumably the case with the physiological anomalies noted above and given the longevity of the spore, is the most likely explanation for the widespread distribution of insect pathogens.

(2). Both *B. thuringiensis* and group IIA *B. sphaericus* are saprophytic bacteria which will grow given conducive surroundings, in particular additional nutrients or moisture. They do not depend on susceptible insect larvae for growth

(3). There may be unknown insect hosts in the soil allowing for the growth of these oppportunistic pathogens. This is unlikely, especially for the very specific toxins of group IIA *B. sphaericus*. These possibilities have been thoughtfully presented by Meadows (1993) and in combination can account for the widespread distribution of these opportunistic pathogens. However, given the ubiquitous distribution of endospore-forming bacteria in general, the first explanation is likely to be a major influence.

4   Obligate Pathogens

Bacillus sphaericus and *B. thuringiensis* are generally considered opportunistic pathogens for two reasons, larval death is largely a toxicosis resulting from ingestion of the crystal and bacterial growth is not dependent on pathogenicity. Both bacteria will grow and sporulate on simple media in a saprophytic fashion. Some other endospore forming bacteria which cause diseases in insects have become much more closely associated with their hosts and may be considered obligate or primary pathogens because the long-term survival of the microbe depends absolutely on its ability to replicate in the host (Falkow, 1997). These organisms do not grow well outside the insect body and sporulate poorly if at all on laboratory media. Thus they cannot complete their life cycle outside the host larva and have evolved a much closer association with their insect hosts than have *B. sphaericus* and *B. thuringiensis*. These bacteria include *P. lentimorbus*, and *P. popilliae*, pathogens of beetles and *Paenibacillus larvae*, a pathogen of honeybees.

4.1   PAENIBACILLUS LENTIMORBUS AND PAENIBACILLUS POPILLIAE

Paenibacillus lentimorbus and *P. popilliae* cause "milky disease" in certain beetles of the Scarabaeidae family, in particular the Japanese beetle, a common pest of turf grass in the USA. The name of the disease derives from the characteristic appearance of infected larvae in which the hemolymph is so loaded with bacteria and spores (equivalent of about 5×10^{10} spores/ml hemolymph) that the normal translucent grub takes on the turbid, milky appearance shown in Figure 4 (Bulla *et al.*, 1978; Klein & Kaya, 1995).

In brief, the beetle larva eats spores of the bacterium while foraging. The spores germinate in the hindgut of the larva and vegetative cells migrate to the midgut where the bacteria penetrate the epithelial cells of the gut probably by phagocytosis. About two days after infection some bacteria enter the hemolymph where they grow vegetatively for a period prior to vegetative growth becoming accompanied by sporulation. After about two weeks, gross septicemia is accompanied by massive sporulation and larval death (Stahly *et al.*, 1992). Although crystal proteins are present in sporulating cells of *P. popilliae* their contribution to milky disease pathology is uncertain since the purified protein is toxic when injected into the grub but not when eaten which is the normal route (Weiner, 1978). The *cry* gene

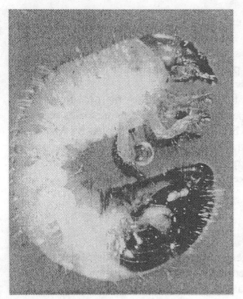

Figure 4.   Healthy (left) and diseased (right) larvae of *Popillia japonica* (Japanese beetle). Note the "milky" appearance of the larva on the right due to the massive growth and sporulation of *Paenibacillus popilliae*. (Photographs courtesy of M. G. Klein).

from a strain of *P. popilliae* has been characterized and codes for a protein of 79 kDa with some 40% sequence identity to the Cry2 polypeptides of *B. thuringiensis* (Zhang *et al.*, 1997). It may be that the crystal protein weakens the gut lining without killing the larva thus enabling the bacterium to invade the hemolymph more easily (Zhang *et al.*, 1997).

This primary disease process defines the habitat of *P. lentimorbus* and *P. popilliae*. These organisms are confined to diseased larvae and the spores are only found in the soil surrounding the larvae. In the biological control context, the obligate/parasitic nature of the infection provides for long-lasting reduction of the Japanese beetle since the spores recycle and re-infect new larvae.

Both *P. lentimorbus* and *P. popilliae* are highly fastidious organisms which do not grow on repeated subculture in nutrient broth and therefore are generally grown on a particularly nutritive medium high in yeast extract called J-medium (Gordon *et al.*, 1973; Stahly *et al.*, 1992). They are facultative anaerobes, although they grow much better aerobically than anaerobically, and lack nitrate reductase. One unusual feature is a lack of catalase. The spores are oval and swell the mother cell. These general features are consistent with an assignment to the genus *Paenibacillus* as is their fatty acid composition which comprises less than 3% unsaturated fattty acids (Kaneda, 1977). The original 16S rRNA sequence analysis recovered these bacteria in rRNA group I very close to *B. subtilis* (Ash *et al.*, 1991a) but this is apparently erroneous, and more recent studies are consistent with the physiological and morphological evidence and support assignment to rRNA group 2 or *Paenibacillus* (Pettersson *et al.*, 1999; see Fig. 2).

There have been suggestions in the past that *P. lentimorbus* and *P. popilliae* may be varieties of a single species but the two taxa are distinct by modern criteria (Rippere *et al.*, 1998). Strains of the two species show about 50-65% DNA reassociation but within species the level is generally greater than 70%. It was thought that crystal proteins were restricted to *P. popilliae* but recent studies have revealed that some strains of *P. lentimorbus* also synthesize a crystal. The distinction of these two species is evident at the molecular level and RAPD clearly separates the two taxa (Rippere *et al.*, 1998). The power of the RAPD approach is shown in Figure 5 which highlights variation within *P. popilliae* strains and shows two subgroups, one based on the type strain ATCC 14706 (lanes 1 and 12-17) and isolates in lanes 2-10 which comprise a second group. This division suggests a clonal population structure to this species with two clones represented here.

The two species were originally associated with slightly different forms of milky disease (Dutky, 1940). *Paenibacillus popilliae* is responsible for type A milky disease and *P. lentimorbus* with type B milky disease. The latter is characterized by the appearance of brown clots which block the circulation of hemolymph in the larva and lead to gangrenous conditions in the affected parts.

A number of isolates from various insects with milky-type diseases have been afforded separate species or varietal status in particular "*B. popilliae* var. *melolonthae*" and

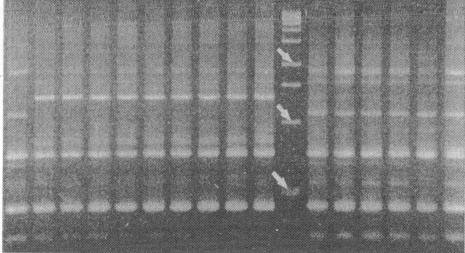

Figure 5. Random amplified polymorphic DNA (RAPD) patterns generated from strains of *Paenibacillus popilliae* using a 10-bp primer. Isolates can be assigned to two major patterns (clones) comprising: Lane 1 ATCC 14706[T] and isolates in lanes 2 to 9, and the isolates shown in lanes 12 to 17. Lane 11 contains molecular weight markers of 2.0 (top arrow), 1.0 and 0.5 kb. (Photograph courtesy of A A Yousten & M T Tran; see Rippere *et al.*, 1998 for further details).

"*B. popilliae* var. *rhopaea*" based on morphological features of the spores and crystals and cultural features (Milner, 1981) The taxonomic positions of these bacteria are largely unknown but DNA reassociation studies and RAPDS suggest that "var. *melolonthae*" may represent a valid subspecies (Rippere *et al.*, 1998). Such molecular techniques will perhaps resolve the status of some of these bacteria when more strains have been isolated and studied. This will be a valuable step forward as it could enable correlation between target pest and variety of bacterium and improvements in biological control of some of these important pests.

A frustrating and impenetrable feature of these bacteria is the inability to obtain sporulation in the laboratory. The spore is the most effective form of the bacterium for biocontrol applications because of the rapid loss of viability of vegetative cells, and consequently the inability to induce sporulation *in vitro* has severley hampered the implementation of *P. popilliae* as a biocontrol agent despite many decades of effort. The reasons for this lack of sporulation may be based on inappropriate nutrition, some physical factor(s) such as oxygen tension or a combination of both (reviewed by Stahly *et al.*, 1992), but until this is solved spores must be prepared *in vivo* and the inordinate costs of the process impede production and implementation.

4.2   PAENIBACILLUS LARVAE SUBSP. LARVAE AND PAENIBACILLUS LARVAE SUBSP. PULVIFACIENS

"*Bacillus larvae*", a pathogen of the honey bee (*Apis mellifera*) is responsible for American foulbrood, a serious infection of honey bee hives. The bacterium has recently been reclassified in the genus *Paenibacillus* (*Bacillus* rRNA group III; Ash *et al.*, 1993) on the basis of 16S rRNA sequence and physiological properties. Like *P. lentimorbus* and *P. popilliae*, this bacterium causes a fatal septicemia of larvae following ingestion of endospores. Infection is restricted to young (below 1.5 days old) larvae in which spores germinate in the midgut and grow. As in milky disease, the vegetative cells traverse the epithelium and enter the hemocoel were they grow to high populations. Death generally occurs at an age of 8 to 11 days and the larvae rapidly decompose as the bacteria sporulate in massive numbers (Stahly *et al.*, 1992). The spores, which can be isolated from comb, honey and wax remain to infect other larvae and the disease is so important that infected hives must be destroyed, generally by burning.

The taxonomy of *P. larvae* is more established than for the milky disease bacteria largely because of its economic importance in honey production. Like these other obligate pathogens, *P. larvae* forms swollen sporangia containing ellipsoidal spores. The bacterium grows poorly, if at all in nutrient broth, but well in complex media such as J agar. Full phenotypic descriptions of *P. larvae* have been published (Jelinski, 1985). The bacterium is catalase negative and sporulates poorly in artificial media although sporulation is rampant in the infected larvae. The placement of this bacterium in *Paenibacillus* is therefore consistent with its physiological and morphological properties and it is comforting that the other obligate pathogens are also members of this phylogentic branch of the endospore-forming bacteria rather than rRNA group 1 as originally supposed.

American foulbrood is one of the most economically significant diseases affecting honeybees and there is therefore a need to be able to type and trace strains in an epidemiological fashion as with any pathogen. DNA restriction endonuclease profiles have proven useful in this respect and revealed a clonal structure to *P. larvae* populations with clones associated with geographically localized regions (Djordjevic *et al.*, 1994). For example, type 1A strains originated from New South Wales and Queensland (Australia) while type 2 strains were from Victoria and South Australia (Fig. 6). In all, 20 isolates were assigned to five types (clones) by this technique with isolates from geographically localized regions showing highest similarity (Djordjevic, *et al.*, 1994). This should lead to opportunities to trace, and hopefully limit, the spread of *P. larvae*.

Paenibacillus pulvifaciens (originally "*Bacillus pulvifaciens*") can be isolated from "powdery scale" within bee hives. This material is light brown in colour with a powdery texture and comprises the dried remnants of dead larvae. It is still not clear if the bacterium is responsible for the condition since reintroduction of the bacteria into larvae does not cause the disease (Gilliam & Dunham, 1977). The bacterium shares numerous phenotypic features with *P. larvae* including lack of catalase, but it grows and sporulates on normal media and therefore cannot be considered an obligate pathogen, particularly given its questionable pathogenicity. A recent polyphasic taxonomic study has shown that *P. pulvifaciens* is closely related to *P. larvae* and has recommended reclassification as a subspecies of *P. larvae*, namely *P. larvae* subsp. *pulvifaciens* (Heyndrickx *et al.*, 1996).

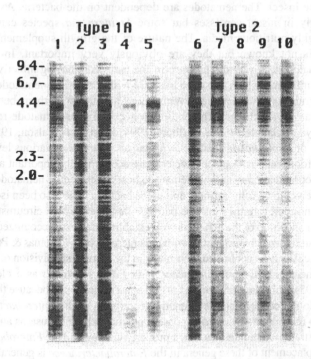

Figure 6. DNA restriction endonuclease profiles of representative strains of *Paenibacillus larvae* subsp. *larvae*. Lanes 1 to 5 represent DNA from strains considered to be members of the same clone (type 1A) as do lanes 6 to 10 (clone type 2). DNA was digested with *Cfo*I, resoved through a 3.5% polyacrylamide gel and stained with silver (Photographs courtesy of M. Hornitzky; for further details see Djordjevic *et al.*, 1994).

5   *Photorhabdus* and *Xenorhabdus*

Photorhabdus and *Xenorhabdus* are Gram-negative bacteria which are the key to a fascinating predatory interaction between entomopathogenic nematodes of the families *Heterorhabditidae* and *Steirnematidae* respectively and their insect prey. The bacteria inhabit the guts of the infective juvenile stages of the nematodes in a mutualistic fashion. The nematodes invade the hemocoel of the insect generally via the intestinal tract and release the bacteria into the hemolymph. Here the bacteria grow producing a fatal septicemia within about 48 hours (Hurlburt, 1994). The nematodes are effectively acting as a delivery device to release the bacteria into the nutritive insect hemolymph. But the interaction goes further because it is now the turn of the nematode. The bacteria produce a variety of antimicrobial compounds during growth in the hemolymph including anthraquinones, hydroxystilbenes, xenocoumacins, xenorhabdins and various indole derivatives (Hurlburt, 1994; Forst & Nealson, 1996). These are active against both Gram-negative and Gram-positive bacteria and protect the development of the nematode in the insect cadaver from proliferation of other, competing bacteria. The bacteria also secrete extracellular enzymes such as lipase(s) and protease(s) which begin the degradation of the insect macromolecules to provide nutrients and growth factors needed by the nematode.

This, then is a true mutualistic relationship between nematode and bacterium at the expense of the insect. The nematodes are dependent on the bacteria. Axenic nematodes grow poorly in insect carcasses but some *Steinernema* species can be grown axenically on highly nutritious media. The nature of the growth supplements provided by the bacteria is not known but they are obviously very important. In the natural situation the nematodes grow and reproduce in the carcass releasing infective juveniles 10-14 days after infection which carry the bacteria with them. The nematodes therefore protect the bacteria from the external environment and transmit them between hosts while the bacteria provide the growth conditions necessary for nematode reproduction (reviewed by Kaya & Gaugler, 1993; Hurlburt, 1994; Forst and Nealson, 1996).

The taxonomy of *Photorhabdus* and *Xenorhabdus* is complex at various levels and yet highly important for several reasons; the bacteria are economically important as biological control agents in combination with their nematode hosts, the specific nematode/bacterium relationships have yet to be fully defined, and these bacteria have also been isolated from human wounds and therefore they may be pathogenic under certain circumstances.

The phylogentic position of the genera is well established. It was recognized at an early stage that they shared many characteristics with the enterobacteria (Thomas & Poinar, 1979) and rRNA sequence analysis confirmed their place in the gamma subdivision of the *Proteobacteria* within the family *Enterobacteriaceae* with *Proteus vulgaris* as a close relative (Rainey *et al.*, 1995). This classification has raised some controversy because the members of these genera do not conform to the phenetic description of the *Enterobacteriaceae* as they are unable to reduce nitrate to nitrite, possess polycrystalline inclusions and, of course in *Photorhabdus* strains are bioluminescent, a property unknown in the *Enterobacteriaceae*. Nevertheless the placement of these genera in the *Enterobacteriacece* is generally accepted and is evidence of how influential bacterial phylogeny can be.

The establishment of two genera for these bacteria is a good example of the difficulties sometimes associated with taxonomic decision-making. Initially, *P. luminescens* was included in the genus *Xenorhabdus* as *X. luminescens* and the non-luminous forms were assigned to several species, notably, *X. beddingii, X. bovienii, X. nematophilus* and *X. poinarii*, each associated with various host nematode species. It was later proposed on the basis of phenotypic characteristics and DNA: DNA reassociation that *X. luminescens* should be given separate generic status and the genus *Photorhabdus* with the sole member *P. luminescens* was established (Boemare *et al.*, 1993).

The particularly low level of DNA reassociation between *P. luminescens* strains and *Xenorhabdus* strains (<20% hybridization) was taken as strong evidence for this new genus. This was, perhaps, misconceived since such low levels of DNA relatedness have little meaning other than indicating that the two strains are not closely related. Such low levels of DNA relatedness are not normally used as evidence for separate generic status (Priest & Austin, 1993). Subsequent 16S rRNA comparisons were unable to substantiate this generic division when only strains of the single species of *Photorhabdus* could be compared with four *Xenorhabdus* species. The bifurcation of the tree around the division between *Xenorhabdus* and *Photorhabdus* was very unstable and dependent on the choice of outgroup organisms (Rainey *et al.*, 1995). From a phylogenetic point of view, separate genera are only indicated when the type strain of each genus is firmly placed outside the radiation of the other genus (Stackebrandt *et al.*, 1997) and this situation was clearly not reached in this instance. When additional representatives of the genus *Photorhabdus* were studied, some of which originated from clinical specimens, the trees became more stable and the division between *Photorhabdus* and *Xenorhabdus* more profound (Szallas *et al.*, 1997). This is a good example of the difficulties sometimes faced when making decisions on the status of taxa when only a few strains and species are available and the extent of the genetic variation within a group of organisms is unrecognized.

Speciation within the two genera is less controversial. Various species of *Xenorhabdus* have been described on DNA reassociation and phenotypic criteria and all are symbionts of *Steinernema* nematodes. These species have been supported by rRNA sequence differences, however, the relationships between bacterial species and nematode species are currently confusing. Each nematode species has a specific natural association with members of a unique *Xenorhabdus* species but a bacterial species may be associated with more than one nematode. For example *X. bovienii* naturally associates with six different *Steinernema* species (Hurlburt, 1994).

The subgeneric structure of *Photorhabdus* is still being determined. The symbiotic strains can be divided into several taxa based on DNA relatedness (Farmer *et al.*, 1989) and 16S rRNA sequence data, although these have yet to be formally named (Szallas *et al.*, 1997). These taxa have been associated with various nematodes of the *Heterorhabiditis* group, but because the identity of the host nematodes is often unsure, no firm correlation between bacterial taxon with host can be made (Brunel *et al.*, 1997). A rapid approach to the identification of both *Photorhabdus* and *Xenorhabdus* taxa relies on the patterns obtained by restriction enzyme analysis of PCR-amplified 16S rRNA genes which gave clear distinction between members of the two genera and allowed identification to the "species" level (Brunel *et al.*, 1997).

Strains of *Photorhabdus* have rather surprisingly been isolated from human clinical specimens (Farmer *et al.*, 1989). No symbiotic association has been demonstrated for these bacteria and it is not clear whether they are free-living opportunistic human pathogens or originated from nematode-bacterium infected material. The symbiotic strains and those from clinical specimens form two distinct DNA hybridization groups which could be assigned separate species status (Akhurst *et al.*, 1996).

Further taxonomic interest in *Photorhabdus* and *Xenorhabdus* lies in the concept of phase variation. Like *Salmonella, Neisseria* and some other pathogens, *Photorhabdus* and *Xenorhabdus* species present two markedly different physiological forms (Boemare & Akhurst, 1988). Phase I (primary form) cells, which are typically those isolated from the nematode host, provide better conditions for reproduction of the nematode in the insect host through synthesis of antibiotics (Sundar & Chang, 1993), extracellular enzymes associated with nutrition of the developing nematode (Wang & Dowds, 1993) as well as metabolic end products which may be nutrients themselves. These physiological variations are accompanied by morphological traits such as distinct and unique colony morphologies, adsorption of dyes and usually the development of large crystalline protein inclusions of unknown function (Bleakey & Nealson, 1988). Phase II colonies are equally pathogenic but are not generally recovered from host nematodes. They lack the properties which enhance nematode reproduction and in the case of *Photorhabdus* are weakly, if at all luminescent. Indeed, the two phases are so different that numerical taxonomic studies are totally skewed by the phenotypic distinction to the extent that phases are recovered in clusters rather than species being distinguished (Akhurst & Boemare, 1988).

Apart from causing disquiet among bacteriologists thinking that they have contaminants in their cultures, what is the likely reason for the development of this dimorphic lifestyle? One plausible explanation is that phase II forms are more adapted to life outside the host nematode. They are physiologically more active, have higher respiratory activity and do not waste energy synthesizing unnecessary antibiotics and metabolites (Smigielski *et al.,* 1994; Rosner *et al.*, 1996). Although this does not answer the baffling question of luminescence, in other aspects it makes good sense.

Photorhabdus and *Xenorhabdus* then, are partly adapted to their host. In one life they are pathogens, and providers of nutrients in a mutualistic relationship, in another they are saprophytes. Other bacteria have abandoned the saprophytic life and become more intimately associated with their insect hosts to the extent that they are dependent parasites.

6   Bacterial Endosymbionts of Insects

The close association of free-living bacteria with insects and arthropods has led to more intimate pairings resulting in endosymbiotic relationships in which the bacteria become dependent on the insect host cell and *vice versa*. One of the principal areas in which the two organisms collaborate is in nutrition. For example the role of saprophytic, intestinal bacteria in honey bee nutrition has been studied extensively because of the commercial implications (Gilliam, 1997). *Enterobacteriaceae, Bacillus* species and Gram-variable pleomorphic bacteria of unknown taxonomy have been implicated in the synthesis of numerous enzymes to aid digestion in the bee gut (Gilliam *et al.*, 1988). Given that symbiotic parasitic relationships have developed quite frequently during evolution, it was inevitable that symbiosis between such distinct organisms as the bacterium and insect resulting in a survival advantage to both should occur. Indeed, such mutualism is particularly common in insects which feed on restricted or single food sources and thus may suffer nutritional deficiencies (Buchner, 1965). Endosymbiosis is a wide ranging topic which cannot be covered in detail here. However, two excellent books on the subject are available (Buchner, 1965; Schwemmler & Gassner, 1989) as well as several reviews which will be mentioned below.

The existence of endosymbiotic bacteria in insect tissues has been recognized since the 19th century but the inability to culture the bacteria outside of the host hampered taxonomic and physiological studies. The introduction of the PCR revolutionized taxonomy in several ways but one important contribution was the ability to amplify specific genes from endosymbiotic bacteria. This provided the opportunity to construct phylogenetic trees of these bacteria from gene sequences in the absence of axenic culture and provided an approach to the classification and identification of bacterial endosymbionts of insects. Two groups of endosymbiotic bacteria will be considered here namely, those involved primarily in nutrition of the insect (genus *Buchnera* and relatives) and some bizarre bacterial endosymbionts which influence the sexual reproduction of various insects (genus *Wolbachia*).

6.1   BUCHNERA

Endosymbiotic bacteria of the genus *Buchnera* are associated with virtually all aphid genera. Aphids feed on plant sap which is a carbohydrate rich diet low in amino acids and protein and one of the contributions the bacterium makes to the insect nutrition is to supply at least some of the essential amino acids to its host. This has been particularly well documented for tryptophan; feeding of aphids with chlortetracycline (to remove the endosymbionts) in a tryptophan-deficient diet quickly results in aphid death whereas fatality could be reversed by the addition of tryptophan to the antibiotic-containing diet (Douglas & Prosser, 1992). The evidence for the supply of other essential amino acids, although not so well established, seems likely.

Buchnera live in membranous vacuoles (symbiosomes) within specialized cells called bacteriocytes. The bacteriocytes in turn aggregate within the body cavity of the aphid as a bacteriome. *Buchnera* are transmitted to eggs and embryos in a vertical fashion during parthenogenetic (development of an ovum without fertilization) reproduction. This is contrasted by the situation in ants where bacteriocytes containing free, intracellular endosymbionts are located in the midgut epithelium and come into direct contact with the lumen of the gut. The same endosymbionts are also found in oocytes, enabling maternal transmission (Schröder *et al.*, 1996). The endosymbionts of tsetse flies are also located in bacteriocytes (mycetocytes) located in the anterior region of the gut in a U-shaped organelle referred to as the mycetome (Buchner, 1965; Aksoy, 1995).

Two questions related to the systematics of endosymbionts of aphids have been addressed by Baumann and his collaborators (Baumann *et al.*, 1995); what are the evolutionary origins of aphid endosymbionts and how do they compare with other eukaryotic cytoplasmic organelles such as chloroplasts and mitochondria? The endosymbionts are bounded by a cell wall similar in structure to a Gram-negative envelope and thus it was consistent that 16S rRNA sequence comparisons placed them in a Gram-negative branch of the bacterial phylogenetic tree in the γ-3 subdivision of the class *Proteobacteria* (Munson *et al.*, 1991a,b) with *E. coli* and other members of the family *Enterobacteriaceae* as close relatives. Endosymbionts of ants of the genus *Camponotus* (Schröder *et al.*, 1996) and

Wigglesworthia glossinidia, the primary endosymbiont of tsetse flies (Aksoy, 1995) also form distinct taxa in this branch of the phylogenetic tree.

Of particular interest in the aphid endosymbionts, where several "species" have been studied from various hosts, is that the molecular phylogeny of *Buchnera* superimposes on the phylogeny of the aphid hosts based on phenotypic features. This indicates vertical evolution of the endosymbionts from a monophyletic source (Moran *et al.,* 1993; Baumann *et al.,* 1995). Thus some 200-250 million years ago an aphid ancestor began an endosymbiotic relationship with a Gram-negative bacterium which has resulted in parallel evolution of insect and endosymbiotic symbiont into the range of genera and species of insect known today each with its distinctive *Buchnera* "species" (Moran *et al.,* 1993). On a nomenclatural note; species of *Buchnera* and *Wigglesworthia* have not been introduced beyond the type species required for establishment of the genera because the diversity within these taxa is based almost entirely on 16S rRNA sequence comparisons and this must be extended by sequences of some protein coding genes and some phenotypic criteria before speciation can be approached systematically. Currently, endosymbionts from different insect species are named after the host insect.

A second question addressed by Baumann and his colleagues was, how does the genomic composition of *Buchnera* compare with that of organellar endosymbionts such as the chloroplast or mitochondrion? *Buchnera aphidicola,* although unable to grow axenically, has a much larger genome than the mitochondrion and has retained much more independent metabolic machinery than cellular organelles (Baumann *et al.,* 1995). Indeed, it is more like a free-living bacterium than an organelle. One explanation for this is that it has been engaged in a symbiotic relationship for much less time than the mitochondrion, 250 million years versus more than 1 billion years, and thus has yet to lose metabolic capacity (Baumann *et al.,* 1995).

Much remains to be learnt about these fascinating endosymbiotic bacteria, but it seems likely that with the development and refinement of molecular techniques more insects will be studied, the endosymbionts will be further characterized, and the diversity and evolution of these relationshps will become more conclusively established.

6.2   WOLBACHIA

A second group of insect endosymbionts is associated with reproduction rather than nutrition. Bacterial endosymbionts in the reproductive systems of insects were first noted in the mosquito *Culex pipiens* and named accordingly as *Wolbachia pipientis* (Hertig, 1936). Since then they have been detected in insects from every major order and some non-insect arthropods. In a recent survey, *Wolbachia* were detected in 26 out of 154 insect species (Werren *et al.,* 1995a) which, if extrapolated to the estimated 10 to 30 million insect species, suggests that at least 2 million insect species are infected with these bacteria making them an extremely abundant group.

The effects of *Wolbachia* on the reproduction of their insect hosts are extensive. Initially endosymbionts were recognized for being responsible for cytoplasmic incompatibility in the mosquito *Culex pipiens.* In this phenomenon, crosses between uninfected females and infected males fail to produce offspring while crosses between infected females and males, whatever their status, result in infected offspring. When infected males are treated with antibiotics to cure them of *Wolbachia* they can reproduce normally (Yen & Barr, 1971). The bacteria live in the gonadal tissues of the insects and are passed vertically through the maternal line in the cytoplasm of the egg cells rather like mitochondria in eukaryotic cells. The abortive matings, induced by defective sperm condensation reactions, reduce the number of offspring from uninfected females and enhance the transmission of the endosymbionts through providing infected females with a reproductive advantage (Hart, 1995).

Thelytokous parthenogenesis in which diploid females carrying *Wolbachia* lay unfertilized eggs which develop into more infected females occurs in certain species of wasp. This result from *Wolbachia* infection enhances spread of the bacterium. Treatment of the female with antibiotics cures the host of the endosymbionts and restores sexuality and males appear in the progeny. These and other effects of *Wolbachia* on its host have been extensively reviewed (Hart, 1995; Rigaud & Rousset, 1996).

The initial suggestion that *Wolbachia* be included in the family *Rickettsiaceae*, since the rickettsiae initially encompassed all intracellular bacteria regardless of any other characteristic proved to be accurate (Drancourt & Raoult, 1994). Sequence comparisons of 16S rRNA genes were able to rationalize the taxonomy of the rickettsias and placed most of these bacteria in the α-subdivision of the class *Proteobacteria* (Breeuwer *et al.*, 1992; O'Neill *et al.*, 1992; Rousset *et al.*, 1992). In particular, *Ehrlichia* species are polyphyletic on these criteria and *Wolbachia pipientis* forms a lineage within the *Ehrlichia* (Drancourt & Raoult, 1994). In order to gain some insight into the taxonomic structure within *Wolbachia*, sequence comparisons of less conserved genes than the 16S rRNA were needed. *ftsZ* is a rapidly evolving bacterial gene involved in cell division and was chosen by Werren *et al.* (1995b) for this purpose. Phylogenetic analysis of the *ftsZ* sequences revealed two major taxa of *Wolbachia* (A and B), which were estimated to have diverged 58 to 67 million years ago. The extensive divergence among *Wolbachia* strains based on *ftsZ* sequences indicates that, as with *Buchnera*, numerous species are evident but the lack of phenotypic features complicates speciation. Therefore *Wolbachia* strains are not given species status at this stage but simply identified by their host arthropod (Werren *et al.*, 1995b).

The *ftsZ* results were important in two other respects. First the A/B division of strains was concordant with trees based on 16S rRNA sequence comparisons supporting the gene trees as organism trees. Second, the *ftsZ*-based *Wolbachia* tree was dissimilar to the phylogeny of the insect hosts strongly supporting horizontal transfer of the endosymbionts. This is different from the situation in *Buchnera* in which vertical transmission is the norm and the bacterial and insect phylogenetic trees correspond (Baumann *et al.*, 1995). A subgroup within the A branch of the *Wolbachia* tree (labelled Adm after *Drosophila melanogaster*) is particularly interesting in this regard since it shows a high degree of *ftsZ* sequence similarity despite being found in host species from different orders. Estimates from sequence comparisons suggest that the bacteria diverged only 0-2.5 million years ago while the insect orders diverged at least 200 million years ago. This indicates a relatively recent expansion of Adm *Wolbachia* into new insect species possibly via transmission between parasitic insects and their insect hosts. It is also conjectured that the recent expansion of Adm *Wolbachia* may result from human activity which has altered the ecology of numerous insect species and brought different species into contact thus enabling transmission of endosymbionts (Werren *et al.*, 1995a,b).

7   Conclusions

Bacterial relationships range from mutualism through opportunistic pathogenicity and obligate pathogencitiy to complete interdependence in endosymbiotic pairings. In all instances where we wish to understand these relationships it is essential to define the systematics of the organisms involved. Without comprehensive phenetic recognition of the bacterium, the relationships with its insect target or host are confusing. For example, in the case of *B. sphaericus* the lack of mosquitocidal toxicity in so many strains was baffling until DNA reassociation studies revealed the enormous diversity in round-spore forming bacilli and the existence of numerous "species" of which only one had insect toxicity (Krych *et al.*, 1980). This realization also facilitated the typing and population genetic studies which are providing information on the evolution of toxicity in this bacterium (Zahner *et al.*, 1998). On a more practical note, the recognition of DNA homology group IIA as the mosquitocidal "species" has simplified selective isolation procedures (Yousten *et al.*, 1985) and enhanced screening programmes for the isolation of novel strains (Liu *et al.*, 1993).

The high affinity between *B. cereus* and *B. thuringiensis* has been revealed in numerous ways, both phenetic and phylogenetic, and in most respects they can be considered the same species (Priest, 1994). Again the implications for selective isolation of new strains for biocontrol purposes are very important and difficulties associated with selective recovery of *B. thuringiensis* are fully explained by the systematics, although here numerical phenetics has currently failed to provide effective selective isolation procedures.

The most profound development in the systematics of bacteria associated with insects has been the introduction of molecular approaches to bacterial classification and identification. With the endospore-forming bacteria this has promoted the rationalization of the genus *Bacillus* and the introduction of new genera such as *Brevibacillus* and *Paenibacillus*. Such divisions based, as in the case of *Paenibacillus,* almost entirely on 16S rRNA sequence analysis have their critics. However, such profound differences in rRNA sequence as demonstrated between *Bacillus* and *Paenibacillus* must be accompanied by distinct phenetic properties and one would expect to see this reflected in pathogenicity attributes. Indeed this is the case, as obligate pathogens such as *P. larvae* and *P. popilliae* display very different virulence properties and disease processes from the opportunistic pathogens such as *B. thuringiensis*. Here taxonomy takes on its informative role and from simple PCR recognition of an isolate as *Paenibacillus* [based on its unique rRNA sequence (Shida, *et al.*, 1997)] aspects of its pathogenicity and lifestyle can be predicted. Similarly, the realization that *B. sphaericus* is phylogenetically distinct from *B. thuringiensis* suggests that there should be little in common between their pathogenicity processes and the molecular analysis of the crystal proteins supports this contention. The phylogenetic classification of "*B. laterosporus*" in *Brevibacillus* similarly suggests that insect pathogenicity in this bacterium may be unique.

Molecular classification was essential for determining the taxonomic relationships of the insect endosymbionts which cannot be cultivated in the laboratory and this is one of the most exciting areas of unknown bacterial diversity. Now that the tools are available for identifying these bacteria based on probes targeted to rRNA and protein-coding DNA sequences, it is possible to ask some of the interesting questions about distribution, co-evolution with the insect host, vertical versus horizontal transmission of the endosymbionts and the physiological interdependency of the two organisms. There is no doubt that this is going to be an area of intense interest which should lead to fascinating evolutionary and ecological insights as well as environmentally friendly approaches to insect pest control.

The diversity of bacteria associated with insects must be enormous given the multitudinous number of insect species. Initial studies on this reservoir of microbial biodiversity has already led to a fascinating range of bacteria from the bioluminescent *Photorhabdus* with its biphasic lifestyle in nematodes and insects to paenibacilli which can only sporulate in hemolymph; from *Wolbachia* endosymbionts which affect insect reproduction to opportunistic pathogens such as *B. thuringiensis* with exquisitely targeted protein toxins and enormous potential for insect control in agriculture. It is clear that there will be more surprises in store as we explore the realms of bacterial insect interactions more extensively.

Acknowledgements

We are grateful to Bertil Pettersson for the preparation of phylogenetic trees and to Michael Hornitzky, Michael Klein and Allan Yousten for provision of material for use in figures. Work from the laboratory of FGP was supported by grants from the Overseas Development Administration and the World Health Organization.

8 References

Akhurst, R.J. & Boemare, N.E. (1988). A numerical taxonomic study of the genus *Xenorhabdus* (*Enterobacteriaceae*) and proposed elevation of the subspecies of *X. nematophilus* to species. *Journal of General Microbiology* 134, 751-761.

Akhurst, R.J., Mourant, R.G., Baud, I. & Boemare, N.E. (1996). Phenotypic and DNA relatedness between nematode symbionts and clinical strains of the genus *Photorhabdus* (*Enterobacteriaceae*). *International Journal of Systematic Bacteriology* 46, 1034-1041.

Aksoy, S. (1995). *Wigglesworthia* gen. nov. and *Wigglesworthia glossinidia* sp. nov., taxa consisting of the mycetocyte-associated, primary endosymbionts of tsetse flies. *International Journal of Systematic Bacteriology* 45, 848-851.

Alexander, B. & Priest, F.G. (1990). Numerical classification and identification of *Bacillus sphaericus* including some strains pathogenic for mosquito larvae. *Journal of General Microbiology* 136, 367-376.

Aquino de Muro, M. & Priest, F.G. (1993). Phylogenetic analysis of *Bacillus sphaericus* and development of an oligonucleotide probe specific for mosquito pathogenic strains. *FEMS Microbiology Letters* 112, 205-210.

Aquino de Muro, M., Mitchell, W.J. & Priest, F.G. (1992). Differentiation of mosquito pathogenic strains of *Bacillus sphaericus* from nontoxic varieties by ribosomal RNA gene restriction patterns. *Journal of General Microbiology* 138, 1159-1166.

Aronson, A.I. (1993). The two faces of *Bacillus thuringiensis:* insecticidal proteins and post-exponential survival. *Molecular Microbiology* 7, 489-496.

Ash, C., Farrow, J.A., Wallbanks, S. & Collins, M.D. (1991a). Phylogenetic heterogeneity of the genus *Bacillus* revealed by comparative analysis of small subunit ribosomal RNA sequences. *Letters in Applied Microbiology* 13, 202-206.

Ash, C., Farrow, J.E., Dorsch, M., Stackebrandt, E. & Collins, M.D. (1991b). Comparative analysis of *Bacillus anthracis, Bacillus cereus* and related species on the basis of reverse transcriptase sequencing of

16S rRNA. *International Journal of Systematic Bacteriology* **41**, 343-346.

Ash, C., Priest, F.G. & Collins, M.D. (1993). Molecular identification of rRNA group 3 bacilli (Ash, Farrow, Wallbanks and Collins) using a PCR probe test. *Antonie van Leeuwenhoek* **64**, 253-260.

Baumann, L. & Baumann, P. (1989). Expression in *Bacillus subtilis* of the 51-kilodalton and 42- kilodalton mosquitocidal toxin genes of *Bacillus sphaericus. Applied and Environmental Microbiology* **55**, 252-253.

Baumann, P., Clark, M.A., Baumann, L. & Broadwell, A.H. (1991). *Bacillus sphaericus* as a mosquito pathogen - properties of the organism and its toxins. *Microbiological Reviews* **55**, 425-436.

Baumann, P., Baumann, L., Lai, C.-Y., Rouhbakhsh, D., Moran, N.A. & Clark, M.A. (1995). Genetics, physiology and evolutionary relationships of the genus *Buchnera*: intracellular symbionts of aphids. *Annual Review of Microbiology* **49**, 55-94.

Bleakey, B. & Nealson, K.H. (1988). Characterization of primary and secondary forms of *Xenorhabdus luminescens* strain HM. *FEMS Microbiology Ecology* **53**, 241-250.

Boemare, N.E. & Akhurst, R.J. (1988). Biochemical and physiological characterization of colony variants in *Xenorhabdus* spp.(*Enterobacteriaceae*). *Journal of General Microbiology* **134**, 1835-1841.

Boemare, N.E., Akhurst, R.J. & Mourant, R.G. (1993). DNA relatedness between *Xenorhabdus* spp. (*Enterobacteriaceae*), symbiotic bacteria of entomopathogenic nematodes, and a proposal to transfer *Xenorhabdus luminescens* to a new genus, *Photorhabdus* gen. nov. *International Journal of Systematic Bacteriology* **43**, 249-255.

Bravo, A. (1997). Phylogenetic relationships of *Bacillus thuringiensis* delta-endotoxin family proteins and their functional domains. *Journal of Bacteriology* **179**, 2793-2801.

Breeuwer, J.A.J., Stouthamer, R., Barns, S.M., Pelletier, D.A., Weisberg, W.G. & Werren, J.H. (1992). Phylogeny of cytoplasmic incompatibility microorganisms in the parasitoid wasp genus *Nasonia* (Hymenoptera, Pteromalidae) based on 16S ribosomal DNA sequences. *Insect Molecular Biology* **1**, 25-36.

Brousseau, R., Saint-Onge, A., Prefontaine, G., Masson, L. & Cabana, J. (1993). Arbitrary primer polymerase chain reaction, a powerful method to identify *Bacillus thuringiensis* serovars and strains. *Applied and Environmental Microbiology* **59**, 114-119.

Brunel, B., Givaudan, A., Lanois, A., Akhurst, R.J. & Boemare, N. (1997). Fast and accurate identification of *Xenorhabdus* and *Photorhabdus* species by restriction analysis of PCR-amplified 16S rRNA genes. *Applied and Environmental Microbiology* **63**, 574-580.

Bucher, C. (1981). Identification of bacteria found in insects. In *Microbial Control of Pests and Plant Diseases 1970-1980*, pp. 7-33. Edited by H. D. Burges. London: Academic Press.

Buchner, P. (1965). *Endosymbiosis of Animals with Plant Microorganisms*. New York: Interscience.

Bulla, L.A., Costilow, R.N. & Sharpe, E.S. (1978). Biology of *Bacillus popilliae. Advances in Applied Microbiology* **23**, 1-18.

Carlson, C.R., Caugant, D.A. & Kolstø, A.-B. (1994). Genotypic diversity among *Bacillus cereus* and *Bacillus thuringiensis* strains. *Applied and Environmental Microbiology* **60**, 1719-1725.

Carlson, C.R., Johansen, T. & Kolstø, A.-B. (1996a). The chromosome map of *Bacillus thuringiensis* subsp. *canadensis* HD224 is highly similar to that of *Bacillus cereus* type strain ATCC 14579. *FEMS Microbiology Letters* **141**, 163-167.

Carlson, C.R., Johansen, T., Lecadet, M.-M. & Kolstø, A.-B. (1996b). Genomic organization of the entomopathogenic bacterium *Bacillus thuringiensis* subsp. *berliner. Microbiology* **142**, 1625-1634.

Charles, J.F., Nielsen Leroux, C. & Delécluse, A. (1996). *Bacillus sphaericus* toxins, molecular biology and mode of action. *Annual Review of Entomology* **41**, 451-472.

Correa, M. & Yousten, A.A. (1995). *Bacillus sphaericus* spore germination and recycling in mosquito larval cadavers. *Journal of Invertebrate Pathology* **66**, 76-81.

Damgaard, P.H. (1995). Diarrhoeal enterotoxin production by strains of *Bacillus thuringiensis* isolated from commercial *Bacillus thuringiensis*-based insecticides. *FEMS Immunology and Medical Microbiology* 12, 245-250.

Davidson, E.W. (1995). Biochemistry and mode of action of the *Bacillus sphaericus* toxins. *Memorias do Instituto Oswaldo Cruz* 90, 81-86.

de Barjac, H. & Frachon, E. (1990). Classification of *Bacillus thuringiensis* strains. *Entomophaga* 35, 233-240.

de Barjac, H., Larget-Thiéry, I., Cosmao Dumanoir, V.C. & Ripouteau, H. (1985). Serological classification of *Bacillus sphaericus* strains in relation to toxicity to mosquito larvae. *Applied Microbiology and Biotechnology* 21, 85-90.

Delécluse, A., Rosso, M.-L. & Ragni, A. (1995). Cloning and expression of a novel toxin gene from *Bacillus thuringiensis* subsp. *jegathesan* encoding a highly mosquitocidal protein. *Applied and Environmental Microbiology* 61, 4230-4235.

Dillon, R.J. & Charnley, A.K. (1995). Chemical barriers to gut infection in the desert locus: *in vivo* production of antimicrobial phenols associated with *Pantoea agglomerans*. *Journal of Invertebrate Pathology* 66, 72-75.

Djordjevic, S., Ho-Shon, M. & Hornitzky, M. (1994). DNA restriction endonuclease profiles and typing of geographically diverse isolates of *Bacillus larvae*. *Journal of Apicultural Research* 33, 95-103.

Douglas, A.E. & Prosser, W.A. (1992). Synthesis of the essential amino acid tryptophan in the pea aphid (*Acrthosiphon pisumi*) symbiosis. *Journal of Insect Physiology* 38, 565-568.

Drancourt, M. & Raoult, D. (1994). Taxonomic position of the rickettsiae: current knowledge. *FEMS Microbiology Reviews* 13, 13-24.

Dutky, S.R. (1940). Two new spore-forming bacteria causing millky diseases of Japanese beetle larvae. *Journal of Agricultural Research* 61, 57-68.

Eisenstein, B.I. (1990). New techniques for microbial epidemiology and the diagnosis of infectious diseases. *Journal of Infectious Diseases* 161, 595-602.

Estruch, J.J., Warren, G.W., Mullins, M.A., Nye, G.J., Craig, J.A. & Koziel, M.G. (1996). Vip3A, a novel *Bacillus thuringiensis* vegetative toxin with a wide spectrum of activity against lepidopteran insects. *Proceedings of the National Academy of Sciences of the United States of America* 93, 5389-5394.

Falkow, S. (1997). What is a pathogen? *ASM News* 63, 359-365.

Farmer, J.J.I., Jorgensen, J.H., Grimont, P.A., Akhurst, R.J., Poinar, G.O., Ageron, E., Pierce, G.E., Smith, J.A., Carter, G.P., K.L., W. & Hickman-Brenner, F.W. (1989). *Xenorhabdus luminescens* (DNA hybridization group) from human clinical specimens. *Journal of Clinical Microbiology* 27, 1594-1602.

Favret, M.E. & Yousten, A.A. (1985). Insecticidal activity of *Bacillus laterosporus*. *Journal of Invertebrate Pathology* 45, 195-203.

Forst, S. & Nealson, K. (1996). Molecular biology of the symbiotic pathogenic bacteria *Xenorhabdus* spp. and *Photorhabdus* spp. *Microbiological Reviews* 60, 21-43.

Gill, S.S. (1995). Biochemistry and mode of action of *Bacillus thuringiensis* toxins. *Memorias do Instituto Oswaldo Cruz* 90, 69-74.

Gilliam, M. (1997). Identification and roles of non-pathogenic microflora associated with honey bees. *FEMS Microbiology Letters* 155, 1-10.

Gilliam, M. & Dunham, D.R. (1977). Recent isolations of *Bacillus pulvifaciens* from powdery scales of honey bee, *Apis mellifera*, larvae. *Journal of Invertebrate Pathology* 32, 222-223.

Gilliam, M., Lorenz, B.J. & Richardson, G.V. (1988a). Digestive enzymes and microorganisms in honey bees, *Apis mellifera*: influence of streptomycin, age, season and pollen. *Microbios* 55, 95-114.

Gilliam, M., Taber, S., III, Lorenz, B.J. & Prest, D.B. (1988b). Factors affecting the development of chalkbrood disease in colonies of honey bees, *Apis mellifera*, fed pollen contaminated with *Ascosphsera apis*.

Journal of Invertebrate Pathology 52, 314-325.

Gordon, R.E., Haynes, W.C. & Pang, C.H.-N. (1973). *The genus Bacillus. Agriculture Handbook no. 427.* Washington, D.C.: United States Department of Agriculture.

Greenberg, B., Kowalski, J.A. & Flowden, M.J. (1970). Factors affecting the transmission of *Salmonella* by flies: natural resistance to colonization and bacterial interference. *Infection and Immunity* 2, 800-809.

Hacker, J., Blum-Oehler, G., Muhidorfer, I. & Tschape, H. (1997). Pathogenicity islands of virulent bacteria: structure, function and impact on microbial evolution. *Molecular Microbiology* 23, 1089-1097.

Hart, S. (1995). When *Wolbachia* invades, insect sex lives get into a spin. *BioScience* 45, 4-6.

Heimpel, A.M. & Angus, T.A. (1960). Bacterial insecticides. *Bacteriological Reviews* 24, 266-288.

Henderson, I., Duggleby, C.J. & Turnbull, P.C.B. (1994). Differentiation of *Bacillus anthracis* from other *Bacillus cereus* group bacteria with the PCR. *International Journal of Systematic Bacteriology* 44, 99-105.

Hertig, M. (1936). The rickettsia, *Wolbachia pipientis* (gen. nov. et sp. nov.) and associated inclusions of the mosquito *Culex pipiens. Parasitology* 28, 453-486.

Heyndrickx, M., Vandemeulebroecke, K., Hoste, B., Janssen, P., Kersters, K., Devos, P., Logan, N.A., Ali, N. & Berkeley, R.C.W. (1996). Reclassification of *Paenibacillus* (formerly *Bacillus*) *pulvifaciens* (Nakamura 1984) Ash *et al.* 1994, a later subjective synonym of *Paenibacillus* (formerly *Bacillus*) *larvae* (White 1906) Ash *et al.* 1994, as a subspecies of *Paenibacillus larvae*, with emended descriptions of *Paenibacillus larvae* as *Paenibacillus larvae* subsp. *larvae* and *Paenibacillus larvae* subsp. *pulvifaciens*. *International Journal of Systematic Bacteriology* 46, 270-279.

Humphreys, M.J. & Berry, C. (1998). Variants of the *Bacillus sphaericus* binary toxins: implications for differential toxicity of strains. *Journal of Invertebrate Pathology* 71, 184-185.

Hurlburt, R.E. (1994). Investigations into the pathogenic mechanisms of the bacterium-nematode complex. *ASM News* 60, 473-478.

Istock, C.A., Duncan, K.E., Ferguson, N. & Zhou, X. (1992). Sexuality in a natural population of bacteria-*Bacillus subtilis* challenges the the clonal paradigm. *Molecular Ecology* 1, 95-103.

Jahnz, U., Fitch, A. & Priest, F.G. (1996). Evaluation of an rRNA-targeted oligonucleotide probe for the detection of mosquitocidal strains of *Bacillus sphaericus* in soils - characterization of novel strains lacking toxin genes. *FEMS Microbiology Ecology* 20, 91-99.

Jarrett, P. & Stephenson, M. (1990). Plasmid transfer between strains of *Bacillus thuringiensis* infecting *Galleria mellonella* and *Spodoptera littoralis. Applied and Environmental Microbiology* 56, 1608-1614.

Jelinski, M. (1985). Some biochemical properties of *Bacillus larvae* White. *Apidologie* 16, 69-76.

Kaji, D.A., Rosato, Y.B., Canhos, V.P. & Priest, F.G. (1994). Characterization by polyacylamide gel electrophoresis of whole cell proteins of some strains of *Bacillus thuringiensis* subsp. *israelensis* isolated in Brazil. *Systematic and Applied Microbiology* 17, 104-107.

Kaneda, T. (1977). Fatty acids of the genus *Bacillus*: an example of branched-chain preference. *Bacteriological Reviews* 41, 391-418.

Kaya, H.K. & Gaugler, R. (1993). Entomopathogenic nematodes. *Annual Review of Entomology* 38, 181-206.

Klein, M.G. & Kaya, H.K. (1995). *Bacillus* and *Serratia* species for scarab control. *Memorias de Instito Oswaldo Cruz* 90, 87-95.

Krych, V., Johnson, J.L. & Yousten, A.A. (1980). Deoxyribonucleic acid homologies among strains of *Bacillus sphaericus. International Journal of Systematic Bacteriology* 30, 476-484.

Kumar, P.A., Sharma, R.P. & Malik, V.S. (1996). The insecticidal proteins of *Bacillus thuringiensis. Advances in Applied Microbiology* 42, 1-43.

Kuo, W-S. & Chak, K-F. (1996). Identification of novel *cry*-type genes from *Bacillus thuringiensis* strains on the basis of restriction fragment length polymorphism of the PCR-amplified DNA. *Applied and*

Environmental Microbiology **62**, 1369-1377.

Lereclus, D., Delécluse, A & Lecadet, M-M. (1993). Diversity of *Bacillus thuringiensis* toxins and genes. In *Bacillus thuringiensis, an Environmental Biopesticide: Theory and Practice*, pp. 37-69. Edited by P.F. Entwistle, J.S. Cory, M.J. Bailey & S. Higgs. Chichester: John Wiley & Sons.

Liu, J.W., Hindley, J., Porter, A. & Priest, F.G. (1993). New high toxicity mosquitocidal strains of *Bacillus sphaericus* lacking a 100-kilodalton gene. *Applied and Environmental Microbiology* **59**, 3470-3473.

Gonzalez, J.M., Jr., Brown, B.J. & Carlton, B.C. (1982). Transfer of *Bacillus thuringensis* plasmids coding for delta-endotoxins among strains of *Bacillus thuringiensis* and *Bacillus cereus*. *Proceedings of the National Academy of Sciences of the United States of America* **79**, 6951-6955.

Martin, P.A.W. & Travers, R.S. (1989). Worldwide abundance and distribution of *Bacillus thuringiensis*. *Applied and Environmental Microbiology* **55**, 2437-2442.

Martin, P.A.W., Haransky, E.B., Travers, R.S. & Reichelfderfer, C.F. (1985). Rapid biochemical resting of large numbers of *Bacillus thuringiensis* isolates using agar dots. *BioTechniques* **3**, 386-392.

Maynard Smith, J. (1995). Do bacteria have population genetics? In *Population Genetics of Bacteria*, pp. 1-12. Edited by S. Baumberg, J.P.W. Young, E.H.M. Wellington & J.R. Saunders. Cambridge: Cambridge University Press.

Meadows, M.P. (1993). *Bacillus thuringiensis* in the environment: ecology and risk assessment. In *Bacillus thuringiensis, an Environmental Biopesticide: Theory and Practice*, pp. 193-220. Edited by P.F. Entwistle, J.S. Cory, M.J. Bailey & S. Higgs. Chichester: John Wiley & Sons.

Milner, R.J. (1981). Identification of the *Bacillus popilliae* group of insect pathogens. In *Microbial Control of Pests and Plant Diseases 1970-1980*, pp. 45-59. Edited by H.D. Burges. London: Academic Press.

Moran, N.A., Munson, M.A., Baumann, P. & Ishikawa, H. (1993). A molecular clock in endosymbiotic bacteria is calibrated using insect hosts. *Proceedings of the Royal Society of London Series B* **253**, 167-171.

Munson, M.A., Baumann, P., Clark, M.A., Baumann, L., Moran, N.A., Voegtlin, D.J. & Campbell, B.C. (1991a). Evidence for the establishment of aphid-eubacterium endosymbiosis is an ancestor of four aphid families. *Journal of Bacteriology* **173**, 6321-6324.

Munson, M.A., Baumann, P. & Kinsey, M.G. (1991b). *Buchnera* gen. nov. and *Buchnera aphidicola* sp. nov., a taxon consisting of the mycetocyte-associated endosymbionts of aphids. *International Journal of Systematic Bacteriology* **41**, 566-568.

Nakamura, L.K. (1994). DNA relatedness among *Bacillus thuringiensis* serovars. *International Journal of Systematic Bacteriology* **44**, 125-129.

Nakamura, L.K. & Jackson, M.A. (1995). Clarification of the taxonomy of *Bacillus mycoides*. *International Journal of Systematic Bacteriology* **45**, 46-49.

Nicolas, L., Regis, L.N. & Rios, E.M. (1994). Role of the exosporium in the stability of the *Bacillus sphaericus* binary toxin. *FEMS Microbiology Letters* **124**, 271-275.

Nielsen, P., Rainey, F.A., Outtrup, H., Priest, F.G. & Fritze, D. (1994). Comparative 16S rDNA sequence analysis of some alkaliphilic bacilli and the establishment of a sixth rRNA group within *Bacillus*. *FEMS Microbiology Letters* **117**, 61-66.

O'Neill (1995). *Wolbachia pipientis:* symbiont or parasite. *Parasitology Today* **11**, 168-169.

O'Neill, S., Giordano, R., Colbert, A.M.E., Karr, T.L. & Robertson, H.M. (1992). 16S rRNA phylogenetic analysis of the bacterial endosymbionts associated with cytoplasmic incompatibility in insects. *Proceedings of the National Academy of Sciences of the United States of America* **89**, 2699-2702.

Petras, S.T. & Casida, L.E. (1985). Survival of *Bacillus thuringiensis* spores in soil. *Applied and Environmental Microbiology* **50**, 1496-1501.

Pettersson, B., Rippere, K.E., Yousten, A.A. & Priest, F.G. (1999). Transfer of *Bacillus lentimorbus* and

Bacillus popilliae to the genus _Paenibacillus_ with emended descriptions of _Paenibacillus lentimorbus_ comb. nov. and _Paenibacillus popilliae_ comb. nov. _International Journal of Systematic Bacteriology_, 49, 531-540.

Priest, F.G. (1994). Systematics and ecology of _Bacillus_. In _Bacillus subtilis and other Gram-positive Bacteria; Biochemistry, Physiology and Molecular Genetics_, pp. 3-16. Edited by J.A. Hoch, A.L. Sonenshein & R. Losick. Washington, D.C.: American Society for Microbiology.

Priest, F.G. & Grigorova, R. (1990). Methods for studying the ecology of endospore-forming bacteria. In _Methods in Microbiology_, pp. 565-591. Edited by R. Grigorova & J.R. Norris. London: Academic Press.

Priest, F.G. & Austin, B.A. (1993). _Modern Bacterial Systematics_, Second Edition. London: Chaman and Hall.

Priest, F.G., Goodfellow, M. & Todd, C. (1988). Numerical classification of the genus _Bacillus_. _Journal of General Microbiology_ 134, 1847-1882.

Priest, F.G., Aquino de Muro, M. & Kaji, D.A. (1994a). Systematics of insect pathogenic bacilli: uses in strain identification and isolation of novel strains. In _Bacterial Diversity and Systematics_, pp. 275-296. Edited by F.G. Priest, A. Ramos-Cormenzana & B.J. Tindall. New York: Plenum Press.

Priest, F.G., Kaji, D.A., Rosato, Y.B. & Canhos, V.P. (1994b). Characterization of _Bacillus thuringiensis_ and related bacteria by ribosomal RNA gene restriction fragment length polymorphisms. _Microbiology_ 140, 1015-1022.

Rainey, F.A., Fritze, D. & Stackebrandt, E. (1994). The phylogenetic diversity of thermophilic members of the genus _Bacillus_ as revealed by 16S rDNA analysis. _FEMS Microbiology Letters_ 115, 205-212.

Rainey, F.A., Ehlers, R.U. & Stackebrandt, E. (1995). Inability of the polyphasic approach to systematics to determine the relatedness of the genera _Xenorhabdus_ and _Photorhabdus_. _International Journal of Systematic Bacteriology_ 45, 379-381.

Rigaud, T. & Rousset, F. (1996). What generates the diversity of _Wolbachia_ - arthropod interactions? _Biodiversity and Conservation_ 5, 999-1013.

Rippere, K.E., Tran, M.T., Yousten, A.A., Hilu, K.H. & Klein, M.G. (1998). _Bacillus popilliae_ and _Bacillus lentimorbus_, bacteria causing milky disease in Japanese beetles and related scarab larvae. _International Journal of Systematic Bacteriology_ 48, 395-402.

Rosner, B.M., Ensign, J.C. & Schink, B. (1996). Anaerobic metabolism of primary and secondary forms of _Photorhabdus luminescens_. _FEMS Microbiology Letters_ 140, 227-232.

Rousset, F., Vautrin, D. & Solignac, M. (1992). Molecular identification of _Wolbachia_, the agent of cytoplasmic incompatibility in _Drosophila simulans_, and variation in relation to host mitochondrial types. _Proceedings of the Royal Society of London Series B_ 247, 163-168.

Russell, B.L., Jelley, S.A. & Yousten, A.A. (1989). Carbohydrate metabolism in the mosquito pathogen _Bacillus sphaericus_ 2362. _Applied and Environmental Microbiology_ 55, 294-297.

Schröder, D., Deppisch, H., Obermayer, M., Krohne, G., Stackebrandt, E., Hölldober, B., Goebel, W. & Gross, R. (1996). Intracellular endosymbiotic bacteria of _Camponotus_ species (carpenter ants): systematics, evolution and ultrastructural characterization. _Molecular Microbiology_ 21, 479-489.

Schwemmler, W. & Gassner, G. (1989). _Insect Endosymbiosis: Morphology, Physiology, Genetics, Evolution_. Boca Raton, Fl.: CRC Press.

Selander, R.K., Cougant, D.A., Ochman, H., Musser, J.M., Gilmour, M.N. & Whittam, T.S. (1986). Methods of multilocus electrophoresis for bacterial population genetics and systematics. _Applied and Environmental Microbiology_ 51, 873-884.

Selander, R.K., Li, J., Boyd, E.F., Wang, F.-S. & Nelson, K. (1994). DNA sequence analysis of the genetic structure of populations of _Salmonella enterica_ and _Escherichia coli_. In _Bacterial Diversity and Systematics_, pp. 17-50. Edited by F.G. Priest, A. Ramos-Cormenzana & B. Tindall. New York: Plenum Press.

Shida, O., Takagi, H., Kadowaki, K. & Komagata, K. (1996). Proposal for two new genera, _Brevibacillus_ gen. nov. and *Aneurinibacillus* gen. nov. *International Journal of Systematic Bacteriology* **46**, 939-946.

Shida, O., Takagi, H., Kadowaki, K., Nakamura, L.K. & Komagata, K. (1997). Transfer of *Bacillus alginolyticus*, *Bacillus chondroitinus*, *Bacillus glucanolyticus*, *Bacillus kobensis* and *Bacillus thiaminolyticus* to the genus *Paenibacillus* and emended description of the genus *Paenibacillus*. *International Journal of Systematic Bacteriology* **47**, 289-298.

Singer, S. (1996). The utility of strains of morphological group II *Bacillus*. *Advances in Applied Microbiology* **42**, 219-261.

Smigielski, A.J., Akhurst, R.J. & Boemare, N.E. (1994). Phase variation in *Xenorhabdus nematophilus* and *Photorhabdus luminescens:* differences in respiratory activity and membrane energization. *Applied and Environmental Microbiology* **60**, 120-125.

Smith, R.A. & Couche, G.A. (1991). The phylloplane as a source of *Bacillus thiuringiensis* variants. *Applied and Environmental Microbiology* **57**, 311-315.

Stackebrandt, E., Ehlers, R.-U. & Rainey, F.A. (1997). *Xenorhabdus* and *Photorhabdus*: are they sister genera or are their members phylogenetically intertwined. *Symbiosis* **22**, 50-65.

Stahly, D.P., Andrews, R. & Yousten, A.A. (1992). The genus *Bacillus*: insect pathogens. In *The Procaryotes*, pp. 1697-1745. Edited by A. Balows, H.G. Trüper, M. Dworkin, W. Harder & K.-H. Schleifer. New York: Springer-Verlag.

Sundar, L. & Chang, F.N. (1993). Antimicrobial activity and biosynthesis of indole antibiotics produced by *Xenorhabdus nematophilus*. *Journal of General Microbiology* **139**, 3139-3148.

Szallas, E., Koch, C., Fodor, A., Burghardt, J., Buss, O., Szentirmai, A., Nealson, K.H. & Stackebrandt, E. (1997). Phylogenetic evidence for the taxonomic heterogeneity of *Photorhabdus luminescens*. *International Journal of Systematic Bacteriology* **47**, 402-407.

te Giffel, M.C., Beumer, R.R., Klijn, N., Wagendorp, A. & Rombouts, F.M. (1997). Discrimination between *Bacillus cereus* and *Bacillus thuringiensis* using DNA probes based on variable regions of 16S rRNA. *FEMS Microbiology Letters* **146**, 47-51.

Thanabalu, T., Berry, C. & Hindley, J. (1993). Cytotoxicity and ADP-ribosylating activity of the mosquitocidal toxin from *Bacillus sphaericus* SSII-1, possible roles of the 27-kilodalton and 70-kilodalton peptides. *Journal of Bacteriology* **175**, 2314-2320.

Thomas, G.M. & Poinar, G.O. (1979). *Xenorhabdus* gen. nov. a genus of entomopathogenic nematophilic bacteria of the family *Enterobacteriaceae*. *International Journal of Systematic Bacteriology* **29**, 352-360.

Travers, R.S., Martin, P.W.A. & Reichelderfer, C.F. (1987). Selective process for efficient isolation of soil *Bacillus* species. *Applied and Environmental Microbiology* **53**, 1263-1266.

Turnbull, P.C., Hutson, R.A., Ward, M.J., Jones, M.N., Quinn, C.P., Finnie, N.J., Duggleby, C.J., Kramer, J.M. & Melling, J. (1992). *Bacillus anthracis* but not always anthrax. *Journal of Applied Bacteriology* **72**, 21-28.

Walsh, J. (1986). River blindness, a gamble pays off. *Science* **232**, 922-925.

Wang, H.Y. & Dowds, B.C.A. (1993). Phase variation in *Xenorhabdus luminescens* - cloning and sequencing of the lipase gene and analysis of its expression in primary and secondary phases of the bacterium. *Journal of Bacteriology* **175**, 1665-1673.

Weiner, B.A. (1978). Isolation and partial characterization of the parasporal body of *Bacillus popilliae*. *Canadian Journal of Microbiology* **24**, 1557-1561.

Werren, J.H., Windsor, D. & Guo, L. (1995a). Distribution of *Wolbachia* among neotropical arthropods. *Proceedings of the Royal Society of London Series* B **262**, 197-204.

Werren, J.H., Zhang, W. & Rong Guo, L. (1995b). Evolution and phylogeny of *Wolbachia:* reproductive parasites of arthropods. *Proceedings of the Royal Society of London Series* B **261**, 55-71.

West, A.W., Burges, H.D., Dixon, T.J. & Wyborn, C.H. (1985). Survival of *Bacillus thuringiensis* and *Bacillus cereus* spore inocula in soil: effects of pH, moisture, nutrient availability and indigenous microorganisms. *Soil Biology and Biochemistry* **17**, 657-665.

Wintzingerode, F.v., Rainey, F.A., Kroppenstedt, R.M. & Stackebrandt, E. (1997). Identification of environmental strains of *Bacillus mycoides* by fatty acid analysis and species-specific 16S rDNA oligonucleotide probing. *FEMS Microbiology Ecology* **24**, 201-209.

Yen, J.H. & Barr, A.R. (1971). New hypothesis on the cause of cytoplasmic incompatibility in *Culex pipiens*. *Nature* **232**, 657-658.

Yousten, A.A., Fretz, S.B. & Jelley, S.A. (1985). Selective medium for insect pathogenic strains of *Bacillus sphaericus*. *Applied and Environmental Microbiology* **49**, 1532-1533.

Zahner, V., Momen, H., Salles, C.A. & Rabinovitch, L. (1989). A comparative study of enzyme variation in *Bacillus cereus* and *Bacillus thuringiensis*. *Journal of Applied Bacteriology* **67**, 275-282.

Zahner, V., Momen, H. & Priest, F.G. (1998). Serotype H5a5b is a major clone within mosquito-pathogenic strains of *Bacillus sphaericus*. *Systematic and Applied Microbiology* **21**, 162-170.

Zhang, J., Hodgman, C., Krieger, L., Schnetter, W. & Schairer, H.U. (1997). Cloning and analysis of the first *cry* gene from *Bacillus popilliae*. *Journal of Bacteriology* **179**, 4336-4341.

8 FUNGAL PATHOGENS AND PARASITES OF INSECTS

RICHARD A. HUMBER

*USDA-ARS Plant Protection Research Unit, US Plant,
Soil and Nutrition Laboratory, Tower Road Ithaca,
New York 14853-2901, U.S.A.*

1 Introduction

Biologists generally agree that insects are the most numerous, ecologically successful, and taxonomically diverse macro-organisms on the face of this planet. Comparatively few of these biologists, however, seem to appreciate that these arthropods are typical animals much like ourselves and, for purposes of this chapter, are attacked by a wide spectrum of fungi as well as by bacteria (see Chapter 7), viruses, and other microbes. Perhaps the major difference between the diseases of insects and those of most mammals is the perspective with which these diseases are held by humans: the diseases of most animals (and, of course, of many of the plants most important to us) are maladies to be prevented or treated within the limits of our technological and fiscal capabilities. The diseases of insects and pest vertebrates are generally regarded as beneficial for humans and are to be encouraged.

Biological pest control methods seek to manipulate the host and/or microbial pathogen to maximize the negative impact of disease upon the target populations. In all fairness, however, there are also many insects such as pollinator bees, other agriculturally beneficial insects - e.g., ladybird beetles (Coccinellidae), parasitic wasps, and phytophagous insects used to control weeds - and spiders, other commercially important species such as silkworms (*Bombyx mori*), and research colonies of insects (whether they are pests or not!) in whose populations microbial diseases are distinctly undesirable, can have major negative economic impacts, and whose control can be extremely difficult (see Humber, 1996).

Nonlethal parasitic diseases may be caused by microsporidia, by latent viruses dormant in insect populations for many generations before causing epizootic disease outbreaks, or by a spectrum of fungi very different from that of the entomopathogens, or by other microbes. Lethal diseases of insects may be caused by pathogens such as fungi, viruses, or bacteria (Tanada & Kaya, 1993). Some sick insects can cure themselves by their own immune defenses (Cooper *et al.*, 1992; Jarosz, 1996; Gillespie *et al.*, 1997) or by modified behaviours ("behavioural fevers") that artificially raise the body temperature enough to kill or debilitate the disease-causing organisms (Carruthers *et al.*, 1992; Watson *et al.*, 1993). Usually, however, diseased insects die as a direct result of their infection.

Pathogens of insects, mites, and spiders occur in nearly every major class of the fungi (Roberts & Humber, 1981), and those who study these fungi are a mix of people specialized in entomology or mycology. Indeed, there are many more entomologists than mycologists

F. G. Priest and M. Goodfellow (eds.), Applied Microbial Systematics, 203-230
© 2000 *Kluwer Academic Publishers. Printed in the Netherlands.*

working on the biology and even the taxonomy and systematics of these fungi. This chapter centers primarily on the entomopathogenic ascomycetes of the family *Clavicipitaceae* *(Pyrenomycetes: Hypocreales)* and their associated conidial states (phylum *Deuteromycota*: class *Hyphomycetes*); these ascomycetes and hyphomycetes comprise the largest and most taxonomically complex group of nearly 1000 taxa of fungal entomopathogens. Secondary emphasis is given to the *Entomophthorales (Zygomycota)* whose 200 species pathogenic for insects and other invertebrates comprise the second largest cluster of entomopathogens.

This chapter focuses on selected aspects of entomopathogenic fungi that are critical for their taxonomy (the naming and classification of taxa) and systematics (the consideration of the larger relationships among these organisms). The great diversity of these fungi becomes comprehensible only through the perspective offered by their systematics. To present an analogy for this situation, the bricks, windows, doors, wiring and plumbing and other parts that brought together to make an enduring edifice will remain just so many unassembled bits and pieces with no meaningful form or relationship unless there is a detailed plan, a solid foundation, and appropriate framing of the superstructure onto which they are placed. The underlying plan, the foundation, and the superstructure for the complex edifice of the entomopathogenic fungi is the systematics and taxonomy of these organisms. At the risk of overextending this analogy, it should also be noted that the way in which the bits and pieces of the edifice come together during construction allow the alert observer to infer much about the final nature of the structure: The observation and correct interpretation of the partially completed structure of a building makes it possible to know without reference to the plans - and without seeing the completed, decorated, and furnished structure - where a kitchen, bathroom, or grand ballroom will be and to visualize their general shape and relationships to the rest of the structure (but offer no clues about the final details of the building's finishing and decoration). In a similar way, a few facts about organisms may, if judiciously applied (indeed, the very theme of this book!), reveal a great deal about their overall biology and relationships.

Many complex systematic considerations derive directly from the life cycles for the clavicipitaceous taxa and their independently self-reproducing conidial states. Thanks to the special nomenclatural rules for morphologically variable fungi, there are usually two or more valid taxonomic names for each organism in this group (a point that is discussed below). The problems of discovering and interpreting the relationships among the sexual and conidial states of these fungi are both fascinating and frustrating; no complete understanding of these fungi is possible without integrating the diverse sorts of information gained from all phases of their life histories. While issues of host specificity may seem at first glance to be rather academic, there are major problems attached to the definition and use of this concept that affect the systematics of these fungi, and which provide useful guideposts to the more complete understanding of the roles that entomopathogenic fungi can play both as biological control agents and as the sources of compounds of potential pharmaceutical interest. Lastly, some consideration is given to the rise and increasing importance of molecular systematic techniques for insect mycology (see Chapter 3 by Nakase and Hamamoto for more detailed discussion on molecular systematics and fungal phylogenetics), and an attempt is made to indicate some probable future directions for research on the systematics of insect fungi.

Unfortunately, this chapter cannot include any discussion of some of the less common but no less important fungi associated with insects. Foremost among the "omitted" fungi may be the only basidiomycetes intimately associated with insects, the many species of *Septobasidium (Teliomycetes: Septobasidiales)* and its relatives; the taxonomy of *Septobasidium* and the amazing ways in which these fungi husband rather than suppress their associated populations of scale insects are discussed in great detail by Couch (1931, 1935, 1938). The biology and systematics of the genus *Coelomomyces (Chytridiomycetes: Blastocladiales)*, the only large group of entomopathogenic watermolds and the only entomopathogens with an obligatory alternation of hosts (the haploid and diploid states are lethal for copepods and mosquitoes, respectively) is thoroughly discussed and illustrated by Couch & Bland (1985).

2   Assembling a Puzzle: Integrating Complex Life Histories

Many diverse organisms have phases in their life histories that are so fundamentally dissimilar that they would seem unlikely to represent a single organism. Examples include the very mobile planktonic larval states of uni- and bivalve molluscs or the tiny, filmy gametophytes of ferns. Among terrestrial invertebrates, the undisputed champions of change among the various stages of life are holometabolous insects, whose larvae pupate and undergo a complete bodily reorganization to the adult form.

An alternation of haploid and diploid generations is an alien concept for zoologists since in most animal life histories the only haploid cells are eggs and sperm. Separate haploid and diploid generations are much more routine for plants and fungi. Distinctly different haploid and diploid phases in the life cycles of mosses, ferns, and many other plant groups may exist but, as for animals and most botanical taxa, the organism can bear only one nomenclaturally valid name. Many fungi not only display distinct differences between the asexual and the sexual phases, but these morphologically distinct phases may lead completely independent existences functioning as entirely distinct organisms separated in space and/or time, with each phase able to grow and reproduce in the complete absence of the other for prolonged or even indefinite periods of time.

Persons with no formal training in mycology often fail to grasp the complexities of fungal life cycles. And this difficulty is not lessened by the fact that Article 59 of the International Code of Botanical Nomenclature (ICBN; Greuter *et al.*, 1994) allows the morphologically and reproductively distinct states of a huge number of ascomycetes and a relatively few basidiomycetes to bear separate and nomenclaturally valid names for the teleomorphic (sexual) state and each associated anamorphic (conidial) state (see Fig. 1). For example, the teleomorph of a widely distributed and important pathogen of whiteflies, aphids, thrips and other insects, *Verticillium lecanii*, is a much less common entomopathogenic ascomycete, *Torrubiella confragosa* (Evans & Samson, 1982). The ICBN allows but does not *require* all anamorphic states to bear names, and an example of this is Gams' (1971) characterization of a "*Verticillium* species" as the "conidial form of *Cordyceps militaris*" equal in status and detail of treatment with all other species in his monograph of this genus.

A. Ascomycete with single anamorphic state

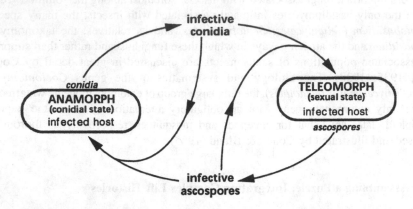

B. Ascomycete with two synanamorphic states

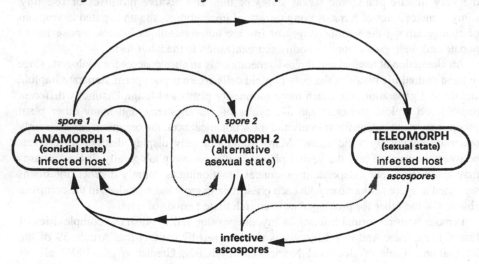

Figure 1. Diagrammatic life cycles of an entomopathogenic ascomycete with a sexual (teleomorphic) state in the *Clavicipitaceae* (*Pyrenomycetes: Hypocreales*) and one or two conidial (anamorphic) states in the *Hyphomycetes*. The conidial state(s) may cycle indefinitely on susceptible host populations, but the lighter paths from ascospores (or their secondary spores) directly back to the teleomorph and for a cycling of Anamorph 2 indicate that these events are probably very rare. Most clavicipitaceous ascomycetes produce only a single conidial form, but when alternative synanamorphic states do occur (B, Anamorph 2), the alternative state is most often a developmental stage of the other morph chlamydospore (some type of environmentally resistant spore) or a comparatively less common conidial form; Anamorph 2 may appear in sequence on a specimen (but not often simultaneously) or in culture but it would be rare for Anamorph 2 to recycle on susceptible hosts as an independent infective state of the fungus. The teleomorph must and each of its associated anamorphic states may bear separate nomenclaturally valid names.

This allowance of two or even more valid names applicable to a single organism makes many zoologists and botanists uncomfortable and may lead them to regard the mycologists who must deal with these multiple names to be either lazy or crazy. The application of powerful molecular approaches to find relationships among fungi - see the discussion by Nakase and Hamamoto in Chapter 3 of this book - has led a significant number of systematic mycologists to believe that the maintenance of separate taxonomic category for anamorphic fungi is superfluous (Taylor, 1995). Nonetheless, the practical reality for the pleomorphic fungi is that separate names for their distinct stages may be indispensable because both the anamorphic (asexual) and the teleomorphic (sexual) states may never or only rarely occur simultaneously. Even if it is possible to find and correctly to associate conidial and sexual states as belonging to a single fungus, these independently named morphs may only be found at completely different times or locations.

The relative importance of the anamorphic states as opposed to their teleomorphic states is underscored by the fact that these conidial forms are *much* more commonly found than are their equivalent sexual states. Some ubiquitous conidial entomopathogens such as *Beauveria bassiana* and *Metarhizium anisopliae* have been known for much more than 100 years but have never been definitively connected with any sexual state; it is thought that many deuteromycetes may have lost the capacity to undergo sexual reproduction and depend on parasexuality (mitotic recombination) or heterokaryosis for gene flow, or that they may even have become completely clonal in their reproduction.

Entomopathogenic fungi, and most especially those ascomycetes in the family *Clavicipitaceae (Pyrenomycetes: Hypocreales)* provide very useful illustrations for a number of the related taxonomic and systematic issues involving morphological convergences, overlapping and confused generic concepts, and biogeographic considerations. Figure 2 illustrates the numerous connections between the main entomopathogenic genera of the *Clavicipitaceae (Cordyceps, Hypocrella, and Torrubiella)* and an unusually large array of anamorphic genera; it also indicates the known pairings of synanamorphic states among these conidial partners to *Cordyceps* species. *Hirsutella*, with its direct connections to *Cordyceps* species and diverse synanamorphic relationships with several other hyphomycete genera, is immediately seen to be a special problem.

The early taxonomic history of *Cordyceps* and its allies focused entirely on the discovery and description of the (sexual) fruiting bodies of these fungi. No consideration was given to the existence of any conidial states for these fungi. Cooke's (1892) remarkable book was the earliest review of insect fungi and is still useful (especially for its treatment of *Cordyceps* species) and fun to read. The modern history of *Cordyceps* taxonomy is almost wholly the work of Yosio Kobayasi (1941, 1982; Kobayasi & Shimizu 1982, 1983) who treated hundreds of *Cordyceps* species, many of which he described although, sadly, Kobayasi's approach to *Cordyceps* and *Torrubiella* paid very little attention to the evidence of any conidial states or to the connections of conidial states with the teleomorphs. Tom Petch's extensive collections of insect fungi in Ceylon led to a tandem monograph (Petch, 1921) of *Hypocrella* and its anamorphic state, *Aschersonia*; both of these genera are pathogenic exclusively to whiteflies and scale insects. No more modern monograph of either of these genera has been produced; in fact, there is lamentable lack of recent and useful monographic treatments for a shockingly large proportion of the most important entomopathogenic fungi.

A key concept about the systematics of the conidial fungi, currently treated as the division *Deuteromycota*, is that these fungi are classified in *form*-genera defined by their general morphological characters. The relative simplicity of the mechanisms for producing conidia makes it likely that phylogenetically unrelated fungi may develop conidiogenetic mechanisms and conidial shapes similar enough for them to be classified together in morphologically based form-genera. Many deuteromycete genera may be monophyletic and, thus, natural taxa according to the current systematic standards. Nonetheless, some important conidial genera are polyphyletic since their species are connected to phylogenetically diverse teleomorphs; *Paecilomyces* and *Verticillium* contain important entomopathogens, and both genera will eventually have to be split with some inevitable confusion and distress for many "consumers" of mycological systematics. The possibility of morphological convergence by phylogenetically disparate fungi also raises the possibility that a single conidial form-species might be connected to more than one teleomorph; to date, this possibility remains hypothetical rather than actual except for taxa such as *Beauveria bassiana* and *Metarhizium anisopliae* that have been shown to be unresolved species complexes.

The classification of conidial fungi has gone through several significant major upheavals. The first major classification for conidial genera was an intellectually satisfying and highly artificial system devised by Saccardo (1899) that depended on readily observed morphological characters such as the colour and morphologies of the conidia and the hyphae bearing the conidia. In a more biological and less architectural approach Vuillemin (1910, 1911) put primary taxonomic emphasis on conidiogenous cells and the mechanism by which they produce conidia. This approach was greatly amplified by Hughes (1953) who added more modes of conidiogenesis and began a sweeping taxonomic overhaul of deuteromycete genera by confirming the mode of conidiogenesis in the type species of these genera and transferring taxa not sharing the same conidiogenetic pattern as the type species to other genera. The Hughesian system and its later modifications placed conidial fungi into the two currently recognized classes, *Hyphomycetes* (with conidiogenous cells borne on exposed surfaces) and *Coelomycetes* (with conidiogenous cells formed inside flask-like fruiting bodies called *pycnidia*). This developmentally based Hughesian classification logically required abandoning any attempt to assign conidial fungi to orders, families or any other taxonomic ranks between the ranks of class and genus. The removal of such standard taxonomic guideposts did little to aid in revealing the phylogenetic relationships among these fungi. It did, however, underscore the artificiality and provisional status of these form-genera while assuming that the underlying phylogenetic relationships among these conidial fungi, and also with their sexual states, would be discovered and incorporated into the classification later. Molecular techniques combined with phylogenetic (cladistic) analyses are now providing the types of "missing" information needed to reveal these relationships and to revise the systematics of the organisms concerned. It is noteworthy that none of these molecular or phylogenetic techniques to gather and to analyse data existed when the Hughesian system was proposed and accepted.

2.1 MORPHOLOGICAL CONVERGENCES AND OVERLAPPING GENERIC CONCEPTS

Both *Verticillium* and *Paecilomyces* have been segregated into two infrageneric groups (as sections rather than as subgenera) that do reflect the taxonomically significant phylogenetic differences among their species. Within *Verticillium* (Gams, 1971), the Section *Verticillium* accommodates species with erect conidiophores that are well differentiated from the vegetative hyphae; these species (which include all the major phytopathogens in *Verticillium*) are connected where known with teleomorphs in *Nectria* and other genera of the *Hypocreaceae (Pyrenomycetes: Hypocreales)*. *Verticillium* section *Prostrata*, which includes all of the insect and nematode pathogenic species, accommodates species with conidiogenous cells borne on prostrate hyphae that are neither erect nor morphologically differentiated into distinct conidiophores, and all teleomorphic connections known for these species are with species also in the order *Hypocreales* but exclusively with the family *Clavicipitaceae* (with species of *Cordyceps* and *Torrubiella*).

As treated by Samson (1974), the phylogenetic divergence of the two sections within *Paecilomyces* is even more dramatic, but is also supported by diagnostic characters that readily separate these sections: Species in *Paecilomyces* Section *Paecilomyces* have mononematous (single) and generally erect conidiophores, form conidia that are mostly olivaceous to dark in colour, and have teleomorphic connections to ascomycetes in the class *Plectomycetes* that produce cleistothecia (more or less spherical, closed fruiting bodies enclosing spherical indehiscent asci), to genera such as *Byssochlamys*; species in *Paecilomyces* Section *Isarioidea* have mononematous or synnematous (fasciculate) conidiophores, form conidia whose colours range from colourless through a series of bright colours, and with teleomorphic connections (where known or suspected) to the genera *Cordyceps* or *Torrubiella* in the *Clavicipitaceae (Pyrenomycetes: Hypocreales)*, fungi whose pear- to flask-shaped fruiting bodies (perithecia) have an apical opening (ostiole) and contain elongated, dehiscent asci.

When *Paecilomyces* and *Verticillium* are finally revised to make them phylogenetically homogeneous genera, it is important to note that the type sections in both genera (those which will continue to bear the familiar generic names) are the ones that do *not* include the entomopathogenic species; the insect pathogens will eventually have to be placed in other genera bearing altogether new generic name (for *Verticillium*) or, possibly for *Paecilomyces*, one resurrected from its current synonymous status.

A glance at Figure 2 suggests that the most taxonomically complicated genus among the anamorphs of *Cordyceps* species is, indeed, *Hirsutella*. All *Hirsutella* species are pathogenic for insects or mites although *H. rhossiliensis* was originally described from nematodes (but does affects mites). *Hirsutella* species do not proliferate in nature as facultative saprobes even though most grow relatively well in axenic culture. Another taxonomic confusion with *Hirsutella* species is that the conidiogenous cells (which have a swollen base and bear one or more elongated, hair-like necks) and, in some instances, the conidia of these fungi can bear a striking morphological similarity to other anamorphs of *Cordyceps* in a wide range of other hyphomycete genera.

A strong resemblance to *Hirsutella* can be seen in species of *Beauveria* (when the conidiogenous cells have produced the first conidia and before the denticulate rachis proliferates), *Paraisaria* and *Syngliocladium* (especially in polyphialidic conidiogenous cells where several conidiogenous sterigmata may form on each conidiogenous cell), *Tolypocladium*, some *Mariannaea* or *Verticillium* species, and even to *Culicinomyces* species (which affect aquatic mosquito and blackfly larvae but have not yet been linked to any teleomorph).

   The taxonomic difficulty with *Hirsutella* is compounded by the lack of any monographic
treatment of the genus; it still awaits a formal synthesis to sort through its taxonomy
and biology, to condense a rich and diverse literature covering more than a century of
research in a myriad of publications in many languages. In other words, the literature for
Hirsutella is typical of most entomopathogenic fungal genera except that it is more
extensive because of its large number of species. The obvious morphological similarities
and synanamorphic relationships with five other genera whose morphologies are not
obviously similar to that of *Hirsutella* underscores the need for a thorough morphological,
developmental, and molecular study of this genus and its diverse interrelationships within
the *Clavicipitaceae*.
   The sheer diversity of associations between *Hirsutella* and so many other conidial
genera suggests that one or more characters of *Hirsutella* fulfill an essential ecological
function for these synanamorphic genera. Humber & Rombach (1987) warned against
the dangers of beginning to synonymize linked (synanamorphic) conidial genera to *Hirsutella*
(in an attempt to begin to combine hyphomycete form-genera into natural genera having a
variety of synanamorphic states); to begin such a process would destroy all morphological
definition of the conidial genera and remove any rational bases for classifying these
fungi *except* for their synanamorphic relationships and linkages to *Cordyceps* species.

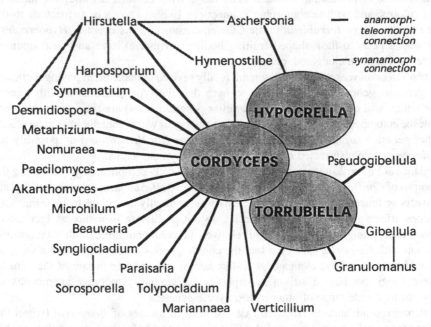

Figure 2. Complex relationships within the entomopathogenic *Clavicipitaceae*. The major teleomorph genera
(*Cordyceps*, *Hypocrella* and *Torrubiella*) genera have known connections to a very diverse array of anamorphic
genera (light type) of the *Hyphomycetes*. Several of these conidial genera have morphologically distinct alternative
conidial states (synanamorphs, indicated with grey lines) classified in other genera. Note the large number of
connections between *Hirsutella* and other anamorphic genera as well as with teleomorphs in *Cordyceps*. Some other
entomopathogenic hyphomycete genera not currently connected to any teleomorphic genus will probably be found
eventually to fit into this complex in the *Clavicipitaceae* either by virtue of the discovery of a teleomorph or by
gene sequence data matching those known for other clavicipitaceous fungi.

Entomogenous *Verticillium* species have long been known to be linked to *Cordyceps* and *Torrubiella*. The conidial state of *Cordyceps militaris* is now treated as a *Verticillium* but was once thought to be a *Paecilomyces*; Brown & Smith (1957) noted that the conidia of this fungus initially form imbricate chains (with conidia lying side by side and at an angle to the axis of the conidial chain) that upon the secretion of slime from the conidiogenous cells then transform into mucoid spore balls. *Mariannaea* (Samson, 1974) was originally described for taxa with exactly this type of sporogenesis, with imbricate conidial chains sliming down into spore balls. Liang (1991) saw this pattern for the anamorph of *Cordyceps pruinosa* and described that fungus as *Mariannaea pruinosa*. This pattern of conidiogenesis also occurs in a large number of isolates from pear thrips (*Taeniothrips inconsequens*; Thysanura, Thripidae) in the northeastern United States; except for the initial production of imbricate conidial chains that soon slime down, these isolates are otherwise wholly similar to *Verticillium lecanii*. Molecular studies have confirmed that these thrips isolates to not have molecular profiles matching those of a wide range of isolates from the *V. lecanii* species complex (R.A. Humber, unpublished).

How the entomopathogenic anamorphs following the *Mariannaea* pattern should be treated depends on the teleomorphic connections: The teleomorph of *Mariannaea elegans*, the type species, is a *Nectria* (Hypocreales: Hypocreaceae) (Samuels & Seifert, 1991); it is probable that the entomopathogenic species of *Mariannaea* and *Verticillium* will all prove to be clavicipitaceous, closely related, and will probably have to be placed together in whatever genus is eventually chosen to accommodate the species of *Verticillium* Section *Prostrata* once the systematics of *Verticillium* is clarified and restricted to species with teleomorphs in the *Hypocreaceae*.

The systematic controversies about the fungi in Figure 2 are not confined to the generic level. At various times the phytopathogens (e.g., *Claviceps*, the genus that causes ergot of rye and other grasses) and the entomopathogens now placed together in the family *Clavicipitaceae* in the *Hypocreales* have been retained in one or two families (with the insect fungi in the *Cordycipitaceae*), and the one or two families placed either in the *Hypocreales* or in a separate order *Clavicipitales*. Recent molecular studies (Spatafora & Blackwell, 1993; Blackwell, 1994) indicate that *Claviceps, Cordyceps* and their allies are a monophyletic group (despite a history of host-shifting among insect, fungal, and plant hosts) occupying a clade nested wholly within the Hypocreales.

Some of the greatest mycological surprises (along the lines of "everything is not what is seems to be") have been linked to the revelation of the correct life histories for some of the smallest fungi associated with insects. Roland Thaxter, a founding father of American mycology, planned to be an entomologist but was diverted to mycology by the repeated annoyances of fungi killing the insects whose life histories he was studying (Thaxter, 1888). Thaxter described and exquisitely illustrated more than 100 genera and nearly 1300 species of the *Laboulbeniales*, perithecial ascomycete ectoparasites on insects and other arthropods. Thaxter's passion for the small fungal ectoparasites on insects led him to describe several genera of sterile or conidial (non-ascomycetous) ectoparasites, most of which produced tiny thalli attached to their hosts' cuticles by blackened holdfasts in a manner much like the *Laboulbeniales*; these fungi became known as the "Thaxteriolae" after the genus *Thaxteriola* named in Thaxter's honour by a major scientific rival, Carlo Spegazzini. The initial emphasis on the morphological descriptions of these fungi without knowing their life cycles has led to some latter-day surprises about the mycological "reality" of some of them. That the "Thaxteriolae" are phylogenetically heterogeneous suggests that there may be strong selective forces favouring the morphological convergences of these reduced, ectoparasites; some of these fungi have yielded unexpected clues toward solving long-standing questions about the origins and of the *Laboulbeniales* themselves (Blackwell, 1994).

The two-celled "thalli" of *Thaxteriola* and the similar genus *Acariniola* (Majewski & Wisniewski, 1978a,b) are now understood to be the insect-borne *ascospores* of fungi of pyrenomycetous ascomycetes in the genus *Pyxidiophora* (Blackwell *et al.*, 1986a, 1986b; 1989; Blackwell & Malloch, 1990). Mature *Pyxidiophora* asci contain eight fully formed "thalli" of *Thaxteriola* or *Acariniola* lined up and awaiting their forcible dispersal from the ascus. Yet another of the Thaxteriolae has been found to be altogether different from what it was thought to be. The ovoid to elongated thalli of *Amphoromorpha* species are one-celled, attach to the substrate by a blackened foot, and produce small spores by the internal cleavage of the distal portion of the cell. Blackwell & Malloch (1989) showed that these "thalli" (and probably those of *Amphoropsis* species as well) are not hyphomycetes but the secondary capilliconidia of *Basidiobolus* species (*Zygomycetes: Entomophthorales*). Further, *Amphoromorpha* species have no real specificity for insects since they may also occur on fungi of the *Mucorales* or *Laboulbeniales*. *Basidiobolus* cells can cleave internally to form several to many small, amoeboid cells (this phenomenon is best known from spores ingested by cold-blooded vertebrates whose faeces are the classic substrate for isolating *Basidiobolus*). Drechsler (1955, and in later papers) demonstrated cleavage of only parts of *Basidiobolus* conidia and illustrated capilliconidia distinctly resembling *Amphoromorpha* thalli.

2.2   BIOGEOGRAPHY AND LIFE CYCLES

Verticillium lecanii is a globally distributed, taxonomically confused hyphomycete pathogen of whiteflies, aphids, thrips, and other insects as well as a mycoparasite on rust fungi or even a saprobe on leaf surfaces and in plant detritus. A culture isolated from ascospores of *Torrubiella confragosa* (*Clavicipitaceae*) on scale insects from the Galápagos Islands produced a *Verticillium lecanii* anamorph (Evans & Samson, 1982); the isolation of cultures from freshly discharged ascospores may be the simplest, fastest, and most reliable way to connect teleomorphic and anamorphic states since such cultures rarely produce the teleomorphic but do usually begin quickly to show the anamorphic conidial state. The molecular characteristics of this Galápagos isolate of *V. lecanii* group it tightly with those of an Indonesian isolate of *V. lecanii* from *Coccus viridis* (*Homoptera: Coccidae*), the site and host from which this species was described differ markedly from several other distinct groups within the *V. lecanii* species complex (R.A. Humber, unpublished).

In addition to the greater diversity of *Cordyceps* and *Torrubiella* species from northeastern Asia and Japan (Table 1), this is the region where teleomorphic states have been found to occur for several universally distributed hyphomycetes for which teleomorphs have never been confirmed to exist elsewhere. *Cordyceps brongniartii* was described by Shimazu *et al.* (1988) with unimpeachable evidence as the teleomorph of *Beauveria brongniartii* (a widely distributed, nearly cosmopolitan pathogen mostly affecting scarab beetle larvae);

REGION	Number of species present [2]		Type Localities [3]	
	Cordyceps	*Hypocrella/Aschersonia*[4]	*Cordyceps*	*Torrubiella*
Japan	31	3	104	24
Northern Asia[5]	20	3	11	4
Southeastern Asia[6]	22	13	10	5
Indonesia/New Guinea/Philippines	23	11	24	2
Australia/New Zealand/Pacific Islands	16	9	18	5
South America	41	15	35	8
Mexico/Central America/Caribbean	15	12	10	1
United States/Canada	29	5	27	4
Europe	18	0	25	3
Africa/Madagascar	7	8	22	3

[1]   Data included here are drawn from Kobayasi (1941) for the numbers of *Cordyceps* species, Kobayasi (1982) for type localities of *Cordyceps* and *Torrubiella* species, and Petch (1921) for the distributions of *Hypocre¹·¹* and *Aschersonia* species. This table does not reflect an attempt to collate all available literature about the presence of particular species of type locations for species described after or omitted from these publications.

[2]   A count of the total number of described species present in this region.

[3]   Type localities: region from which the type specimen was collected.

[4]   Each known teleomorph/anamorph pair or each species of *Aschersonia* or *Hypocrella* without a known teleomorph or anamorph is counted as a single species.

[5]   Includes Korea, China, Taiwan, and Tibet.

[6]   Includes Vietnam, Cambodia, Burma, Thailand, Malay Peninsula, India, Sri Lanka and islands in the western Indian Ocean.

C. brongniartii appears to be very rare and has been found so far only in Japan. Sung (1996) found a *Beauveria* species produced by cultures isolated from ascospores of Korean collections of a fungus identified as *Cordyceps scarabaeicola*. In China, Liang *et al.* (1991) described *Cordyceps taii*, a new species whose anamorph is *Metarhizium taii* (which is indistinguishable from *M. anisopliae* except that a very few conidia have a single transverse septum); this is the first credible teleomorphic connection for any *Metarhizium* species.

That teleomorphic states for species of *Beauveria* and *Metarhizium*, the entomopathogens with the most cosmopolitan distributions and diverse host ranges of any insect fungi, have not been found elsewhere remains a puzzlement. The possibility is that the high diversity of clavicipitaceous taxa in eastern Asia combined with the presence of these long-sought teleomorphs suggests the possibility that these fungi originated there and that their capacity for sexual reproduction has been lost in other parts of their geographic ranges. Nothing is yet known about the homothallic or heterothallic nature of sexual reproduction in *Cordyceps* or any of the other entomopathogens of the *Clavicipitaceae*. Methods devised by Sung (1996) to obtain abundant *in vitro* production of ascomata by several *Cordyceps* species open the possibility of solving many intriguing questions about the biology of *Cordyceps* species and their anamorphic states.

It is possible to do a bit better, however, when considering the distributions of species within some genera. *Cordyceps, Hypocrella*, and *Torrubiella* are reported from throughout the world in the northern and southern hemispheres, in the tropics and temperate zones. Nonetheless, data from a few essential publications on the regional occurrence of these species and of the type localities for their species (Table 1) provide useful information. These geographical data do suggest that the diversity of *Cordyceps* and *Torrubiella* is greater in Japan and northeastern Asia than elsewhere. Petch's (1921) data on *Hypocrella* and its tightly associated *Aschersonia* anamorphic state clearly indicates centres of diversity for these genera in southern Asia and the northern countries of South America as well as throughout Central America and Caribbean islands. Because Kobaysi was a widely travelled and capable collector with an outstanding ability to find clavicipitaceous pathogens of insects, the high diversity of species of *Cordyceps* and *Torrubiella* reported for Japan (Kobayasi, 1982) cannot be discounted as an artifact. The increase in the recorded diversity of *Cordyceps* in Japan between Koyabasi's 1941 monograph and his 1982 listing of species and their type localities exemplifies the intensity of his prolific studies on this genus. This high species diversity in Japan - and in Korea (see Sung, 1996) - also suggests the extraordinary suitability of habitats and/or of the insect fauna in northeastern Asia to sustain such a rich flora of the entomopathogenic *Clavicipitaceae*.

3   Specificity

The taxonomic characters traditionally used to circumscribe and to classify fungi include all aspects of their morphology as well as their vegetative and reproductive developmental patterns. For fungal parasites and pathogens, some of the next most important diagnostic characters tend to deal with their specificity. This seems proper since "specificity" is a character that integrates much of the genome and many aspects of the individual biologies of a parasite or pathogen and its hosts as well as the whole range of possible interactions between the invading organism and its host, geography and other abiotic factors. Among all of the factors contributing to the observed specificity, at least two of them, the host range (those organisms actually affected by a parasite or pathogen) and the geographic range in which an organism occurs, are taxonomically significant; many other factors such as those that determine whether a propagule of any specific fungus will or will not be able to germinate, to grow and to establish itself on or inside a potential host play no part in the taxonomy of the host or its pathogens and parasites despite their inherent biological interest.

The use of the terms "parasite" and "pathogen" in this chapter already reflects a significant amount of information about the nature of relationship between two organisms. Many people seem to use "parasite" and "pathogen" interchangeably, but to do so blurs a real distinction that should be preserved: Parasites live on or in their host and derive their nutrition from the host; they may cause disease or discomfort for the host or may not even be much noticed, but the result of their association with the host will *not* be the death of the host. Pathogens also cause disease but, on the other hand, do cause the host's death in at least the majority of successfully established infections. For most fungal pathogens or parasites associated with insects and other invertebrates the specificity for the hosts is a basic character at one or more taxonomic ranks.

It is easy to confirm the activity of a virulent pathogenic fungus but it can be maddeningly difficult to confirm the activity of a weak or facultative pathogen by using bioassays to test Koch's postulates (requiring the putative pathogen to infect a host successfully and then to be recovered after the host's death). A bioassay is a very artificial system in which to handle both a target host and a potential pathogen; the more stringent the bioassay conditions, the greater the likelihood that the increased stress on the host and/or fungus may artificially increase the incidence of facultative infections that may have a much lower likelihood of occurring in a natural setting. This situation is biologically analogous to the Heisenberg Uncertainty Principle, a fundamental paradox describing the decreasing likelihood of observing subatomic particles as the exactitude of the observation technique increases.

The understandable expectation that a fungus growing on insect hosts is most probably a pathogen of that host can occasionally be suspect or wrong. A true fungal pathogen of an insect should produce a final biomass that is smaller than that of the affected host. If the total mass of the fungus exceeds that of the affected host(s), then a close look at the nature of the fungus-host relationship is in order. Three *Hypocrella* species (*Hypocreales: Clavicipitaceae*) produce stromata that are hundreds of times larger than the scale insects that this genus routinely affects (Hywel-Jones & Samuels, 1998). Two possible relationships between these three species and their hosts are suggested: Either these species do not kill their hosts outright but utilize them as a continuing source of nutrients (much as *Septobasidium* species use their scale insect hosts; Couch, 1938) to be transmitted to the fungus from the host plant through the body of the living scale insect. Alternatively, the fungus is able to absorb a significant portion of its nutrients directly from the plant on which the fungus/scale combination occurs. Either of these possibilities is a complete break from the usual relationship expected between *Hypocrella* and scale insects. Confirmatory anatomical studies are not available to suggest which of these possibilities may be correct, but the latter scenario (that of a facultative phytoparasitism) is intriguing in view of the apparently closer relationship of *Hypocrella* species to the phytopathogens than to the entomopathogens of the *Clavicipitaceae* (J.W. Spatafora, unpublished).

3.1   HOST RANGE

With virtually every parasitic or pathogenic organism, the spectrum of hosts that can be affected is recognized almost instinctively to be a quintessentially important character. The host range is not just taxonomically significant, but it colors and shapes nearly every approach to the biology of the parasite or pathogen and to its relationship with its host. A few entomopathogens such as *Beauveria bassiana* and *Metarhizium anisopliae* appear to have ubiquitous distributions and can successfully attack nearly any insect they contact; Veen, (1968), Li (1988), and Humber, (1992) provide partial listings of the vast host ranges for these species.

It is probably more characteristic for entomopathogenic fungi to show some degree of specificity, usually to a single family or cluster of related families of host insects; *Nomuraea rileyi* (*Hyphomycetes*) is found globally as a pathogen of hosts in the *Noctuidae* (*Lepidoptera*) but may also occasionally attack hosts from other families in the super-family *Noctuoidea*. Another example of this very common level of "familial" specificity is *Pandora neoaphidis* (*Entomophthorales: Entomophthoraceae*), a globally distributed, relatively common pathogen of a wide range of aphids (*Homoptera: Aphididae*). A closely related fungus, *Pandora delphacis*, affects homopterans mainly in the *Delphacidae* and *Cicadellidae* but is known from laboratory tests to be able to infect aphids. This situation is similar to the opportunistic bacterial entomopathogens such as *Bacillus sphaericus* and *Bacillus thuringiensis* which generally have limited host ranges (see Chapter 7).

The most extreme sorts of specificities for insect pathogens may be those, again, of other entomophthoraleans: Species of *Massospora* are highly specific pathogens of gregarious cicadas (Soper, 1974); each *Massospora* species affects only a single cicada species, and no cicada species is known to be affected by two or more *Massospora* species. Another genus of the *Entomophthoraceae* has two species with distinctly different host preferences: *Strongwellsea castrans* affects a wide range of adult flies in the family *Anthomyiidae* while *S. magna* is known only from adults of *Fannia* species (*Fanniidae*) from the United States (Humber, 1976) and Denmark (Eilenberg & Michelsen, 1999). The ectoparasitic *Laboulbeniales* can show an even more extreme form of specificity: These fungi do show strong specificities for their host arthropods, but some of them also have specificities in which they may occur on the host body depending upon the sex of the host when fungi are transmitted from one individual to another during mating (Benjamin, 1965).

Similar to the obligate bacterial pathogens such as *Paenibacillus larvae* (which is specific for honeybee larvae) or *Paenibacillus popilliae* (which infects larvae of the Japanese beetle, *Popillia japonica*), many fungi and bacteria show some degree of pathogenicity during laboratory host range studies to host species that may never be found to be infected by that pathogen under field conditions. It is possible that such seemingly anomalous infections obtained during bioassay studies of host ranges may reflect the situation noted above that the very conditions of a bioassay test may exaggerate the apparent infectivity by a test fungus. Non-target host safety testing protocols required by the U.S. Environmental Protection Agency routinely include honeybees, *Apis mellifera*, because of their obvious importance for agriculture. There are some published reports of some very common fungal entomopathogens killing honeybees in laboratory bioassays that are not known to be natural mortality factors in wild populations of bees (Goerzen *et al.*, 1990; Vandenberg, 1990; Alves *et al.*, 1996). Any differences involving non-target beneficial arthropods between observed ("ecological") and experimental ("laboratory") host ranges of a pathogen need to be evaluated carefully by researchers and regulators alike before significant field use of the pathogen should be undertaken.

Except for a few species of *Nectria* and their *Fusarium* anamorphs (Booth, 1971; Hajek *et al.*, 1993), the *Hypocreales* do not include entomopathogens except those in the *Clavicipitaceae*. The entomopathogenicity of *Fusarium* species exemplifies the issues for fungi whose pathogenicity for insects is weak or facultative. Booth (1971) recognized *Fusarium* Section *Coccophilum* for a few primary pathogens of scale insects. However, some other saprobic or normally phytopathogenic species may occasionally affect insects; the question is whether any of these "unusual" insect mycoses reflect primary infections or facultative attacks of hosts that are weakened or injured so as to protect the *Fusarium* inoculum from exposure to the host's usual defenses; the fungus might also be found in some instances as a necrobe growing on insect cadavers.

Surprisingly few fungal pathogens affect wild populations of honeybees or other pollinators such as leafcutting bees (*Hymenoptera: Megachilidae*). The safety of fungal (and other microbial) biocontrol agents for nontarget insects, such as honey bees or endangered insect species, is obviously a critical concern for the applied use of any entomopathogenic biocontrol agent. Stringent safety tests of several fungi against *A. mellifera* show that the susceptibility of bees to major entomopathogenic fungal biocontrol agents varies from low to significant (as evidenced by increased mortality) with *Beauveria bassiana* or *Metarhizium anisopliae* (Vandenberg, 1990; Alves *et al.*, 1996) to undetectable with *Culicinomyces clavisporus* (Cooper *et al.*, 1984; a mosquito-pathogen) or *Paecilomyces fumosoroseus* (Sterk *et al.*, 1995; a pathogen affecting many diverse terrestrial insects). Even though *Beauveria* and *Metarhizium* have some negative impact on bees in bioassay studies, these same fungi are *not* known to infect wild populations of bees in any significant way and may be functionally safe for bees exposed during the applied use of these biocontrol agents in the field (Alves *et al.*, 1996).

Despite the apparent general safety for bees of the most common fungal entomopathogens, a few fungi do cause very specific and sometimes economically significant diseases of bees. *Aspergillus flavus* and/or *A. parasiticus* cause stonebrood, a disease that kills and mummifies larvae in honeybee brood combs. The bee-pathogenic *Aspergillus* species are unusual since very few species of this huge genus affect insects. *Aspergillus* species do seem to be very rare as pathogens of wild insect populations but may cause significant mortality among insects in stressed and crowded conditions (e.g., in beehives or commercial silkworm rearings). Even though *Aspergillus* species are unusual pathogens of most insects, these fungi produce a diverse range of compounds that can be potent insect toxins and potentially useful as biorational pesticidal compounds (Wicklow & Dowd, 1989; DeGurman *et al.*, 1993; Dowd *et al.*, 1994).

More important than the aspergilli, however, are the species of *Ascosphaera* (Skou, 1972, 1982; Bissett *et al.*, 1996), ascomycetes whose ascospores form in spherical spore-cysts and which are the causal agents of chalkbrood in larval bees. It is worth noting that the paucity of natural fungal pathogens affecting bees may be less a matter of their low susceptibility to various entomopathogens than to these insects' hygienic behaviours that tend to contain or to eliminate pathogens from a hive before epizootics can break out (Spivak & Gilliam, 1993). The *Ascosphaerales* includes one fungus that appears to have undergone a major host-shift: These cleistothecial fungi are associated exclusively with bees, and while all *Ascosphaera* species are pathogens, *Bettsia alvei* (Skou, 1972, 1975), a monotypic genus, occurs in hives as a saprobe on pollen grains in the honey combs rather than as a pathogen on bee larvae in the brood combs.

3.2   BIOGEOGRAPHIC CONSIDERATIONS

Many of the most important hyphomycete entomopathogens have cosmopolitan distributions; the teleomorphs of these fungi, however, may be much more geographically restricted (with the extreme examples noted above of the teleomorphs of *Beauveria* and *Metarhizium* species having been found only very recently and, so far, only in Japan, Korea, and China). The biotic and abiotic conditions that stimulate or allow the clavicipitaceous teleomorphs to develop remain unknown, but it is sure that *Cordyceps* fruiting bodies are often widely dispersed and only rarely or never very numerous in any given collection trip to a site where they are known to occur. No ecological studies have been made to plot the relative incidences on a site of the populations of susceptible insect hosts, of the available inoculum of the fungal pathogen, nor of the incidence of infections on the site resulting in the production of either the anamorphic or the teleomorphic state of the fungus. Once some of these factors are understood for one or two model systems, some of the disparities in the geographical distributions of anamorphs and their associated teleomorphs might be better understood.

Some *Cordyceps* species do seem to be geographically restricted (Sung, 1996). *C. sinensis*, a species much esteemed for its medicinal properties, affects hepialid moth larvae on montane sites (above 3000 m) in southern and western China, Tibet, and Nepal; this species has not been found in Korea or Japan despite years of diligent collecting. Similarly, the highly diverse flora of *Torrubiella* species in Japan (Table 1) may be related in some way to the diversity or population sizes of the arachnid and homopteran hosts of these fungi, or to some climatic factors that are especially suitable, or to other, wholly different factors that in some way account for the taxonomic proliferation of this genus throughout that island nation.

Decisions to segregate some taxa are often supported by differing geographic distributions for the entities being recognized. For example, Humber & Feng (1991) restored *Entomophthora chromaphidis* (*Entomophthorales*) from synonymy with *Entomophthora planchoniana* (*Entomophthorales*), a major aphid pathogen, on the basis of morphological differences that were also supported by the fact that *E. planchoniana* is known only from Europe while *E. chromaphidis* was known from North America and Australia but not Europe.

Geographic isolation of populations can result in speciation events, but such effects are probably stronger for sexual organisms in which outcrossing occurs than for those organisms in which reproduction is strictly clonal. The *Entomophthorales* appears to be a highly unusual example of the latter case, of a rather large, globally distributed, and eminently successful group that has undergone radiation into many families, genera, and species as (functionally) clonally reproducing fungi. Sexual reproduction may occur in this order (McCabe *et al.*, 1984) but it is homothallic in all taxa; neither do *any* vegetative cell fusions occur in these fungi except during zygosporogenesis, so no genetic variability can be introduced by heterokaryosis or parasexuality.

3.3   APPARENT SPECIFICITY AS A SAMPLING PROBLEM

The interpretation of host specificity for entomopathogens must always be regarded as a problem in population sampling. To find a taxon only once, and especially if it is new, leaves the impression that an entomopathogen is rare and may have extremely narrow host and /or geographic ranges. With entomopathogenic fungi this impression may be correct since many of these fungi have rather widespread distributions and are found with some regularity. As more collections of a pathogen are made in a greater variety of locations, one or more subsequent collections from the same or different hosts begin to give a different impression of the host and geographical ranges but even these new impressions may also be misrepresentative.

It is easy to forget that the cumulative knowledge about locations and hosts in which parasitic or pathogenic fungi may occur is wholly limited to the activities of one or more collectors. Entomogenous fungi do occur in places and at times where and when nobody has specifically sought them. They occur in places that have never been explored for such fungi. And they certainly may affect hosts at some time or place that have never been seen and noted. Such unseen, unrecorded fungus-host-site associations are data that may remain forever "missing" until they are collected by a suitably trained or observant collector.

Many of the most notable contributors to the knowledge of the flora of fungi associated with insects or other invertebrates are those persons who spent the most time collecting them. Tom Petch spent nearly thirty years in Sri Lanka at the beginning of the 20th century and made an impressive number of collections of entomogenous fungi during that time, but primarily in the few locations where he spent the majority of his time; the resulting string of publications describing numerous new taxa of entomopathogenic fungi appeared primarily in the *Transactions of the British Mycological Society* from the 1920s through the 1940s.

In the United States, Charles Drechsler had extraordinary success in finding and describing a huge number of taxonomically diverse fungal pathogens of nematodes, tardigrades and other soil-dwelling invertebrates largely because he had enough under-standing and patience to allow cultures to develop on an open-ended basis as small ecosystems long past the point where most mycologists would throw them out as hopelessly overridden with bacteria, mites, nematodes, and other "nasties" that would cause most mycologists to consign these dishes to the autoclave. And during the course of his studies, Drechsler also became the principal student of the soil-dwelling and mostly saprobic entomophthoralean fungi in the genus *Conidiobolus* (see King, 1976, 1977) when he discovered that these fungi can be routinely recovered by a special isolation technique that depends on the forcible discharge of their conidia (Drechsler, 1952). Most of Drechsler's collecting activity was confined to the southeastern United States.

Another series of publications describing numerous *Conidiobolus* species from soil and plant detritus in southern India (publications mostly by Narasimhan, Srinivasan, and Thirumalachar; see King, 1977) and a much less intensive effort to examine the flora of *Conidiobolus* in England (Waters & Callaghan, 1989) suggest that many new taxa of

Conidiobolus await description throughout the world wherever and whenever the basic effort is made to collect them.

Strongwellsea (*Entomophthorales*), which forms gaping holes in the abdomen of adult flies and disperses its conidia from a cup-like hymenium inside the hole, again offers another useful example. Batko & Weiser (1965) described this genus from two slides comprising 10 histological sections made from an infected fly collected in Wisconsin. They concluded from what they saw in the slides that the fungus prevented ovarial development in the affected female seed corn maggot (*Delia platura*; *Diptera: Anthomyiidae*) and thus described *Strongwellsea castrans* as a new (monotypic) genus. Batko & Weiser did not see other sections from the holotype fly or from the same series of infected flies showing normal ovaries with fully developed eggs or testes with normal spermatozoa; they were correct, however, about the parasitic castration since infected females are unable to deposit their eggs (Humber, 1976).

A later discovery of *Strongwellsea* from lesser houseflies (*Fannia canicularis*; Diptera: Fanniidae) collected in Berkeley, California, was maintained *in vivo* on the insect host and described as a new species, *S. magna* (Humber, 1976), based on differences in host specificity and the size and shape of the primary conidia. While there were reports of fungi identifiable as *S. castrans* from anthomyiid flies from sites throughout the northern hemisphere, the Berkeley collection of *S. magna* remained unique for more than twenty years. Recent intensive collections of fanniid flies in Denmark (Eilenberg & Michelsen, 1999) indicate that *S. magna* does occur there on several *Fannia* species. The Danish collections do reinforce the known specificity of *S. magna* for *Fannia* but this fungus is probably even more widely distributed than this new information indicates.

Clearly, the known host and geographic ranges of the huge number of entomopathogenic fungi that have been collected only once or very infrequently are subject must be regarded as "works in progress." Narrow specificities should be accepted as "real" only when host specificity has been confirmed by studies with a wide range of potential hosts or when a geographic (or host) restriction of a taxon is confirmed by its repeated association with a few specific sites (or hosts) and an accumulation of substantial quantities of negative data from other sites or hosts.

3.4   SPECIFICITY AND SPECIES COMPLEXES

Commonly occurring, globally distributed species that affect a very diverse spectrum of hosts and harbour an appreciable range of morphological and genetic variability should be treated as probable species complexes. The use of random amplified polymorphic DNA (RAPD) or restriction fragment length polymorphism (RFLP) analyses is now the fastest and, in many ways, the easiest way to confirm the existence and to speed the resolution of taxa within these complexes. Molecular studies (including the use of allozyme polymorphisms in addition to more direct genomic examination by RAPD or other techniques) co firm the existence of significant genetic variability within *Beauveria bassiana* and *Metarhizium anisopliae* (Mugnai *et al.*, 1989; St. Leger *et al.*, 1992, 1993; Bridge *et al.*, 1993; Cobb & Clarkson, 1993; Fegan *et al.*, 1993) that will eventually become the basis for the taxonomic splitting of these species into smaller taxa whose overall biological characters will, hopefully, be more homogeneous.

4 Molecular Systematics and Fungal Entomopathogens

Recent decades have witnessed a revolution in the techniques used for taxonomic research. The invention of the polymerase chain reaction (PCR) that earned Kary Mullis the 1993 Nobel Prize for Chemistry has been the single most significant technological advance for taxonomic research and has brought the use of molecular systematics techniques into general use. Before the advent of PCR techniques, most molecular systematics research was confined to such approaches as comparisons of fatty acid spectra, to chromatographic techniques using pigments or protein extracts, the electrophoretic comparison of enzyme polymorphisms (isozymes), and immunological approaches to determine whether organisms shared antigens.

Once armed with the PCR reaction and thermocyclers, the analyses of restriction fragment length polymorphisms, of random amplified polymorphic DNA profiles, of gene sequences (most notably for fungi, several portions of nuclear genes for ribosomes; see Chapter 4), and several other newer techniques has enabled many diverse and often imaginative applications of molecular systematic techniques: PCR approaches have been used to solve questions of taxonomy and phylogeny, to track the establishment and dispersal of biocontrol agents in the field, to undertake general diagnostics in some specialized situations, and to fingerprint organisms for regulatory, patent, and other commercial purposes.

The diverse biochemical and molecular techniques available are not uniformly useful at all levels of the taxonomic hierarchy (see Fig. 3). Each has its strengths and weaknesses, and there is no reason that any of these approaches should be abandoned in favour of any other. The level of confidence in taxonomic conclusions increases when two or, preferably, more wholly different technological approaches support the same conclusions.

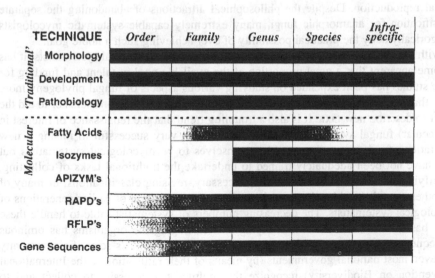

Figure 3. Diagrammatic presentation of taxonomic ranks at which the most common taxonomic techniques and approaches are useful for taxonomic and systematics studies of entomopathogenic fungi. The lighter the bar, the less applicable the particular technique is likely to be for the particular taxonomic rank.

The "traditional" (classical, descriptive, alpha-taxonomic) methods for fungal systematics have been useful at nearly every level of the taxonomic hierarchy and must remain the foundation on which fungal systematics forever rests. It is reassuring that traditional taxonomies may be refined after molecular studies, but that they are generally validated rather than overthrown by the newer techniques. Up to now, the most common biochemical and molecular systematic techniques have been used most successfully at the lower ranges of the taxonomic hierarchy (usually at the generic level or below). Phylogenetic studies, which usually depend on gene sequences, may be the most successful of the molecular approaches for studies above the rank of genus (see Chapter 2). RAPD studies are excellent vehicles for studying specific problems below the level of genus, but suffer from a lower level of reproducibility than a more specifically targeted approach to the amplification of genomic segments of the genome such as RFLP.

The common use of molecular systematic techniques, especially those comparing the sequences of ribosomal genes, have spawned a powerful enthusiasm among many mycological systematists that these sequences can, in the absence of any other information, be sufficiently informative to place an isolate among its closest relatives if not to provide an exact identification. Indeed, it has been argued that such technological powers can render the state in which a fungus may be collected, observed, or cultured irrelevant. And it has also been argued that such technical capabilities remove any need to retain separate taxonomies for anamorphic and teleomorphic states or even to retain the Division *Deuteromycota* (for the conidial or anamorphic fungi) as a distinct taxonomic entity. Such a radical view of fungal systematics, that it is now technically possible to dispense with Article 59 of the ICBN and to integrate all anamorphic fungi into a single phylogenetically "correct" systematic scheme would bring all fungi into line with the general systematic practices requiring animals and plants to be classified according to their modes of sexual reproduction. Despite the philosophical attractions of abandoning the separate classification for anamorphic fungi, many extremely capable systematic mycologists categorically deny the practical possibility of ever achieving such a noble goal.

With the rapid adoption of molecular approaches by mycological systematists, it has become apparent that a great proportion of the available pool of talent and funding for these studies has been expended on studying various aspects of fungal phylogeny more often than on using these powerful techniques to solve taxonomic problems or to aid the effort to describe new taxa. It is also a troubling trend that the renaissance of interest in (molecular) fungal systematics seems to have been very successful in raising a new generation of scientists who consider themselves to be mycological systematists but who have not been adequately trained to undertake the traditional tasks of collecting, identifying, and describing new taxa as necessary, revising classifications, or many of the other central activities that have been the main occupation of earlier generations of mycological systematists. The decreasing numbers of systematists able to handle these most basic, essential tasks of identifying and cataloging the organisms has ominous consequences, especially at this time when much of the world's scientific community and even most national governments (by means of their ratification of the International Convention on Biodiversity) recognize the multifaceted necessity to collect and to describe the diversity of living organisms. In the meantime, molecularly oriented systematists are gaining the numerical majority within the community of systematists and also seem to attract the majority of the limited funding that is available for systematic and taxonomic research.

The final solution to these conflicts amongst systematists must be for the molecular and "traditional" communities to embrace their mutual interdependence. The systematist who is interested solely in phylogenetic studies on a particular group of fungi, for example, is absolutely dependent upon the traditionally trained systematist to suggest which taxa should be studied and compared, and to be sure that the cultures and specimens examined by the molecular specialist are, in fact, identified correctly (because the results of studies using improperly identified materials is both chaotic and counterproductive). On the other hand, no responsible laboratory studying the systematics of fungi or other organisms can now ignore the invaluable inputs available from molecular techniques, but it is also necessary to appreciate the strengths and weaknesses of each of the available techniques and to know when molecular or traditional taxonomic techniques should or must be used or avoided. Not all problems can be approached in the same manner, and common sense must be allowed to prevail.

5 And a Glance into the Future

The progress of systematics and taxonomy for entomopathogenic fungi is very much like that in other groups of fungi. There is no direct linear progress towards the goal of improving the taxonomy, classification, and systematics of these fungi so long as there are any two people with differing opinions on any of these matters. "Progress" in systematics and taxonomy may be compared fairly to the slow motions of a plasmodial slimemold in which subtle differences in the net duration and direction of the reversible streaming of the cytoplasm within the plasmodium results in the slow translocation of the plasmodium across the substrate towards food, light, or any other preferred stimulus or location. The plasmodial cytoplasm is always under dynamic tension being pulled first one direction and then another; those who must deal with the organisms whose classifications are studied and, seemingly, tinkered with perpetually by the taxonomists must choose to use whichever among the current conflicting classifications they perceive to be the most correct or most suitable scheme; there is nothing to stop anybody's allegiances from flipping from one classification or taxonomy to a competing scheme. It is through this fundamentally democratic process that a consensus about the preferred, most acceptable current taxonomy emerges as pre-eminent, and, if active research on a particular group is being pursued, there should probably be a progressive evolution of the systematics towards an increasingly effective and biologically predictive classification that will be expected to be divided into monophyletic taxa.

Among the fungal entomopathogens there are at least two important examples - the *Entomophthorales* and the taxonomy of species within the genus *Metarhizium* - for which taxonomic concepts have long been controversial but for which the taxonomies now appear to be stabilizing. The taxonomy of the *Entomophthorales* (Zygomycetes) has been controversial ever since the description of the first species in this group in 1855. Of note here, however, are the events in the wake of new generic classification proposed by Batko (1964a-d, 1966; Batko & Weiser, 1965) that began to splinter the long-standing use of *Entomophthora* as a large and very heterogeneous genus accommodating nearly all entomophthoralean entomopathogens. The Batko classification languished without published acceptance, rejection, or even substantial criticism until Remaudière & Hennebert (1980) and Remaudière & Keller (1980) published an extensive but altogether morphologically based revision of this classification. Humber (1981, 1982, 1984) argued that the revised classification was based on faulty taxonomic premises and also needed a more evolutionarily realistic basis.

A more detailed classification of these fungi by Ben-Ze'ev & Kenneth (1982a,b) incorporated the taxonomic criteria suggested by Humber (1981) and differed significantly from the Remaudière/Hennebert/Keller classification. The subsequent additions of new taxa, especially that of a new family (Ben-Ze'ev *et al.*, 1987), led Humber (1989) to a new analysis of the taxonomy of the *Entomophthorales* and to a recognition of two new families, a new genus, the invalidation of the possible use of *Erynia* as a substitute name for *Zoophthora* Batko, and the raising of the four subgenera of *Zoophthora* Batko (1966) to generic status. Some controversy still remains about the treatment of the zoophthoroid genera, and the latest monograph of these fungi (Balazy, 1993) adopted some ultra-conservative approaches to the taxonomy of the families and genera while describing many new species (including segregates from *Zoophthora radicans*, a species widely believed to be a species complex) with classical taxonomic criteria unaided by any of the sorts of molecular data that would now be considered to be essential.

Metarhizium has long been recognized to be one of the most important and common of all genera of fungal entomopathogens, but the taxonomy of its few species has tended to be surprisingly unstable. The first significant overall treatment of this genus (Tulloch, 1976) recognized only three taxa - *M. anisopliae* var. *anisopliae*, *M. anisopliae* var. *major* (a nomenclaturally incorrect varietal name that was later corrected to "*majus*"), and *M. flavoviride*. Later studies recognized a second variety of *M. flavoviride* (Rombach *et al.*, 1986) and resurrected *M. album* from synonymy with *M. anisopliae* (Rombach *et al.*, 1987).

Extensive studies in Africa with the use of *Metarhizium* for the biocontrol of locusts led to a recognition that the morphologies of *M. anisopliae* and *M. flavoviride* sometimes were extremely difficult to separate despite the relatively easily interpreted criteria proposed by Rombach *et al.* (1987) to separate these taxa. A series of biochemical studies (Riba *et al.*, 1986; St. Leger *et al.*, 1992; Bridge *et al.*, 1993; Cobb and Clarkson, 1993; Fegan *et al.*, 1993; Bidochka *et al.*, 1994; Curran *et al.*, 1994) confirmed the obvious presence of a great deal of morphological variation in this genus, especially within *M. anisopliae*, and uniformly found support for the separation of *M. anisopliae* from *M. flavoviride*. Despite this molecular support for the broad outlines of the specific taxonomy of *Metarhizium* that had been developed over time, the understandable frustrations caused by the difficulty in assigning many isolates (especially those from the locusts) to one species or another culminated in a suggestion by Milner *et al.* (1994) that it might be appropriate to disregard all of this history of taxonomic work and to treat *Metarhizium* as a monotypic genus typified by a single, highly plastic species, *M. anisopliae*, with a number of varieties. Mercifully, this suggestion attracted no serious following. The continued molecular studies by several of those who suggested such an unjustifiable simplification of *Metarhizium* systematics are now being used as the basis of a much more elaborate taxonomy of *Metarhizium* based on gene sequences in which several new varieties within *M. anisopliae* and *M. flavoviride* are recognized and some support for retaining *M. album* is offered. It seems likely that this latest study, which has yet to be published at the time this chapter was written, will probably substantially stabilize the taxonomy of this genus for the foreseeable future.

It is possible to make several predictions about the future prospects for taxonomic and systematic research on the fungi that are pathogenic to insects. As with other points made in this chapter, these comments are generally applicable for all work with fungal systematics even though some of the particulars in these comments are tailored very specifically to work with entomopathogens.

- There can be no question that molecular systematic techniques will increasingly be used to solve important systematic and taxonomic issues for fungal entomopathogens. The resolution of the taxonomies of the *B. bassiana* and *M. anisopliae* species complexes, and the unraveling of the tangled taxonomy of the genus *Cordyceps* are but three obvious examples.

- There must be a stronger effort than ever before to use all available technologies to establish connections between the anamorphic and teleomorphic parts of the life histories of the entomopathogenic *Hyphomycetes* and their sexual states in the Clavicipitaceae (species of *Cordyceps*, *Torrubiella*, and *Hypocrella* as well as other less prominent genera). The completion of our understanding of the life cycles of these fungi will provide means for the considerable clarification of the overall systematics of these fungi, will guide their use for biological control much more effectively, and will make the exploitation of these fungi through "bioprospecting" for compounds useful in pest control or medicine just that much more efficient.

- Much more exploration of the world's flora of fungal entomopathogens is needed. This is certainly true in the tropics and those other regions suffering from the greatest rates of habitat destruction, but it is also true for undisturbed and minimally disturbed sites in the temperate regions of the world, too. And more effort needs to be made to explore the microbial flora of extreme or unusual habitats (with special emphasis, perhaps, on fungi able to infect and to sporulate in very cool temperatures). The collection, culturing, and classification of these fungi from the widest possible range of habitats will be important not only in the effort to catalogue the earth's biodiversity, but such efforts with entomopathogenic fungi will, again, serve the needs of biological pest control and the discovery through "bioprospecting" of metabolites with possible pharmaceutical or other important commercial utilities.

- In comparison to other sorts of fungi, there may be some real reason to believe that an unusually large proportion of the world's entomopathogenic fungi is already known and described. Further, an unusually large proportion (ca. 20-25%) of the taxa of entomopathogenic fungi have already been cultured and preserved in various culture collections. Nonetheless, there is a pressing need to use traditional and molecular approaches to revise the systematics of most of the major genera of fungal entomopathogens; none may be in greater need of a complete taxonomic overhaul than *Cordyceps*.

- No matter how comparatively well known the entomopathogenic fungi may now be, such a small proportion of the world's fungal flora has been described that it is clear that there will be no lack of potential work for descriptive mycological systematists for many generations to come. However, the magnitude of funding available is almost certainly going to be much smaller than will be needed to accomplish the vast amount of basic descriptive taxonomy required. This situation will automatically force systematists to exercise a great degree of care in selecting those taxa that will receive their attentions.

This chapter has only been able to skim the surface of some of the issues touching on the fungal pathogens and parasites of insects. Despite the seemingly narrow ecological specializations of these fungi, the issues raised and the methods needed to elucidate and to revise their systematics and taxonomy are quite universal. Despite all of their unique features, these fungi can be understood, in the end, to be very typical fungi whose needs for critical taxonomic and systematic research and the means to meet those needs are, in fact, representative of all fungi.

6   References

Alves, S.B., Marchini L.C., Pereira R.M. & BaumGratz L.L. (1996). Effects of some insect pathogens on the africanized honey bee, *Apis mellifera* L. (Hym., Apidae) *Journal of Applied Entomology* 120, 559-564.

Balazy, S. (1993). *Entomophthorales*. Flora of Poland (Flora Polska), Fungi (Mycota) 24, 1-356. Krakow, Polish Academy of Sciences, W. Szafer Institute of Botany.

Batko, A. (1964a). Remarks on the genus *Entomophthora* Fresenius 1856 non Nowakowski 1883 (Phycomycetes: Entomophthoraceae). *Bulletin de l'Académie Polonaise des Sciences, Série des Sciences Biologiques* 12, 319-321.

Batko, A. (1964b). On the new genera: *Zoophthora* gen. nov., *Triplosporium* (Thaxter) gen. nov., and *Entomophaga* gen. nov. (Phycomycetes: Entomophthoraceae). *Bulletin de l'Académie Polonaise des Sciences, Série des Sciences Biologiques* 12, 323-326.

Batko, A. (1964c). Remarks on the genus *Lamia* Nowakowski 1883 vs. *Culicicola* Nieuwland 1916 (Phycomycetes: Entomophthoraceae). *Bulletin de l'Académie Polonaise des Sciences, Série des Sciences Biologiques* 12, 399-402.

Batko, A. (1964d). Some new combinations in the fungus family Entomophthoraceae (Phycomycetes). *Bulletin de l'Académie Polonaise des Sciences, Série des Sciences Biologiques* 12, 403-406.

Batko, A. (1966). On the subgenera of the fungus genus *Zoophthora* Batko 1964 (Entomophthoraceae). *Acta Mycologica* 2, 15-21.

Batko, A. & Weiser, J. (1965). On the taxonomic position of the fungus discovered by Strong, Wells, and Apple: *Strongwellsea castrans* gen. et sp. nov. (Phycomycetes: Entomophthoraceae). *Journal of Invertebrate Pathology* 7, 455-463.

Ben-Ze'ev, I. & Kenneth, R.G. (1982a). Features-criteria of taxonomic value in the Entomophthorales: I. A revision of the Batkoan classification. *Mycotaxon* 14, 393-455.

Ben-Ze'ev, I. & Kenneth, R.G. (1982b). Features-criteria of taxonomic value in the Entomophthorales: II. A revision of the genus *Erynia* Nowakowski 1881 (= *Zoophthora* Batko 1964). *Mycotaxon* 14, 456-475.

Ben-Ze'ev, I., Kenneth, R.G. & Uziel, A. (1987). A reclassification of *Entomophthora turbinata* in *Thaxterosporium* gen. nov., Neozygitaceae fam. nov. (Zygomycetes: Entomophthorales). *Mycotaxon* 28, 313-326.

Benjamin, R.K. (1965). Study in specificity. *Natural History* 74, 42-49.

Bidochka, M.J., McDonald, M.A., St. Leger, R.J. & Roberts, D.W. (1994). Differentiation of species and strains of entomopathogenic fungi by random amplification of polymorphic DNA (RAPD). *Current Genetics*

25, 107-113.

Bissett, J., Duke, G. & Goettel, M. (1996). *Ascosphaera acerosa* sp. nov. isolated from the alfalfa leafcutting bee, with a key to the species of *Ascosphaera*. *Mycologia* 88, 797-803.

Blackwell, M. (1994). Minute mycological mysteries: the influence of arthropods on the lives of fungi. *Mycologia* 86, 1-17.

Blackwell, M., Bridges, J.R., Moser, J.C. & Perry, T.J. (1986a). Hyperphoretic dispersal of a *Pyxidiophora* anamorph. *Science* 232, 993-995.

Blackwell, M. & Malloch, D. (1989). Similarity of *Amphoromorpha* and secondary capilliconidia of *Basidiobolus*. *Mycologia* 81, 735-741.

Blackwell, M. & Malloch, D. (1990). Discovery of a *Pyxidiophora* with *Acariniola*-type ascospores. *Mycological Research* 94, 415-417.

Blackwell, M., Moser, J.C. & Wisniewski, J. (1989). Ascospores of *Pyxidiophora* on mites associated with beetles in trees and wood. *Mycological Research* 92, 397-403.

Blackwell, M., Perry, T.J., Bridges, J.R. & Moser, J.C. (1986b). A new species of *Pyxidiophora* and its *Thaxteriola* anamorph. *Mycologia* 78, 605-612.

Booth, C. (1971). *The genus* Fusarium. Kew, Commonwealth Mycological Institute.

Bridge, P.D., Williams, M.A., Prior, C. & Paterson, R.R.M. (1993). Morphological, biochemical, and molecular characteristics of *Metarhizium anisopliae* and *Metarhizium flavoviride*. *Journal of General Microbiology* 139, 1163-1169.

Brown, A.H.S. & Smith, G. (1957). The genus *Paecilomyces* Bainier and its perfect stage *Byssochlamys* Westling. *Transactions of the British Mycological Society* 40, 17-89.

Carruthers, R.I., Larkin, T. S., Firstencel, H. & Feng, Z. (1992). Influence of thermal ecology on the mycosis of rangeland grasshoppers. *Ecology* 73, 190-204.

Cobb, B.D. & Clarkson, J.M. (1993). Detection of molecular variation in the insect pathogenic fungus *Metarhizium* using RAPD-PCR. *FEMS Microbiology Letters* 112, 3119-3124.

Cooke, M.C. (1892). *Vegetable Wasps and Plant Worms: a Popular History of Entomogenous Fungi, or Fungi Parasitic upon Insects*. London: Society for Promoting Christian Knowledge.

Cooper, E.L., Rinkevich, B., Uhlenbruck, G. & Valembois, P. (1992). Invertebrate immunity: another viewpoint. *Scandinavian Journal of Immunology* 35, 247-266.

Cooper, R., Hornitzky, M. & Medcraft, B.E. (1984). Non-susceptibility of *Apis mellifera* to *Culicinomyces clavisporus*. *Journal of the Australian Entomological Society* 23, 173-174.

Couch, J.N. (1931). The biological relationship between *Septobasidium retiform* (B. & C.) Pat. and *Aspidiotius osborni*. New. and Ckll. *Quarterly Journal of Microscopic Science* 74, 383-437.

Couch, J.N. (1935). *Septobasidium* in the United States. *Journal of the Elisha Mitchell Scientific Society* 51, 1-77.

Couch, J.N. (1938). *The genus* Septobasidium. Chapel Hill: University of North Carolina Press.

Couch, J.N. & Bland, C.E. (eds.) (1985). *The genus* Coelomomyces. Orlando: Academic Press.

Curran, J., Driver, F., Ballard, J.W.O. & Milner, R.J. (1994). Phylogeny of *Metarhizium* analysis of ribosomal DNA sequence data. *Mycological Research* 98, 547-552.

DeGurman, F.S., Dowd, P.H., Gloer, J.B. & Wicklow, D.T. (1993). Cycloechinulin anti-insectan metabolite. U.S.D.A. Patent 5,196,420. (23 March 1993).

Dowd, P.F., Wicklow, D.T., Gloer, J.B. & TePaske, M.R. (1994). Leporin A, and anti-insectan fungal metabolite. U.S.D.A. Patent 5,281,609 (25 January 1994).

Drechsler, C. (1952). Widespread distribution of *Delacroixia coronata* and other saprophytic Entomophthoraceae in plant detritus. *Science* 115, 575-576.

Drechsler, C. (1955). A southern *Basidiobolus* forming many sporangia from globose and from elongated adhesive conidia. *Journal of the Washington Academy of Sciences* **45**, 49-56.

Eilenberg, J. & Michelsen, V. (1999). Host range and prevalence of the genus *Strongwellsea* (Zygomyrota: Entomophthorales in Denmark. *Journal of Invertebrate Pathology* **73**, 189-198.

Evans, H.C. & Samson, R.A. (1982). Entomogenous fungi from the Galápagos Islands. *Canadian Journal of Botany* **60**, 2325-2333.

Fegan, M., Manners, J.M., MacLean, D.J., Irwin, J.A.G., Samuels, K.D.Z., Holdom, D.G. & Li, D.P. (1993). Random amplified polymorphic DNA markers reveal a high degree of genetic diversity in the entomopathogenic fungus *Metarhizium anisopliae* var. *anisopliae*. *Journal of General Microbiology* **139**, 2075-2081.

Gams, W. (1971). Cephalosporium-*artige Schimmelpilze (Hyphomycetes)*. Stuttgart: Gustav Fischer Verlag.

Gillespie, J.P., Kanost, M.R. & Trenszek, T. (1997). Biological mediators of insect immunity. *Annual Review of Entomology* **42**, 611-643.

Goerzen, D.W., Erlandson, M.A. & Moore, K.C. (1990). Effect of two insect viruses and two entomopathogenic fungi on larval and pupal development in the alfalfa leafcutting bee, *Megachile rotundata* (Fab.) (Hymenoptera: Megachilidae) *Canadian Entomologist* **122**, 1039-1040.

Greuter, W., Barrie, F.R., Burdet, H.M., Chaloner, W.G., Demoulin, V., Hawksworth, D.L., Jørgensen, P.M., Nicholson, D.H., Silva, P.C., Trehane, P. & McNeill, J. (1994). *International Code of Botanical Nomenclature (Tokyo Code)*. Königstein: Koeltz Scientific Books.

Hajek, A.E., Nelson, P.E., Humber, R.A. & Perry, J.L. (1993). Two *Fusarium* species pathogenic to gypsy moth, *Lymantria dispar*. *Mycologia* **85**, 937-940.

Hughes, S.J. (1953). Conidiophores, conidia and classification. *Canadian Journal of Botany* **31**, 577-659.

Humber, R.A. (1976). The systematics of the genus *Strongwellsea* (Zygomycetes: Entomophthorales). *Mycologia* **68**, 1042-1060.

Humber, R.A. (1981). An alternative view of certain taxonomic criteria used in the Entomophthorales (Zygomycetes). *Mycotaxon* **13**, 191-240.

Humber, R.A. (1982). *Strongwellsea* vs. *Erynia*: the case for a phylogenetic classification of the Entomophthorales (Zygomycetes). *Mycotaxon* **15**, 167-184.

Humber, R.A. (1984). Foundations for an evolutionary classification of the Entomophthorales (Zygomycetes). In *Fungus/Insect Relationships: Perspectives in Ecology and Evolution*, pp. 166-183. Edited by Q. Wheeler & M. Blackwell. New York: Columbia University Press.

Humber, R.A. (1989). Synopsis of a revised classification of the Entomophthorales (Zygomycotina). *Mycotaxon* **34**, 441-460.

Humber, R.A. (1992). Collection of Entomopathogenic Fungi: Catalog of Strains 1992. *USDA-ARS Publications* **110**, 1-177.

Humber, R.A. (1996). Fungal pathogens of the Chrysomelidae and prospects for their use in biological control. In *Chrysomelidae Biology*, Vol. 2, *Ecological Studies*, pp. 93-115. Edited by P.H.A. Jolivet & M.L. Cox. Amsterdam: SPB Academic Publishing bv.

Humber, R. A. & Feng, M.G. (1991). *Entomophthora chromaphidis* (Entomophthorales): the correct identification of an aphid pathogen in the Pacific Northwest and elsewhere. *Mycotaxon* **41**, 497-504.

Humber, R.A. & Rombach, M. C. (1987). *Torrubiella ratticaudata* sp. nov. (Pyrenomycetes: Clavicipitales) and other fungi from spiders on the Solomon Islands. *Mycologia* **79**, 375-382.

Hywel-Jones, N.L. & Samuels, G.J. (1998). Three species of *Hypocrella* with large stromata pathogenic on scale insects. *Mycologia* **90**, 36-46.

Jarosz, J. (1996). Strategies of insect immune defences and counter defence systems of insect bacterial pathogens and parasites. *Wiadomosci Parazytologiczne* **42**, 3-27. [in Polish].

King, D.S. (1976). Systematics of *Conidiobolus* (Entomophthorales) using numerical taxonomy. II. Taxonomic considerations. *Canadian Journal of Botany* **54**, 1285-1296.

King, D.S. (1977). Systematics of *Conidiobolus* (Entomophthorales) using numerical taxonomy. III. Descriptions of recognized species. *Canadian Journal of Botany* **55**, 718-729.

Kobayasi, Y. (1941). The genus *Cordyceps* and its allies. *Science Reports of the Tokyo Bunrika Daigaku, Section B* **5**, 53-260.

Kobayasi, Y. (1982). Keys to the taxa of the genera *Cordyceps* and *Torrubiella*. *Transactions of the Mycological Society of Japan* **23**, 329-364.

Kobayasi, Y. & Shimizu, D. (1982). Monograph of the genus *Torrubiella*. *Bulletin of the National Science Museum, Series B (Botany)* **8**, 43-78.

Kobayasi, Y. & Shimizu, D. (1983). *Iconography of Vegetable Wasps and Plant Worms*. Osaka: Hoikusha Publ. Co., Ltd.

Li, Z.Z. (1988). A list of insect hosts of *Beauveria bassiana*. In *Study and Application of Entomogenous Fungi in China*, pp. 241-255. Edited by Y.W. Li, Z.Z. Li, Z.Q. Liang, J.W. Wu, Z.K. Wu & Q.F. Xi. Beijing: Academic Periodical Press.

Liang, Z.Q. (1991). Verification and identification of the anamorph of *Cordyceps pruinosa* Petch. *Acta Mycologica Sinica* **10**, 104-107.

Liang, Z.Q., Liu, A.Y. & Liu, J.L. (1991). A new species of the genus *Cordyceps* and its *Metarhizium* anamorph. *Acta Mycologica Sinica* **10**, 257-262.

Majewski, T. & Wisniewski, J. (1978a). New species of parasitic fungi occurring on mites (Acarina). *Acta Mycologica* **14**, 3-12.

Majewski, T. & Wisniewski, J. (1978b). Records of parasitic fungi of the "Thaxteriolae" group on subcortical mites. *Mycotaxon* **7**, 508 510.

McCabe, D.E., Humber, R.A. & Soper, R.S. (1984). Observation and interpretation of nuclear reductions during maturation and germination of entomophthoralean resting spores. *Mycologia* **76**, 1104-1107.

Milner, R.J., Driver, F., Curran, J., Glare, T.R., Prior, C., Bridge, P.D. & Zimmermann, G. (1994). Recent problems with the taxonomy in the genus *Metarhizium*, and a possible solution. Proc. VIth Internat. Colloquium on Invertebrate Pathology and Microbial Control, 28 August - 2 September 1994, Montpellier, France, vol. 2 (Abstracts), 109-110.

Mugnai, L., Bridge, P.D. & H.C. Evans, H.C. (1989). A chemotaxonomic evaluation of the genus *Beauveria*. *Mycological Research* **92**, 199-209.

Petch, T. (1921). Studies in entomogenous fungi: II. The genera *Hypocrella* and *Aschersonia*. *Annals of the Royal Botanic Gardens, Peradeniya* **7**, 167-277.

Remaudière, G. & Hennebert, G.L. (1980). Révision systématique de *Entomophthora aphidis* Hoffm. in Fres. Description de deux nouveaux pathogènes d'aphides. *Mycotaxon* **11**, 269-321.

Remaudière, G. & Keller, S. (1980). Révision systématique des genres d'Entomophthoracae à potentialité pathogène. *Mycotaxon* **11**, 323-338.

Riba, G., Bouvier-Fourcade, I. & Caudal, A. (1986). Isoenzymes polymorphism in *Metarhizium anisopliae* (Deuteromycotina: Hyphomycetes, entomogenous fungi). *Mycopathologia* **96**: 161-170.

Roberts, D.W. & Humber, R.A. (1981). Entomogenous fungi. In *Biology of Conidial Fungi*, vol. 2, pp. 201-236. Edited by G.T. Cole & W.B. Kendrick. New York: Academic Press.

Rombach, M.C., Humber, R.A. & Evans, H.C. (1987). *Metarhizium album* Petch, a fungal pathogen of leaf- and planthoppers of rice. *Transactions of the British Mycological Society* **88**, 451-459.

Rombach, M.C., Humber, R.A. & Roberts, D.W. (1986). *Metarhizium flavoviride* var. *minus*, var. nov., a pathogen of plant- and leafhoppers on rice in the Philippines and Solomon Islands. *Mycotaxon* **27**, 87-92.

Saccardo, P. (1899 et seq.). *Sylloge Fungorum Omnium Hucusque Cognitorum*. 14 volumes. Published by the author. Pavia, Italy.

Samson, R.A. (1974). *Paecilomyces* and some allied Hyphomycetes. *Studies in Mycology* 6, 1-119.

Samuels, G.J. & Seifert, K.A. (1991). Two new species of *Nectria* with *Stilbella* and *Mariannaea* anamorphs. *Sydowia Annales Mycologici* 43, 249-263.

Shimazu, M., Mitsuhashi, W. & Hashimoto, H. (1988). *Cordyceps brongniartii* sp. nov., the teleomorph of *Beauveria brongniartii*. *Transactions of the Mycological Society of Japan*. 29, 323-330.

Skou, J.P. (1972). Ascosphaerales. *Friesia* 10, 1-24.

Skou, J.P. (1975). Two new species of *Ascosphaera* and notes on the conidial state of *Bettsia alvei*. *Friesia* 11, 62-74.

Skou, J.P. (1982). Ascosphaerales and their unique ascomata. *Mycotaxon* 15, 487-499.

Soper, R.S. (1974). The genus *Massospora*, entomopathogenic for cicadas, Part, I, Taxonomy of the genus. *Mycotaxon* 1, 13-40.

Spatafora, J.W. & Blackwell, M. (1993). Molecular systematics of unitunicate perithecial ascomycetes: The Clavicipitales-Hypocreales connection. *Mycologia* 85, 912-922.

Spivak, M. & Gilliam, M. (1993). Facultative expression of hygienic behaviour of honey bees in relation to disease resistance. *Journal of Agricultural Research* 32, 147-157.

Sterk, G., Bolckmans, K., Jonghe, R. de, Wael, L. de & Vermeulen, J. (1995). Side-effects of the microbial insecticide PreFeRal WG (*Paecilomyces fumosoroseus*, strain Apopka 97). On *Bombus terrestris*. *Mededelingen Faculteit Landbouwkundige en Toegepaste Biologische Wetenschappen Universiteit Gent* 60, 713-717.

St. Leger, R.J., Allee, L.L., May, B., Staples, R.C. & Roberts, D.W. (1993). World-wide distribution of genetic variation among isolates of *Beauveria* spp. *Mycological Research* 96, 1007-1015.

St. Leger, R.J., May, B., Allee, L.L., Frank, D.C., Staples, R.C. & Roberts, D.W. (1992). Genetic differences in allozymes and in formation of infection structures among isolates of the entomopathogenic fungus *Metarhizium anisopliae*. *Journal of Invertebrate Pathology* 60, 89-101.

Sung, J.M. (1996). *The Insects-Born Fungus of Korea in Color.* Seoul: Kyohak Publishing Co., Ltd.

Tanada, T. & Kaya, H.K. (1993). *Insect Pathology.* London: Academic Press.

Taylor, J.W. (1995). Making the Deuteromycota redundant: a practical integration of mitosporic and meiosporic fungi. *Canadian Journal of Botany* 73 (Supplement 1, Section E-H), S754-S759.

Thaxter, R. (1888). The Entomophthoreae of the United States. *Memoirs of the Boston Society of Natural History* 4, 133-201.

Tulloch, M. (1976). The genus *Metarhizium*. *Transactions of the British Mycological Society* 66, 407-411.

Vandenberg, J.D. (1990). Safety of four entomopathogens for caged adult honey bees (Hymenoptera: Apidae). *Journal of Economic Entomology* 83, 755-759.

Veen, K.H. (1968). Recherches sur la Maladie due à *Metarrhizium anisopliae* chez le Criquet Pélerin. Ph.D. :dissertation, University of Wageningen: Veenman & Zonen (Netherlands).

Vuillemin, P. (1910). Les conidiospores. *Bulletin de la Société Scientifique de Nancy* 11, 129-172.

Vuillemin, P. (1911). Les aleurospores. *Bulletin de la Société Scientifique de Nancy* 12, 151-175.

Waters, S.D. & Callaghan, A.A. (1989). *Conidiobolus iuxtagenitus*, a new species with discharged elongate repetitional conidia and conjugation tubes. *Mycological Research* 93, 223-226.

Watson, D.W., Mullens, B.A. & Petersen, J.J. (1993). Behavioral fever response of *Musca domestica* (Diptera: Muscidae) to infection by *Entomophthora muscae* (Zygomycetes: Entomophthorales). *Journal of Invertebrate Pathology* 61, 10-16.

Wicklow, D.T. & Dowd, P.F. (1989). Entomotoxigenic potential of wild and domesticated yellow-green aspergilli: toxicity to corn earworm and fall armyworm larvae. *Mycologia* 81, 561-566.

IV. ENVIRONMENT AND ITS EXPLOITATION

9 TAXONOMY OF EXTREMOPHILES

JAKOB K. KRISTJANSSON [1], GUDMUNDUR O. HREGGVIDSSON [1]
and WILLIAM D. GRANT [2]

[1] Institute of Biology, University of Iceland, Liftaeknihus, Keldnaholt, 112
Reykjavik, Iceland
[2] Department of Microbiology and Immunology, University of Leicester,
PO Box 138, University Road, Leicester LE1 9HN, England

1   Introduction

Extremophiles have received ever increasing interest over the last two decades and there
has been almost an explosive increase in the international research effort in this field.
The only thing that all extremophiles have in common is that they are regarded as some
sort of microbial outsiders. They are defined as extremophiles based on the ability to grow
optimally under extreme conditions, in which most other organisms are not able to grow.

A range of physical and chemical environmental factors defines a biotope for an
organism and an extremophile lives optimally outside of the "normal" range for at least
one such factor. We can make a scale for each factor and define high and low ("extreme")
values on the scale. At the ends it is usually possible to observe that the species diversity
is lower than in the middle, which is then the "normal range" for that environmental factor.
Much emphasis in extremophile research has been on discovering new and unusual
organisms and microbiologist, have been like the early discoverers, travelling to remote and
exotic places like smoking volcanos and deep-sea hydrothermal vents to isolate previously
unknown organisms (Prieur, 1997). Because extremophiles are so unusual they are very
interesting to both scientists and the general public and the consequent research effort
has resulted in a flurry of new species that have been isolated and described from ever
more unusual extreme habitats.

From the simple definition above, it is clear that for most environmental factors it is
possible to find extreme conditions and therefore many kinds of organisms could be
termed extremophiles. However, the most commonly termed extremophiles are organisms
that are able to grow at the ends of the scales for temperature, pH and salinity
(Kristjansson & Hreggvidsson, 1995). Thermophiles are by far the most studied
extremophiles but alkaliphiles and halophiles have also been extensively studied in

F. G. Priest and M. Goodfellow (eds.), Applied Microbial Systematics, 231-291 recent years. This chapter is focused on these three groups. However, it is clear that many of the points discussed here also apply to other extremophiles. Acidophiles have also been much studied and their role in bioleaching is covered in Chapter 10 Studies are also increasing on psychrophiles but other extremophiles have been studied to a much lesser degree (DeLong, 1997).

2   Extremophiles in Biotechnology

The special properties of many extremophiles that allow them to thrive under harsh conditions also make them interesting for possible exploitation in industry (Kristjansson, 1989; Hoyle, 1998). The present applications of extremophiles are numerous, with probably the best known being the use of thermostable enzymes in industry and in research or diagnostics (Kristjansson, 1989). One of the most popular applications is the PCR method that is based on thermostable DNA polymerases from various thermophilic bacteria. Several other thermostable enzymes are now used for molecular biology applications, such as DNA ligases, restriction enzymes and alkaline phosphatases (Hjörleifsdottir et al., 1996; 1997). The use of amylases in the sugar industry, proteases in detergents and xylanases in pulp bleaching are probably the best known industrial applications.

As a consequence of this interest, much effort has been devoted to the isolation of new strains of many extremophiles and several large strain collections have been accumulated. Such collections are used for selecting organisms for screening programs to look for specific properties. If such collections are to be used to their fullest potential and in the most economic manner it is very important that the isolated strains have been characterized and classified as much as possible before they are admitted to the collection (Cheetham, 1987; Bull et al., 1992). This helps to avoid re-examination or re-isolation of similar strains and also indicates groups for further screening. Moreover, it is necessary to be able to define strains well for eventual patenting or licensing purposes (see Chapter 16). With this in mind it is clear that taxonomy in all its aspects, is very important for the biotechnology industry that is trying to exploit the vast potential that can be harnessed from extremophiles.

3   The Habitats

Before examining the systematics of extremophilic bacteria, it is important to define the habitats in which they live. The most studied extreme habitats are characterized by high temperature, high or low pH and high salinity. Some biotopes can also be extreme in more than one of these three factors (Kristjansson & Hreggvidsson, 1995).

3.1   HIGH TEMPERATURE ENVIRONMENTS

Geothermal areas of the world are the main natural habitats of thermophilic microorganisms. Natural geothermal areas are associated with tectonic activity and found in all parts of the globe but are usually concentrated in small areas. The best known geothermal areas and most studied biologically are in Iceland, Italy, Japan, New Zealand, North America (Yellowstone National Park) and Russia. Geothermal areas are of two main types, depending on the geological heat source, which then gives characteristic surface chemistry, resulting in either high or low pH (Brock, 1978; Kristjansson & Stetter, 1992).

The low pH type are the solfatara fields characterized by acidic soils, H_2S, and sulfur with acidic hot springs and boiling mud pots on the surface (Kristjansson & Stetter, 1992). These areas are always located within active volcanic zones and have a deep magma chamber or molten lava inserts as heat sources. These are usually characterized by water temperatures of 150 to 350°C at 500 to 3000 m depth and by emissions of steam and volcanic gases on the surface. The gas is primarily CO_2 and N_2 but H_2 and H_2S can be up to 10% each of total gases. Also traces of ammonia, carbon monoxide and methane are often found. Because of the weak acids, CO_2 (pK = 6.3) and H_2S (pK = 7.2), the pH of the sub-surface steam is near neutrality.

On the surface the H_2S is oxidized, chemically and biologically first to sulfur and then to sulfuric acid. This lowers the pH, causing corrosion of the surrounding rocks and formation of the typical acidic mud of solfatara fields. The pH is often stabilized around 2 to 2.5 with sulfuric acid (pK_2 = 1.92) as the effective buffering agent. Because of the high temperature there is little water that comes to the surface and the "hot springs" are usually in the form of steam holes or fumaroles. These areas are usually unstable and the individual openings often disappear or move to another location within the geothermal field.

The other main type of geothermal area, is characterized by freshwater hot springs and geysers of neutral to alkaline pH. These areas are mainly located outside active volcanic zones and are heated by extinct deep lava flows or by dead magma chambers. The water temperature at 500 to 3000 m depths is usually below 150°C. Groundwater percolates into these hot areas, warms up and returns to the surface containing dissolved minerals such as silica and some dissolved gases, mainly CO_2. Usually there is little H_2S in such fluids. The sub-surface pH is near neutrality but there is usually a lot of water and little sulfide so surface oxidation has no effect on the pH. However, on the surface the CO_2 is blown away and the silica precipitates, resulting in increased pH, often stabilized at around 9, where silicate (pK_n = 9.7) and carbonate (pK_2 = 10.25) start to act as the effective buffering agents. Since these areas are located outside the volcanically active zone, they are geologically rather stable and the individual hot springs are very constant both in temperature and water-flow. However, they are often disrupted or new ones created during earthquake activities.

Other naturally occurring hot places are normally more transient, such as solar-heated ponds and soils or composts. The organisms found in these transient ecosystems are mainly rapidly growing, endospore-formers (Edwards, 1990). Many man-made, long-term hot environments of the neutral-alkaline type now exist. These include hot water pipelines in homes and factories, district heating systems and thermophilic waste treatment plants. Also many processes, for example in food and chemical industries, use aqueous evaporation or extraction processes which run at high temperatures. All such systems that have been studied are inhabited by thermophiles (Pask-Hughes & Williams, 1975; Håskå & Nystrand, 1982; Perttula *et al.*, 1991, Kristjansson *et al.*, 1994).

3.2   HIGH SALT ENVIRONMENTS

The best known high salt environments are the great inland salt lakes, such as the Dead Sea and the Great Salt Lake (Oren, 1988; Gilmour, 1990; Grant, 1991). These lakes are found in sub-tropical or tropical areas and are subject to high rates of evaporation due to generally high temperature and high light intensity. Such hypersaline lakes are formed when evaporation exceeds the input of freshwater from rivers or rain. Transient hypersaline biotopes are formed naturally on the seashore due to rapid evaporation and similar biotopes of more intermediate duration result from human activity in the production of salt from salterns or evaporation ponds near the seashore.

The pH in the hypersaline lakes is determined by the ionic composition of the brine, primarily the relative abundance of Ca^{2+} and Mg^{2+} (Grant & Ross, 1986; Oren, 1988; Gilmour, 1991; Grant & Horikoshi, 1992). The ionic composition of the Great Salt Lake resembles that of seawater. It is slightly alkaline (pH 7.7) and low in Ca^{2+} but high in Mg^{2+} whereas the Dead Sea is slightly acidic (pH 5.9 to 6.3) and extremely high in both Ca^{2+} and Mg^{2+}, about 0.4 and 1.8 M respectively (Oren, 1988; Grant, 1991). The organisms growing in these habitats usually require high concentrations of these ions in the medium, when cultivated in the laboratory.

3.3   HIGHLY ALKALINE ENVIRONMENTS

The main alkaline environments are the soda lakes in the Rift Valley of Kenya and similar lakes found in a few other places on earth. The main difference between the soda lakes and the highly saline lakes is the pH and ionic composition of the brine, which seems to depend on the relative abundance of Ca^{2+} and Mg^{2+} as discussed above. They are highly alkaline with pH values of 11 to 12 (Tindall et al., 1980; Shiba, 1991). However, these lakes are also highly saline and the resident organisms are therefore normally both alkaliphilic and halophilic. The soda lakes are characterized by very low concentrations of Ca^{2+} and Mg^{2+}. Na^+ is normally the main cation with Na_2CO_3 the major source of alkalinity. Alkaline pH also precipitates selectively other important cations, such as iron. The organisms that are found in these alkaline environments therefore require little Ca^{2+} or Mg^{2+} in their growth medium (Tindall et al., 1980; 1984).

Other types of alkaline environments are rare but carbonate-rich springs and alkaline soils are found in many places. Also decaying proteins and in particular hydrolysis of urea can cause high pH due to release of high concentrations of ammonia.

4   Ecology and Population Structure of Extremophiles

Extreme habitats are unique in many aspects and many of their characteristic features make them interesting, and perhaps ideal, model systems for microbial ecology studies. They are ecosystems that are often solely occupied by microbes that are very different from those of the surrounding areas. They are also usually confined to relatively small areas and they can therefore be considered as islands in the ecological sense. On a global scale, most natural extreme habitats are rare but evenly distributed over the globe. Consequently there are large geographical dispersal barriers for extremophilic microorganisms.

This can result in great founder effects in microbial colonization with the subsequent generation of unique species interactions and compositions. Similar niches may be occupied by different microbes in different geographic regions thereby creating opportunities for very isolated and unique adaptations. This can lead to accelerated evolutionary divergence and different speciation at distant locations. Due to the nature of many extreme environments, like for example hot springs, they may have limited lifetimes since they can suddenly appear or disappear due to earthquakes or volcanic activities.

So far the emphasis in the microbiology of extremophiles has been on the search for new species often with biotechnological applications in mind. This has resulted in the accumulation of highly diverse and fascinating extremophilic microorganisms (Horikoshi, 1996; Stetter, 1996b) which in turn has stimulated systematic evaluation.

The first extremophilic microorganisms that were isolated, such as *Sulfolobus* and *Thermus,* were described using the techniques of the time, namely traditional phenotypic and chemotaxonomic methods (Brock & Freeze, 1969; Brock *et al.*, 1972). Molecular methods have since become very important in the development of the taxonomy of most extremophile groups, particularly in revealing phylogenetic relationships and in validating new species descriptions (Vandamme *et al.*, 1996).

Not surprisingly, 16S rRNA sequence analysis has played a pivotal role and the results have contributed significantly towards answering major questions regarding the origin and evolution of life. The findings have challenged old evolutionary scenarios and led to the proposal of new, more radical ones, such as the hypothesis of the thermophilic origin of life and that respiration may be the most ancient metabolic pathway (Woese, 1987; Olsen & Woese, 1993; Stetter, 1996a; Schäfer *et al.*, 1996).

An extensive metabolic diversity is found among extremophilic prokaryotes. Apart from the "major" pathways of aerobic respiration, fermentation (various archaea, *Clostridium,* and *Thermotoga,*), photosynthesis (*Chloroflexus,* cyanobacteria) and various other metabolic pathways are represented among extremophiles. These include activities such as denitrification, methanogenesis (in archaea), nitrate respiration, sulfate respiration and sulfur and hydrogen oxidation, (Kristjansson & Stetter, 1992).

5   Thermophiles

Brock (1986) defined organisms that grow optimally at or above the "thermophile boundary" of 55 to 60°C, as thermophiles. Since only prokaryotes are capable of this, all thermophiles are prokaryotes. The highest growth temperature known for any organisms is, on the other hand, 113°C (Blöchl *et al.*, 1997). The thermobiotic range is therefore very wide and several sub-divisions have been suggested, such as "extreme thermophile", or "hyperthermophile" for those growing above 85°C and even "pyrophile" for those growing above 100°C (Stetter, 1986; Blöchl *et al.*, 1997). The current, validly accepted species of thermophiles that grow at or above 65°C are diverse and distributed across 36 genera while the thermophilic archaea cover 23 genera (Table 1). In the next sections, the contributions of the various taxonomic approaches to the establishment and characterization of this diversity are reviewed with an emphasis on the value of the different techniques and how they can be used to discover new microorganisms with interesting new properties.

5.1   THERMOPHILIC BACTERIA

5.1.1 *Phenotypic methods*
Phenotypic methods were very important in the initial discovery and description of new thermophiles. For example, the original description of *Thermus aquaticus*, one of the first bacteria to be specifically associated with hot geothermal habitats, was in fact based on a few convenient charcteristics that distinguished it from Gram-positive bacteria found under the same environmental conditions (Brock & Freeze, 1969). Soon after its discovery several similar bacteria were isolated in different parts of the world but only *T. ruber* (now *Meiothermus ruber*) could be unequivocally accepted as a new species since it was phenotypically distinct in being red pigmented and had a lower maximum growth temperature (Loginova *et al.*, 1984). The failure of phenotypic methods to resolve the speciation of the yellow pigmented strains related to *T. aquaticus* was highlighted by DNA reassociation studies which revealed several different genomic species, namely *T. aquaticus, T. brockianus, T. filiformis, T. oshimai* and *T. thermophilus* (Oshima & Imahori, 1974; Hudson *et al.*, 1987; Williams *et al.*, 1995; 1996).

Simple phenotypic definitions of species have several limitations and reliance on single or few phenotypic traits can be misleading. A notable example is the species *T. filiformis*. It was described as a new species on the basis of the distinctive formation of stable filaments in liquid cultures but subsequent 16S rRNA analysis and DNA:DNA hybridization studies showed that several non-filamentous strains from the same region could be classified genotypically with *T. filiformis* (Georganta *et al.*, 1993; Saul *et al.*, 1993). Similarly, the hydrogen oxidizing obligate chemoautothroph, *Hydrogenobacter thermophilus* and related bacteria are phenotypically very similar (Kristjansson *et al.*, 1985; Aragno, 1992) but little DNA: DNA homology is evident between geographically distant groups of isolates (Aragno, 1992). These examples underline the importance of screening genomically diverse organisms which may be phenotypically similar for particular enzymes or other products since certain genomic species may have unique and valuable properties.

5.1.2 *Numerical taxonomy*
The extension of simple phenotypic descriptions into numerical taxonomy as a systematic tool in the study of thermophilic microorganisms has been largely restricted to the aerobic heterotrophic bacteria (Hudson *et al.*, 1986; Santos *et al.*, 1989; White *et al.*, 1993; Marteinsson *et al.*, 1995a). This reflects the ease of cultivating and studying the phenotypes of these organisms compared to anaerobes or obligate chemolithoautotrophs.

Early studies on *Thermus* strains showed great phenotypic diversity in isolates from the same site and surprisingly no correlation was observed between physicochemical conditions at a sample site and phenotype of the isolates (Cometta *et al.*, 1982; Alfredsson *et al.*, 1985). Later studies with various methods, but in particular DNA:DNA homology, confirmed that the phenetic relationships based on phenotypic similarities of *Thermus* isolates were superficial and did not reflect true genotypes. This is well demonstrated by strains of *T. filiformis* that share high DNA:DNA homolgy but have no known distinguishing phenotypic characteristics (Williams, 1989; Georganta *et al.*, 1993; Manaia *et al.*, 1994). Similarly a phenetic study of 40 strains of *T. aquaticus* and *T. brockianus* strains from Yellowstone Park, based on 71 characteristics, showed that the only distinguishing

TABLE 1. Validly described species of thermophiles with T_max > 65°C

Taxon/species	Growth conditions		Metabolism	Electron donor	acceptor[1]	Reference
	T_{max}	pH_{opt}				
ARCHAEA						
Acidianus ambivalens	87	2.5	aerobic	H_2	O_2/S^o	Zillig et al. (1986)
A. brierley	75	2.0	aerobic	heterotrophic/H_2/S^o	O_2/S^o	Segerer et al. (1991)
A. infernus	95	2.0	aerobic	heterotrophic/H_2/S^o	O_2/S^o	Segerer et al. (1991)
Aeropyrum pernix	100	7.0	aerobic	heterotrophic	O_2	Sako et al. (1996b)
Archaeoglobus fulgidus	92	5.5-7.5	anaerobic	heterotrophic/H_2	SO_4	Stetter (1988)
A. profundus	92	6.0	anaerobic	heterotrophic	SO_4	Burggraf et al. (1990b)
Desulfurococcus amylolyticus	97	6.4	anaerobic	heterotrophic	f/So	Bonch-Osmolovskaya et al. (1985)
D. mobilis	95	5.5-6.0	aerobic	heterotrophic	f/So	Zillig et al. (1982)
D. mucosus	97	6.0	anaerobic	heterotrophic	f/So	Zillig et al. (1982)
D. saccharovorans	97	6.5	anaerobic	heterotrophic	f/So	Stetter (1996b)
Hyperthermus butylicus	108	7.0	anaerobic	heterotrophic	S^o	Zillig et al. (1991)
Metallosphaera prunae	80	1-4.5	aerobic	heterotrophic/H_2	O_2	Fuchs et al. (1995)
M. sedula	80	3	aerobic	heterotrophic	O_2	Huber et al. (1989a)

TABLE 1. continued

Taxon/species	Growth conditions		Metabolism	Electron donor	acceptor[1]	Reference
	T_{max}	pH_{opt}				
ARCHAEA						
Methanobacterium						
thermoautotrophicum	75	7.4	anaerobic	H_2	CO_2	Kotelnikova et al. (1993a)
M. thermoflexum	65	6.5–7.0	anaerobic	H_2	CO_2	Kotelnikova et al. (1993b)
M. thermophilum	65	7.5	anaerobic	H_2	CO_2	Laurinavichyus et al (1988)
Methanococcus igneus	91	5.7	anaerobic	H_2	CO_2	Burggraf et al. (1990a)
M. jannaschii	86	6.0	anaerobic	H_2	CO_2	Jones et al. (1983)
M. thermolithotrophicus	70	7.0	anaerobic	H_2	CO_2	Huber et al. (1982)
Methanopyrus kandleri	110	6.5	anaerobic	H_2	CO_2	Kurr et al. (1991)
Methanothermus fervidus	97	6.5	anaerobic	H_2	CO_2	Stetter et al. (1981)
M. sociabilis	97	6.5	anaerobic	H_2	CO_2	Lauerer et al. (1986)
Picrophilus oshimae	65	0.7	anaerobic	heterotrophic	f	Schleper et al (1995)
P. torridus	65	0.7	anaerobic	heterotrophic	f	Schleper et al (1995)
Pyrobaculum aerophilum	104	7.0	aerobic	heterotrophic/H_2	O_2/NO_3	Völkl et al. (1993)
P. islandicum	103	6.0	anaerobic	heterotrophic/H_2	$S°$	Huber et al. (1987)
P. organotrophum	103	6.0	anaerobic	heterotrophic/H_2	$S°$	Huber et al. (1987)

TABLE 1. continued

Taxon/species	Growth conditions		Metabolism	Electron donor	Electron acceptor[1]	Reference
	T_max	pH_opt				
ARCHAEA						
Pyrococcus furiosus	103	7.0	anaerobic	heterotrophic	f	Fiala & Stetter (1986)
P. woesei	103	6.2	anaerobic	heterotrophic	f	Zillig et al. (1987)
Pyrodictium abyssii	110	5.5	anaerobic	H₂	S°	Pley et al. (1991)
P. brockii	110	5.5	anaerobic	H₂	S°	Stetter et al. (1983)
P. occultum	110	5.5	anaerobic	H₂	S°	Stetter et al. (1983)
Pyrolobus fumarii	113	5.5	anaerobic	heterotrophic	NO₃/S₂O₃/O₂	Blöchl et al. (1997)
Staphylothermus marinus	98	6.5	anaerobic	heterotrophic	f	Fiala et al. (1986)
Stygioglobus azoricus	89	1-5.5	anaerobic	H₂	S°	Takayanagi et al. (1996)
Sulfolobus acidocaldarius	85	2.5	aerobic	heterotrophic/S°	O₂/Fe(III)	Brock et al. (1972)
S. hakonensis	80	3	aerobic	heterotrophic	O₂	Takayanagi et al. (1996)
S. metallicus	75	1-3.5	aerobic	heterotrophic	O₂	Huber & Stetter (1991)
S. shibatae	86	3	aerobic	heterotrophic	O₂	Grogan et al. (1990)
S. solfataricus	87	4.5	aerobic	heterotrophic	O₂	Zillig et al. (1980)

TABLE 1.   continued

Taxon/species	Growth conditions		Metabolism	Electron donor	acceptor[1]	Reference
	T_{max}	pH$_{opt}$				
ARCHAEA						
Thermococcus alkaliphilus	90	9.0	anaerobic	heterotrophic	f	Keller et al. (1995)
T. celer	93	5.8	anaerobic	heterotrophic	f/S°	Zillig et al. (1983b)
T. chitonophagus	93	6.7	anaerobic	heterotrophic	f	Huber et al. (1996b)
T. fumicolans	103	8.0	anaerobic	heterotrophic	f	Godfroy et al. (1996)
T. hydrothermalis	100	6.0	anaerobic	heterotrophic	f	Godfroy et al. (1997)
T. littoralis	98	7,2	anaerobic	heterotrophic	f	Neuner et al. (1990)
T. peptonophilus	100	6.0	anaerobic	heterotrophic	f	Gonzales et al. (1996)
T. profundus	90	7.5	anaerobic	heterotrophic	f/S°	Kwak et al. (1995)
T. stetteri	98	6.5	anaerobic	heterotrophic	f/S°	Miroshnichenko et al. (1989)
Thermodiscus maritimus	98	5.5	anaerobic	hetcrotrophic	f	Stetter (1996b)
Thermofilum librum	95	6.0	anaerobic	heterotrophic	S°	Stetter (1996b)
T. pendens	95	5.0	anaerobic	heterotrophic	S°	Zillig et al. (1983a)
Thermoplasma acidophilum	63	1.5	anaerobic	heterotrophic	f	Darland et al. (1970)
T. volcanicum	67	2.0	anaerobic	heterotrophic	f	Segerer et al. (1988)
Thermoproteus neutrophilus	97	neutral	anaerobic	H_2	S°	Stetter (1996b)
T. tenax	97	5	anaerobic	heterotrophic/H_2	S°	Zillig et al. (1981)
T. uzoniensis	97	neutral	anaerobic	unkown	S°	Bonch-Osmolovskaya et al. (1990)

TABLE 1. continued

Taxon/species	Growth conditions		Metabolism	Electron donor	Electron acceptor[1]	Reference
	T_{max}	pH_{opt}				
BACTERIA						
Alicyclobacillus acidocaldarius	70	2	aerobic	heterotrophic	O_2	Darland & Brock (1971), Wisotzkey *et al.* (1992)
Ammonifex degensii	77	7.5	anaerobic	H_2	NO_3	Huber *et al.* (1996a)
Aquifex pyrophilus	95	6.8	aerobic	S^o/H_2	O_2	Huber *et al.* (1992)
Bacillus infernus	65	7.3-7.8	aerobic	heterotrophic	$O_2/Fe(III)$	Boone *et al.* (1995)
B. schlegelii	80	6.5	aerobic	heterotrophic/S/H_2	O_2	Schenk & Aragno (1981)
B. stearothermophilus	75	7	aerobic	heterotrophic	O_2	Claus & Berkeley (1986)
B. thermocatenulatus	78	NR^2	aerobic	heterotrophic	O_2	Golovacheva *et al.* (1975)
B. thermocloacae	70	8-9	aerobic	heterotrophic	O_2	Demharter & Hensel (1989)
B. thermoglucosidasius	69	6.5	aerobic	heterotrophic	O_2	Suzuki *et al.* (1983)
B. thermolevorans	70	6.2-7.	aerobic	heterotrophic	O_2	Zarilla & Perry (1987)
B. tusciae	65	4.5	aerobic	heterotrophic/S	O_2	Bonjour & Aragno (1985)
Calderobacterium hydrogenophilum	82	6-7	aerobic	S^o/H_2	O_2	Kryukov *et al.* (1983)
Caldicellulosiruptor saccharolyticus	80	7	anaerobic	heterotrophic	f	Rainey *et al.* (1994a)
Kristjanssonii	82	7	anaerobic	heterotrophic	f	Bredholt *et al.* (1999)

TABLE 1. continued

Taxon/species	Growth conditions		Metabolism	Electron donor	acceptor[1]	Reference
	T_{max}	pHopt				
BACTERIA						
Caloramator fervidus	80	7-7.5	anaerobic	heterotrophic	f	Patel et al. (1987)
C. indicus	75	7	anaerobic	heterotrophic	f	Collins et al. (1994)
C. proteoclasticus	68	7-7.5	anaerobic	heterotrophic	f	Tarlera et al. (1994)
Carboxydothermus hydrogenoformans	78	6.8-7	anaerobic	CO	HO	Svetlitchnii et al. (1991)
Chloroflexus aurantiacus	70	8	aerobic	heterotrophic/photosynthetic	O_2	Pierson & Castenholz (1974)
Clostridium thermocellum	68	5.7	anaerobic	heterotrophic	f	McBee (1954)
C. thermolacticum	70	7	anaerobic	heterotrophic	f	Le Ruyet et al. (1985)
C. thermopapyrolyticum	66	NR	anaerobic	heterotrophic	f	Méndez et al. (1991)
C. thermosuccinogenes	91	NR	anaerobic	heterotrophic	f	Drent et al. (1991), Collins et al. (1994)
Coprothermobacter proteolyticus	70	7.5	anaerobic	heterotrophic	f	Ollivier et al. (1985)
Desulfotomaculum australicum	74	7-7.4	anaerobic	heterotrophic/H_2	SO_4	Love et al. (1993)
D. kuznetsovii	85	7	anaerobic	heterotrophic	S°	Nazina et al. (1989)
D. nigrificans	70	7	anaerobic	heterotrophic/H_2	SO_4	Starkey (1938)
D. thermoacetooxidans	65	NR	anaerobic	heterotrophic	S°	Min & Zinder (1990)
D. thermobenzoicum	70	7.2	anaerobic	heterotrophic/H_2	SO_4	Tasaki et al. (1991)

TABLE 1. continued

Taxon/species	Growth conditions		Metabolism	Electron donor	acceptor[1]	Reference
	Tmax	pHopt				
BACTERIA						
Desulfurella acetivorans	70	6.4-6.8	anaerobic	heterotrophic/H$_2$	S°	Miroshnichenko et al. (1994)
D. multipotens	77	6-7.2	anaerobic	heterotrophic/H$_2$	S°	Miroshnichenko et al. (1994)
Dictyoglomus thermophilus	80	7.0	anaerobic	heterotrophic	f	Saiki et al. (1985)
D. turgidus	86	7.0-7.1	anaerobic	heterotrophic	f	Svetlichnii & Svetlichnaya (1988)
Fervidobacterium gondwanense	80	7.0	anaerobic	heterotrophic	f	Andrews & Patel (1996)
F. islandicum	80	7.2	anaerobic	heterotrophic	f	Huber et al. (1990)
F. nodosum	80	7.0	anaerobic	heterotrophic	f	Patel et al. (1985)
Hydrogenobacter acidophilus	70	3-4	aerobic	S/H$_2$	O$_2$	Shima & Suzuki (1993)
H. thermophilus	77	7	aerobic	S/H$_2$	O$_2$	Kawasumi et al. (1984)
Meiothermus ruber	70	7.2	aerobic	heterotrophic	O$_2$	Longinova et al. (1984)
M. silvanus	65	8-8.5	aerobic	heterotrophic	O$_2$	Tenreiro et al. (1995a)
Moorella glycerini	65	6.3-6.5	anaerobic	heterotrophic	f	Slobodkin et al. (1997a)
M. thermoacetica	65	6.6-6.8	anaerobic	heterotrophic/H$_2$	f/CO$_2$	Fontaine et al. (1942)
M. thermoautotrophicum	70	5.7	anaerobic	heterotrophic/H$_2$	f/CO$_2$	Wiegel et al. (1982)
Rhodothermus marinus	72	6.5	aerobic	heterotrophic	O$_2$	Alfredsson et al. (1988)
R. obamensis	85	7	aerobic	heterotrophic	O$_2$	Sako et al. (1996a)

TABLE 1. continued

| Taxon/species | Growth conditions | | Metabolism | Electron | | acceptor[1] | Reference |
	T_{max}	pH_{opt}		donor			
BACTERIA							
Synechococcus lividus	73	8	anaerobic	photosynthesis		f	Meeks & Castenholz (1971)
Thermoanaerobacter acetoethylicus	80	NR	anaerobic	heterotrophic		f	Ben-Bassat et al. (1981)
T. brockii	75	6.5-7.5	anaerobic	heterotrophic		f	Cayol et al. (1995)
T. ethanolicus	78	5.8-8.5	anaerobic	heterotrophic		f	Wiegel & Ljungdahl (1981)
T. kivui	75	6.2-6.8	anaerobic	heterotrophic/H_2		f/CO_2	Collins et al. (1994)
T. thermopriae	70	6.5-7.3	anaerobic	heterotrophic		f	Collins et al. (1994)
T. thermohydrosulfuricum	78	6.9-7.5	anaerobic	heterotrophic		f	Lee et al. (1993)
Thermoanaerobacterium aotearoense	66	5.2	anaerobic	heterotrophic		f	Liu et al. (1996)
T. saccharolyticum	70	6	anaerobic	heterotrophic		f	Lee et al. (1993)
T. thermosulfurigenes	75	5.5-6.5	anaerobic	heterotrophic		f	Lee et al. (1993)
T. xylanolyticum	70	6	anaerobic	heterotrophic		f	Lee et al. (1993)
Thermobrachium celere	75	8.2	anaerobic	heterotrophic		f	Engle et al. (1996)
Thermocrinis ruber	89	8	aerobic	S/H_2/heterotrophic		O_2	Huber et al. (1998)
Thermodesulfobacterium commune	85	7	anaerobic	heterotrophic		f/SO_4	Zeikus et al. (1983a)
T. mobile	85	7.5	anaerobic	heterotrophic		f/SO_4	Rozanova & Pivovarova (1988)

TABLE 1.   continued

Taxon/species	Growth conditions		Metabolism	Electron donor	Electron acceptor[1]	Reference
	T_{max}	pH_{opt}				
BACTERIA						
Thermodesulforhabdus norvegicus	74	6.9	anaerobic	heterotrophic	SO_4	Beeder et al. (1995)
Thermodesulfovibrium yellowstonii	70	6.8-7	anaerobic	heterotrophic	f/SO_4	Henry et al. (1994)
Thermomicrobium roseum	80	8.3	aerobic	heterotrophic	O_2	Jackson et al. (1973)
Thermonema lapsum	70	6.5	aerobic	heterotrophic	O_2	Hudson et al. (1989)
T. rossianum	65	7-7.5	aerobic	heterotrophic	O_2	Tenreiro et al. (1997)
Thermooleophilum album	70	7.3	aerobic	heterotrophic	O_2	Zarilla & Perry (1984)
T. minutum	70	6.8	aerobic	heterotrophic	O_2	Zarilla & Perry (1986)
Thermoterrabacterium ferrireducens	74	6-6.2	anaerobic	heterotrophic/H	$f/Fe(III)/S^o$	Slobodkin et al. (1997b)
Thermosipho africanus	77	7.2	anaerobic	heterotrophic	f	Huber et al. (1989b)
T. melanesiensis	80	6.5-7.5	anaerobic	heterotrophic	f	Antoine et al. (1997)
Thermosyntropha lipolytica	70	8.1-8.9	anaerobic	heterotrophic	f	Svetlichnii et al. (1996)

TABLE 1. continued

Taxon/species	Growth conditions		Metabolism	Electron donor	acceptor[1]	Reference
	T_{max}	pH_{opt}				
BACTERIA						
Thermothrix azorensis	86	7-7.5	aerobic	S°	O_2	Odintsova et al. (1996)
T. thiopara	80	6.8	aerobic	heterotrophic/S	O_2	Caldwell et al. (1976)
Thermotoga elfii	72	7.5	anaerobic	heterotrophic	f/S$^\circ$	Ravot et al. (1995)
T. hypogea	90	7.3-7.4	anaerobic	heterotrophic	f	Fardeau et al. (1997)
T. maritima	90	6.5	anaerobic	heterotrophic	f	Huber et al. (1986)
T. neapolitana	90	7.0	anaerobic	heterotrophic	f	Jannasch et al. (1988)
T. subterranea	75	7.0	anaerobic	heterotrophic	f	Jeanthon et al. (1995)
T. thermarum	84	7.0	anaerobic	heterotrophic	f	Windberger et al. (1989)
Thermus aquaticus	80	7.5	aerobic	heterotrophic	O_2	Brock & Freeze (1969)
T. brockianus	80	NR	aerobic	heterotrophic	O_2	Williams et al. (1995)
T. filiformis	80	7.3	aerobic	heterotrophic	O_2	Hudson et al. (1987)
T. oshimai	80	NR	aerobic	heterotrophic	O_2	Williams et al. (1996)
T. scotoductus	80	7.5	aerobic	heterotrophic	O_2	Kristjansson et al. (1994)
T. thermophilus	85	7.5	aerobic	heterotrophic	O_2	Oshima et al. (1974)

[1] f; fermentative; [2] Not recorded

characters for the two species, was a paler yellow pigmentation and spreading morphology on yeast tryptone agar for *T. brockianus* (Williams *et al.*, 1995). Characters such as growth on galactose, fructose and trehalose are commonly found among strains of other *Thermus* species and are not diagnostic (Kristjansson, unpublished).

The geographical isolation of thermophilic habitats may also influence numerical taxonomic analyses. For example, *Thermus* strains of widely different origin cluster according to geographic origin or physiological adaptations to local conditions rather than affinities by genotype (Hudson *et al.*, 1986; 1987; Santos *et al.*, 1989). In other words, while different *Thermus* genospecies from the same geothermal region may be distinguished by numerical taxonomy, it does not necessarily group together isolates of the same genospecies, from different geographic locations.

However, numerical taxonomy and molecular methods both indicate that *Rhodothermus* strains of different origin form a very homogeneous group (Nunes *et al.*, 1992) and its behaviour is therefore very different from that of the genus *Thermus* perhaps reflecting a comparatively recent adaptation to thermophily.

Numerical taxonomy has been used successfully on thermophilic *Bacillus* to reveal substantial diversity and many more taxa than the commonly recognized *B. stearothermophilus*. Many of these strains make industrially attractive amylolytic and proteolytic enzymes. Different studies indicated that the phenotypic clustering correlates with genotype grouping by DNA:DNA hybridization (White *et al.*, 1993). The various clusters were also largely supported by 16S rRNA sequence analysis (Rainey *et al.*,1994b).

A comparative numerical taxonomic study on 51 strain of thermophilic, Gram-positive, anaerobic, cellulolytic bacteria based on 92 phenotypic characters with partial 16S rRNA sequences from 16 of the isolates revealed high diversity by both methods and a fairly good correlation between the results of the phenetic and the phylogenetic study (Rainey *et al.*, 1993b). However, although numerical analysis of phenotypic properties may be useful in some cases for defining taxonomic groups or distinguishing between thermophilic bacteria these methods have largely been superseded by chemotaxonomic and more particularly molecular approaches.

5.1.3 Chemotaxonomic methods

Various specific chemical markers have been identified in thermophilic bacteria that are potentially useful for discriminating between taxa at different levels but little use has been made of these techniques.

Many chemotaxonomic methods have only slight discriminatory power in studies of thermophilic microorganisms, below the level of genus. A few studies have been undertaken to examine the applicability of membrane lipids and fatty acid composition in the classification of *Rhodothermus* and *Thermus* (Tenreiro *et al.*, 1995; Moreira *et al.*, 1996). Williams (1989) discussed the biochemical characterization of *Thermus* and its potential use for classification. All isolates belonging to different *Thermus* species were shown to have menaquinone as the predominant respiratory quinone, and similar peptioglycan containing ornithine, a diamino acid which is relatively rare in Gram-negative bacteria.

In some instances, lipids are valuable phylogenetic markers. Within the family *Bacillicaeae* the thermoacidophilic bacilli belonging to the genus *Alicyclobacillus* was established on the basis of a unique type of lipid, ω-alicyclic fatty acid (Wisotzkey *et al.*, 1992).

Moreover, ether-linked lipids made of alkyl chains with one subterminal methyl branch (*anteiso* branching) have been found to be characteristic for the deeply branching sulfate-reducing bacteria of the genus *Thermodesulfobacterium* (Langworthy *et al.*, 1983).

Whole cell protein (WCP) analysis by SDS PAGE and multilocus enzyme electrophoresis (MLEE) analysis by native PAGE differ from other chemotaxonomic methods in that ~v relate directly to stable genomic characteristics (Selander *et al.*, 1986; Goodfellow & O'Donnell, 1993). WCP analysis is based on comparing size distribution of proteins without knowing their identity or function. It gives taxonomic resolution that reflects genomic relationships within the family at species and sometimes subspecies level. MLEE, which is based on the distribution of different alleles of a number of enzymes, indicates relationships of a clonal nature at the subspecies level (Selander *et al.*, 1986; Vandamme *et al.*, 1996).

WCP analysis correlated well with DNA:DNA hybridization for delineating different species of *Thermus* (Williams, 1989). It has also been used with *Thermococcus* strains (Marteinsson *et al.*, 1995b) and with anaerobic fermentative bacteria for establishing species status in conjunction with 16S rRNA analysis (Engle *et al.*, 1996). We have also recently found MLEE to be very useful for taxonomic studies of *Rhodothermus* (Petursdottir *et al.* unpublished) and for establishing sub-specific groups within species of *Thermus* (see Fig. 1) (Hreggvidsson *et al.*, unpublished).

5.1.4 Nucleic acids

Thermophilic microbiology has almost from the beginning been based on phylogeny (Table 1). It is evident from published taxonomic and phylogenetic studies on thermophiles that 16S rRNA sequencing and DNA:DNA hybridization are the critical methods for delineating taxa and establishing species status. These methods allow classification and delineation of a taxon independent of the evolution of phenotypic traits and regardless of biochemical or physiological properties.

Small subunit (16S) rRNA gene sequencing is now accepted as the universal method of establishing hierarchical relationships of extant organism at and above species level and thus enabling natural phylogenetic taxonomy (Woese, 1987; Goodfellow & O'Donnell, 1993). It provides the necessary framework for evolutionary studies, such as tracing the evolutionary history of organisms or the origin and evolution of particular metabolic or cellular machinery (Woese, 1987; Olsen & Woese, 1993). Taxonomic structures of large groups of thermophilic bacteria have been recently reconsidered on the basis of 16S rRNA sequences analysis, resulting in proposals for new genera and species. This has, for example, been accomplished for Gram-positive, fermentative anaerobes (Rainey *et al.*, 1994b; Collins *et al.*, 1994) and the family *Thermaceae* (Nobre *et al.*, 1996; Rainey *et al.*, 1997) (see Table 1).

Other methods covering a more limited taxonomic range than 16S rRNA gene sequence analysis but allowing finer distinctions between organisms are more suitable for different aspects of microbial systematics for example in the field of population genetics and ecology. Available methods show different levels of sensitivity, and they also vary in ease of application. Many of the recently developed genotypic methods, such as restriction fragment lengh polymorphisms (RFLP), random amplified polymorphic DNA (RAPD-PCR), pulsed field gel electrophoresis-RFLP (PFGE-RFLP) and ribotyping are

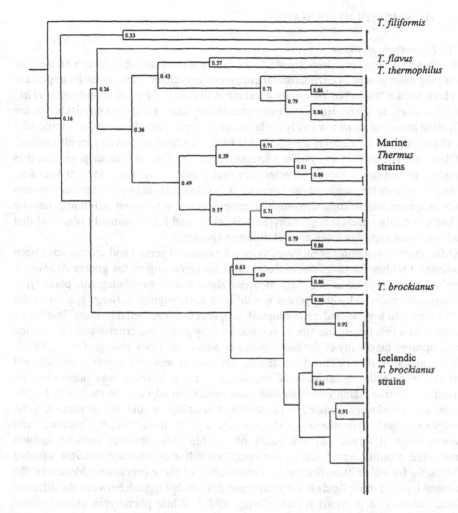

Figure 1. Dendrogram based on MLEE analysis of Icelandic *Thermus* isolates including type strains for the species *T. filiformis*, *T. thermophilus* and *T. brockianus*. *Thermus flavus* belongs to *T. thermophilus*. The dendrogram is based on allelic distribution of 7 enzymes. The study contained a subset of 23 marine strains and 36 terrestrial strains, most of which had been presumed to be *T. brockianus*. The MEE analysis agrees with marked groups and the type strain groups as expected from other phenotypic and phylogenetic studies. The numbers are similarity values at branching points.

particularly suitable for processing many samples and therefore ideal in ecological studies. These methods have different sensitivities but cover taxonomic ranges at and below the level of species. They have been used sporadically in thermophilic microbiology (Rodrigo *et al.*, 1994; Kwak *et al.*, 1995; Ferris *et al.*, 1996; Ward *et al.*, 1997) and are particularly useful for identifying and tracing individual strains (Ronimus, 1997) or for determining the fine structures in geographical distribution (Moreira *et al.* 1996; 1997)

5.2   THERMOPHILIC ARCHAEA

5.2.1   *Phenotypic methods*
Some phenotypes are very characteristic for certain groups of archaea and can be used for identifying those groups in environmental samples or in enrichment cultures. Examples are the characteristic "blebs" of *Thermotoga* and the disk shape of *Pyrodictium* (Stetter *et al.*, 1983; Huber *et al.*, 1986). However, many archaea are also fastidious organisms that are difficult to grow and show few easily evaluated phenotypic traits. In these cases molecular and chemotaxonomic methods are invaluable tools for classification and identification.

Obviously phenotypic analysis is a necessary part of a species description and it is important to determine those characteristics that would allow for easy identification. However, as with the thermophilic bacteria, it is often difficult to distinguish between different species within some thermophilic genera and even between different genera on the basis of easily observable phenotypic traits alone and it is commonly observed that one basic phenotype can mask several different species.

Difficulties of assigning phenotypes to specific archaeal genera and species have been encountered within the order *Sulfolobales*. Species belonging to the genera *Acidianus, Metallospaera, Sulfolobus* and *Sulforococcus* share many morphological, phenotypic and metabolic traits and within genera it is difficult to distinguish between species on the basis of morphological and physiological properties. More reliable classification of species within this family has since been obtained by genotypic criteria and this has led to an improved taxonomy of the *Sulfolobales* (Fuchs *et al.* 1996; Burggraf *et al.*, 1997).

Marteinsson *et al.* (1995b) discussed the difficulties and shortcomings of different methods used in the systematics of archaeal cocci. Their study was undertaken on anaerobic, sulfur-utilizing cocci isolated from hydrothermal vents in the South Pacific Ocean and previously isolated type strains belonging to the six genera namely, *Desulfurococcus, Hyperthermus, Pyrococcus, Pyrodictium, Staphylothermus* and *Thermococcus*. It turned out that easily observable characteristics such as, growth temperature, motility, nutritional requirements and sulfur dependence were too variable to be useful for either classification or identification of these organisms. Moreover, the structural types of ether lipids in the membrane did not distinguish between the different genera, contrary to previous reports (Zillig, 1992). While phenotypic characteristics were of limited value for distinguishing these taxa, cell morphology could be used to differentiate between thermococci and the other archaeal cocci and members of the genera *Hyperthermus* and *Pyrodictium* can also be excluded from the others by butanol formation which is characteristic of the former two genera. The G+C ratio of the DNA alone did not give any conclusive results, but in conjunction with other data it could be used to distinguish between *Pyrococcus* and *Thermococcus* isolates. Resolution at the genomicspecies level and lower can be obtained by DNA:DNA hybridization and WCP profile studies.

5.2.2   *Chemotaxonomic methods*
Many chemotaxonomic methods are well suited for discriminating between higher taxa, above the level of species. Some of these clearly distinguish archaea from bacteria. For example, the cell wall of archaea does not contain peptidoglycan while different

peptidoglycan structures are found in most groups of bacteria. Thermophilic archaea, unlike bacteria, also contain characteristic isoprenoid glycerol ether-linked lipids. Variation of lipid structure based on the ether-linked, branched theme has been reported to discriminate between some archaeal groups (Ross *et al.*, 1985).

Other specific chemicals can also be associated with certain groups such as calditoglycerocaldarcaeol in members of the order *Sulfolobales* and specific coenzymes in methanogens (Balch *et al.*, 1979; Takayanagi *et al.*, 1996).

5.2.3 *Nucleic acids*

Nucleic acid analyses have revealed significant diversity among thermophilic archaea which, as discussed above, may not be evident from phenotypic data. For example, most of the easily determined phenotypic traits of *Sulfolobus* are similar for different species but genetic divergence is great. The largest phylogenetic distance based on 16S rRNA gene sequences is close to 14% between S. *metallicus* and other members of the genus (Fuchs *et al.*, 1996; Burggraf *et al.*, 1997). Such a large distance would usually warrant splitting the group into at least two genera and emphasizes that in the search for useful enzymes and metabolites the full diversity of genera should be considered. Furthermore, the species S. *acidocaldarius* and "S. *thuringiensis*" exhibit only 60% DNA:DNA similarity in spite of having identical 16S rRNA sequence (Fuchs *et al.*, 1996).

Burggraf *et al.* (1997) pointed out that there are clear examples of hyperthermophiles belonging to the kingdom *Crenarchaeaota* that form distinct phylogentic clusters but which are also well separated by a variety of phenotypic characters. However, the phylogenetic branches are much shorter than those observed in mesophiles. The authors argue that the high G+C content in the rRNAs of hyperthermophiles may result in a lower rate of sequence changes in hyperthermophiles than in mesophiles. This would mean that the rate of evolution in thermophiles is generally lower than that in mesophiles and consequently the phylogenetic depth, based on 16S rRNA divergence, would be less for members of thermophilic taxa than is observed for mesophilic taxa.

6   Halophiles

Although bacteria have been categorized according to salt tolerance/requirement (Table 2) there are organisms that do not fit exclusively into any one category (Kushner, 1978; 1993). The term halophile is generally restricted to those organisms that require NaCl for growth. The red-pigmented archaeal types, which require at least 1.5 M NaCl for growth, are the most halophilic organisms known, but the salt requirements for the slight halophiles and moderate halophiles may be considerably less unequivocal and may vary depending on environmental conditions. Organisms capable of growing over a range of salt concentrations, but capable of maximal growth in the absence of salt are halotolerant, whereas the unusual group growing over the whole concentration range from zero to saturation with maximal growth in the presence of salt are referred to as haloversatile (James *et al.*, 1990) or euryhaline (Vreeland, 1987).

TABLE 2.   Classification of halophiles

Category	Salt Concentration (M)	
	Range	Optimum
Non-halophile	0-1.0	<0.2
Slight halophile	0.2-2.0	0.2-0.5
Moderate halophile	0.4-3.5	0.5-2.0
Borderline extreme halophile	1.4-4.0	2.0-3.0
Extreme halophile	2.0-5.2	>3.0
Halotolerant	0->1.0	<0.2
Haloversatile	0->3.0	0.2-0.5

6.1   HALOPHILIC ARCHAEA

The term 'halobacteria' refers to the red-pigmented, extremely halophilic archaea, members of the order *Halobacteriales*. Most halobacteria require at least 1.5 M NaCl for growth and to retain the structural integrity of their cells. In recent years halophilic methanogens have also been described, but the halobacteria remain the most halophilic organisms known (Table 3). Halobacteria can be readily differentiated from halophilic bacteria by the presence of characteristic archaeal ether-linked lipids (Ross *et al.*, 1981). Halophily has been acquired by more than one methanogen group and the systematics of these organisms is firmly based around the methanogen phenotype rather than the halophilic phenotype - this section is therefore devoted to the halobacteria.

6.1.1   *Phenotypic, chemotaxonomic and molecular analyses of halophilic archaea*
Phenotypic characterization has proven relatively unsuccessful in the taxonomy of halobacteria since many are biochemically inactive although the original descriptions of the genera *Halobacterium* and *Halococcus* were based on phenotypic features. In more recent times, the genera *Haloarcula* and *Haloferax* were also defined largely by a numerical taxonomic analysis (Torreblanca *et al.*, 1986). Current isolates comprise two physiological groups:
(a) Isolates that grow only at high pH (8.5-11.0) with a very low Mg^{2+} requirement (<1 mM). These halobacteria, classified in the genera *Natronobacterium* and *Natronococcus* are restricted to soda lakes and soda deserts.
(b) Isolates that grow at neutrality or close to neutrality with a significant Mg^{2+} requirement (5-50 mM). These organisms are currently classified in the genera *Haloarcula*, *Halobacterium*, *Halococcus*, *Halobaculum*, *Haloferax*, *Halorubrum* and *Natrialba*.

Current genera are best defined by chemotaxonomic analyses, notably polar lipid composition. Minimal standards for description of new taxa have now been published (Oren *et al.*, 1997). All halobacteria have phytanyl (C_{20}) or sesterterpanyl (C_{25}) derivatives of phosphatidyl glycerol (PG) and methylated phosphatidyl glycerol phosphate (PGP).

TABLE 3.   Characteristics of halophilic archaea

Species	Polar lipid signature	Salt range (w/v %)	Salt optimum (w/v%)	Reference
Halobacteria				
Haloarcula	PG, PGP, PGS, S-TGD-2, (C_{20},C_{20})			
Haloarcula				
argentinensis		15-30	22	Ihara *et al.* (1997)
hispanica		15-30	20	Torreblanca *et al.* (1986)
japonica		15-30	20	Takashina *et al.* (1990)
marismortui		15-30	20	Oren *et al.* (1990)
mukohataei 1		5-30	20	Ihara *et al.* (1997)
vallismortis		15-30	20	Torreblanca *et al.* (1986)
Halobacterium	PG, PGP, PGS, S-TGD-1, S-TeGD (C_{20},C_{20})			
Halobacterium salinarum		15-30	25	Grant & Larsen (1989) Ventosa & Oren (1996)
Halobaculum	PG, PGP, S-DGD, (C_{20},C_{25})			
Halobaculum gomorrense		15-30	20	Oren *et al.* (1995)
Halococcus	PG, PGP, S-DGD, (C_{20},C_{25})			
Halococcus				
morrhuae		15-30	20	Kocur & Hodgkiss (1973)
saccharolyticus		15-30	20	Montero *et al.* (1989)
salifodinae		15-30	20	Denner *et al.* (1994)
turkmenicus		15-30	20	Oren *et al.* (1997)
Haloferax	PG, PGP, S-DGD (C_{20},C_{20})			
Haloferax				
denitrificans		10-30	15	Tomlinson *et al.* (1986), Grant & Larsen (1989)
gibbonsii		10-30	15	Torreblanca *et al.* (1986),
mediterranei		10-30	15	Torreblanca *et al.* (1986)
volcanii		10-30	15	Torreblanca *et al.* (1986)

TABLE 3.   continued

Species	Polar lipid signature	Salt range (w/v %)	Salt optimum (w/v%)	Reference
Halorubrum	PG, PGP, PGS, S-DGD (C_{20},C_{20})			
Halorubrum				
coriense		15-30	20	Oren & Ventosa (1996)
distributum		15-30	20	Oren & Ventosa (1996)
lacusprofundii		15-30	20	McGenity & Grant (1995)
saccharovorum		15-30	20	McGenity & Grant (1995)
sodomense		10-25	15	McGenity & Grant (1995)
trapanicum		15-30	20	McGenity & Grant (1995)
Natronobacterium	PG, PGP, (C_{20},C_{25})			
Natronobacterium				
gregoryi		15-30	20	Tindall *et al.* (1984)
magadii		15-30	20	Tindall *et al.* (1984)
pharaonis		15-30	20	Tindall *et al.* (1984)
vacuolata		15-30	20	Mwatha & Grant (1993)
Natronococcus	PG, PGP, (C_{20},C_{20})			
Natronococcus				
amylolyticus		10-30	20	Kanai *et al.* (1995)
occultus		10-30	20	Tindall *et al.* (1984)
Natrialba	PG, PGP, PGS, S2-DGD, (C_{20},C_{25})	15-30	22	Kamekura & Dyall-Smith (1995)
Methanogens				
Halomethanococcus				
doii		2-15	12	Yu & Kawamura (1987)
Methanohalophilus				
halophilus		2-15	7-9	Wilharm *et al.* (1991)
mahii		3-20	6-15	Paterek & Smith (1988)
portucalensis		2-25	3-12	Boone *et al.* (1993)
zhilinae		1-12	4	Mathrani *et al.* (1988)
Methanohalobium				
evestigatum		15-30	25	Zhilina & Zavarzin (1987)

Abbreviations

PG, phosphatidyl glycerol; PGP, methylated phosphatidyl glycerol phosphate; PGS, phosphatidyl glycerol sulphate; (S)-DGD, (sulphated) diglycosyl diether; (S)-TGD, (sulphated) triglycosyl diether; (S)-TeGD, (sulphated) tetraglycosyl diether; S2-DGD, disulphated diglycosyl diether. The structures of the glycolipids are given in Kamekura (1993) and Kates (1996).

Most also contain phosphatidyl glycerol sulphate (PGS). A family of glycolipids and sulphated glycolipids is also present derived from a basic (α-D) glucosyl phytanyl or sesterterpanyl core. A summary of the lipid signatures of the various genera is given in Table 3. Features like cell morphology, motility, pigmentation, requirement for salt to prevent cell lysis, optimum NaCl and $MgCl_2$ concentration for growth, temperature and pH range for growth, ability to grow on single carbon sources, hydrolysis of starch and sensitivity to antibiotics enable further taxonomic discrimination (Oren *et al.*, 1997).

Thin layer chromatography (TLC) of polar lipids offers a simple and rapid way of assigning isolates to genera. Lipids are extracted from freeze dried cells by a modified Bligh and Dyer procedure (Ross *et al.*, 1985) and separated by two dimensional TLC. A range of spray reagents may be used to discriminate between different classes of lipids. There is still some dispute over the precise structure of some of the glycolipids (Kamekura & Dyall-Smith, 1995; McGenity & Grant, 1995).

16S rRNA/DNA hybridization analyses support groups obtained by lipid analysis (Ross & Grant, 1985). Moreover, 16S rRNA gene sequence analyses confirm the conclusions of the hybridization and polar lipid compositions (Fig. 2), (but see later under alkaliphiles) with most generic groups sharing only around 90% sequence identify, although *Natrialba, Natronobacterium* and *Natronococcus* are more closely related at 94-95% identity (Kamekura *et al.*, 1997). Care has to be exercised over the way that the sequence analysis is carried out since most halobacteria have more than one 16S RNA gene. Generally, the genes are virtually identical and normal cloning or PCR procedures are adequate.

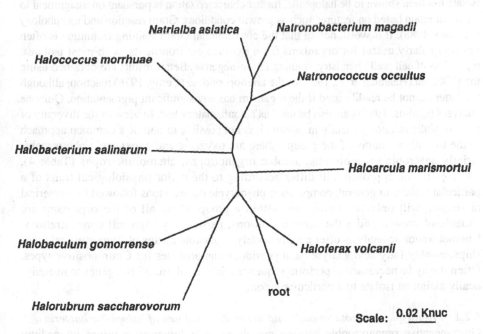

Figure 2. A maximum likelihood tree of 16S rRNA gene sequences of extremely halophilic archaea. The tree was rooted by the outgroup *Methanospirillum hungatei*. A preliminary sequence alignment was obtained using CLUSTAL V, which was manually optimized. Aphylogenetic tree was drawn using the DNAML maximum likelihood program in PHYLIP 3.4.

However, gene copies in *Haloarcula* spp. in particular, may show some 70 or 80 base differences (i.e. around 5% of the 16S rRNA gene) localized mainly in two variable regions (Mylvaganam & Dennis, 1992; unpublished results). Nevertheless, *Haloarcula* spp. still constitute a distinct phylogenetic line when these multiple 16S rRNA gene sequences are included in the analyses (unpublished results). It is worth noting that a difference of 5% in 16S rRNA sequence is normally regarded as indicating a distinct species! Figure 2 includes one of the two dissimilar gene sequences from *Haloarcula marismortui*. Signature sequences between positions 151-200 are available for a limited number of the generic groups (Benlloch *et al.*, 1996), and these may have potential as probes for use in screening programmes. In view of the limited array of discriminatory phenotypic tests that are available, and the unreactive nature of many isolates, DNA/DNA reassociation determinations may be the only way of establishing identity at the species level. Validly published species are listed in Oren *et al.* (1997), together with guidelines for phenotypic characterization.

6.2   HALOPHILIC BACTERIA

Examples of halophilic bacteria are listed in Table 4. Inevitably, given the definition of halophiles based on what is effectively detected as a single phenotypic characteristic, there is great phylogenetic diversity amongst the group. It is evident from Figure 3 that the halophilic phenotype has evolved in most of the major phylogenetic lines, but also that some halophiles e.g. *Flexistipes sinusarabadici* may represent novel lineages. Once an isolate has been shown to be halophilic, further characterization is pursuant on assignment to an initial group based on features such as growth conditions, Gram reaction and morphology (Ventosa, 1988). It is worth noting that the conventional Gram staining technique is often not particularly useful for organisms from extreme environments, with most isolates, regardless of cell wall chemistry, staining Gram-negative. Better results are obtained using the KOH test (Halebian *et al.*, 1981) or the aminopeptidase (Cerny, 1976) reaction, although the latter cannot be readily used if the organism has any significant pigmentation. Quinone analyses (Collins, 1985) can also be used as a confirmatory test. In view of the diversity of the halophilic isolates presently in culture, it is not possible to outline a common approach to the overall taxonomy of the group - there are oxygenic and anoxygenic phototrophs, strictly anaerobic organotrophs, aerobic organotrophs, chemolithotrophs (Table 4), and clearly, the approach will differ according to the major physiological traits of a particular isolate. In general, comparative phenotypic comparisons followed by numerical taxonomy, will only be appropriate within a group where all of the organisms are capable of growth under the same conditions, particularly high salt concentrations. Chemotaxonomic analysis offer a more widely applicable means of supporting relationships, notably fatty acid analyses and peptidoglycan analyses for Gram-positive types. Often it may be necessary to perform sequence analyses of 16S rRNA genes to unequivocally assign an isolate to a particular taxon.

6.2.1 *Phenotypic, chemotaxonomic and molecular analyses of halophilic bacteria*
Gram-negative organotrophic bacteria are abundant in hypersaline brines of medium salinity. Originally classified in several genera such as *Deleya, Halomonas, Halovibrio, Pseudomonas, Paracoccus,* and *Volcaniella* (Ventosa, 1994), these have now been

reclassified into a single generic grouping as *Halomonas* spp. on the basis of 16S rRNA sequence analyses (Dobson & Franzmann, 1996). The relative coherence of the group can be seen in Figure 3, which also supports the suggestion that the genus *Chromohalobacter* might be included (Dobson & Franzmann, 1996). The organisms cannot be distinguished on the basis of fatty acid or polar lipid profiles and all contain ubiquinone 9 (Franzmann & Tindall, 1990). Numerical taxonomy has not proved successful in clearly distinguishing the original genera, although it has been of value in discriminating individual species. Polar lipid patterns and fatty acid analyses that define the group are listed in Franzmann & Tindall (1990). There are also limited 16S rRNA signature sequences available for the group (Franzmann *et al.*, 1988; Dobson & Franzmann, 1996).

Of the other *Proteobacteria* listed in Table 4 and shown in Figure 3, *Vibrio costicola*, probably the best studied of all moderate halophiles, shows little relationship with other *Vibrio* spp. and has recently been reclassified as *Salinivibrio* spp, a new phylogenetic line (Mellado *et al.*, 1996). *Flavobacterium gondwandsense* and *Flavobacterium salegens* from Antarctic hypersaline lakes are members of the *Flavobacterium - Bacteroides* line.

Ollivier *et al.* (1994) have recently reviewed the systematics and biology of the anaerobic Gram-negative rods (Table 4). These organisms, apart from *Acetohalobium arabaticum*, ferment carbohydrates. Fermentation patterns describe the genera established so far, a variety of different carbon sources being used by these obligate anaerobes. These organisms represent a separate lineage on the basis of 16S rRNA sequence analyse¬ (Fig. 3).

Gram-positive organisms are less common in hypersaline environments, but Gram-positive cocci in particular have been recovered. In early numerical taxonomic studies *Micrococcus halobius* clustered in association with other non-halophilic *Micrococcus* spp. (Ventosa, 1988). Other Gram-positive cocci have since been reclassified as *Marinococcus* and *Salinicoccus* spp. (Marquez *et al.*, 1990; Ventosa, 1994) on the basis of peptidoglycan, isoprenoid quinone, polar lipid and fatty acid data. Gram-positive rods are less common, although moderately halophilic *Bacillus* and *Clostridium* species have been described (Fendrich *et al.*, 1990; Ventosa *et al.*, 1994; Garabito *et al.*, 1997). Recently, Spring *et al.* (1996) showed that other halophilic aerobic, spore forming rods formed a group together with *Sporosarcina halophila* on the periphery of the *Bacillus* spectrum, and proposed a new genus *Halobacillus* for these organisms. It is clear that with the exception of *Micrococcus halobius*, which is a member of the high % G+C line of the Gram-positive phylum, the other types are distinct lineages within the low % G+C Gram-positive lineage (Fig. 3).

The phototrophic and sulphate-reducing halophilic bacteria (Fig. 3) are included in the γ1, α and δ subclasses of the class *Proteobacteria*. The phototrophic bacteria, in particular, are extremely diverse and proposals for new genera to separate halophilic representatives from non-halophilic types has occurred (Kawasaki *et al.*, 1993; Imhoff & Suling, 1996). The non-phototroph, *Arhodomonas oleiferhydrens* is clearly a member of the γ1 subgroup.

In general, it is likely that as more detailed phylogenetic analyses of other types become available, halophilic examples in major phyla will be shown to represent separate sublines within particular major phyla, meriting separate generic descriptions as has already happened with some of the Gram-positive types.

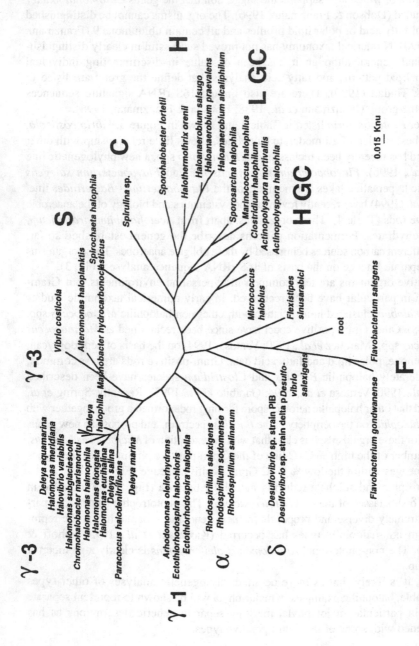

Figure 3. A maximum likelihood tree of 16S rRNA gene sequences of halophilic bacteria. The root was imposed by the archaeon outgroup *Haloferax volcanii*. Sequences were accessed from the Ribosomal Database Project. A preliminary sequence alignment was obtained using CLUSTAL W, which was manually optimized, then a phylogenetic tree drawn using the DNAML maximum likelihood program in PHYLIP 3.4. Evolutionary lines are indicated: γ-3, γ-1, α, δ proteobacteria; S spirochaetes; C cyanobacteria; H *Haloanaerobiaceae*; LGC, low % G+C Gram-positives; HGC, high % G+C Gram-positives; F, Flavobacteria *Flexistipes sinusarabici* does not associate with any known group.

TABLE 4. Characteristics of halophilic bacteria

Species	Habitat	Salt Range	Salt Optimum	Reference
Cyanobacteria				
Aphanothece halophytica	Salt lakes	6-12	3-24	Javor (1989)
Dactylococcopsis salina	Dead Sea Sabka	5-15	3-8	Walsby *et al.* (1983)
Spirulina platensis	Soda lakes	3-12	8	Javor (1989)
Gram-negative anaerobic rods				
Acetohalobium arabaticum	Lake Sivash, saline lagoons, microbial mats	10-25	15-18	Zhilina & Zavarzin (1990)
Haloanaerobium alcaliphilum	Great Salt Lake, sediment	3-15	10	Tsai *et al.* (1995)
praevalens	Great Salt Lake, sediment	3-25	13	Zeikus *et al.* (1983b)
salsugo	Saline oil field brine	6-24	9	Bhupathiraju *et al* (1994).
Halobacteroides acetoethylicus	Subsurface brines	6-20	10	Rengpipat *et al.* (1988a, 1988b)
halobius	Dead Sea sediment	8-17	9-15	Oren (1992)
lacunaris	Chockrack Lake	10-30	15-18	Zhilina *et al.* (1991)
Haloanaerobacter chitinovorans	Saltern	3-30	11-18	Liaw & Mah (1992)
Halocella cellulolytica	Lake Sivash, saline lagoons	5-20	15	Simankova *et al.* (1993)
Haloincola saccharolytica	Lake Sivash	3-30	10	Zhilina *et al.* (1992)
Halothermothrix orenii	Tunisian salt lake sediment	4-20	5-10	Cayol *et al.* (1994)
Sporohalobacter lortetii	Dead Sea, sediment	6-12	8 - 9	Oren *et al.* (1987)
marismortui	Dead Sea,	3-17	3-12	Oren *et al.* (1987)
Gram-positive (clostridia)				
Clostridium halophilum	Various saline sediments	1.5-10	6	Fendrich *et al.* (1990)

TABLE 4. continued

Species	Habitat	Salt Range	Salt Optimum	Reference
Gram-positive cocci (aerobic/microaerophilic)				
Marinococcus				
albus	Saline soils	2-25	10	Ventosa (1993)
halophilus	Saline soils, salterns, salted fish	2-25	10	Novitsky & Kushner (1976), Ventosa (1993)
Micrococcus				
halobius	Solar salt	2-25	10	Onishi & Kamekura (1972), Ventosa (1993)
Pediococcus halophilus	Salted anchovies	0-15	6.5-10	Villar et al. (1985)
Salinicoccus				
hispanicus	Saline soils, salterns	0.5-25	10	Marquez et al. (1990), Ventosa (1993)
roseus	Saltern	0.9-25	10	Ventosa et al. (1990), Ventosa (1993)
Sporosarcina halophila (*Halobacillus* halophila)	Saline soils, salterns, salted fish	1-15	3-5	Claus et al. (1983) Spring et al. (1996)
Gram-positive rods (aerobic)				
Bacillus	Rotted wood,			
halophilus	Pacific Ocean	3-30	15	Ventosa et al. (1989a)
salexigens	Hypersaline	7-20	10	Garabito et al. (1997)
Halobacillus				
litoralis	Great salt lake	5-20	10	Spring et al. (1996)
trueperi	Great salt lake	5-20	10	Spring et al. (1996)
Gram-positive (actinomycetes)				
Actinopolyspora				
halophila	Contaminant	10-32	20	Gochnauer et al. (1975)
iraqiensis	Iraqi saline soil	5-30	10-25	Ruan et al. (1994)
mortivallis	Death Valley soil	5-30	10-15	Yoshida et al. (1991)
Distinct lineage				
Flexistipes sinusarabidici	Red Sea, deep brine (2000 m)	3-18		Fiala et al. (1990)
Flavobacteria				
Flavobacterium gondwanense	Antarctic saline lake	0-20	5	Dobson et al. (1993), James et al. (1994)
salegens	Antarctic saline lake	0-20	5	Dobson et al. (1993), James et al. (1994)

TABLE 4.   continued

Species	Habitat	Salt Range	Salt Optimum	Reference
Proteobacteria, sulphate reducers				
Desulfovibrio halophilus	Solar lake, microbial mat	3-18	6-7	Caumette *et al.* (1991)
Desulfohalobium retbaense	Retba lake, Senegal	0.2-24	10	Ollivier *et al.* (1991)
Proteobacteria, rhodospirilla				
Rhodospirillum				
mediosalinum	Hot spring	2-10	6	Kompantseva & Gorlenko (1984)
salexigens	Saltern	5-20	6-8.5	Drews (1981)
salinarum	Saltern	2-25	6-12	Nissen & Dundas (1984)
sodomense	Dead Sea	6-21	12	Mack *et al.* (1993)
Proteobacteria, purple sulphur bacteria				
Chromatium salexigens	Salt ponds	4-20	10	Caumette *et al.* (1988)
Ectothiorhodo- spira marismortui	Salt lake	3-15	3-8	Oren *et al.* (1989)
Ectothiorhodospira			*(Halorhodospira)*	
abdelmalekii	Soda lake	5-30	15	Imhoff & Trüper (1981)
halochloris	Soda lake	10-37	14-27	Imhoff & Trüper (1977)
halophila	Salt lake	5-30	18	Raymond & Sistron (1969)
Thiocapsa halophila	Salt lake	3-20	8	Caumette *et al.* (1992)
Proteobacteria, halomonads				
Chromohalobacter (*Halomonas*?) *marismortui*	Dead Sea, saltern	1-30	10	Ventosa *et al.* (1989b)
Deleya (*Halomonas*)				
halophila	Saline soils, salterns	2-30	7.5	Quesada *et al.* (1984)
salina	Saline soils,	2.5-20	5	Valderrama *et al.* (1991)
Proteobacteria, halomonads				
Halomonas				
canadensis	Saline soils	0-20	3	Huval *et al.* (1995)
desiterata	Sewage	0-20	3	Berendes *et al.* (1996)
elongata	Salterns	0-32	3.5-8	Vreeland *et al.* (1980)
halodurans	Estuary	0-20	3-5	Herbert & Vreeland (1987)
halmophila	Dead Sea	3-25	7.5-12	Franzmann *et al.* (1988)
israelensis	Saline soils	0-20	3	Huval *et al.* (1995)

TABLE 4.   continued

Species	Habitat	Salt Range	Salt Optimum	Reference
Proteobacteria, halomonads				
Halomonas				
meridiana	Antarctic saline lake	0.5-20	1-3	James *et al.* (1990)
pantelleriense	Lake sand	0-20	3	Romano *et al.* (1996)
subglaciescola	Antarctic saline lake salterns, sea	0.5-20	3-5	Franzmann *et al.* (1987)
Halovibrio (*Halomonas*) variabilis	Great Salt Lake (N. arm)	7-29	9	Fendrich (1988)
Paracoccus (*Halomonas*) halodenitrificans	Bacon curing brine	0.5-25	4.4-8.8	Robinson & Gibbons (1952)
Pseudomonas (*Halomonas*) halosaccharolytica	Crude salt	3-27	12	Ventosa (1988)
Pseudomonas (*Halomonas*) halophila	Great salt lake (N. arm)	0.1-19	5	Fendrich, (1988)
Volcaniella (*Halomonas*) eurihalina	Saline soils	1.5-25	7.5	Quesada *et al.* (1990)
Proteobacteria, vibrios				
Vibrio (*Salinivibrio*) costicola	Salterns, cured meats	0.5-20	10	Garcia *et al.* (1987)
Proteobacteria, other lineage				
Marinobacter hydrocarbono-clasticus	Mediterranean	0.5-20	3.5	Gauthier *et al.* (1992)
Thiobacillus halophilus	Saline lake	0.5-22	5	Wood & Kelly (1991)
Proteobacteria, other lineage				
Arhodomonas aquaeolei	Petroleum reservoir production fluid	6-20	15	Adkins *et al.* (1993)
Spirochaetes				
Spirochaeta halophila	Saline pond	0.5-10	4	Greenberg & Canale-Parolé (1976)
Dichotomicrobium thermohalophilum	Solar Lake	0.8-22	8-14.2	Hirsch & Hoffmann (1989)

7   Alkaliphiles

Organisms that can grow optimally at a pH value greater than 8 are referred to as alkaliphiles. Obligate alkaliphiles generally have a pH optimum for growth between 9 and 10 and are incapable of growth at neutrality. The group is, like the halophiles, based around what is essentially a single phenotypic trait, and as such, the isolates in pure culture represent considerable phylogenetic diversity, including archaeal types. The validly published species of alkaliphilic bacteria are listed in Table 5. The alkaliphilic archaea were considered with the halophiles (Table 4).

Much of the impetus for the isolation and characterization of alkaliphilic bacteria has arisen from their application in the enzymes industry, the extracellular carbohydrases and proteases are used in laundry detergents and the xylanases in paper maunfacture. This has resulted in a large number of new species and strains in recent years (Table 5). For example, systematic screening of alkaliphilic *Bacillus* strains for novel proteases resulted in a diverse collection of isolates representing at least nine new species several of which produce commercially valuable extracellular proteases (Nielsen *et al.*, 1995).

The most alkaliphilic organisms known are also halobacteria and have been discussed earlier. The group is chemotaxonomically and phenotypically coherent, but confusingly, a recent study exclusively based on 16S rRNA sequence data has concluded that the rod-shaped types are not monophyletic and represent four separate lineages (Kamekura *et al.*, 1997). To date, this is probably the only example known where phylogenetic placement at the genus level is not supported by any chemotaxonomic or phenotypic data.

TABLE 5.   Characteristics of alkaliphilic bacteria

Species	Habitat	pH optimum	Reference
Cyanobacteria			
Spirulina platensis	Soda lakes	8.5	Javor (1989)
Proteobacteria			
Desulfonatronovibrio hydrogenovorans	Soda lakes	9.5	Zhilina *et al.* (1997)
"*Halomonas pantelleriense*"	Lake sand	9.0	Romano *et al.* (1996)
desiderata	Sewage	9.5	Berendes *et al.* (1996)
Halomonas "spp."	Soda lakes	9.5	Duckworth *et al.* (1996)
Pseudomonas spp."	Soda lakes	9.5	Duckworth *et al.* (1996)
Vibrio/Aeromonas/enteric spp.	Soda lakes	9.5	Duckworth *et al.* (1996)
Gram-negative anaerobic rods			
Haloanaerobium alcaliphilum	Salt lake	5-10	Tsai *et al.* (1995)
Natrioniella acetigena	Soda lake	9.7	Zhilina *et al.* (1996b)
Thermosyntropha lipolytica	Soda lake	8.5	Svetlitshnyi *et al.* (1996)

TABLE 5.   continued

Species	Habitat	pH optimum	Reference
Gram-positive anaerobic rods			
Clostridium			
paradoxum	Sewage digester	9.5	Li *et al.* (1993)
thermoalkaliphilium	Sewage digester	9.5	Li *et al.* (1994)
Gram-positive aerobic rods			
"*Arthrobacter*" spp.	Soda lakes	9.5-10	Duckworth *et al.*(1996)
Bacillus			
agaradherens	Soil	9.5	Nielsen *et al.* (1995)
alcalophilus	Soil	9.5	Vedder (1934)
clausii	Soil	9.5	Nielsen *et al.* (1995)
clarkii	Soil	9.5	Nielsen *et al.* (1995)
gibsonii	Soil	9.5	Nielsen *et al.* (1995)
halmapalus	Soil	9.5	Nielsen *et al.* (1995)
halodurans	Soil	9.5	Nielsen *et al.* (1995)
horikoshii	Soil	9.5	Nielsen *et al.* (1995)
pseudoalcalophilus	Soil	9.5	Nielsen *et al.* (1995)
pseudofirmis	Soil	9.5	Nielsen *et al.* (1995)
vedderi	Bauxite waste	9.5	Agnew *et al.* (1995)
Bacillus spp."	Various soils	9.5-10	Horikoshi (1992)
Bacillus spp."	Soda lakes	9.5	Duckworth *et al.* (1996)
Bogoriella caseilytica	Soda lakes	9-10	Groth *et al.* (1997)
Dietzia spp."	Soda lakes	9.5-10	Duckworth *et al.* (1996)
Exiguobacterium *auriantiacum*	Potato waste	9.5	Gee *et al.* (1980)
Lactobacillus spp.	Soda lakes	9.5-10	Duckworth *et al.* (1996)
Phototrophic bacteria			
Ectothiorhodospira			
marismortui	Salt lake	9.0	Imhoff & Siling (1996)
mobilis	Soda lake	9.5	Imhoff & Siling (1996)
shaposhinikovii	Salt lake	9.5	Imhoff & Siling (1996)
vacuolata	Soda lake	9.5	Imhoff &Siling (1996)
Ectothiorhodospira (*Halorhodospira*)			
abdelmalekii	Soda lake	9.5-10	Imhoff & Siling (1996)
halochloris	Soda lake	9.5-10	Imhoff & Siling (1996)
halophila	Soda lake	9.5-10	Imhoff & Siling (1996)

TABLE 5. continued

Species	Habitat	pH optimum	Reference
Gram-positive anaerobic rods			
Spirochaetes			
Spirochaeta			
africana	Soda lake	8-9	Zhilina *et al.* (1996)
alkalica	Soda lake	8-9	Zhilina *et al.* (1996)
asiatica	Soda lake	8-9	Zhilina *et al.* (1996)
Others			
"*Thermotoga* spp."	Soda lake	9	Duckworth *et al.* (1996)

7.1 PHENOTYPIC, CHEMOTAXONOMIC AND MOLECULAR ANALYSES OF ALKALIPHILIC BACTERIA

Vedder (1934) was the first to isolate an alkaliphile, in this case an alkaliphilic spore forming rod classified as *Bacillus alcalophilus*. Very many alkaliphile spore formers have been isolated often from normal soils (Fritze *et al.*, 1990). The organisms were originally assigned to the genus *Bacillus* on the basis of morphology, spore position and phenotypic characteristics (Nielsen *et al.*, 1994, 1995). Recent phylogenetic placement on the basis of 16S rRNA sequence analyses (Nielsen *et al.*, 1994; Agnew *et al.*, 1995; Duckworth *et al.*, 1996) have indicated that the alkaliphilic bacilli probably represent at least two separate evolutionary lines amongst the complex *Bacillus* spectrum.

Most isolates have come from highly alkaline African soda lakes, notably the Kenyan soda lakes and the Egyptian Wadi Natrun group. The lakes harbour cyanobacteria such as *Spirulina* spp. and especially, anoxygenic phototrophic bacteria as primary producers. The halophilic and alkaliphilic anoxygenic phototrophs, originally all classified as *Ectothiorhodospira* spp. have, on the basis of 16S rRNA gene sequence analyses, been divided into alkaliphilic halophilic types (*Halorhodospira*) and less halophilic types (*Ectothiorodospira*) (Imhoff & Suling, 1996).

Numerous aerobic soda lake organotrophs have been isolated - several hundred strains were subjected to a preliminary numerical taxonomic analysis for more than one hundred characters after preliminary sorting on the basis of the Gram reaction (Jones *et al.*, 1994). Gram-negative strains were recovered in six clusters, only one of which was related to any of the known organisms included in the analysis. Chemotaxonomic data including isoprenoid quinone, polar lipid and fatty acid analyses supported the phenotypic clustering. The analysis of a much fewer number of Gram-positive isolates produced less satisfactory clusters, with some association of certain clusters with *Bacillus* spp.

Phylogenetic analysis of Gram-negative isolates showed that all were members of the γ-3 sub-branch of the *Proteobacteria*, with notably, most isolates clustering within the *Halomonas/Deleya* line previously discussed in the halophile section. However, the alkaliphiles were distinct in three subclusters (Duckworth *et al.*, 1996). There have since been reports of other alkaliphilic halomonads (Berendes *et al.*, 1996; Romano *et al.*, 1996). Other isolates were members of the *Aeromonas/Vibrio/Enterobacteria* line, with all others related to the pseudomonads.

Phylogenetic analysis of Gram-positive isolates revealed a dominant association with *Bacillus* spp., with three groups, one containing the original *B. alcalophilus*, and another containing mainly halophilic and alkaliphilic types (Duckworth *et al.*, 1996). One further isolate was found to associate most closely with *Listeria*. Other Gram-positive isolates were associated with the *Arthrobacter/Terrabacter* part of the high % G+C part of the Gram-positive lineage. There has been a recent report of another isolate that is also associated with *Terrabacter* (Groth *et al.*, 1997). Two isolates were associated with the organism recently named as *Dietzia maris* (Rainey *et al.*, 1995).

The anaerobic alkaline environment has received less attention. Alkaliphilic thermophilic anaerobic *Clostridium* spp. were isolated from anaerobic digesters (Li *et al.*, 1993; 1994) and an alkaliphilic *Haloanaerobium* spp. has been described from the Great Salt Lake (Tsai *et al.*, 1995). Strictly anaerobic alkaliphilic spirochaetes were isolated from Kenyan and Russian soda lakes (Zhilina *et al.*, 1996a). *Natrionella acetigena*, isolated from Lake Magadi is another member of the *Haloanaerobiales* (Zhilina *et al.*, 1996b). Recently the elusive alkaliphilic sulphate reducers have been cultured for the first time (Zhilina *et al.*, 1997). Other alkaliphilic anaerobes described include a member of the *Themotogales* (Duckworth *et al.*, 1996) and, other phylogenetically distinct *Clostridium* spp. (Owenson & Grant, unpublished).

8   The Implications of Taxonomy in the Extremophile Field

8.1   PHYLOGENY OF EXTREMOPHILES

Taxonomy has had an enormous effect on the development of extremophile biology in the last two decades. Indeed, the development of molecular taxonomy has been a key to the discovery, characterization and classification of most of the newly discovered extremophiles and with this an appreciation of the metabolic and morphological diversity of these remarkable microorganisms. This important role of molecular taxonomy has not diminished the value of the more traditional methods, which have provided the essential phenetic descriptions, but has provided the development of a natural phylogeny for the extremophiles (Goodfellow & O'Donnell, 1993; Vandamme *et al.*, 1996).

The phylogeny of extremophiles has been very closely linked to questions on the origin and early evolution of life on earth (Woese, 1987; Olsen & Woese, 1993; Stetter, 1996a). It is interesting that many extremophiles, and in particular thermophiles, seem to be very "old" lineages. This is particularly evident among the archaeal thermophiles which are very deeply rooted in the phylogenetic tree (Fig. 4). It is also very striking that seven out of the twelve major bacterial phyla contain thermophiles. Four of these, the *Aquifex* - *Hydrogenobacter*, the *Thermotoga*, the Green non-sulfur bacteria, including *Thermus*, and the *Thermodesulfobacter* group, comprise the deepest phylogenetic branches in the bacterial tree and thermophiles almost invariably represent the deepest branching groups in their corresponding phylogenetic branches. These findings indicate that thermophiles may represent the most ancient forms of life now present on earth (Woese, 1987; Stetter, 1996a).

Since identical or similar metabolic processes are found in diverse and distantly related organisms, questions of monophyletic versus polyphyletic origin of corresponding metabolic pathways and convergent evolution among these bacteria are some of the more fascinating issues to be addressed; the phylogenetic frameworks provided by 16S rRNA and protein-coding gene sequence analyses will help resolve such issues.

8.2   GLOBAL DISTRIBUTION OF EXTREMOPHILES

The global distribution of most extremophiles is not known especially at the species and subspecies levels. It is also not known to what extent geographically distinct extreme ecosystems may differ. These different sites may each have a unique composition of global and endemic genera, species or subspecies but this cannot be determined without a robust and refined taxonomic database, especially at the subspecies or population level.

Some thermophilic bacteria appear to be ubiquitous. Thus, the thermophilic anaerobic alkaliphile, *Thermobrachium celere*, has been isolated on three continents from very diverse habitats ranging from hot springs to compost (Engle *et al.*, 1996). Indeed 40 isolates showed very similar WCP patterns by SDS-PAGE and a phylogenetic analysis of a subset of these strains revealed that the 16S rRNA sequences were nearly identical. Moreover, members of phenotypically relatively homogeneous thermophilic genera such as *Bacillus, Clostridium, Hydrogenobacter, Thermoplasma, Thermus* and *Sulfolobus* are found in terrestrial thermal habitats all over the world. However, the methods of molecular systematics are now beginning to reveal, at the finer taxonomic levels of species and subspecies, that the distribution of extremophiles in many cases appears to show clear endemic patterns (Aragno, 1992; Moreira *et al.* 1996; 1997).

On the other hand, some species have been isolated only at one location and sometimes only once. This is the case with the aerobic, heterotrophic bacterium, *Thermomicrobium roseum*, that was isolated from a sample taken near the source of Toadstole spring in Yellowstone National Park (Jackson *et al.*, 1973). The cyanobacterium *Synechococcus* (Tmax 72°C) is abundant and easily visible in American hot springs but has not been detected in Icelandic hot springs (Brock, 1978). Similarly, the archaeum *Methanothermus* has only been found at one location in Iceland (Lauerer *et al.*, 1986). The distribution of other thermophilic fermentative anaerobes seems to be much influenced by geography. Thus some deep phylogenetic clusters corresponding to genomic species of thermophilic anaerobic fermentative bacteria appear to be confined to New Zealand, while other clusters are more widespread (Rainey *et al.*, 1993).

Distinct differences in species composition are being revealed between marine and terrestrial thermal habitats (Prieur, 1997). Hydrothermal vents appear to have characteristic species compositions. While isolates from these sites all belong to genera previously found in coastal or terrestrial habitats, such as *Archeaoglobus, Desulforococcus, Methanococcus, Methanopyrus, Pyrodictium, Pyrococcus, Staphylothermus, Thermococcus* and *Thermodiscus*, phylogenetic analyses have revealed that most isolates so far analysed are new species (Prieur, 1997). The exceptions are *Methanopyrus kandlerii, Staphylothermus marinus* and *Thermodiscus maritimus* that have also been isolated from coastal areas (Fiala *et al.* 1986; Kurr *et al.* 1991). In contrast, representatives of some other archaeal genera such as *Hyperthermus, Pyrobaculum, Thermofilum, Thermoproteus* and the thermoacidophiles *Acidianus* and *Sulfolobus* have not been found in samples from deep-sea hydrothermal vents (Marteinsson, 1997).

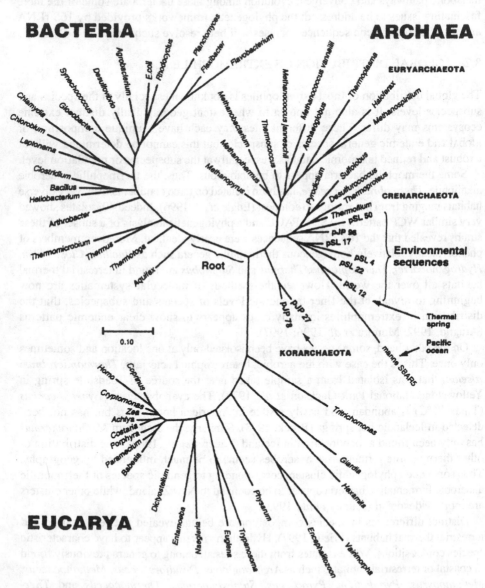

Figure 4. Universal unrooted phylogenetic tree obtained by maximum likelihood analysis, showing positions of major groups including uncultivated sequences from a Yellowstone hot spring. Bar indicates 0.1 changes per site. Redrawn from Barns *et al.* (1996).

The importance of molecular systematics in defining the distribution of extremophiles in different sites was evident in a study of 24 *Thermaplasma* strains obtained from solfatara fields in the Azores, Iceland, Indonesia, Italy and the United States. Little phenotypic distinction was observed between the isolates and they could have been considered a homogeneous taxon but DNA:DNA hybridization revealed that the samples consisted of two genomic species, *Thermoplasma acidophilum* and *Thermoplasma volcanium* (Segerer *et al.* 1988). *Thermoplasma acidophilum* was the only species found in two solfatara sites in Naples, on the volcanic island Volcano in Italy and in the Azores. It was not found in Iceland nor in Indonesia. These isolates exhibited 100% DNA homology with the type strain previously isolated from self-heated coal piles in Southern Indiana and Western Pennsylvania, USA. *Thermoplasma volcanium* could be divided into three distinct DNA homology groups, v1, v2 and v3, which showed a clear geographic distribution with v1 only found in the Vulcano island, v2 restricted to Indonesia and v3 found only in Iceland and Yellowstone Park.

Similarly, a number of species belonging to the genus *Thermococcus* have been isolated from isolated hydrothermal vents. Despite phenotypic or physiological similarity and apparent similar ecological roles, different isolates have been assigned to different genomic species (Godfroy *et al.*, 1996; 1997, Gonzales *et al.*, 1995; Huber *et al.*, 1996b). The same situation occurs with *Thermus* isolates classified by 16S rRNA sequencing and DNA:DNA hybridization (Fig. 5).The *T. thermophilus* cluster circumscribed an ecologically distinct population as the isolates were found mainly in marine or coastal hot springs (Manaia *et al.* 1994). The species has a world wide distribution and is found in coastal or submarine hot springs in the Azores, Iceland, Fiji, Japan, New Zealand and Portugal. The genotypic distinctiveness of *T. thermophilus* is also evident by DNA:DNA hybridization studies where the marine isolates cluster apart form other species found in hot springs on land. The only two phenotypic traits that distinguish these strains from strains of other *Thermus* species found in hot springs on land is growth at 80°C and in 3% NaCl.

The predominantly colourless *Thermus scotoductus* isolates have been isolated from subterranean or man-made enclosed thermal habitats, including outlets from hot water systems all over the world. Thermal habitats shielded from the sun may therefore be the preferred habitat of *T. scotoductus* (Kristjansson *et al.*, 1994). DNA:DNA homology studies have shown that *T. scotoductus* has a wide distribution in the northern hemisphere (Tenreiro *et al.*, 1995b).

The *Thermus filiformis* cluster is phylogenetically homogeneous as estimated by 16S rRNA analysis and this population appears to be confined to New Zealand (Saul *et al.*, 1993). Close relatives have not been found elsewhere. The strains were isolated from locations wide apart in New Zealand and are phenotypically very different (Hudson *et al.*, 1989). geological isolation of New Zealand may explain the phylogenetic uniqueness of the *T. filiformis* cluster.

Thermus aquaticus, *T. brockianus* and *T. oshimai* also form distinct DNA:DNA homology clusters (Williams *et al.* 1995; 1996). *Thermus aquaticus* has only been found in the USA, but representatives of *T. brockianus* have been found at widely separated sites in Iceland and the USA and *T. oshimai* in Iceland and Portugal (Williams *et al.* 1995; 1996).

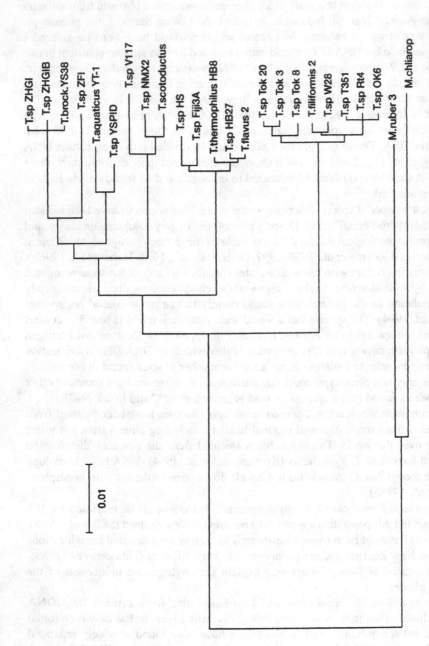

Figure 5. A phylogenetic tree of the *Thermus* family indicating considerable influence of geography on the grouping of strains. Strains ZHGI, ZHGIB, ZFI and the *T. scotoductus* type strain are from Iceland. Strains YS38 (type strain for *T. brockianus*), YT-1 (type strain for *T. aquaticus*), YSPID and NMX2 are from North-America. Strain V117 is from Portugal, strain Fiji3A is from Fiji Islands, and strains HB8 (type strain for *T. thermophilus*), HB27 and *T. flavus* 2 are from Japan; all of the *Thermus* strains are from New Zealand. The two *Meiothermus* strains were used as outgroups. Scale bar indicates 1 substitution per 100 nucleotides. Redrawn from Radax et al. (1998).

8.3 DIVERSITY OF UNCULTURED EXTREMOPHILES

Many bacteria may be difficult to grow or to select for against a background of more abundant and more easily cultured species. Their presence is therefore not easily confirmed. Such bacteria however may, be both abundant and important in the ecosystem and of value to the biotechnology industry as a source of varied genomic material for exploitation by genetic and protein engineering techniques. It is becoming evident that traditional isolation techniques are inadequate for describing the true diversity of any given habitat. It is therefore likely that the known extremophilic species represent only a small fraction of the actual diversity of microorganisms living under various extreme conditions. Results from several environments indicate that only a fraction (0.001-0.3%) of naturally occurring microorganisms are cultivated with standard techniques (Amann *et al.*, 1995). Frequently microorganisms are observed to grow in considerable quantities at very high temperatures in hot springs but have proved difficult to grow in the laboratory. These organisms include the pink filaments in the outflow of hot springs in Yellowstone Park at temperature as high as 84°C (Reysenbach *et al.*, 1994) and the so called Blue threads found at high temperatures in Icelandic hot springs (Skirnisdottir *et al.*, unpublished) which have been demonstated by PCR techniques to be largely composed of members of the *Aquifecalis*. Recently one of them has been isolated and named *Thermocrinis ruber* (Huber *et al.* 1998).

If the distribution of 16S rRNA of a particular uncultured species correlates to certain environmental conditions, this knowledge may prove useful if the corresponding organism is to be isolated. Similarly, the phylogenetic type of the 16S rRNA may give useful information concerning the possible metabolic phenotype of the organism (Amann *et al.* 1995). For example, Huber *et al.* (1995) have used the knowledge of an environmental 16S rRNA sequence obtained by PCR of a previously unknown thermophile to obtain the organism in pure culture. They combined the visual recognition of single cells by phylogenetic staining and retrieval by optical tweezers. Moreover, a "spectacular wealth of novel archaea" was detected in a hot spring sediment at 74°C using PCR-based rRNA analyses (Barns *et al.*, 1994). Two of these sequences branch very deeply and possibly represent a new and ancient kingdom of archaea, provisionally named "Korarchaeota" (Barns *et al.*, 1996). Similarly impressive new diversity was also found among the eubacteria, where even new phyla were discovered in a Yellowstone hot spring (Hugenholtz *et al.*, 1998).

Such diversity is not restrcted to archaea. Weller *et al.* (1991;1992) have assessed the microbial populations in hot spring cyanobacterial mats at 55°C by cloning and analysing cDNA synthesized from rRNA from biomass samples. Sequences representing many novel species of bacteria were retrieved, belonging to different phyla such as the cyanobacteria, proteobacteria, green non-sulfur bacteria and the spirochaete group.

Methods that do not depend on cultivation will undoubtedly improve in refinement and applications. They will not only be important for detecting known organisms under various environmental conditions but also for studying the unknown microbial diversity. With more knowledge on the vast microbial diversity that has yet to be detected, there will be ever more pressing attempts to isolate and/or to find ways of actually studying the *in situ* activity of these "new" organisms (Staley *et al.*, 1997; Ward *et al.*, 1997; Hugenholtz *et al.*, 1998).

The continued taxonomic analysis of extremophiles should help assess their diversity and indicate taxa for future exploitation. But extremophiles are interesting in many other ways, their biochemistry, ecology, physiology, phylogeny and population genetics, constrained and developed by the peculiar habitats in which the organisms live, offer distinctive systems and stuctures which can aid in understanding of more "normal" organisms. These studies will also benefit from a sound and extensive taxonomic base.

9   Acknowledgements

We acknowledge the financial support of EC Generic Project "Biotechnology of Extremophiles" Contract BIO-CT93-O2734, the Icelandic Ministry of Education, the Icelandic National Research Council (Grant no. 93235) and the Icelandic Science Foundation (Grant no. 94-L-105).

10   References

Adkins, J.P., Madigan, M.T., Mandelco, L., Woese, C.R. & Tanner, R.S. (1993). *Arhodomonas aquaeolei* gen. nov. sp. nov., an aerobic halophilic bacterium isolated from a sub-terranean brine. *International Journal of Systematic Bacteriology* 43, 514-520.

Agnew, M.D., Koval, S.F. & Jarrell, K.F. (1995). Isolation and characterisation of novel alkaliphiles from bauxite-processing waste and description of *Bacillus vedderi* sp. nov. *Systematic and Applied Microbiology* 18, 221-230.

Alfredsson, G.A, Baldursson, S. & Kristjansson, J.K. (1985). Nutritional diversity among *Thermus* spp. isolated from Icelandic hot springs. *Systematic and Applied Microbiolology* 6, 308-311.

Alfredsson, G.A. and Kristjansson, J.K. (1995). Ecology, distribution and isolation of *Thermus*. In *Thermus Species: Biotechnology Handbooks*, pp. 43-66. Edited by R.J. Sharp & R.A.D Williams. London: Plenum Press.

Alfredsson, G.A., Kristjansson, J.K., Hjörleifsdottir, S. & Stetter, K.O. (1988). *Rhodothermus marinus*, gen. nov., sp. nov., a thermophilic, halophilic bacterium from submarine hot springs in Iceland. *Journal of General Microbiology* 134, 299-306.

Amann, R.I., Ludwig, W. & Schleifer, K.-H. (1995). Phylogenetic identification and *in situ* detection of individual microbial cells without cultivation. *Microbiological Reviews* 59, 143-169.

Andrews, K.T. & Patel, B.K. (1996). *Fervidobacterium gondwanense* sp. nov., a new thermophilic anaerobic bacterium isolated from nonvolcanically heated geothermal waters of the Great Artesian Basin of Australia. *International Journal of Systematic Bacteriology* 46, 265-269.

Antoine, E., Cilia, V., Meunier, J.R., Guezennec, J., Lesongeur, F. & Barbier, G. (1997). *Thermosipho melanesiensis* sp. nov., a new thermophilic anaerobic bacterium belonging to the order *Thermotogales*, isolated from deep-sea hydrothermal vents in the southwestern Pacific Ocean. *International Journal of Systematic Bacteriology* 47, 1118-1123.

Aragno, M. (1992). The aerobic, chemolithoautrophic, thermophilic bacteria. In *Thermophilic Bacteria* pp. 77-103. Edited by J.K. Kristjansson. Boca Raton: CRC Press Inc.

Balch, W.E., Fox, G.E., Magrum, L.J., Woese, C.R. & Wolfe, R.S. (1979). Methanogens: reevaluation of a unique biological group. *Microbiological Reviews* 43, 260-296.

Barns, S.M., Fundyga, R.E., Jeffries, M.W. & Pace, N.R. (1994). Remarkable archaeal diversity detected in a Yellowstone National Park hot spring environment. *Proceedings of the National Academy of Sciences of the United States of America* **91**, 1609-1613.

Barns, S.M., Delwiche, C.F., Palmer, J.D. & Pace, N.R. (1996). Perspectives on archaeal diversity, thermophily and monophyly from environmental rRNA sequences. *Proceedings of the National Academy of Sciences of the United States of America* **93**, 9188-9193.

Beeder, J., Torsvik, T. & Lien, T. (1995). *Thermodesulforhabdus norvegicus* gen. nov., sp. nov., a novel sulfate-reducing bacterium from oil field water. *Archives of Microbiology* **164**, 331-336.

Ben-Bassat, A. & Zeikus, J. G. (1981). *Thermobacteroides acetoethylicus* gen. nov. and sp. nov., a new chemoorganotrophic, anaerobic, thermophilic bacterium. *Archives of Microbiology* **128**, 365-370.

Benlloch, S., Acinas S.G., Martinez-Murcia A.J. & Rodriguez-Valera, F. (1996). Description of prokaryote diversity along the salinity gradient of a multipond solar saltern by direct PCR amplification of 16S rRNA. *Hydrobiologia* **329**, 19-31.

Berendes, F., Gottschalk, G., Hemedobbernack, E., Moore, E.R.B. & Tindall, B.J. (1996). *Halomonas desiderda* sp. nov., a new alkaliphilic, halotolerant and denitrifying bacterium from a municipal sewage works. *Systematic and Applied Microbiology* **19**, 158-167.

Bhupathiraju, V.K., Oren, A., Sharma, P.K., Tanner, R.S. & Woese, C.R. (1994). *Haloanaerobium salsugo* sp. nov. a moderately halophilic anaerobic bacterium from a subterranean brine. *International Journal of Systematic Bacteriology* **44**, 565-572.

Blöchl, E., Rachel, R., Burggraf, S., Hafenbradl, D., Jannasch, H. W. & Stetter, K.O. (1997). *Pyrolobus fumarii*, gen. and sp. nov., represents a novel group of archaea, extending the upper temperature limit for life to 113°C. *Extremophiles* **1**, 14-21.

Bonch-Osmolovskaya, E.A., Slesarev, A.I., Miroshnichenko, M.L., Svetlichnaya, T.P. & Alexeyev, V.A. (1985). Characteristics of *Desulfurococcus amylolyticus* n. sp., a new extreme thermophilic archaebacterium from hot volcanic vents of Kamchatka and Kunashir. *Microbiologyia* **57**, 78-85.

Bonch-Osmolovskaya, E.A., Miroshnichenko, M.L., Kostrikina, N.A., Chernych, N.A. & Zavarzin, G.A. (1990). *Thermoproteus uzoniensis* sp. nov., a new extremely thermophilic archaebacterium from Kamchatka continental hot springs. *Archives of Microbiology* **154**, 556-559.

Bonjour, F. & Aragno, M. (1984). *Bacillus tusciae*, a new species of thermoacidophilic, facultatively chemolithoautotrophic, hydrogen oxidizing sporeformer from a geothermal area. *Archives of Microbiology* **139**, 397-401.

Boone, D.R., Mathrani, I.M., Liu, Y, Meanaia, J.A., Mah, R.A. & Boone, J.E. (1993). Isolation and characterisation of *Methanohalophilus portugalensis* sp. nov. and DNA reassociation study of the genus *Methanohalophilus*. *International Journal of Systematic Bacteriology* **43**, 430-437.

Boone, D.R., Liu, Y., Zhao, Z.J., Balkwill, D.L., Drake, G.R., Stevens, T.O. & Aldrich, H.C. (1995). *Bacillus infernus* sp. nov., an Fe(III)- and Mn(IV)-reducing anaerobe from Deep Terrestrial Subsurface. *International Journal of Systematic Bacteriology* **45**, 441-448.

Bredholt, S., Sonne-Hansen, J., Nielsen, P., Mathrani I.M. & Ahring, B.K. (1999). *Caldicellulosiruptor kristjanssonii* sp. nov., a cellulolytic, extremely thermophilic, anaerobic bacterium. *International Journal of Systematic Bacteriology* **49**, 991-996.

Brock, T.D. (1978). *Thermophilic Microorganisms and Life at High Temperatures*. New York: Springer-Verlag.

Brock, T. D. (1986). Introduction: an overview of the thermophiles. In, *Thermophiles: General, Molecular and Applied Microbiology*, pp. 1-16. Edited by T. D. Brock, New York: John Wiley & Sons.

Brock, T.D. & Freeze, H. (1969). *Thermus aquaticus* gen. n., sp. n., a nonsporulating extreme thermophile. *Journal of Bacteriology* **98**, 289-297.

Brock, T.D., Brock, K.M., Belly, R.T. & Weiss, R.L. (1972). *Sulfolobus*: a new genus of sulfur-oxidizing bacteria living at low pH and high temperature. *Archives of Microbiology* **84**, 54-68.

Bull, A.T., Goodfellow, M. & Slater, J.H. (1992). Biodiversity as a source of innovation in biotechnology. *Annual Review of Microbiology* **46**, 219-252.

Burggraf, S., Fricke, H., Neuner, A., Kristjansson, J.K., Rouvier, P., Mandelco, L., Woese, C. R. & Stetter, K.O. (1990a). *Methanococcus igneus* sp. nov., a novel hyperthermophilic methanogen from a shallow submarine hydrothermal system. *Systematic and Applied Microbiology* **13**, 263-269.

Burggraf, S., Jannasch, H.W., Nicolaus, B. & Stetter, K.O. (1990b). *Archaeglobus profundus* sp. nov. represents a new species within the sulfate-reducing archaebacteria. *Systematic and Applied Microbiology* **13**, 24-28.

Burggraf, S., Huber, H. & Stetter, K.O. (1997). Reclassification of the crenarchael orders and families in accordance with 16S rRNA sequence data. *International Journal of Systematic Bacteriolology* **47**, 657-660.

Caldwell, D.E., Caldwell, S.J. & Laycock, P. (1976). *Thermothrix thioparus* gen. et sp. nov. a facultatively anaerobic facultative chemolithotroph living at neutral pH and high temperature. *Canadian Journal of Microbiology* **22**, 1509-1517.

Caumette, P., Baulaigue, R. & Matheron, R. (1988). Characterization of *Chromatium salexigens* sp. nov. a halophilic *Chromatiaceae* isolated from Mediterranean salinas. *Systematic and Applied Microbiology* **10**, 284-292.

Caumette, P., Cohen, Y. & Matheson, R. (1991). Isolation and characterisation of *Desulfovibrio halophilus* gen. nov., sp. nov., a halophilic sulphate-reducing bacterium from Solar Lake (Sinai). *Systematic and Applied Microbiology* **14**, 33-38.

Caumette, P., Baulaigue, R. & Matheron, R. (1992). *Thiocapsa halophila* sp. nov. a new halophilic phototropic bacterium. *Archives of Microbiology* **55**, 70-73.

Cayol, J.L., Ollivier, B., Patel, K.C., Prensier, G., Guezennec, J. & Garcia, J-L. (1994). Isolation and characterisation of *Halothermothrix orenii* gen. nov. sp. nov a halophilic, thermophilic fermentative strictly anaerobic bacterium. *International Journal of Systematic Bacteriology* **44**, 534-543.

Cayol, J.-L., Ollivier, B., Patel, B.K.C., Ravot, G., Magot, M., Ageron, E., Grimont, P.A.D. & Garcia, J.-L. (1995). Description of *Thermoanaerobacter brockii* subsp. *lactiethylicus* subsp. nov., isolated from a deep subsurface french oil well. A proposal to reclassify *Thermoanaerobacter finnii* as *Thermoanaerobacter brockii* subsp. *finnii* comb. nov., and an emended description of *Thermoanaerobacter brockii*. *International Journal of Systematic Bacteriology* **45**, 783-789.

Cerny, G. (1976). Method for the distinction of Gram-negative from Gram-positive bacteria. *European Journal of Applied Microbiology* **3**, 223-225.

Cheetham, P.S.J. (1987). Screening for novel biocatalysts. *Enzyme and Microbial Technology* **9**, 194-213.

Claus, D., Fahny, F., Rolf, H.J. & Tosunoglu, N. (1983). *Sporosarcina halophila* sp. nov. an obligate halophilic bacterium from salt marsh soils. *Systematic and Applied Microbiology* **4**, 496-506.

Claus, D. & Berkeley, R.C.W. (1986). Genus *Bacillus* Cohn 1872, 174[AL]. In *Bergey's Manual of Systematic Bacteriology*, vol. 2, pp. 1105-1139. Edited by P.H.A. Sneath, N.S. Mair, M.E. Sharpe, J.G. Holt. Baltimore: Williams & Wilkins.

Collins, M.D. (1985). Isoprenoid quinone analysis - bacterial classification and identification. In: *Chemical Methods in Bacterial Systematics*, pp. 267-288. Edited by M. Goodfellow & D. Minnikin. London: Academic Press.

Collins, M.D., Lawson, P.A., Willems, A., Cordoba, J.J., Fernandez-Garayzabal, J., Garcia, P., Cai, J., Hippe, H. & Farrow, J.A.E. (1994). The phylogeny of the genus *Clostridium:* proposal of five new genera and eleven species combinations. *International Journal of Systematic Bacteriology* **44**, 812-826.

Cometta, S., Sonneleitner, B. & Fiechter, A. (1982). The growth behaviour of *Thermus aquaticus* in continuous cultivation. *Journal of Applied Microbiology and Biotechnology* **15**, 69-74.

Darland, G., Brock, T.D., Samsonoff, W. & Conti, S.F. (1970). A thermophilic, acidophilic mycoplasma isolated from a coal refuse pile. *Science* **170**, 1416-1418.

Darland, G. & Brock, T.D. (1971). *Bacillus acidocaldarius* sp. nov., an acidophilic thermophilic spore-forming bacterium. *Journal of General Microbiology* **67**, 9-15.

DeLong, E.F. (1997). Marine microbial diversity: the tip of the iceberg. *Trends in Biotechnology* **15**, 203-207.

Demharter, W. & Hensel, R. (1989). *Bacillus thermocloaceae* sp. nov., a new thermophilic species from sewage sludge. *Systematic and Applied Microbiology* **11**, 272-276.

Denner, E., Busse, H.J., McGenity, T.J., Grant, W.D., Wanner, G. & Stan-Lotter, H. (1994). *Halococcus salifodinae* sp. nov. an archaeal isolate from an Austrian salt mine. *International Journal of Systematic Bacteriology* **44**, 774-780.

Dobson, S.J. & Franzmann, P.D. (1996). Unification of genera *Deleya, Halomonas, Halovibrio* and the species *Paracoccus halodenitrificans* into a single genus, *Halomonas*, and the placement of the genus *Zymomonas* in the family *Halomonadaceae*. *International Journal of Systematic Bacteriology* **46**, 550-558.

Dobson, S.J., Colwell, R.R., McMeekin, T.A. & Franzmann, P.D. (1993). Direct sequencing of the polymerase chain reaction amplified 16S rRNA gene of *Flavobacterium gondwanense* sp. nov. and *Flavobacterium salegens* sp. nov., two new species from a hypersaline Antarctic Lake. *International Journal of Systematic Bacteriology* **43**, 77-83.

Drent, W.J., Lahpor, G.A., Wiegant, W.M. & Gottschal, J.C. (1991). Fermentation of inulin by *Clostridium thermosuccinogenes* sp. nov., a thermophilic anaerobic bacterium isolated from various habitats. *Applied and Environmental Microbiology* **57**, 455-462.

Drews, G. (1981). *Rhodospirillum salexigens* sp. nov., an obligately halophilic phototrophic bacterium. *Archives of Microbiology* **130**, 325-327.

Duckworth, A.W., Grant, W.D., Jones, B.E. & Van Steenbergen, R. (1996). Phylogenetic diversity of soda lake alkaliphiles. *FEMS Microbiology Ecology* **19**, 181-191.

Edwards, C. (1990). Thermophiles. In *Microbiology of Extreme Environments*, pp. 1-32. Edited by C. Edwards. Milton Keynes: Open University Press.

Engle, M., Li, Y., Rainey, F., DeBlois, S., Mai, V., Reichert, A., Mayer, F., Messner, P. & Wiegel, J. (1996). *Thermobrachium celere* gen. nov., sp. nov., a rapidly growing thermophilic, alkalitolerant, and proteolytic obligate anaerobe. *International Journal of Systematic Bacteriology* **46**, 1025-1033.

Fardeau, M.L., Ollivier, B., Patel, B.K., Magot, M., Thomas, P., Rimbault, A., Rocchiccioli, F. & Garcia, J.L. (1997). *Thermotoga hypogea* sp. nov., a xylanolytic, thermophilic bacterium from an oil-producing well. *International Journal of Systematic Bacteriology* **47**, 1013-1019.

Fendrich, C. (1988). *Halovibrio variabilis* gen. nov. sp. nov. *Pseudomonas halophila* sp. nov. and a new aerobic coccoid eubacterium from Great Salt Lake Utah. *Systematic and Applied Microbiology* **11**, 36-43.

Fendrich, C., Hippe, H. & Gottschalk, G. (1990). *Clostridium halophilum* sp. nov. and *C. litorale* sp. nov., an obligate halophilic and marine species degrading betaine in the Stickland reaction. *Archives of Microbiology* **154**, 127-132.

Ferris, M.J., Muyzer, G. & Ward, D.M. (1996). Denaturing gradient gel electrophoresis profiles of 16S rRNA-defined populations inhabiting a hot spring microbial mat community. *Applied and Environmental Microbiology* **62**, 340-346.

Fiala, G. & Stetter, K.O. (1986). *Pyrococcus furiosus* sp. nov. represents a novel genus of marine heterotrophic archaebacteria growing optimally at 100°C. *Archives of Microbiology* **145**, 56-61.

Fiala, G., Stetter, K.O., Jannasch, H.W., Langworthy, T.A. & Madon, J. (1986). *Staphylothermus marinus* sp. nov. represents a novel genus of extremely thermophilic submarine heterotrophic archaebacteria growing up to 98°C. *Systematic and Applied Microbiology* **8**, 106-113.

Fiala, G., Woese, C.R., Langworthy, T.A. & Stetter, K.O. (1990). *Flexistipes sinusarabidici* a novel genus and species of eubacteria occurring in the Atlantis II deep brines of the Red Sea. *Archives of Microbiology* **154**, 120-126.

Fontaine, F. E., Peterson, W. H., McCoy, E., Johnson, M. J. & Ritter, G. J. (1942). A new type of glucose fermentation by *Clostridium thermoaceticum* n. sp. *Journal of Bacteriology* **43**, 701-715.

Franzmann, P.D. & Tindall, B.J. (1990). A chemotaxonomic study of members of the family *Halomonadaceae. Systematic and Applied Microbiology* **13**, 142-147.

Franzmann, P.D., Burton, H.R. & McMeekin, T.A. (1987). *Halomonas subglaciescola*, a new species of halotolerant bacteria isolated from Antarctica. *International Journal of Systematic Bacteriology* **37**, 27-34.

Franzmann, P.D., Wehmeyer, U. & Stackebrandt, E. (1988). *Halomonadaceae* fam. nov. a new family of the class *Proteobacteria* to accommodate the genera *Halomonas* and *Deleya*. *Systematic and Applied Microbiology* **11**, 16-19.

Fricke, H., Giere, O., Stetter K.O., Alfredsson, G.A., Kristjansson, J.K., Stoffers, P. & Svavarsson, J. (1989). Hydrothermal vent communities at the shallow subpolar Mid-Atlantic ridge. *Marine Biology* **102**, 425-429.

Fritze, D., Flossdorf, J. & Claus, D. (1990). Taxonomy of alkaliphilic *Bacillus* strains. *International Journal of Systematic Bacteriology* **40**, 300-307.

Fuchs, T., Huber, H., Teiner, K., Burggraf, S. & Stetter, K.O. (1995). *Metallosphaera prunae* sp. nov., a novel metal-mobilizing, thermoacidophilic Archaeum, isolated from a uranium mine in Germany. *Systematic and Applied Microbiology* **18**, 560-566.

Fuchs, T., Huber, H., Burggraf, S. & Stetter, K.O. (1996). 16S rDNA-based phylogeny of the archaeal order *Sulfolobus* and reclassification of *Desulfurolobus ambivalens* as *Acidianus ambivalens* comb. nov. *Systematic and Applied Microbiology* **19**, 56-60.

Garabito, M.J., Arahal, D.R., Millado, E., Marquez, M.M. & Ventosa, A. (1997). *Bacillus salexigens* sp. nov., a new moderately halophilic *Bacillus* sp. *International Journal of Systematic Bacteriology* **47**, 735-741.

Garcia, M.T., Ventosa, A., Ruiz-Berraquero, F. & Kocur, M. (1987). Taxonomic study and emended description of *Vibrio costicola. International Journal of Systematic Bacteriology* **37**, 251-256.

Gauthier, M.J., LaFay, B., Christen, R., Fernandez, L., Acquaviva, M., Boninn, P. & Bertrand, J-C. (1992). *Marinobacter hydrocarbonoclasticus* gen. nov. sp. nov., an extremely halotolerant hydrocarbon-degrading marine bacterium. *International Journal of Systematic Bacteriology* **42**, 568-576.

Gee, J.M., Lund B.M., Metcalf, G. & Peel, J.L. (1980). Properties of a new group of alkaliphilic bacteria. *Journal of General Microbiology* **117**, 9-17.

Georganta, G., Smith, K.E. & Williams, R.A.D. (1993). DNA:DNA homology and cellular components of *Thermus filiformis* and other strains of *Thermus* from New Zealand hot springs. *FEMS Microbiology Letters* **107**, 145-150.

Gilmour, D. (1990). Halotolerant and halophilic microorganisms. In *Microbiology of Extreme Environments*, pp. 147-177. Edited by C. Edwards. Milton Keynes: Open University Press.

Gochnauer, M.B., Leppard, G.G., Komaratat, P., Kates, M., Novitsky, T. & Kushner, D.J. (1975). Isolation and characterisation of *Actinopolyspora halophila* gen. et. sp. nov., an extremely halophilic actinomycete. *Canadian Journal of Microbiology* **21**, 1500-1511.

Godfroy, A. Meunier, J.R, Guezennec, J., Lesongeur, F., Raguenes, G., Rimbault, A. & Barbier, G. (1996). *Thermococcus fumicolans* sp. nov., a new hyperthermophilic archaeon isolated from a deep-sea hydrothermal vent in the North Fiji Basin. *International Journal of Systematic Bacteriology* **46**, 1113-1119.

Godfroy, A., Lesongeur, F., Raguenes, G., Querellou, J., Antoine, E., Meunier, J., Guezennec, J., Rimbault, A. & Barbier, G. (1997). *Thermococcus hydrothermalis*, sp. nov., a new hydrothermophilic archaeon isolated from a deep-sea hydrothermal vent. *International Journal of Systematic Bacteriology*

47, 622-626.

Golovacheva, R.S., Loginova, L.G., Salikhov, T.A., Kolesnikov, A.A. & Zaitseva, G.N. (1975). A new thermophilic species, *Bacillus thermocatenulatus* nov. spec. *Mikrobiologiya* 44, 265-268.

Gonzales, J.M., Kato, C. & Horikoshi, K. (1995). *Thermococcus peptonophilus* sp. nov., a fast growing, extreme thermophilic archaebacterium isolated from deep-sea hydrothermal vents. *Archives of Microbiology* 164, 159-164.

Goodfellow, M. & O'Donnell, A.G. (1993). Roots of bacterial systematics. In *Handbook of New Bacterial Systematics*, pp. 3-54. Edited by M. Goodfellow and A.G. O'Donnell. London: Academic Press Ltd.

Grant, W.D. (1991). General view of halophiles. In *Superbugs. Microorganisms in Extreme Environments*, pp. 15-37. Edited by K. Horikoshi & W.D. Grant. Berlin: Springer-Verlag.

Grant, W. D. & Horikoshi, K. (1992). Alkaliphiles: ecology and biotechnological applications. In *Molecular Biology and Biotechnology of Extremophiles*, pp. 143-162. Edited by R.A. Herbert & R.J. Sharp. London: Blackie & Son.

Grant, W.D. & Larsen, H. (1989). Extremely halophilic archaebacteria. In *"Bergey's Manual of Systematic Bacteriology"*, Vol. 3, pp. 2216-2219. Edited by J.T. Staley, M.P. Bryant, M.P., N. Pfennig, and J.G. Holt. Baltimore: Williams and Wilkins.

Grant, W.D. & Ross, H.N.M. (1986). The ecology and taxonomy of halobacteria. *FEMS Microbiology Reviews* 39, 9-15.

Greenberg, S. & Canale-Parolé, G. (1976). *Spirochaeta halophila* sp. nov., a facultative anaerobe from a high salinity pond. *Archives of Microbiology* 110, 185-194.

Grogan, D.W. (1989). Phenotypic characterization of the archaebacterial genus *Sulfolobus*: Comparison of five wild-type strains. *Journal of Bacteriology* 171, 6710-6719.

Grogan, D., Palm, P. & Zillig, W. (1990). Isolate B12, which harbours a virus-like element, represents a new species of the archaebacterial genus *Sulfolobus*, *Sulfolobus shibatae*, sp. nov. *Archives of Microbiology* 154, 594-599.

Groth, I., Schumann, P., Rainey, F., Martin, K., Schuetze, B. & Augsten, K. (1997). *Bogoriella caseilytica* gen. nov. sp. nov. a new alkaliphilic actinomycete from a soda lake in Africa. *International Journal of Systematic Bacteriology* 47, 788-794.

Halebian, S., Harris, B., Finegold, S.M. & Rolfe, A.D. (1981). Rapid method that aids in distinguishing Gram-positive from Gram-negative bacteria. *Journal of Clinical Microbiology* 13, 444-448.

Haskå, G. & Nystrand, R. (1982). Size and activity of the microflora in beet sugar extraction. *La Sucrerie Belge* 101, 131-144.

Henry, E.A., Devereux, R., Maki, J.S., Gilmour, C.C., Woese, C.R., Mandelco, L., Schauder, R., Remsen, C.C. & Mitchell, R. (1994). Characterization of a new thermophilic sulfate-reducing bacterium. *Thermodesulfovibrio yellowstonii*, gen. nov. and sp. nov.: its phylogenetic relationship to *Thermodesulfobacterium commune* and their origins deep within the bacterial domain. *Archives of Microbiology* 161, 62-69.

Herbert, A.M. & Vreeland, R.H. (1987). Phenotypic comparison of halotolerant bacteria: *Halomonas halodurans* sp. nov., nom. rev. comb. nov. *International Journal of Systematic Bacteriology* 37, 347-350.

Hirsch, P. & Hoffmann, B. (1989). *Dichotomicrobium thermohalophilum* gen. nov. sp. nov. budding prosthecate bacteria from Solar Lake (Sinai) and some related strains. *Systematic and Applied Microbiology* 11, 291-301.

Hjörleifsdottir, S., Peturdottir, S.K., Korpela, J., Torsti, A.-M., Mattila, P. & Kristjansson, J.K. (1996). Screening for restriction endonucleases in aerobic, thermophilic eubacteria. *Biotechnology Techniques* 10, 13-18.

Hjörleifsdottir, S., Ritterbusch, W., Petursdottir, S.K. & Kristjansson, J.K. (1997). Thermostabilities of DNA ligases and DNA polymerases from four genera of thermophilic eubacteria. *Biotechnology Letters*

19, 147-150.

Holt, J.G., Krieg, N.R., Sneath, P.H.A., Staley J.T. and Williams, S.T. (Eds..) (1994). *Bergey's Manual of Determinative Bacteriology* 9th Ed., p. **741**. Baltimore: Williams & Wilkins.

Horikoshi, K. (1992). *Microorganisms in Alkaline Environments.* Kodangha: VCH Verlags GmbH.

Horikoshi, K. (1996). Alkaliphiles, from an industrial point of view. *FEMS Microbiology Reviews* **18**, 259-270.

Hoyle, R. (1998). In hot pursuit of extremophiles. *Nature Biotechnology* **16**, 312.

Huber, G. & Stetter, K.O. (1991). *Sulfolobus metallicus*, sp. nov., a novel strictly chemolithotrophic thermophilic archaeal species of metal-mobilizers. *Systematic and Applied Microbiology* **14**, 372-378.

Huber, H., Thomm, M., König, H., Thies, G. & Stetter, K.O. (1982). *Methanococcus thermolithotrophicus*, a novel thermophilic lithotrophic methanogen. *Archives of Microbiology* **132**, 47-50.

Huber, R., Langworthy, T.A., König, H., Thomm, M., Woese, C.R., Sleytr, U.B. & Stetter, K.O. (1986). *Thermotoga maritima* sp. nov. represents a new genus of unique extremely thermophilic eubacteria growing up to 90° C. *Archives of Microbiology* **144**, 324-333.

Huber, R., Kristjansson, J.K. Stetter, K.O. (1987). *Pyrobaculum* gen. nov., a new genus of neutrophilic, rod-shaped archaebacteria from continental solfataras growing optimally at 100°C. *Archives of Microbiology* **149**, 95-101.

Huber, G., Spinnler, C., Gambacorta, A. & Stetter, K.O. (1989a). *Metallosphaera sedula* gen. and sp. nov. represents a new genus of aerobic, metal-mobilizing, thermoacidophilic archaebacteria. *Systematic and Applied Microbiology* **12**, 38-47.

Huber, R., Woese, C. R., Langworthy, T. A., Fricke, H. & Stetter, K.O. (1989b). *Thermosipho africanus* gen. nov., represents a new genus of thermophilic eubacteria within the "Thermotogales". *Systematic and Applied Microbiology* **12**, 32-37.

Huber, R., Woese, C. R., Langworthy, T. A., Kristjansson, J.K. & Stetter, K.O. (1990). *Fervidobacterium islandicum* sp. nov., a new extremely thermophilic eubacterium belonging to the "Thermotogales". *Archives of Microbiology* **154**, 105-111.

Huber, R., Wilharm, T., Huber, D., Trincone, A., Burggraf, S., König, H., Rachel, R., Rockinger, I., Fricke, H. & Stetter, K.O. (1992). *Aquifex pyrophilus* gen. nov. sp. nov., represents a novel group of marine hyperthermophilic hydrogen-oxidizing bacteria. *Systematic and Applied Microbiology* **15**, 340-351.

Huber, R., Burggraf, S., Mayer, T., Barns, S.M., Rossnagel, P. & Stetter, K.O. (1995). Isolation of a hyperthermophilic archaeum predicted by *in situ* RNA analysis. *Nature* **376**, 57-58.

Huber, R., Rossnagel, P., Woese, C. R., Rachel, R., Langworthy, T. A. & Stetter, K.O. (1996a). Formation of ammonium from nitrate during chemolithoautotrophic growth of the extremely thermophilic bacterium *Ammonifex degensii* gen. nov., sp. nov. *Systematic and Applied Microbiology* **19**, 40-49.

Huber, R., Stöhr, J., Hohenhaus, S., Rachel, R., Burggraf, S., Jannasch, H. W. & Stetter, K.O. (1996b). *Thermococcus chitonophagus* sp. nov., a novel, chitin-degrading, hyperthermophilic archaeum from a deep-sea hydrothermal vent environment. *Archives of Microbiology* **164**, 255-264.

Huber, R., Eder, W., Hedwein, S., Wanner, G., Huber, H., Rachel, R. & Stetter, K.O. (1998). *Thermocrinis ruber* gen. nov., sp. nov., a pink-filament-forming hyperthermophilic bacterium isolated from Yellowstone National Park. *Applied and Environmental Microbiology* **64**, 3576-3583.

Hudson, J.A., Morgan, H.W. & Daniel, R.M. (1986). A numerical classification of some *Thermus* isolates. *Journal of General Microbiology* **132**, 531-540.

Hudson, J.A., Morgan, H.W. & Daniel, R.M. (1987). *Thermus filiformis* sp. nov., a filamentous caldoactive bacterium. *International Journal of Systematic Bacteriology* **37**, 431-436.

Hudson, J.A., Schofield, K.M., Morgan, H.W. & Daniel, R.M. (1989). *Thermonema lapsum* gen. nov., sp. nov., a thermophilic gliding bacterium. *International Journal of Systematic Bacteriology* **39**, 485-487.

Hugenholtz, P., Pitulle, C., Hersberger, K. L. & Pace, N. R. (1998). Novel division level bacterial diversity in a Yellowstone hot spring. *Journal of Bacteriology* **180**, 366-376.

Huval, J.H., Latta, R., Wallace, R., Kushner, D.J. & Vreeland, RH. (1995). Description of two new species of *Halomonas, Halomonas israelensis* and *Halomonas canadensis*. *Canadian Journal of Microbiology* **41**, 1124-1131.

Ihara, K., Watanabe, S. & Tamura, T. (1997). *Halvarcula argentinensis* and *Halarcula mukohataei*, two new extremely halophilic archaea collected in Argentina. *International Journal of Systematic Bacteriology* **47**, 73-77.

Imhoff, J.F. & Suling, J. (1996). The phylogenetic relationship among *Ectothiorhodospiraceae* a re-evaluation of their taxonomy on the basis of of 16S rRNA analysis. *Archives of Microbiology* **165**, 106-113.

Imhoff, J.F. & Trüper, H.G. (1977). *Ectothiorhodospira halochloris* sp. nov., a new extremely halophilic phototrophic bacterium containing bacteriochlorophyll *b*. *Archives of Microbiology* **114**, 115-121.

Imhoff, J.F. & Trüper, H.G. (1981). *Ectothiorhodospira abdelmalekii* sp. nov, a new halophilic and alkaliphilic phototrophic bacteria. *Zentralblatt für Bakteriologie, Mikrobiologie und Hygiene* C2, 228-234.

Jackson, T.J., Ramaley, R.F. & Meinschein, W.G. (1973). *Thermomicrobium,* a new genus of extremely thermophilic bacteria. *International Journal of Systematic Bacteriology* **23**, 28-36.

James, S.R., Dobson, S.J., Franzmann, P.B. & McMeekin, T.A. (1990). *Halomonas meridiana* a new species of extremely halotolerant bacteria isolated from Antarctic saline lakes. *Systematic and Applied Microbiology* **13**, 270-278.

James, S.R., Burton, H.R., McMeekin, T.A. & Mancuso, C.A. (1994). Seasonal abundance of *Halomonas meridipha, Halomonas subglasciescola, Flavobacterium gondwanense* and *Flavobacterium salegens* in four antarctic lakes. *Journal of Antarctic Science* **6**, 325-332.

Jannasch, H.W., Huber, R., Belkin, S. & Stetter, K.O. (1988). *Thermotoga neapolitana* sp. nov. of the extremely thermophilic, eubacterial genus *Thermotoga*. *Archives of Microbiology* **150**, 103-104.

Javor, B. (1989). *Hypersaline Environments*. Berlin: Springer-Verlag.

Jeanthon, C., Reysenbach, A.-L., L'Haridon, S., Gambacorta, A., Pace, N.R., Glenat, P. & Prieur, D. (1995). *Thermotoga subterranea* sp. nov., a new thermophilic bacterium isolated from a continental oil reservoir. *Archives of Microbiology* **164**, 91-97.

Jones, B.E., Grant, W.D., Collins, N.C. and Mwatha, W.E. (1994). Alkalikphiles: diversity and identification. In, *Bacterial Diversity and Systematics*, pp. 195-230. Edited by F.G. Priest, B.J. Tindall & A. Ramos-Cormenzana. New York: Plenum Press.

Jones, W.J., Leigh, J.A., Mayer, F., Woese, C.R. & Wolfe, R.S. (1983). *Methanococcus jannaschii* sp. nov., an extremely thermophilic methanogen from a submarine hydrothermal vent. *Archives of Microbiology* **136:**, 254-261.

Kamekura, M. (1993). Lipids of extreme halophiles. In *Biology of Halophilic Bacteria*, pp. 135-161. Edited by R. Vreeland & L. Hochstein. Boca Raton: CRC Press.

Kamekura, M. & Dyall-Smith, M.L. (1995). Taxonomy of the *Halobacteriaceae* and the description of two new genera, *Halorubrobacterium* and *Natrialba*. *Journal of General and Applied Microbiology* **41**, 333-350.

Kamekura, M., Dyall-Smith, M.L., Upasani, V., Ventosa, A. & Kates, M. (1997). Diversity of alkaliphilic halobacteria: Proposals for transfer of *Natronobacterium vacuolatum, Natronobacterium magadii* and *Natronobacterium pharaonis* to *Halorubrum, Natrialba* and *Natronomonas* gen. nov. *International Journal of Systematic Bacteriology* **47**, 853-857.

Kanai, H., Kobayashi, T., Aono, R. & Kudo T. (1995). *Natrococcus amylolyticus* sp. nov. a haloalkaliphilic archaeon. *International Journal of Systematic Bacteriology* **45**, 762-766.

Kates, M. (1996). Structural analysis of phospholipids and glycolipids in extremely halo- philic archaebacteria. *Journal of Microbiological Methods* **25**, 113-128.

Kawasaki, H., Hoshino, Y. & Yamasoto, K. (1993). Phylogenetic diversity of phototrophic purple non-sulphur bacteria. *FEMS Microbiology Letters* **112**, 61-66.

Kawasumi, T., Igarashi, Y., Kodama, T. & Minoda, Y. (1984). *Hydrogenobacter thermophilus* gen. nov. sp. nov., an extremely thermophilic, aerobic, hydrogen-oxidizing bacterium. *International Journal of Systematic Bacteriology* **34**, 5-10.

Keller, M., Braun, F-J., Dirmeier, R., Hafenbradl, D., Burggraf, S., Rachel, R. & Stetter, K.O. (1995). *Thermococcus alkaliphilus* sp. nov., a new hyperthermophilic archaeum growing on polysulfide at alkaline pH. *Archives of Microbiology* **164**, 390-395.

Kocur, M. & Hodgkiss, W. (1973). Taxonomic status of the genus *Halococcus*. *International Journal of Systematic Bacteriology* **23**, 151-156.

Kompantseva, E.I. & Gorlenko, V.M. (1984). A new species of moderately halophilic purple bacteria *Rhodospirillum mediosalinarum*. *Microbiology* **53**, 954-961.

Kotelnikova, S. V., Obraztsova, A. Y., Bloetvogel, K.-H. & Popov, I.N. (1993a). Taxonomic analysis of thermophilic strains of the genus *Methanobacterium*: reclassification of *Methanobacterium thermoalcaliphilum* as a synonym of *Methanobacterium thermoautotrophicum*. *International Journal of Systematic Bacteriology* **43**, 591-596.

Kotelnikova, S.V., Obraztsova, A.Y., Gongadze, G.M. & Laurinavichius, K.S. (1993b). *Methanobacterium thermoflexum* sp. nov. and *Methanobacterium defluvii* sp. nov., thermophilic rod-shaped methanogens isolated from anaerobic digestor sludge. *Systematic and Applied Microbiology* **16**, 427-435.

Kristjansson, J.K. (1989). Thermophilic bacteria as a sources of thermostable enzymes. *Trends in Biotechnology* **7**, 349-353.

Kristjansson, J.K. (1992). *Thermophilic Bacteria*. Boca Raton: CRC Press.

Kristjansson, J.K. & Alfredsson, G.A. (1986). Distribution of *Thermus* spp. in Icelandic hot springs and a thermal gradient. *Applied and Environmental Microbiology* **45**, 1785-1789.

Kristjansson, J.K. & Alfredsson, G.A. (1992). The heterotrophic, thermophilic genera *Thermomicrobium*, *Rhodothermus*, *Saccharococcus*, *Acidothermus* and *Scotothermus*. In *Thermophilic Bacteria*, pp. 63-76. Edited by J.K. Kristjansson. Boca Raton: CRC Press.

Kristjansson, J.K. & Hreggvidsson, G.O. (1995). Ecology and habitats of extremophiles. *World Journal of Microbiology and Biotechnology* **11**, 17-25.

Kristjansson, J.K. & Stetter, K.O. (1992). Thermophilic bacteria. In *Thermophilic Bacteria*, pp. 1-18. Edited by J.K. Kristjansson. Boca Raton: CRC Press.

Kristjansson, J.K. Ingason, A. & Alfredsson, G.A. (1985). Isolation of a thermophilic obligately autotrophic hydrogen-oxidizing bacteria, similar to *Hydrogenobacter thermophilus*, from Icelandic hot springs. *Archives of Microbiology* **140**, 321-325.

Kristjansson, J.K., Hreggvidsson G.O. & Alfredsson G.A. (1986). Isolation of halotolerant *Thermus* spp. from submarine hot springs in Iceland. *Applied and Environmental Microbiology* **52**, 1313-1316.

Kristjansson, J.K., Hjörleifsdóttir, S., Marteinsson, V.Th. & Alfredsson, G.A. (1994). *Thermus scotoductus*, sp.nov., a pigment-producing thermophilic bacterium from hot tap water in Iceland and including *Thermus* sp. X-1. *Systematic and Applied Microbiology* **17**, 44-50.

Kryukov, V.R., Saveleva, N.D. & Pusheva, M.A. (1983). *Calderobacterium hydrogenophilum* nov. gen., nov. sp., an extreme thermophilic bacterium and its hydrogenase activity. *Mikrobiologiya* **52**, 611-618.

Kurr, M., Huber, R., König, H., Jannasch, H. W., Fricke, H., Trincone, A., Kristjansson, J.K. & Stetter, K.O. (1991). *Methanopyrus kandleri*, gen. and sp. nov. represents a novel group of hyperthermophilic methanogens, growing at 110° C. *Archives of Microbiology* **156**, 239-247.

Kushner, D.J. (1978). Life in high salt and solute concentrations halophilic bacteria. In *Microbial Life in*

Extreme Environments, pp. 317-368. Edited by. D.J. Kushner. London: Academic Press..

Kushner DJ. (1993). Growth and nutrition of halophilic bacteria. In *The Biology of Halophilic Bacteria*, pp. 87-103. Edited by R. H. Vreeland, R.H. & L. Hochstein. Boca Raton: CRC Press.

Kwak, Y.S., Kobayashi, T., Akiba, T., Horikoshi, K. & Kim, Y.B. (1995). A hyperthermophilic sulfur-reducing archaebacterium, *Thermococcus* sp. DT1331, isolated from a deep-sea hydrothermal vent. *Bioscience, Biotechnology and Biochemistry* 59, 1666-1669.

Langworthy, T.A., Holzer, G., Zeikus, J.G. & Tornabene, T.G. (1983). *Iso-* and *anteiso*-branched glycerol diethers of the thermophilic anaerobe *Thermodesulfobacterium commune*. *Systematic and Applied Microbiology* 2, 1-17.

Lauerer, G., Kristjansson, J.K., Langworthy, T. A., König, H. & Stetter, K.O. (1986). *Methanothermus sociabilis* sp. nov., a second species within the Methanothermaceae growing at 97°C. *Systematic and Applied Microbiology* 8, 100-105.

Laurinavichyus, K.S., Kotelnikova, S.V. & Obraztsova, A.Y. (1988). New species of thermophilic methane-producing bacteria *Methanobacterium thermophilum*. *Microbiology* 57, 832-838. (English translation of *Mikrobiologiya*).

Le Ruyet, P., Dubourguier, H. C., Albagnac, G. & Prensier, G. (1985). Characterization of *Clostridium thermolacticum* sp. nov., a hydrolytic thermophilic anaerobe producing high amounts of lactate. *Systematic and Applied Microbiology* 6, 196-202.

Lee, Y.-E., Jain, M.K., Lee, C.Y., Lowe, S.E. & Zeikus, J.G. (1993). Taxonomic distinction of saccharolytic thermophilic anaerobes: description of *Thermoanaerobacterium xylanolyticum* gen. nov., sp. nov., and *Thermoanaerobacterium saccharolyticum* gen. nov., sp. nov.; reclassification of *Thermoanaerobium brockii, Clostridium thermosulfurogenes*, and *Clostridium thermohydrosulfuricum* E100-69 as *Thermoanaerobacter brockii* comb. nov., *Thermoanaerobacterium thermosulfurigenes* comb. nov., and *Thermoanaerobacter thermohydrosulfuricus* comb.nov., respectively; and transfer of *Clostridium thermohydrosulfuricum* 39E to *Thermoanaerobacter ethanolicus*. *International Journal of Systematic Bacteriology* 43, 41-51.

Leigh, J.A. & Wolfe, R.S. (1981). *Acetogenium kivui* a new hydrogen-oxidizing thermophilic, acetogenic bacterium. *Archives of Microbiology* 129, 275-280.

Li, Y, Mandelco, L. & Wiegel, J. (1993). Isolation and characterisation of a moderately thermophilic anaerobic alkaliphile, *Clostridium paradoxum* sp. nov. *International Journal of Systematic Bacteriology*. 43, 450-466.

Li, Y., Engle, M., Weiss, N., Mandelco, L. & Wiegel, J. (1994). *Clostridium thermoalcaliphilium* sp. nov. an anaerobic and thermotolerant facultative alkaliphile. *International Journal of Systematic Bacteriology* 44, 111-118.

Liaw, H.J. & Mah, R.A. (1992). Isolation and characterisation of *Haloanaerobacter chitinovorans* gen. nov. sp. nov., a halophilic, anaerobic chitinolytic bacterium from a solar saltern. *Applied and Environmental Microbiology* 58, 260-266.

Liu, S.-Y., Rainey, F.A., Morgan, H.W., Mayer, F. & Wiegel, J. (1996). *Thermoanaerobacterium aotearoense* sp. nov., a slightly acidophilic, anaerobic thermophile isolated from various hot springs in New Zealand, and emendation of the genus *Thermoanaerobacterium*. *International Journal of Systematic Bacteriology* 46, 388-396.

Loginova, L.G., Egorova, L.A., Golovacheva, R.S. & Seregina, L.M. (1984). *Thermus ruber* sp. nov., nom. rev. *International Journal of Systematic Bacteriology* 34, 498-499.

Love, C.A., Patel, B.K.C., Nichols, P.D. & Stackebrandt, E. (1993). *Desulfotomaculum australicum*, sp. nov., a thermophilic sulfate-reducing bacterium isolated from the great artesian basin of Australia.

Systematic and Applied Microbiology 16, 244-251.

Mack, E.E., Mandelco, L., Woese, C.R. & Madigan, M.T. (1993). Rhodospirillum sodomense, a Dead Sea rhodospirillim species. Archives of Microbiology 160, 363-371.

Manaia, C.M., Hoste, B., Gutierrez, M.C., Gillis, M., Ventosa, A., Kersters, K. & Da Costa, M.S. (1994). Halotolerant Thermus strains from marine and terrestrial hot springs belong to Thermus thermophilus (ex Oshima and Imahori, 1974) nom. rev emend. Systematic and Applied Microbiology 17, 526-532.

Marquez, M.C., Ventosa, A. & Ruiz-Berraquero, F. (1990). Marinococcus hispanicus, a new species of moderately halophilic Gram-positive cocci. International Journal of Systematic Bacteriology 40, 165-169.

Marteinsson, V.T. (1997). Isolement et charactérisation de micro-organismes extrêmophiles originiaires de sites hydrothermaux océaniques Influence du couple température-pression sur la physiologie. Ph.D. thesis. University of Bretagne Occidentale, Brest, France.

Marteinsson, V.T., Birrien J.-L, Kristjansson, J.K. & Prieur D. (1995a). First isolation of thermophilic aerobic non-sporulating heterotrophic bacteria from deep-sea hydrothermal vents. FEMS Microbiology Ecology 18, 163-174.

Marteinsson, V.T., Watrin, L., Prieur, D., Caprais, J.-C., Raguénés, G. & Erauso, G. (1995b). Phenotypic chracterization, DNA similarities, and protein profiles of twenty sulfur-metabolizing hyperthermophilic anaerobic archaea isolated from hydrothermal vents in the Southwestern Pacific Ocean. International Journal of Systematic Bacteriology 45, 623-632.

Mathrani, I.M., Boone, D.R., Mah, R.A., Fox G.E. & Lav, P.P. (1988). Methanohalophilus zhilinae sp. nov., an alkaliphilic, halophilic, methylotrophic methanogen. International Journal of Systematic Bacteriology 38, 139-142.

McBee, R. H. (1954). The characteristics of Clostridium thermocellum. Journal of Bacteriology 67, 505-506.

McGenity, T.J. & Grant, W.D. (1995). Transfer of Halobacterium saccharovorum, Halobacterium sodomense, Halobacterium trapanicum NRC 34021 and Halobacterium lacusprofundii to the genus Halorubrum gen. nov. Systematic and Applied Microbiology 18, 237-241.

Meeks, J.C. & Castenholz, R.W. (1971). Growth and photosynthesis in extreme thermophile, Synechococcus lividus (Cyanophyta). Archives of Microbiology 78, 25-41.

Mellado, E., Moore, E.R.B., Nieto, J.J. & Ventosa, A. (1996). Analysis of 16S ribosomal RNA genes sequences of Vibrio costicola strains - description of Salinivibrio costicola gen. nov. comb. nov. International Journal of Systematic Bacteriology. 46, 817-821.

Méndez, B.S., Pettinari, M.J., Ivanier, S.E., Ramos, C.A. & Sineriz, F. (1991). Clostridium thermopapyrolyticus sp. nov., a cellulolytic thermophile. International Journal of Systematic Bacteriology 41, 281-283.

Min, H. & Zinder, S. H. (1990). Isolation and characterization of a thermophilic sulfate-reducing bacterium Desulfotomaculum thermoacetoxidans sp. nov. Archives of Microbiology 153, 399-404.

Miroshnichenko, M.L., Bonch-Osmolovskaya, E.A., Neuner, A., Kostrikina, N.A.Chernych, N.A. & Alekseev, V.A. (1989). Thermococcus stetteri sp. nov., a new extremely thermophilic marine sulfur-metabolizing archaebacterium. Systematic and Applied Microbiology 12, 257-262.

Miroshnichenko, M.L., Gongadze, G.A., Lysenko, A.M. & Bonch-Osmolovskaya, E.A. (1994). Desulfurella multipotens sp. nov., a new sulfur-respiring thermophilic eubacterium from Raoul Island (Kermadec archipelago, New Zealand). Archives of Microbiology 161, 88-93.

Montero, C.G., Ventosa, A., Rodriguez-Valera, F., Kates, M., Moldoveanu, N. & Ruiz-Berraquero, F. (1989). Halococcus saccharolyticus sp. nov., a new species of extremely halophilic non-alkaliphilic cocci. Systematic and Applied Microbiology 12, 167-171.

Moreira, L., Nobre, M.F., Sa-Correia, I. & Da Costa, M.S. (1996). Genomic typing and fatty acid composition of Rhodothermus marinus. Systematic and Applied Microbiology 19, 83-90.

Moreira, L.M., Da Costa, M.S. & Sá-Correia, I. (1997). Comparative genomic analysis of isolates belonging to the six species of the genus *Thermus* using pulsed-field gel electrophoresis and ribotyping. *Archives of Microbiology* 168, 92-101.

Mwatha, W.E. & Grant, W.D. (1993). *Natronobacterium vacuolata* sp. nov. a haloalkaliphilic archaeon isolated from Lake Magadi Kenya. *International Journal of Systematic Bacteriology* 43, 401-404.

Mylvaganam, S. & Dennis, P.P. (1992). Sequence heterogeneity between the two genes encoding 16S rRNA from the halophilic archaebacterium *Haloarcula marismortui*. *Genetics* 130, 399-410.

Nazina, T. N., Ivanova, A. E., Kanchaveli, L. P. & Rozanova, E. P. (1989). A new sporeforming thermophilic methylotrophic sulfate-reducing bacterium, *Desulfotomaculum kuznetsovii* sp. nov. *Microbiology* 57, 659-663 (English Translation of Mikrobiologiya).

Nielsen, P., Rainey, F., Outtrup, H., Priest, F.G. & Fritze, D. (1994). Comparative 16S rDNA sequence analysis of some alkaliphilic bacilli and the establishment of a sixth rRNA group within the genus *Bacillus*. *FEMS Microbiology Letters* 117, 61-66.

Nielsen, P., Fritze, D. & Priest, F.G. (1995). Phenetic diversity of alkaliphilic *Bacillus* strains: proposal for nine new species. *Microbiology* 141, 1745-1761.

Nissen, H. & Dundas, I.D. (1984). *Rhodospirillum salinarum* sp. nov., a halophilic photosynthetic bacterium isolated from a Portuguese saltern. *Archives of Microbiology* 138, 251-256.

Nobre, M. F., Trüper, H. G. & Da Costa, M. S. (1996). Transfer of *Thermus ruber* (Loginova *et al.* 1984), *Thermus silvanus* and *Thermus chliarophilus* to *Meiothermus* gen. nov. as *Meiothermus ruber* comb. nov., *Meiothermus silvanus* comb. nov., and *Meiothermus chliarophilus* comb. nov., respectively, and emendation of the genus *Thermus*. *International Journal of Systematic Bacteriology* 46, 604-606.

Novitsky, T.J. & Kushner, D.J. (1976). *Planococcus halophilus* sp. nov., a facultatively halophilic coccus. *International Journal of Systematic Bacteriology* 26, 53-57.

Nunes, O.C., Donato, M.M. & Da Costa, M.S. (1992). Isolation and characterization of *Rhodothermus* strains from S. Miguel, Azores. *Systematic and Applied Microbiology* 15, 92-97.

Odintsova, E.V., Jannasch, H.W., Mamone, J.A. & Langworthy, T.A. (1996). *Thermothrix azorensis* sp. nov., an obligately chemolithoautotrophic, sulfur-oxidizing, thermophilic bacterium. *International Journal of Systematic Bacteriology* 46, 422-428.

Ollivier, B.M., Mah R.A., Ferguson, T.J., Boone, D.R. Garcia, J.L. & Robinson, R. (1985). Emendation of the genus *Thermobacteroides*: *Thermobacteroides proteolyticus* sp. nov., a proteolytic acetogen from a methanogenic enrichment. *International Journal of Systematic Bacteriology* 35, 425-428.

Ollivier, B., Hatchikian, C.E., Prensier, G. & Garcia, J.-L. (1991). *Desulfohalobium retbaense* gen. nov., sp. nov., a halophilic sulphate - reducing bacterium from the sediments of a hypersaline lake in Senegal. *International Journal of Systematic Bacteriology* 41, 74-81.

Ollivier, B., Caumette, P., Garcia, J.-L. & Mah, R. (1994). Anaerobic bacteria from hypersaline environments. *Microbiological Reviews* 58, 27-38.

Olsen, G.J. & Woese, C.R. (1993). Ribosomal RNA: a key to phylogeny. *FASEB Journal* 7, 113-123.

Onishi, H. & Kamekura, M. (1972). *Micrococcus halobius* sp. nov. *International Journal of Systematic Bacteriology* 22, 233-235.

Oren, A. (1988). The microbiology of the Dead Sea. *Advances in Microbial Ecology* 10, 193-230.

Oren, A. (1992). The genera *Haloanaerobium*, *Halobacteroides* and *Sporohalobacter*. In *The Prokaryotes*, pp. 1893-1900. Edited by A. Balows, H. G. Trüper, M. Dworkin, W. Harder, K-H. Scheifer. Berlin: Springer Verlag.

Oren, A. & Ventosa, A. (1996). A proposal for the transfer of *Halrubrobacterium distributum* and *Halorubrobacterium distributum* to the genus *Halorubrum*. *International Journal of Systematic Bacteriology* 46, 1180.

Oren, A., Pohla, H. & Stackebrandt, E. (1987). Transfer of *Clostridium lortetii* to a new genus *Sporohalobacter* gen. nov. as *Sporohalobacter lortetii* comb. nov. and description of *Sporohalobacter marismortui* sp. nov. *Systematic and Applied Microbiology* 9, 239-246.

Oren, A., Kessel, M. & Stackebrandt, E. (1989). *Ectothiorhodospira marismortui* sp. nov., an obligately halophilic purple sulfur bacterium from a hypersaline sulfur spring on the shore of the Dead Sea. *Archives of Microbiology* 151, 524-529.

Oren, A., Ginzburg, M., Ginzburg, B.Z., Hochstein, L.I. & Volcani, B.E. (1990). *Haloarcula marismortui* (volcani) sp. nov. nom. rev., an extremely halophilic bacterium from the Dead Sea. *International Journal of Systematic Bacteriology* 40, 209-210.

Oren, A., Guevich, P., Gemmell, R.T. & Teske, A. (1995). *Halobaculum gomorrense* gen. nov., sp. nov., a novel extremely halophilic archaeon from the Dead Sea. *International Journal of Systematic Bacteriology* 45, 747-754.

Oren, A., Ventosa, A. & Grant, W.D. (1997). Proposed minimal standards for description of new taxa in the order *Halobacteriales*. *International Journal of Systematic Bacteriology* 47, 233-238.

Oshima, T. & Imahori, K. (1974). Description of *Thermus thermophilus* (Yoshida and Oshima) comb. nov., a nonsporulating thermophilic bacterium from a Japanese thermal spa. *International Journal of Systematic Bacteriology* 24, 102-112.

Pask-Hughes, R.A. & Williams, R.A.D. (1975). Extremely thermophilic gram-negative bacteria from hot tap water. *Journal of General Microbiology* 88, 321-328.

Patel, B.K.C., Morgan, H.W. & Daniel, R.M (1985). *Fervidobacterium nodosum* gen. nov. and spec. nov., a new chemoorganotrophic, caldoactive, anaerobic bacterium. *Archives of Microbiology* 141, 63-69.

Patel, B.K.C., Monk, C., Littleworth, H., Morgan, H.W. & Daniel, R.M. (1987). *Clostridium fervidus* sp. nov., a new chemoorganotrophic acetogenic thermophile. *International Journal of Systematic Bacteriology* 37, 123-126.

Paterek, J.R. & Smith, P.H. (1988). *Methanohalophilus mahii* gen. nov. sp. nov. a methylotrophic methanogen. *International Journal of Systematic Bacteriology* 38, 122-127.

Perttula, M., Pere, J., Konradsdottir, M., Kristjansson, J.K. & Viikari, L. (1991). Removal of acetate from NSSC sulphite pulp mill condensates using thermophilic bacteria. *Water Research* 25, 599-604.

Pierson, B.K. & Castenholz, R.W. (1974). A phototrophic gliding filamentous bacterium of hot springs, *Chloroflexus aurantiacus*, gen. and sp. nov. *Archives of Microbiology* 100, 5-24.

Pley, U., Schipka, J., Gambacorta, A., Jannasch, H. W., Fricke, H., Rachel, R. & Stetter, K.O. (1991). *Pyrodictium abyssi* sp. nov. represents a novel heterotrophic marine archaeal hyperthermophile growing at 110°C. *Systematic and Applied Microbiology* 14, 245-253.

Prieur, D. (1997). Microbiology of deep-sea hydrothermal vents. *Trends in Biotechnology* 15, 242-244.

Quesada, E., Ventosa, A., Ruiz-Berraquero, F. & Ramos-Cormenzana, A. (1984). *Deleya halophila*, a new species of moderately halophilic bacteria. *International Journal of Systematic Bacteriology* 34, 287-292.

Quesada, E., Valderrama M.J., Bejar, V., Ventosa, A., Gutierrez, M.C., Ruiz-Berraquero, F. & Ramos-Cormenzana, A. (1990). *Volcaniella eurihalina* gen. nov. sp. nov., a moderately halophilic non-motile Gram-negative rod. *International Journal of Systematic Bacteriology* 40, 261-267.

Radax, C., Sigurdsson, O., Hreggvidsson, G.O., Aichinger, N., Gruber, G., Kristjansson, J.K. & Stan-Lotter, H. (1998). F and V-ATPases in the genus *Thermus* and related species. *Systematic and Applied Microbiology* 21, 12-22.

Rainey, F.A., Ward, N.L., Morgan, H.W., Toalster, R. & Stackebrandt, E. (1993a). Phylogenetic analyses of anaerobic thermophilic bacteria: aid for their reclassification. *Journal of Bacteriology* 175, 4772-4779.

Rainey, F.A., Jansen, P.H., Morgan, H.W. & Stackebrandt, E. (1993b). A biphasic approach to the determination of the phenotypic and genotypic diversity of some anaerobic, cellulolytic thermophilic, rod-shaped bacteria. *Antonie van Leeuwenhoek* **64**, 341-355.

Rainey, F.A., Donnison, A.M., Janssen, P.H., Saul, D., Rodrigo, A., Berguist, P.I., Daniel, R.M., Stackebrandt, E. & Morgan, H.W. (1994a). Description of *Caldicellulosiruptor saccharolyticus* gen. nov., sp. nov.: An obligately anaerobic, extremely thermophilic, cellulolytic bacterium. *FEMS Microbiology Letters* **120**, 263-266.

Rainey, F.A., Fritze, D. & Stackebrandt, E. (1994b). The phylogenetic diversity of thermophilic members of the genus *Bacillus* as revealed by 16S rDNA analysis. *FEMS Microbiology Letters* **115**, 205-211.

Rainey, F.A., Nobre, M.F., Schumann, P., Stackebrandt, E. & Da Costa, M.S. (1997). Phylogenetic diversity of the deinococci as determined by 16S ribosomal DNA sequence comparison. *International Journal of Systematic Bacteriology* **47**, 510-514.

Ravot, G., Magot, M., Fardeau, M.-L., Patel, B.K.C., Prensier, G., Egan, A., Garcia, J.-L. & Ollivier, B. (1995). *Thermotoga elfii* sp. nov., a novel thermophilic bacterium from an African oil-producing well. *International Journal of Systematic Bacteriology* **45**, 308-314.

Raymond, J.C. & Sistrom, W.R. (1969). *Ectothiorhodospira halophila*: a new member of the genus *Ectothiorhodospira*. *Archives of Microbiology* **69**, 121-126.

Rengipat, S., Langworthy T.A. & Zeikus, J.G. (1988a). *Halobacteroides acetoethylicus* sp. nov. a new obligately anaerobic halophile isolated from deep subsurface hypersaline environments. *Systematic and Applied Microbiology* **11**, 28-35.

Rengipat, S., Lowe S.E. & Zeikus, J.G. (1988b). Effect of extreme salt concentrations on the physiology and biochemistry of *Halobacteroides acetoethyticus*. *Journal of Bacteriology* **170**, 3065-3071.

Reysenbach, A.L., Wickham, G.S. & Pace, N.R. (1994). Phylogenetic analysis of the hyperthermophilic pink filaments community in Octopus Spring, Yellowstone National Park. *Applied and Environmental Microbiology* **60**, 2113-2119.

Robinson, J. & Gibbons, N. (1952). The effect of salts on the growth of *Paracoccus halo- denitificans*. *Canadian Journal of Botany* **30**, 147-154.

Rodrigo, A.G., Borges, K.M. & Bergquist, P.L. (1994). Pulsed-field gel electrophoresis of genomic digests of *Thermus* strains and its implications for taxonomic and evolutionary studies. *International Journal of Systematic Bacteriology* **44**, 547-552.

Romano, I., Nicolaus, B., Lana, L., Manca M.C. & Gambacorta, A. (1996). Characterisation of a haloalkaliphilic strictly aerobic bacterium from Pantelleria Island. *Systematic and Applied Microbiology* **19**, 326-333.

Ronimus, R.S., Parker, L.E. & Morgan, H.W. (1997). The utilization of RAPD-PCR for identifying thermophilic and mesophilic *Bacillus* species. *FEMS Microbiology Letters* **147**, 75-79.

Ross, H.M.N. & Grant, W.D. (1985). Nucleic acid studies on halophilic archaebacteria. *Journal of General Microbiology* **131**, 165-173.

Ross, H.N.M., Collins, M.D., Tindall, B.J. & Grant, W.D. (1981). A rapid method for the detection of archaebacterial lipids in halophilic bacteria. *Journal of General Microbiology* **123**, 75-80.

Ross, H.M.N., Grant, W.D. & Harris, J.E. (1985). Lipids in archaebacterial taxonomy. In, *Chemical Methods in Bacterial Systematics*, pp. 289-300. Edited by M. Goodfellow & D.E. Mimnikin. London: Academic Press.

Rozanova, E.P. & Pivovarova, T.A. (1988). Reclassification of *Desulfovibrio thermophilus* (Rozanova, Khudyakova, 1974). *Microbiology* **57**, 85-89 (English translation of *Mikrobiologiya*).

Ruan, J.S., Al-Tai A.M., Zhou, Z.H. & Qu, L.H. (1994). *Actinopolyspora iraqiensis* sp. nov. a new halophilic actinomycete isolated from soil. *International Journal of Systematic Bacteriology* **44**, 759-763.

Rönkä, J., Hjörleifsdóttir, S., Tenkanen, T., Pitkänen, K., Mattila, P. & Kristjansson, J.K. (1991). *Rma*I, a type II restriction endonuclease from *Rhodothermus marinus* which recognizes 5'CTAG 3'. *Nucleic Acids Research* 19, 2789.

Saiki, T., Kobayashi, Y., Kawagoe, K. & Beppu, T. (1985). *Dictyoglomus thermophilus* gen. nov., sp. nov, a chemoorganotrophic, anaerobic, thermophilic bacterium. *International Journal of Systematic Bacteriology* 35, 253-259.

Sako, Y., Takai, K., Ishida, Y., Uchida, A. & Katayama, Y. (1996a). *Rhodothermus obamensis* sp. nov., a modern lineage of extremely thermophilic marine bacteria. *International Journal of Systematic Bacteriology* 46, 1099-1104.

Sako, Y., Nomura, N., Uchida, A., Ishida, Y., Morii, H., Koga, Y., Hoaki, T. & Maruyama, T. (1996b). *Aeropyrum pernix* gen. nov., a novel aerobic hyperthermophilic archaeon growing at temperatures up to 100°C. *International Journal of Systematic Bacteriology* 46, 1070-1077.

Santos, M.A., Williams, R.A.D. & Da Costa, M.S. (1989). Numerical taxonomy of *Thermus* isolates from hot springs in Portugal. *Systematic and Applied Microbiology* 12, 310-315.

Saul, D.J., Rodrigo, A.G., Reeves, R.A., Williams, L.C., Borges, K.M., Morgan, H.W. & Berquist, P.L. (1993). Phylogeny of twenty *Thermus* isolates constructed from 16S rRNA gene sequence data. *International Journal of Systematic Bacteriology* 43, 754-760.

Schäfer, G., Purschke, W. & Schmidt, C.L. (1996). On the origin of respiration: electron transport proteins from archaea to man. *FEMS Microbiology Reviews* 18, 173-188.

Schenk, A. & Aragno, M. (1979). *Bacillus schlegelii*, a new species of thermophilic facultatively chemolithoautotrophic bacterium oxidizing molecular hydrogen. *Journal of General Microbiology* 115, 333-341.

Schleper, C., Puehler, G., Holz, I., Gambacorta, A., Janekovic, D., Santarus, U., Klenk, H. & Zillig, W. (1995). *Picrophilus* gen. nov., fam. nov.: a novel aerobic heterotrophic thermoacidophilic genus and family comprising archaea capable of growing around pH 0. *Journal of Bacteriology* 177, 7050-7059.

Segerer, A., Neuner, A., Kristjansson, J.K. & Stetter, K.O. (1986). *Acidianus infernus* gen. nov., sp. nov., and *Acidianus brierleyi* comb. nov.: facultatively aerobic, extremely acidophilic thermophilic sulfur metabolizing archaebacteria. *International Journal of Systematic Bacteriology* 36, 559-564.

Segerer, A., Langworthy, T.A. & Stetter, K.O. (1988). *Thermoplasma acidophilum* and *Thermoplasma volcanium* sp. nov. from solfatara fields. *Systematic and Applied Microbiology* 10, 161-171.

Segerer, A.H., Trincone, A., Gahrtz, M. & Stetter, K.O. (1991). *Stygioglobus azoricus* gen. and sp. nov., represents a novel genus of anaerobic, extremely thermoacidophilic archaea of the order *Sulfolobales*. *Journal of Bacteriology* 41, 495-501.

Selander, R.K., Caugant, D.A., Ochman, H., Musser, J.M., Gilmour, M.N. & Whittam, T.S. (1986). Methods of multilocus enzyme electrophoresis for bacterial population genetics and systematics. *Applied and Environmental Microbiology* 51, 873-884.

Shiba, H. (1991). Anaerobic halophiles. In *Superbugs. Microorganisms in Extreme Environments*, pp. 191-211. Edited by K. Horikoshi & W. D. Grant. Berlin: Springer-Verlag.

Shima, S. & Suzuki, K-I. (1993). *Hydrogenobacter acidophilus* sp. nov., a thermoacidophilic, aerobic, hydrogen-oxidizing bacterium requiring elemental sulfur for growth. *International Journal of Systematic Bacteriology* 43, 703-708.

Simankova, M.V., Chernych, N.A., Osipov, G.A. & Zavarzin, G.A. (1993). *Halocella cellulolytica* gen. nov. sp. nov. a new obligately anaerobic halophilic cellulolytic bacterium. *Systematic and Applied Microbiology* 16, 385-389.

Slobodkin, A., Reysenbach, A.-L., Mayer, F. & Wiegel, J. (1997a). Isolation and characterization of the homoacetogenic thermophilic bacterium *Moorella glycerini* sp. nov. *International Journal of*

Systematic Bacteriology **47**, 969-974.

Slobodkin, A., Reysenbach, A.-L., Strutz, N., Dreier, M. & Wiegel, J. (1997b). *Thermoterrabacterium ferrireducens* gen. nov., sp. nov., a thermophilic anaerobic dissimilatory Fe(III)-reducing bacterium from a continental hot spring. *International Journal of Systematic Bacteriology* **47**, 541-547.

Sneath, P.H.A. (1989). Analysis and interpretation of sequence data for bacterial systematics: The view of a numerical taxonomist. *Systematic and Applied Microbiology* **12**, 15-31.

Spring, S., Ludwig, W., Marquez M.C., Ventosa, A. & Schleifer, K-H. (1996). *Halobacillus* gen. nov. with descriptions of *Halobacillus litoralis* sp. nov. and *Halobacillus trueperi* sp. nov. and transfer of *Sporosarcina halophila* to *Halobacillus* comb. nov. *International Journal of Systematic Bacteriology* **46**, 492-496.

Staley, J.T., Castenholz, R.W., Colwell, R.R., Holt, J.G., Kane, M.D., Pace, N.R. Salyers, A.A. & Tiedje, J.M. (1997). *The Microbial World: Foundation of the Biosphere*. Washington: American Society for Microbiology.

Starkey, F.L. (1938). A study of spore formation and other morphological characteristics of *Vibrio desulfuricans*. *Archives für Mikrobiology* **9**, 268-278.

Stetter, K.O. (1988). *Archaeglobus fulgidus* gen. nov., sp. nov.: a new taxon of extremely thermophilic archaebacteria. *Systematic and Applied Microbiology* **10**, 172-173.

Stetter, K.O. (1996a). Hyperthermophiles in the history of life. *Ciba Foundation Symposium* **202**, 1-10.

Stetter, K.O. (1996b). Hyperthermophilic procaryotes. *FEMS Microbiology Reviews* **18**, 149-158.

Stetter, K.O., Thomm, M., Winter, J., Wildgruber G., Huber, H., Zillig, W., Janecovic, D., König, H. & Palm, P. (1981). *Methanothermus fervidus* sp. nov., a novel extremely thermophilic methanogen isolated from an Icelandic hot spring. *Zentralblatt für Bakteriologie und Hygiene* C **2**, 166-178.

Stetter, K.O., König, H. & Stackebrandt, E. (1983). *Pyrodictium*, a new genus of submarine disc-shaped sulfur reducing archaebacteria growing optimally at 105°C. *Systematic and Applied Microbiology* **4**, 535-551.

Suzuki, Y., Kishigami, T., Inoue, K., Mizoguchi, Y., Eto, N., Takagi, M. & Abe, S. (1983). *Bacillus thermoglucosidasius* sp. nov., a new species of obligately thermophilic bacilli. *Systematic and Applied Microbiology* **4**, 487-495.

Svetlichnii, V. A., Sokolova, T. G., Gerhardt, M., Ringpfeil, M., Kostrikina, N.A. & Zavarzin, G. A. (1991). *Carboxydothermus hydrogenoformans* gen. nov., sp. nov., a CO - utilizing thermophilic anaerobic bacterium from hydrothermal environments of Kunashir Island. *Systematic and Applied Microbiology* **14**, 254-260.

Svetlichnii, V.A. & Svetlichnaya, T.P. (1988). *Dictyoglomus turgidus* sp. nov., a new extremly thermophilic eubacterium isolated from hot springs of the Uzon volcano crater. *Mikrobiologiya* **57**, 435-441.

Svetlinshky, V., Rainey, F. & Wiegel, J. (1996). *Thermosyntropha lipolytica* gen nov. sp. nov. A lipolytic anaerobic alkalitolerant thermophilic bacterium utilizing short chain and long chain fatty acids in syntrophic coculture with a methogenic archaeon. *International Journal of Systematic Bacteriology* **46**, 1131-1137.

Takashina, T., Hamamoto, T., Kiyataka, O., Grant, W.D. & Horikoshi, K. (1990). *Haloarcula japonica* sp. nov., a new triangular archaebacterium. *Systematic and Applied Microbiology* **13**, 177-181.

Takayanagi, S., Kawasaki, H., Sugimori, K., Yamada, T., Sugai, A., Ito, T., Yamasato, K. & Shioda, M. (1996). *Sulfolobus hakonensis* sp. nov., a novel species of acidothermophilic archaeon. *International Journal of Systematic Bacteriology* **46**, 377-382.

Tarlera, S., Muxi, L. Soubes, M. & Stams, A J.M. (1997). *Caloramator proteoclasticus* sp. nov., a new moderately thermophilic anaerobic proteolytic bacterium. *International Journal of Systematic Bacteriology* **47**, 651-656.

Tasaki, M., Kamagata, Y., Nakamura, K. & Mikami, E. (1991). Isolation and characterization of a thermophilic benzoate-degrading, sulfate-reducing bacterium, *Desulfotomaculum thermobenzoicum* sp. nov.

Archives of Microbiology 155, 348-352.

Tenreiro, S., Nobre, M.F. & Da Costa, M.S. (1995a). *Thermus silvanus* sp. nov. and *Thermus chliarophilus* sp. nov., two new species related to *Thermus ruber* but with lower growth temperatures. *International Journal of Systematic Bacteriology* 45, 633-639.

Tenreiro, S., Nobre, M.F., Hoste, B., Gillis, M., Kristjansson, J.K. & Da Costa, M.S. (1995b). DNA:DNA hybridization and chemotaxonomic studies of *Thermus scotoductus. Research in Microbiology* 146, 315-324.

Tenreiro, S., Nobre, M.F., Rainey, F.A., Miguel, C. & Da Costa, M.S. (1997). *Thermonema rossianum* sp. nov., a new thermophilic and slightly halophilic species from saline hot springs in Naples, Italy. *International Journal of Systematic Bacteriology* 47, 122-126.

Tindall, B.J., Mills, A.A. & Grant, W.D. (1980). An alkalophilic red halophilic bacterium with a low magnesium requirement from a Kenyan soda lake. *Journal of General Microbiology* 116, 257-260.

Tindall, B.J., Ross, H.N.M. & Grant, W.D. (1984). *Natronobacterium* gen. nov. and *Natronococcus* gen. nov. two new genera of haloalkaliphilic archaebacteria. *Systematic and Applied Microbiology* 5, 41-47.

Tomlinson, G.A., Jahnke, L.L. & Hochstein, L.I. (1986). *Halobacterium denitrificans* sp. nov., an extremely halophilic denitrifying bacterium. *International Journal of Systematic Bacteriology* 36, 66-70.

Torreblanca, M., Rodriguez-Valera, F., Juez, G., Ventosa, A., Kamekura, M. & Kates, M. (1986). Classification of non-alkaliphilic halobacteria based on numerical taxonomy and polar lipid composition and description of *Haloarcua* gen. nov. and *Haloferax* gen. nov. *Systematic and Applied Microbiology* 8, 89-99.

Tsai, C.R., Garcia, J.L., Patel, B.K.C., Baresi, L. & Mah, R. (1995). *Haloanaaerobium alcaliphilium* sp. nov. an anaerobic moderate halophile from the sediments of the Great Salt Lake. *International Journal of Systematic Bacteriology* 45, 301-307.

Valderrama, M.J., Quesada, E., Bejar, V., Ventosa, A., Gutierrez, M.C., Ruiz-Berraquero, F. & Ramos- Cormenzana, A. (1991). *Deleya salina* sp. nov., a moderately halophilic Gram- negative bacterium. *International Journal of Systematic Bacteriology* 41, 377-384.

Vandamme, P., Pot, B., Gillis, M., De Vos, P., Kersters, K. & Swings, J. (1996). Polyphasic taxonomy, a consensus approach to bacterial systematics. *Microbiological Reviews* 60, 407-438.

Vedder, A. (1934). *Bacillus alcalophilus* n. sp., benevens enkele ervaringen met alcalische voedingsbodems. *Antonie Van Leeuwenhoek* 1:141-147.

Ventosa, A. (1988). Taxonomy of moderately halophilic heterotrophic bacteria. In *Halophilic Bacteria* Vol. 1, pp. 71-84. Edited by F. Rodriguez-Valera. Boca Raton: CRC Press.

Ventosa, A. (1993). Molecular taxonomy of Gram-positive moderately halophilic cocci. *Experientia* 49, 1055-1059.

Ventosa, A. (1994). Taxonomy and phylogeny of moderately halophilic bacteria. In *Bacterial Diversity and Systematics*, pp. 231-242 (Edited by F.G. Priest, B.J. Tindall & A. Ramos-Cormenzana). New York: Plenum Press.

Ventosa, A. & Oren, A. (1996). *Halobacterium salinarum* non-corrig., a name to replace *Halobacterium salinarium* and to include *Halobacterium halobrium* and *Halobacterium antrirubrum. International Journal of Systematic Bacteriology* 46, 347.

Ventosa, A., Garcia, M.T., Kamekura, M., Onishi, H. & Ruiz-Berraquero, F. (1989a). *Bacillus halophilus* sp. nov., a moderately halophilic *Bacillus* sp. *Systematic and Applied Microbiology* 12, 162-166.

Ventosa, A., Gutierrez, M.C., Garcia, M.T. & Ruiz-Berraquero, F. (1989b). Classification of *"Chromobacterium mortismortui"* in a new genus *Chromohalobacter* gen. nov. as *Chromohalobacter marismortui* comb. nov. nom. rev. *International Journal of Systematic Bacteriology* 39, 382-386.

Villar, M., de Ruiz Holgado, A.P., Sanchez, J.J., Trucco, R.E. & Oliver, G. (1985). Isolation and characterisation of *Pediococcus halophilus* from salted anchovies. *Applied and Environmental Microbiology* 49, 664-666.

Vreeland, R.H. (1987). Mechanism of halotolerance in microorganisms. CRC *Critical Reviews in Microbiology* 14, 311-355.

Vreeland, R.H., Litchfield, C.D., Martin, E.L. & Elliot, E. (1980). *Halomonas elongata*, a new genus and species of extremely salt tolerant bacteria. *International Journal of Systematic Bacteriology* 30, 485-495.

Völkl, P., Huber, R., Drobner, E., Rachel, R., Burggraf, S., Trincone, A. & Stetter, K.O. (1993). *Pyrobaculum aerophilum* sp. nov., a novel nitrate-reducing hyperthermophilic archaeum. *Applied and Environmental Microbiology* 59, 2918-2926.

Walsby, A.E., van Rijon, J. & Cohen, Y. (1983). The biology of a new gas-vacuolate cyanobacterium *Dactylococcopsis salina* sp. nov. in Solar Lake. *Proceedings of the Royal Society B* 217, 417-447.

Ward, D.M., Santegoeds, C.M., Nold, S.C., Ramsing, N.B., Ferris, M.J. & Bateson, M.M. (1997). Biodiversity within hot spring microbial mat communities: molecular monitoring of enrichment cultures. *Antonie van Leeuwenhoek* 71, 143-150.

Weller, R., Weller, J.W. & Ward, D.M. (1991). 16S rRNA sequences of uncultivated hot spring cyanobacterial mat inhabitants retrieved as randomly primed cDNA. *Applied and Environmental Microbiology* 57, 1146-1151.

Weller, R., Bateson, M.M., Heimbuch, B.K., Kopczynski, E.D. & Ward, D.M. (1992). Uncultivated cyanobacteria *Chloroflexus*-like inhabitants and spirochaete-like inhabitants of a hot spring microbial mat. *Applied and Environmental Microbiology* 58, 3964-3969.

White, D., Sharp, R.J. & Priest, F.G. (1993). A polyphasic taxonomic study of thermophilic bacilli from a wide geographical area. *Antonie van Leeuwenhoek* 64, 357-386.

Wiegel, J. & Ljungdahl, L. G. (1981). *Thermoanaerobacter ethanolicus* gen. nov., spec. nov., a new, extreme thermophilic, anaerobic bacterium. *Archives of Microbiology* 128, 343-348.

Wiegel, J., Kuk, S.-U. & Kohring, G. W. (1989). *Clostridium thermobutyricum* sp. nov., a moderate thermophile isolated from a cellulolytic culture, that produces butyrate as the major product. *International Journal of Systematic Bacteriology* 39, 199-204.

Wilharm, T., Zhilina T.N. & Hummel, P. (1991). DNA:DNA hybridization of methylotrophic halophilic methanogenic bacteria and transfer of *Methanococcus halophilus* to the genus *Methanohalophilus*. *International Journal of Systematic Bacteriology* 41, 558-562.

Williams, R.A.D. (1989). Biochemical taxonomy of the genus *Thermus*. In *Microbiology of Extreme Environments and its Potential for Biotechnology*, pp. 82-97. Edited by M.S. Da Costa, J.C. Duarte & R.A.D. Williams. London: Elsevier.

Williams, R.A.D., Smith, K.E., Welch, S.G., Micallef, J. & Sharp, R.J. (1995). DNA relatedness of *Thermus* strains, description of *Thermus brockianus* sp. nov., and proposal to reestablish *Thermus thermophilus* (Oshima and Imahori). *International Journal of Systematic Bacteriology* 45, 495-499.

Williams, R.A.D., Smith, K.E., Welch, S.G. & Micallef, J. (1996). *Thermus oshimai* sp. nov., Isolated from hot springs in Portugal, Iceland, and the Azores, and comment on the concept of a limited geographical distribution of *Thermus* species. *International Journal of Systematic Bacteriology* 46, 403-408.

Windberger, E., Huber, R., Trincone, A., Fricke, H. & Stetter, K.O. (1989). *Thermotoga thermarum* sp. nov. and *Thermotoga neapolitana* occuring in African continental solfataric springs. *Archives of Microbiology* 151, 506-512.

Wisotzkey, J.D., Jurtshuk, P.Jr., Fox, G.E., Deinhard, G. & Poralla, K. (1992). Comparative sequences analyses on the 16S rRNA (rDNA) of *Bacillus acidocaldarius*, *Bacillus acidoterrestris*, and *Bacillus cycloheptanicus* and proposal for creation of a new genus, *Alicyclobacillus* gen. nov. *International Journal of Systematic Bacteriology* 42, 263-269.

Woese, C.R. (1987). Bacterial evolution. *Microbiological Reviews* 51, 221-271.

Woese, C.R., Kandler, O. & Wheelis, M.L. (1990). Towards a natural system of organisms: Proposal for the domains Archaea, Bacteria and Eucarya. *Proceedings of the Nataional Academy of Sciences of the United States of America* **87**, 4576-4579.

Wood, A.P. & Kelly, D.P. (1991). Isolation and characterisation of *Thiobacillus halophilus* sp. nov., a sulphur oxidising autotrophic bacterium from a Western Australian hypersaline lake. *Archives of Microbiology* **156**, 277-280.

Yoshida, A., Matsubara, K., Kudo, T. & Horikoshi, K. (1991). *Actinopolyspora mortivallis* sp. nov., a moderately halophilic actinomycete. *International Journal of Systematic Bacteriology* **41**, 15-20.

Yu, I.K. & Kawamura, F. (1987). *Halomethococcus doii* gen. nov. sp. nov., an obligate halophilic methanogenus bacterium from solar salt ponds. *Journal of General and Applied Microbiology* **33**, 303-310.

Zarilla, K.A. & Perry, J.J. (1984). *Thermooleophilum album* gen. nov. and sp. nov., a bacterium obligate for thermophily and n-alkane substrates. *Archives of Microbiology* **137**, 286-290.

Zarilla, K.A. & Perry, J.J. (1986). Deoxyribonucleic acid homology and other comparisons among obligately thermophilic hydrocarbonoclastic bacteria with a proposal for *Thermooleophilum minutum* sp.nov. *International Journal of Systematic Bacteriology* **36**, 13-16.

Zarilla, K.A. & Perry, J.J. (1987). *Bacillus thermoleovorans*, sp.nov., a species of obligately thermophilic hydrocarbon utilizing andospore-forming bacteria. *Systematic and applied Microbiology* **9**, 258-264.

Zeikus, G., Hegge, T.E., Thompson, T.J., Phelps, T.J. & Langworthy, T.A. (1983a). Isolation and description of *Haloanaerobium praevalens* gen. nov. sp. nov., an obligately anaerobic halophile common to Great Salt Lake sediments. *Current Microbiology* **9**, 225-234.

Zeikus, J.G., Dawson, M.A., Thompson, T.E., Ingvorsen, K. & Hatchikian, E.C. (1983b). Microbial ecology of volcanic sulphidogenesis: isolation and characterization of *Thermodesulfotobacterium commune* gen. nov. and sp. nov. *Journal of General Microbiology* **129**, 1159-1169.

Zhilina, T.N. & Zavarzin, G.A. (1987). *Methanohalobium evestigatum* gen. nov. sp. nov. an extremely halophilic methane-producing archaebacterium. *Doklady Akademic Nauk* **293**, 464-468.

Zhilina, T.N. & Zavarzin, G. (1990). Extremely halophilic, methylotrophic, anaerobic bacteria. *FEMS Microbiology Reviews* **87**, 315-322.

Zhilina, T.N., Miroshnikova, L.V., Osipov, G.A. & Zavarzin, G.A. (1991). *Halobacteroides lacunaris* sp. nov., a new saccharolytic anaerobic extremely halophilic bacteria from a hypersaline Chockrack Lake. *Microbiology* **60**, 704-714.

Zhilina, T.N., Zavarzin, G.A., Bulygina, E.S., Kevbrin, V.V., Osipov, G.A. & Chumakov, K.M. (1992). Ecology, physiology and taxonomy studies on a new taxon of *Haloanaerobiaceae* gen. nov. sp. nov. *Haloincola saccharolytica* gen. nov. sp. nov. *Systematic and Applied Microbiology* **15**, 275-284.

Zhilina, T.N., Zavarsin, G.A., Rainey, F., Feubrin, V.V., Kostrikina, N.A. & Lysenko, A.M. (1996a). *Spirochaeta alkalica* sp. nov. *Spirocheata africana* sp. nov. and *Spirochaeta asiatica* sp. nov. alkaliphilic anaerobes from the Continental Soda Lakes in Central Asia and the East African Rift. *International Journal of Systematic Bacteriology* **46**, 305-312.

Zhilina, T.N., Zavarsin, G.A., Detkova, E.N. & Rainey, F. (1996b). *Natrionella acetigena* gen. nov. sp. nov., an extremely haloalkaliphilic homoacetic bacterium. A new member of the *Haloanaerobiaceae. Current Microbiology* **32**, 320-326.

Zhilina, T.N., Zavarsin, G.A., Rainey, F., Pikuta, E.N., Osipov, G.A. & Kostrikina, N.A. (1997). *Desulfonatronoribrio hydrogenovorans* sp. nov. an alkaliphilic sulfate-reducing bacterium. *International Journal of Systematic Bacteriology* **47**, 144-149.

Zillig, W. (1992). The order *Thermococcales*. In *The Prokaryotes*, 2nd Edition. pp. 702-706. Edited by A. Barlows, H.G. Trüper, M. Dworkin, W. Harder, and K.L. Schleifer. Berlin: Springer-Verlag.

Zillig, W., Stetter, K.O., Wunderl, S., Schulz, W., Priess, H. & Scholz, I. (1980). The *Sulfolobus-*

"Caldariella" group: Taxonomy on the basis of the structure of DNA-dependent RNA polymerases. *Archives of Microbiology* **125**, 259-269.

Zillig, W., Stetter, K.O., Schäfer, W., Janecovic, D., Wunderl, S., Holz, I. & Palm, P. (1981). *Thermoproteales*: a novel type of extremely thermoacidophilic anaerobic archaebacteria isolated from Icelandic solfataras. *Zentralblatt für Bakteriologie und Hygiene* C2, 205-227.

Zillig, W., Stetter, K.O., Prangishvilli, C., Schäfer, W., Wunderl, S., Janecovic, D., Holz, I. & Palm, P. (1982). *Desulfurococcaceae*, the second family of the extremely thermophilic, anaerobic, sulfur-respiring *Thermoproteales*. *Zentralblatt für Bakteriologie und Hygiene* C3,304-317.

Zillig, W., Gierl, A., Schreiber, G., Wunderl, S., Janekovic, D., Stetter, K.O. & Klenk, H.P. (1983a). The archaebacterium *Thermofilum pendens* represents a novel genus of the thermophilic, anaerobic sulfur respiring *Thermoproteales*. *Systematic and Applied Microbiology* 4, 79-87.

Zillig, W., Holz, I., Janekovic, D., Schäfer, W. & Reiter, W. D. (1983b). The archaebacterium *Thermococcus celer* represents a novel genus within the thermophilic branch of the archaebacteria. *Systematic and Applied Microbiology* 4, 88-94.

Zillig, W., Yeats, S., Holz, I., Böck, A., Rettenberger, M., Gropp, F. & Simon, G. (1986). *Desulfurolobus ambivalens*, gen. nov., sp. nov., an autotrophic archaebacterium facultatively oxidizing or reducing sulfur. *Systematic and Applied Microbiology* 8, 197-203.

Zillig, W., Holz, I., Klenk, H. P., Trent, J., Wunderl, S., Janekovic, D., Insel, E. & Haas, B. (1987). *Pyrococcus woesei*, sp. nov., an ultra-thermophilic marine archaebacterium, representing a novel order, *Thermococcales*. *Systematic and Applied Microbiology* 9, 62-70.

Zillig, W., Holz, I. & Wunderl, S. (1991). *Hyperthermus butylicus* gen. nov., sp. nov., a hyperthermophilic, anaerobic, peptide-fermenting, facultatively H_2S-generating archaebacterium. *International Journal of Systematic Bacteriology* 41, 169-170.

10   ACIDOPHILES IN BIOMINING

BRETT M. GOEBEL [1], PAUL R. NORRIS [2] and NICOLAS P. BURTON [2]

[1] Department of Plant and Microbial Biology, University of California,
Berkeley, CA 94720-3102 U.S.A.
[2] Department of Biological Sciences, University of Warwick, Coventry
CV4 7AL U.K.

1   Introduction

Acidophilic microorganisms that oxidize mineral sulfides are of obvious biogeochemical significance in their natural habitats and are increasingly being utilized commercially in mineral sulfide processing. However, few studies have focused on their taxonomy and phylogeny, hence there are unresolved questions in these areas. For example, the organism that appears to be the major catalyst in the dissolution of gold-bearing minerals in all but one of the commercial bioreactors used for this purpose, "*Leptospirillum ferrooxidans*", has not been validly named and speculation concerning its phylogeny is only recent. Those unaware of current opinions in biohydrometallurgy are more likely to be familiar with the mineral sulfide-oxidizing *Thiobacillus ferrooxidans*, an extremely well-studied organism in comparison to "*L. ferrooxidans*". For *T. ferrooxidans*, phylogenetic analysis is well established and actually indicates a case for its renaming outside of the *Thiobacillus* genus.

Commercial reactors that involved neither "*L. ferrooxidans*" nor *T. ferrooxidans*, because of an operating temperature beyond the range of the activity of these mesophiles, contained a mixture of moderate thermophiles; but the iron-oxidizing species in these reactors were not defined. Moderately thermophilic, iron-oxidizing acidophiles that have been found in many acidic, natural environments have been placed in two genera, *Acidimicrobium* and *Sulfobacillus*, and novel species of *Sulfobacillus*-like bacteria have already been cultured and others predicted to exist from analyses of environmental 16S rDNA sequences.

A likely byproduct of the commercial interest in proprietary culture definition and in the discovery of novel species with useful characteristics for biomining will be an increase in the representation of acidophiles on the microbial phylogenetic tree and a better appreciation of their diversity. In return, the bioleaching industry should gain from improved systematics for the acidophiles. Iron- and sulfur-oxidizing organisms are already recognized to span the bacterial domain (Lane et al., 1992) and are well represented among the *Sulfolobus*-like thermoacidophiles of the Crenarchaeota (Fuchs et al., 1996). Only a few of the thermoacidophilic archaea have so far been investigated with a view to high-temperature bioleaching

F. G. Priest and M. Goodfellow (eds.), Applied Microbial Systematics, 293-314
© 2000 Kluwer Academic Publishers. Printed in the Netherlands.

in reactors. The recent isolation of previously undescribed, mineral sulfide-oxidizing *Thermoplasma*-like members of the Euryarchaeota is supportive of the expectation that other metal-mobilizing organisms will be found nesting in evolutionary branches where they have been hitherto unsuspected. The rRNA sequence data base on which the major evolutionary tree is based also provides the information for design of specific molecular probes. These should be particularly useful for industry for identification of strains and in molecular ecological studies in relation to the mixed cultures that are active in mineral processing bioreactors and in bioleaching ore heaps, particularly as some of the key organisms in bioleaching are not readily isolated on solid media. A consequent improvement in the understanding of the microbiology of bioleaching should facilitate optimization of the industrial processes.

All of these themes are expanded in this review with emphasis on the phylogeny of key organisms as determined by 16S rRNA sequence analysis, following a brief outline of bioleaching of mineral sulfides.

2   Mineral Sulfide Oxidation by Acidophiles

As well as the desire to improve metal extraction rates in industrial bioleaching processes, the study of mineral sulfide dissolution by microorganisms also concerns the need to understand and reduce pollution resulting from acid mine drainage. Microbially-catalysed dissolution of metal sulfides, particularly pyrite (FeS_2), is a cause of acid and metal pollution. Exposure of metal sulfides to the atmosphere through mining leads to their oxidative dissolution and the subsequent acidification of aquifers and river systems (Evangelou & Zhang, 1995; Nordstrom & Southam, 1997). The natural solubilization of copper from minerals has been observed and exploited for centuries in small scale operations in many countries (Hiskey, 1994). Recent developments have seen the technology develop through leaching of low grade ores and waste in dumps to heap leaching processes which involve stacking copper and gold ores over aeration pipes to improve leaching rates. The first commercial bioreactor for oxidation of gold-bearing concentrates of arsenopyrite began operation in 1986 and several plants are currently in operation in Africa, Australia and South America, the largest treating about 1,000 tonnes of mineral per day (see Brierley, 1997), for a review of mining biotechnology; one of a collection of reviews (Rawlings, 1997) covering most aspects of biomining.

Discussion of the mechanism of microbial dissolution of mineral sulfides (e.g. Kelly *et al.*, 1979) has often revolved around the "direct" and "indirect" processes proposed by Silverman & Ehrlich (1964). A direct mechanism was suggested to involve intimate contact of organisms with the mineral surface and enzymatic oxidation of the sulfide moiety leading to capture of electrons by bacterial respiratory chains. The indirect mechanism referred to the capacity of *T. ferrooxidans* and other iron-oxidizing bacteria to catalyse the oxidation of soluble ferrous iron to ferric iron, which would act as the primary, chemical oxidant of sulfide minerals. Recent considerations of the process of mineral sulfide oxidation have led to a proposal that attached cells create a micro-environment in which indirect leaching by chemical oxidation with ferric iron is stimulated by selective accumulation of the iron in uronic-acid-containing capsules (Gehrke *et al.*, 1995; Sand *et al.*, 1995; Schippers *et al.*, 1996). The surface chemistry of bacterial attachment to mineral sulfides and the electrochemistry of mineral sulfide dissolution has been discussed in detail elsewhere (Crundwell, 1997). The few words on the subject here merely serve to emphasize that the taxonomy and phylogeny of interest concerns both iron- and sulfide/sulfur-oxidizing organisms, the oxidation processes being found together in some individual acidophiles, but also separately in virtually unrelated organisms which can collaborate in the mineral degradation. The most studied iron- and sulfur-oxidizing acidophiles are lithoautotrophs but heterotrophic acidophiles are also relevant to bioleaching (Johnson & Roberto, 1997) since some can oxidize iron and others can remove organic compounds which inhibit the autotrophs.

3 Mesophilic and Thermotolerant Acidophiles

3.1 THE "THIOBACILLI"

The acidophilic, Gram-negative thiobacilli are distributed throughout the α-, β-, and γ-subdivisions of the class *Proteobacteria* (Lane *et al.*, 1992). All are capable of autotrophic growth using reduced sulfur compounds as sole energy sources. Many species will acidify their environment to below pH 4, e.g. *Thiobacillus delicatus*, *Thiobacillus neopolitanus*, *Thiobacillus perometabolis* and *Thiobacillus thioparus*, without being strictly acidophilic (Kuenen *et al.*, 1992). Two true acidophiles associated with bioleaching, *Thiobacillus ferrooxidans* and *Thiobacillus thiooxidans*, have been extensively studied.

Thiobacillus ferrooxidans (Temple & Colmer, 1951) - *Ferrobacillus ferrooxidans* (Leathen *et al.*, 1956) and *Ferrobacillus sulfooxidans* (Kinsel, 1960) were re-classified as *T. ferrooxidans* by Tuovinen & Kelly (1972) - is by far the most studied species of the acidophilic, metal-mobilizing bacteria. An abundance of publications and reviews have highlighted *T. ferrooxidans* as the most important organism associated with bioleaching environments. This perception was universal before the recognition of "*Leptospirillum ferrooxidans*", but was not based on quantitative ecological studies and probably reflected the ease with which *T. ferrooxidans* could be recovered from environmental and industrial samples (although this does not preclude its actual predominance in certain acidic habitats).

There appears to be some variety among the acidophilic, obligately chemolithoautotrophic isolates that have been described as *T. ferrooxidans* principally because they are capable of oxidative growth on reduced sulfur compounds and ferrous iron. The heterogeneity of this taxon was demonstrated by Harrison (1982), who separated 23 isolates identified as *T. ferrooxidans* into seven DNA-DNA hybridization groups, with members of each group sharing less than 70% DNA-DNA similarity.

Group 1 strains were shown to belong to the genus "*Leptospirillum*" (Harrison, 1984). Group 7 contained only the single, non-sulfur oxidizing strain m-1 which is only distantly related to other *T. ferrooxidans* strains according to comparative 16S rRNA sequence analysis (see Fig. 1). However, this leaves five other DNA-DNA similarity groups indicating that the use of only a few simple diagnostic characteristics to identify acidophilic, mesophilic, iron- and sulfur-oxidizing, rod-shaped bacteria might have led to the incomplete or inaccurate speciation of these isolates. Because of the low DNA-DNA similarity values between *T. ferrooxidans* groups (Harrison, 1982), it is predicted that this taxon could be separated into a number of species, or even genera. However, it is presently not possible to foresee how, or even if, subdivision into a number of different species is achievable, as supporting phenotypic data are currently not available (Ursing *et al.*, 1995).

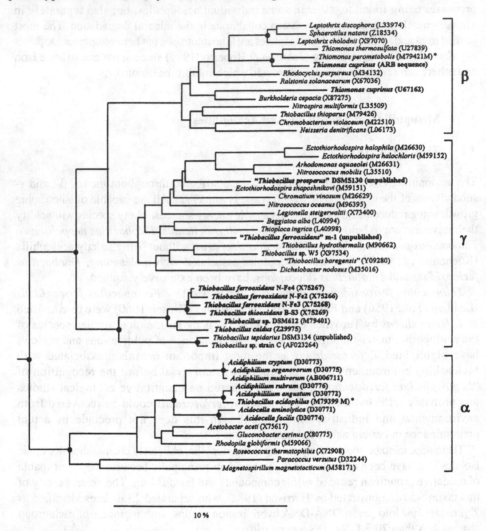

Figure 1. Evolutionary distance dendrogram showing the relative positions of acidophiles (shown in bold type) within the class *Proteobacteria*. GenBank accession numbers are shown in brackets. Species marked with an asterisk were inserted into the tree using a parsimony insertion tool (Strunk *et al.*, unpublished) as these sequence were less than 1 kb in length. A solid circle at selected branches indicates clades which were supported (greater than 70% bootstrap support) in a large number of analyses using different taxa and treeing methods, including distance, parsimony and maximum likelihood algorithms. The scale bar represents 10% sequence divergence.

Thiobacillus thiooxidans was described by Waksman & Joffe (1922). *Thiobacillus concretivorans*, "*Thiobacillus crenatus*", "*Thiobacillus kabobis*", "*Thiobacillus lobatus*", "*Thiobacillus thermitanus*" and "*Thiobacillus umbonatus*" are all considered synonyms (Kuenen *et al.*, 1992) for this sulfur-oxidizing acidophile, which is ubiquitous in acidic, mineral-rich environments. The principle phenotypic difference between *T. ferrooxidans* and *T. thiooxidans* is the inability of the latter to utilize ferrous iron as an electron donor. There are reports of leaching of metals by *T. thiooxidans* (Lizama & Suzuki, 1987), but it is generally accepted that it does not oxidize key mineral sulfides such as pyrite and chalcopyrite in pure culture.

Thiobacillus albertis is an acidophilic, strictly autotrophic organism originally isolated from an acidic soil sample from Alberta, Canada (Bryant *et al.*, 1983). It has many similarities to *T. thiooxidans* but possesses a glycocalyx and a significantly higher mol % G+C content (61.5%) than *T. thiooxidans* (50-52%). *Thiobacillus caldus* (Hallberg & Lindström, 1994) also closely resembles *T. thiooxidans* in physiology, apart from an optimum temperature for growth that qualifies it as thermotolerant or moderately thermophilic. Thermotolerant, sulfur-oxidizing acidophiles have long been known to inhabit hot acid soils and springs (e.g. Fliermans & Brock, 1972) and appear likely to be more significant than *T. thiooxidans* in industrial bioreactor mineral-processing where the temperature is in their favour (see section 6).

A number of facultatively organotrophic, metal-mobilizing strains have been isolated from solfatara fields in Iceland and a uranium mine in the Federal Republic of Germany. These isolates were named *Thiobacillus cuprinus* to indicate their particular and apparently specific preference for the leaching of copper from chalcopyrite (Huber & Stetter, 1990). Moreira & Amils (1997) have proposed, based primarily on 16S rRNA sequence analysis of the type strain Hö5, that this species be transferred to the new genus *Thiomonas* and renamed *Thiomonas cuprinus*. The transfer of *T. intermedius*, *T. perometabolis* and *Thiobacillus thermosulfatus* to the genus *Thiomonas* was also recommended by these authors.

Thiobacillus prosperus (Huber & Stetter, 1989), isolated from a geothermally heated sea bed in Italy, is a halotolerant (up to 6 % w/v NaCl), obligately chemolithoautotrophic, iron- and sulfur-oxidizing species which shares negligible DNA-DNA similarity with other thiobacilli. Almost identical physiological profiles were found for three strains studied in detail. However, strain VC15, which was the only strain which would grow on synthetic FeS, shared less than 36% DNA-DNA similarity with the others and could represent a novel species.

Thiobacillus plumbophilus, isolated by Drobner *et al.* (1992), displays the unusual characteristic of the energetic oxidation of galena (PbS) but of no other metal sulfides so far tested. The authors indicated that the likely reason for this is the low tolerance of this species to soluble metal ions. Anglesite ($PbSO_4$), the oxidative by-product of galena (PbS) oxidation by *T. plumbophilus*, is only weakly soluble in sulfate solutions.

Some mesophilic thiobacilli that are capable of heterotrophic growth under acidic conditions have also been described. Guay & Silver (1975) isolated *T. acidophilus* from a culture of *T. ferrooxidans*. Neither the facultatively autotrophic *T. acidophilus* nor an apparently similar organism, "*Thiobacillus organoparus*" (Markosyan, 1973) appeared on the Approved Lists of Bacterial Names (Skerman *et al.*, 1980), but the species name *T. acidophilus* was revived by Harrison (1983). "*Thiobacillus organoparus*" is now considered a synonym of *T. acidophilus* (Kuenen *et al.*, 1992). *Thiobacillus acidophilus* grows on a range of reduced sulfur compounds (Norris *et al.*, 1986; Mason & Kelly, 1988; Meulenberg *et al.*, 1992) and certain organic acids (e.g. formate and pyruvate) can serve as sole carbon and energy substrates in substrate-limited chemostats (Pronk *et al.*, 1990). The presence of *T. acidophilus* in laboratory cultures of *T. ferrooxidans* has been well established (Guay & Silver, 1975; Johnson & Kelso, 1983; Harrison *et al.*, 1984) but it is has rarely been recovered directly from acidic mineral environments. However, Berthelot *et al.* (1997) reported that eight of the thirty-six heterotrophs they isolated from uranium mine environments were capable of growth both heterotrophically as well as on sulfur and concluded that these isolates were strains of *T. acidophilus*. Further studies, particularly the sequencing of 16S rRNA genes, are required to confirm this observation.

3.2 "LEPTOSPIRILLUM FERROOXIDANS"

The name *"Leptospirillum ferrooxidans"* was proposed by Markosyan (1972) for an acidophilic, spiral-shaped, iron-oxidizing autotroph, but the name did not appear on the Approved Lists of Bacterial Names (Skerman *et al.*, 1980), nor has it subsequently been validated by publication in the *International Journal of Systematic Bacteriology*. There is, however, significant physiological (Balashova *et al.*, 1974; Hallman *et al.*, 1992), genetic (Harrison, 1982, 1986; Harrison & Norris, 1985; Hallman *et al.*, 1992), chemotaxonomic (Blake *et al.*, 1993; Goebel & Stackebrandt, 1995) and phylogenetic (Lane *et al.*, 1992; Goebel & Stackebrandt, 1995) information to support the valid description of this taxon. DNA-DNA similarity data (Harrison, 1982; Hallman *et al.*, 1992) and subsequent phylogenetic inference based on 16S rRNA sequence analyses (Lane *et al.*, 1992; Goebel & Stackebrandt, 1995) suggest that strain L15 (also designated Z1 and Z2), the original isolate of *"Leptospirillum ferrooxidans"*, represents an atypical isolate compared with other leptospirilla subsequently isolated and characterized. It is therefore possible that, like *T. ferrooxidans*, isolates identified as *"L. ferrooxidans"* represent a number of different species or genera. The physiological capabilities of these bacteria have not been explored to the same extent as for *T. ferrooxidans*. Growth on reduced sulfur compounds (except mineral sulfides such as pyrite) has not been obtained.

"Leptospirillum ferrooxidans" isolates have been reported to oxidize elemental sulfur (forming ferrous and bisulfite ions) and possess a hydrogen sulfide:ferric ion oxidoreductase which uses ferric ions as an electron acceptor (Sugio *et al.*, 1992). Analysis of sulfur compounds present during pyrite degradation by *"L. ferrooxidans"*, however, has cast doubt on the significance of any such activity in the oxidation of the sulfide moiety (Schippers *et al.*, 1996). Further studies by Sugio and co-workers (1994) suggested that a low level activity of sulfite:ferric ion oxidoreductase in *"L. ferrooxidans"* (the enzyme responsible for the oxidation of the toxic bisulfite ion formed during sulfur oxidation) could prevent its growth on elemental sulfur due to the rapid accumulation of the bisulfite ion to toxic levels in the growth medium. *"L. ferrooxidans"* was originally thought to be incapable of growth in pure culture on mineral sulfides (Balashova *et al.*, 1974) but growth in pure culture on pyrite was subsequently demonstrated (Norris & Kelly, 1982; Norris, 1983). *"Leptospirillum ferrooxidans"* is frequently the dominant iron-oxidizing mesophile in

pyrite-oxidizing mixed cultures (Helle & Onken, 1988; Norris *et al.*, 1988), probably because of its higher affinity for ferrous iron and greater tolerances of ferric iron and acidity in comparison with *T. ferrooxidans*. This dominance extends to the commercial bioreactors used for gold extraction (see section 6). Leptospirilla are also at least as abundant as *T. ferrooxidans* in some heap leaching habitats, especially at temperatures above 20°C (Sand *et al.*, 1992).

Isolation of a thermotolerant "Leptospirillum" strain from an acidic, hydrothermal spring on the island of Kunashir, Russia, was reported by Golovacheva *et al.* (1993). Named "*Leptospirillum thermoferrooxidans*", this organism appeared to be morphologically and physiologically similar to the mesophilic "*L. ferrooxidans*", except for its higher temperature optimum of 45-50°C and a maximum temperature for growth approaching 60°C, but adequate taxonomic data are lacking.

3.3   ACIDOPHILIC HETEROTROPHS

Barros *et al.* (1984) described a *T. ferrooxidans* strain (FD1) which was capable of heterotrophic growth on glucose. However, this strain shared only 39% DNA-DNA similarity with the type strain (*T. ferrooxidans* ATCC 19377), which is significantly less than the 70% DNA similarity figure recommended by Wayne *et al.* (1987) to represent a phylogenetically-coherent species.

Several early reports of heterotrophic growth of other *T. ferrooxidans* strains were misleading because of the presence of heterotrophic contaminants in the cultures (see Harrison, 1984, for documentation). The persistent appearance of heterotrophic acidophiles and facultatively autotrophic bacteria such as *T. acidophilus* in serial subcultures of *T. ferrooxidans* is reportedly the result of organic material added as contamination in the preparation of growth media (Harrison, 1984) and/or by organic acids and other cell debris released from *T. ferrooxidans* cells during growth (Tuttle *et al.*, 1977).

Simple organic acids (e.g. pyruvate) can inhibit the growth of *T. ferrooxidans* (Tuttle & Dugan, 1976) so their removal by acidophilic heterotrophs can benefit the autotroph. The efficient scavenging of organic material from their surroundings by the heterotrophs has been employed successfully in the development of solid media for isolating acidophilic autotrophs (Butler & Kempton, 1987; Johnson & McGinness, 1991). Incorporation of heterotrophs in an underlay of agarose-gelled medium promotes growth of the autotrophs. In natural habitats and bioleaching heaps similar interactions between heterotrophic and autotrophic acidophiles could occur.

The most diverse group of acidophilic heterotrophs described to date belongs to the genus *Acidiphilium*. This genus currently consists of four validly described species: *A. angustum* and *A. rubrum* (Wichlacz *et al.*, 1986), *A. cryptum* (Harrison, 1982) and *A. organovorum* (Lobos *et al.*, 1986). *Acidiphilium facilis* (Wichlacz *et al.*, 1986) and *A. aminolytica* (Kishimoto *et al.*, 1993) were recently transferred to the genus *Acidocella* gen. nov., based principally on comparative 16S rRNA sequence analysis (Kishimoto *et al.*, 1995b). Two other species of *Acidiphilium* have also been proposed: *A. multivorum* (Wakao *et al.*, 1994) and *A. symbioticum* (Bhattacharyya *et al.*, 1991). While the latter study utilized all validly described species known at the time of publication, and included DNA-DNA similarity data (although no 16S rRNA analysis), the former included only

A. *cryptum* as a comparative strain, leaving a relationship to A. *organovorum*, for example, requiring further resolution.

A number of forms of the photo-active compound bacteriochlorophyll *a* and carotenoid pigments are reportedly produced by A. *angustum*, A. *cryptum*, A. *rubrum* (Wakao *et al.*, 1993) and A. *organovorum* (Kishimoto *et al.*, 1995a). This observation suggests that a possible photosynthetic route for energy production may be present in these species. Detailed studies with A. *rubrum* indicated $^{14}CO_2$ was indeed incorporated into cells, especially when cultures were incubated aerobically in the light (Kishimoto *et al.*, 1995a). However, growth in the absence of an organic substrate was not observed. Kishimoto *et al.* (1995a) concluded that the maintenance of a pH gradient across the cell membrane (necessary to maintain a neutral cytoplasm) could be achieved by the use of this light energy, especially under extreme oligotrophic conditions.

Kishimoto and co-workers (1991) proposed the name *Acidobacterium capsulatum* for a group of atypical, obligately acidophilic, mesophilic and heterotrophic isolates which produced MK-8 as the major respiratory quinone. The isolates, all originating from acidic mineral environments (Kishimoto Tano, 1987), shared very low DNA:DNA homology (less than 18%) with members of the genera *Acidomonas*, *Acidiphilium* and *Deinobacter* (Kishimoto *et al.*, 1991).

Recently, a number of obligately acidophilic, filament-forming, heterotrophic strains capable of the non-energetic oxidation of ferrous ion have been isolated from acidic streams associated with disused mines in Wales and the United States (Johnson *et al.*, 1992, 1995). These strains (represented by strain CCH7) differ from other sheath-producing iron oxidizers (i.e. the *Sphaerotilus-Leptothrix* group) by their obligate acidophily. 16S rRNA sequence data are currently needed for this group of organisms. Non-filamentous, heterotrophic, acidophilic mesophiles which appear to utilize energy obtained from ferrous ion oxidation and solubilize mineral sulfides have also been described (Johnson & Roberto, 1997). These strains (proposed species "*Ferromicrobium acidophilus*") appear to be widely distributed in acidic metal-leaching habitats and form a sister clade within the Actinobacteria division (Fig. 2) with the moderately thermophilic, iron-oxidizing species *Acidimicrobium ferrooxidans*.

3.4   OVERVIEW OF THE PHYLOGENY OF MESOPHILIC ACIDOPHILES

Lane and co-workers (1985, 1992) showed that the sulfur-oxidizing phenotype is a poly-phyletic trait, with species of the genus *Thiobacillus* occurring in a number of evolutionary lines of descent within the class *Proteobacteria*. *Thiobacillus caldus*, *T. ferrooxidans*, *T. thiooxidans* and the neutrophilic *Thiobacillus tepidarius* form a monophyletic cluster that branches deep within the γ-*Proteobacteria* (Fig. 1). Phylogenetic trees, which are available from the Ribosomal Database Project (Maidak *et al.*, 1997) and the ARB database (Strunk *et al.*, unpublished, website URL-http:/www.mikro.biologie.tu-muenchen.de) and which were generated from very large 16S rRNA data sets, indicate that this group of acidophilic thiobacilli branch deep within the γ-*Proteobacteria* and outside of the β-pro-teobacterial radiation. Phylogenetic analysis presented by McDonald and co-workers (1997) also supported this conclusion, although a relatively small dataset of 16S rRNA sequences was used and outgroup sequences were not included. The specific position of

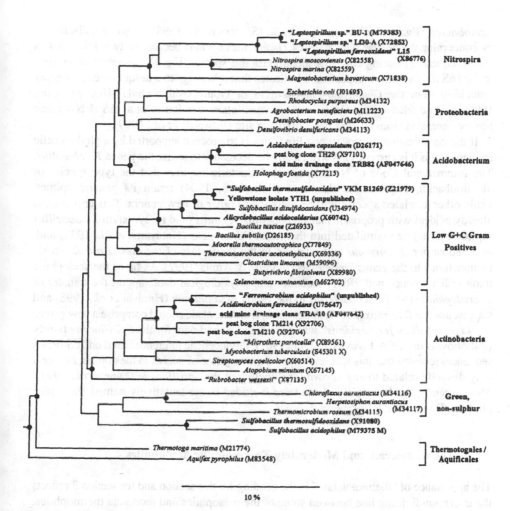

Figure 2. Evolutionary distance dendrogram showing the relative positions of acidophiles (shown in bold type) within the domain *Bacteria*. GenBank accession numbers are shown in brackets. A filled circle at a branch indicates clades which were supported (greater than 70% bootstrap support) in a large number of analyses using different taxa and treeing methods, including distance, parsimony and maximum likelihood algorithms. The scale bar represents 10% sequence divergence.

this clade can be considered a moot point, however, as the β-proteobacterial lineage emerges from within the γ-*Proteobacteria* and it is currently not clear at which point the β-*Proteobacteria* remain monophyletic to the exclusion of the γ-*Proteobacteria*.

The 16S rRNA sequence of the type strain of *Thiobacillus prosperus* (as determined by P. Durand, unpublished data) places this acidophile within the γ-*Proteobacteria* whereas *Thiobacillus acidophilus* groups in the α-1 subclass of the *Proteobacteria* (Fig. 1) along with members of the genera *Acidiphilium* (Lane *et al.*, 1992) and *Acidocella* (Kishimoto *et al.*, 1995b), which form a distinct acidophilic clade with members of the genera

Acetobacter, Gluconobacter and Rhodopila (Sievers et al., 1994). There is a discrepancy concerning the precise phylogeny of Thiomonas cuprinus because of two different 16S rRNA sequences (sharing only 90% identity) for this species (Fig. 1). Comparative analysis of the 16S rDNA from A. capsulatum indicates that it belongs to a unique lineage, deeply branching from the Chlamydia-Planctomyces or Gram-positive phyla (Hiraishi et al., 1995). A close relative of Acidobacterium capsulatum was found in a 16S rRNA clone library generated from an acid mine drainage site (Edwards et al., 1999).

If the classification of prokaryotic life forms is to proceed supported by a phylogenetic framework (as it is currently), it will clearly be necessary to revise the genus Thiobacillus. The International Code of Nomenclature of Bacteria requires that the type species of the thiobacilli, Thiobacillus thioparus (Beijerinck, 1904) retain its generic epithet, while other unrelated species will be assigned to other or new genera. This approach is already evident with proposals that Thiobacillus versutus (and its synonym Thiobacillus rapidicrescens) be assimilated into the genus Paracoccus (Katayama et al., 1995) and, as noted earlier, T. curpinus and relatives in the β-subclass of the class Proteobacteria be transferred to the genus Thiomonas (Moreira & Amils, 1997). As far as the acidophilic thiobacilli are concerned, rRNA sequence and physiological data support the transfer of T. acidophilus (α-1 Proteobacteria) to the genus Acidiphilium (Hiraishi et al., 1998) and suggest that the T. ferrooxidans/T. thiooxidans/T. caldus clade should comprise a new genus.

"Leptospirillum ferrooxidans" was considered a candidate for the Spirillaceae family (Balashova et al., 1974; Pivovarova et al., 1981). Phylogenetic inference based on 16S rRNA sequences revealed that this species formed a primary line of decent in the domain Bacteria only distantly related to any known taxa, including the Spirillaceae (Lane et al., 1992). More recently, the leptospirilla have been included in the tentatively-named Nitrospira phylum (Ehrich et al., 1995; Fig. 2).

4   Thermotolerant and Moderately Thermophilic Acidophiles

The appearance of "thermotolerant" in the heading for this section and for section 3 reflects the uncertain dividing line between some of the mesophiles and moderate thermophiles. Thiobacillus caldus was introduced earlier although its upper temperature limit for growth is about 55°C: this bacterium and some of the iron-oxidizing moderate thermophiles may have optimum temperatures for growth between 45 and 50°C but can be almost as active as T. thiooxidans and T. ferrooxidans respectively at 30-35°C. Two distinct types of ferrous iron- and mineral sulfide-oxidizing moderately thermophilic, acidophilic bacteria have been described, Acidimicrobium ferrooxidans and those of the Sulfobacillus genus. Moderately thermophilic, acidophilic archaea have not previously been reported to oxidize mineral sulfides, but Thermoplasma-like strains with this capacity have recently been isolated from geothermal environments.

4.1   IRON-OXIDIZING SPECIES

Rod-shaped, iron-oxidizing moderate thermophiles were studied as "Thiobacillus-like" organisms (see Brierley & Brierley, 1986) before the first species was named Sulfobacillus *thermosulfidooxidans* (Golovacheva & Karavaiko, 1979). These are nutritionally versatile, Gram-positve, endospore-forming acidophiles. Growth can be autotrophic on ferrous iron, mineral sulfides and sulfur, heterotrophic on yeast extract or mixotrophic with utilization of glucose and carbon dioxide during oxidation of ferrous iron. A range of *Sulfobacillus*-like isolates from many countries were readily separated into only two species on the basis of DNA:DNA relatedness (determined by hybridization). These species, *Sulfobacillus acidophilus* and *Sulfobacillus thermosulfidooxidans*, also differed in their mol % G+C content (55-57 and 48-50 respectively), their morphology and their capacities for oxidation of ferrous iron and sulfur (Norris *et al.*, 1996). During autotrophic growth, *S. thermosulfidooxidans* strains oxidized iron more readily and extensively (before end-product inhibition by ferric iron) than *S. acidophilus* strains. Although both species oxidized elemental sulfur during growth in the presence of yeast extract, only *S. acidophilus* strains grew autotrophically on sulfur. However, these observations were based on a comparative study of the first strains to be available in pure culture which implies that there was a selection for those most readily forming colonies on solid media, a variable trait among iron-oxidizing acidophiles. These strains were also isolated using a restricted range of enrichment culture conditions.

It is clear that the diversity of *Sulfobacillus*-like isolates is greater than that represented by the two named species. For example, enrichment culture of samples from an acidic, geothermal spring on the island of Montserrat with ferrous iron and yeast extract at 60°C, which is above the optimum growth temperature of the named species, has yielded strain M6b (with an approximate 60 mol % G+C content) that grows at up to 65°C. This raises the maximum temperature known to be tolerated by such organisms by several degrees (Norris, unpublished data). Distinctive phenotypic features have yet to be confirmed for other strains that can be placed within the *Sulfobacillus* cluster (based on rRNA sequence analysis), such as strain YTF5 from Yellowstone National Park hot springs (see Fig. 3). Furthermore, examination of *Sulfobacillus*-like isolates from coal spoils in England and geothermal sites in New Zealand has revealed strains of *S. acidophilus* and *S. thermosulfidooxidans* which do not oxidize iron and sulfur with the same relative facility as the type strains, thus already obscuring one of the species-distinguishing phenotypic features noted above.

Acidimicrobium ferrooxidans was first described almost 20 years ago (Brierley, 1978) but named relatively recently (Clark & Norris, 1996). Although known only from one site for many years, a copper leach dump in New Mexico, USA, the widespread distribution of the organism is now becoming apparent with further isolations from geothermal areas in Iceland and New Zealand. The organism possesses a CO_2-uptake system, in contrast to the *Sulfobacillus* species so far studied, which allows it to grow well autotrophically under air in contrast to the more extensively studied *Sulfobacillus* species (Clark & Norris, 1996), and potentially raises its biotechnological potential for mineral processing.

Enrichment cultures for moderately thermophilic acidophiles that might tolerate higher salt concentrations than *Acidimicrobium* and *Sulfobacillus* species have recently yielded *Thermoplasma*-like organisms with the capacity for iron- and mineral sulfide-oxidation (Norris & Burton, unpublished data). 16S rRNA gene sequence analysis has confirmed their close affiliation with the acidophilic Euryarchaeota, *Picrophilus* and *Thermoplasma*. The sources were geothermal sites, but if similar organisms are found in bioleaching heaps or coal spoils (the original source of the non-iron-oxidizing *Thermoplasma acidophilum*;

Darland *et al.*, 1970), a direct role for them in environmental metal-mobilization may be inferred.

4.2   HETEROTROPHIC, MODERATE THERMOPHILES

Heterotrophic bacilli of the *Alicyclobacillus* genus (Wisotzkey *et al.*, 1992) could fill the *Acidiphilium* niche in habitats too warm for this mesophile. The reported pH ranges for growth of the named *Alicyclobacillus* species (*A. acidocaldarius*, *A. acidoterrestris* and *A. cycloheptanicus*; Wisotzkey *et al.*, 1992) suggest they might not be active at the high levels of acidity (less than pH 2) in some bioleaching and geothermally-heated sites. However, an un-named *Alicyclobacillus*-like strain that has been found in natural geothermal sites such as on the island of Montserrat (strain M8b on Fig. 3) and in self-heating coal spoils in England grows well at pH 1.5 and optimally at about pH 2 (Norris, unpublished data). Confirmation of this organism as a member of *Alicyclobacillus* would rest on fatty acid analysis and the demonstration of its possession of the unusual ω-alicyclic fatty acids.

4.3   OVERVIEW OF MODERATE THERMOPHILE PHYLOGENY

Acidimicrobium ferrooxidans is the only named species of the *Acidmicrobiales*, a new order of the recently proposed class *Actinobacteria* (Stackebrandt *et al.*, 1997). However, the iron-oxidizing, heterotrophic, mesophilic acidophile "*Ferromicrobium acidophilus*" appears to be a close relative (Fig. 2) and further study of acidic environments is certain to reveal previously undescribed members of this order. Analysis of 16S rDNA genes isolated from acidic, coal spoil samples in England has revealed sequences most closely related to *A. ferrooxidans* and "*F. acidophilus*" but with only 90% similarity to either (Burton & Norris, unpublished data).

The evolutionary affiliations of *Sulfobacillus* species are uncertain. The sharing of acidophily and endospore formation with *Alicyclobacillus* species appears to lend phenotypic support for the affiliation with these heterotrophs that is indicated by comparative analysis of some sets of 16S rRNA sequences (Fig. 3). However, a distinct lineage has been proposed for *S. thermosulfidooxidans* on the basis of its 5S rRNA sequence (Karavajko *et al.*, 1990). Furthermore, the inclusion of different 16S rRNA sequences in tree construction and consideration of signature nucleotides provisionally places *Sulfobacillus* closest to green non-sulfur bacteria rather than within the low GC Gram-positive bacterial group (Fig. 2; Goebel & Stackebrandt, 1994a).

The un-named *Sulfobacillus*-like strains noted earlier, strains YTF5 and M6b, appear to group closer to *S. acidophilus* than *S. thermosulfidooxidans* (Fig. 3). The grouping of other sequences from "*Sulfobacillus*" species among those from *Alicyclobacillus* species (Fig. 3) suggests that a heterotrophic contaminant of *S. thermosulfidooxidans* was sequenced in one case (Z21979; previously noted by Durand, 1996 and Norris *et al.*, 1996) and that a novel species has been placed in the wrong genus in the case of *S. disulfidooxidans* (Dufresne *et al.*, 1996). The placement of sequences from un-named acidophilic heterotrophs (strain YTH15 from Yellowstone National Park, USA, and strain M8c from Montserrat; Fig. 3) indicates further diversity within the *Alicyclobacillus* cluster.

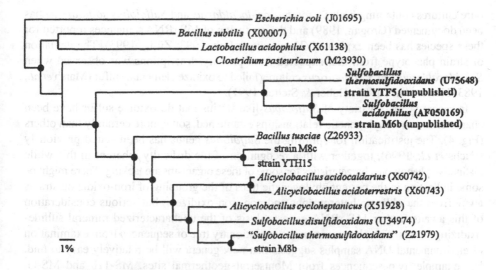

Figure 3. Evolutionary distance dendrogram illustrating relationships among moderately thermophilic, acidophilic bacteria. Strains M6b, M8b and M8c are isolates from Montserrat geothermal sources (Norris and Burton, unpublished data) and YT strains are from Yellowstone National Park, USA (F. Roberto and D. B. Johnson, unpublished data). Iron-oxidizing strains are indicated in bold type. Analysis of 1083 nucleotide positions used the Jukes-Cantor algorithm. Bootstrap values of greater than 70% (for distance and parsimony methods) are indicated by filled circles at selected branches. The scale bar represents 1% sequence divergence.

5 *Sulfolobus*-like Archaea: a Bioleaching Application in Development and a Genus in Need of Re-Organization

A comprehensive account of the mineral sulfide-oxidizing, thermoacidophilic archaea is not presented here. High-temperature mineral processing with thermoacidophilic archaea has not reached commercialization, although several pilot plants have been operated with the aim of developing a bioreactor process for extraction of copper from chalcopyrite ($CuFeS_2$). Early work concerning bioleaching by thermoacidophilic, *Sulfolobus*-like archaea was reviewed by Brierley & Brierley (1986). Prior to the isolation of strains capable of mineral sulfide oxidation above 80°C (Norris & Owen, 1993), the most studied organisms in this context were *Acidianus brierleyi*, *Metallosphaera sedula* and *Sulfolobus metallicus*. In retrospect, *S. metallicus* (Huber & Stetter, 1991) has been the most studied since work in several laboratories from 1981 onwards used a strain, referred to as *Sulfolobus* strain BC, that was only later confirmed as an isolate of *S. metallicus* (see Norris, 1997). The strains that are capable of mineral oxidation at about 80°C provide the most efficient extraction of copper from chalcopyrite. These strains have sufficient differences from each other to warrant the establishment of several new genera (Norris, unpublished data).

There have been two confusions previously in attributing defining characteristics to species of *Sulfolobus* and similar thermoacidophilic archaea. These concern the assignment of 16S rRNA sequences and the capacity for sulfur oxidation. The incidence of supposedly pure cultures containing both *Sulfolobus acidocaldarius* and *Sulfolobus solfataricus* has been documented (Grogan, 1989) and the confusion of 16S rRNA sequences reported for these species has been explained (Kurosawa & Itoh, 1993; Zillig, 1993). The deviation of strain phenotype from that expected from primary descriptions was observed when *S. acidocaldarius* and *S. solfataricus* cultures failed to oxidize elemental sulfur (Marsh *et al.*, 1983; Norris *et al.*, 1986; Huber & Stetter, 1991).

In recent years, a variety of *Sulfolobus*-like strains that do oxidize sulfur have been characterized and phylogenetic relationships established, some more certainly than others (Fig. 4). The justification for splitting the *Sulfolobus* genus has been stated previously (Fuchs *et al.*, 1996), together with a recognition of the difficulty inherent in this while distinctive phenotypic characteristics for some of these organisms are lacking. There might be some indication in the current phylogenetic trees of the grouping of iron-oxidizing strains away from the obligate heterotrophs and non-iron-oxidizers, but serious consideration of this awaits further 16S rRNA gene analysis of the uncharacterized mineral sulfide-oxidizing strains. The placement on the evolutionary tree of sequences from examination of environmental DNA samples suggests that new genera will be relatively easy to find. For example, two sequences from Montserrat geothermal sites, MS-11b and MS-G, are included in Figure 4, but whether the source organisms are metal-mobilizers like their nearest relatives awaits the isolation of these novel strains.

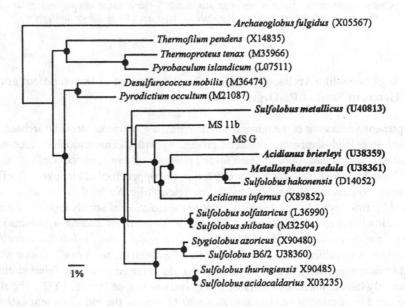

Figure 4. Evolutionary distance dendrogram (based on comparison of 1251 nucleotides) illustrating relationships among Sulfolobus-like organisms within the Archaea. Bootstrap values of over 60% are indictaed by filled circles at selected branches. The scale bar represents 1% sequence divergence. Sequences MS-11b and MS-G are from Montserrat geothermal site environmental samples (Buton and Norris, unpublished data) and strains which have been confirmed as iron-oxidizers are indicated in bold type.

6 Applications of Molecular Ecological Work with Bioleaching Acidophiles

The retrieval, cloning and comparative sequence analysis of 16S rRNA genes from various environmental samples has generally indicated a much greater *in situ* microbial diversity than that revealed by the range of organisms readily cultured from the sources of samples. In contrast, examination of samples from a continuous bioreactor processing zinc sulfide revealed only three sequences which matched those of the culturable organisms (Goebel & Stackebrandt, 1994a). It is noteworthy that one of the clonal types (affiliated with the moderately thermophilic *Sulfobacillus* species) was not represented in the initial organism enrichments, thus indicating the value of the molecular diversity assessment in the case of this undefined, mixed culture, even though it was of low diversity because of the restrictive conditions of acidic, high rate, mineral processing. "*Leptospirillum ferrooxidans*" and a *T. caldus*-like strain were the source of the other clonal types from the bioreactor (Goebel and Stackebrandt, 1994a, b).

Restriction enzyme analysis of 16S rRNA genes isolated from samples from bioreactors processing gold-bearing arsenopyrite (AsFeS) indicated that "*L. ferrooxidans*" and *T. thiooxidans/T. caldus* were again the principal strains in a mixed culture, to the exclusion of *T. ferrooxidans* (Rawlings, 1995). Immunofluorescent staining showed that *T. ferrooxidans* was a significant component of mixed populations in a commercial bioleaching plant treating gold-bearing arsenopyrite, but was generally outnumbered by "*L. ferrooxidans*" and a sulfur-oxidizer described as *T. thiooxidans* (Lawson, 1997). A molecular phylogenetic analysis of the organisms, or of directly extracted rRNA genes in this latter case, would have been interesting with regard to the identity of the principal sulfur-oxidizing organism because *T. caldus* should be favoured over *T. thiooxidans* at the reported temperature (about 40°C) of the bioreactors. The development of PCR-mediated detection of bioleaching bacteria has been described and used to indicate that "*L. ferrooxidans*" was also the principal organism to develop from a mixed culture-inoculum during column leaching of chalcopyrite (de Wulf-Durand *et al.*, 1997).

The application of molecular methods based on rRNA gene probes has been extended to the analysis of the microflora in acid mine drainage. The relative contributions of "*L. ferrooxidans*" and *T. ferrooxidans* to the dissolution of the mineral sulfide source of the acid has been discussed in the light of fluorescence *in situ* hybridization work (Schrenk *et al.*, 1998): a predominance of "*L. ferrooxidans*" over *T. ferrooxidans* in regions of higher acidity and higher temperature appeared to confirm earlier reports (cited in section 3.2) concerning the characteristics that determine the outcome of 'competition' between these organisms.

The application of RNA gene probing has the potential to shed light on many of the questions concerning the nature of mixed microbial populations in bioleaching. The phylogenetic affiliations of the bioleaching organisms described in this review indicates that many of the organisms that can be closely associated in bioleaching processes are not at all closely related, so design of distinctive group- or species-specific 16S rRNA-targeted probes or of primers for PCR-mediated detection can be relatively straightforward. For example, mixed cultures of moderate thermophiles active in mineral processing comprise the Gram-negative proteobacterium *T. caldus*, the putative low GC group, Gram-positive *S. thermosulfidooxidans* and the actinobacterium *A. ferrooxidans*.

Continuous assessment of such mixed populations in bioreactors has begun and the promised revelation of the culture dynamics in concert with physiological studies of key strains should facilitate process understanding and optimization. Design of strain-specific probes, for example to distinguish between *T. ferrooxidans* strains (Goebel & Stackebrandt, 1994b), will be more difficult. Finally, some of the acidophiles noted in this review, including the potentially useful but incompletely characterized strains "*T. ferrooxidans*" m1 and "*Sulfobacillus*" M6b, have been isolated only once: molecular phylogenetic analyses of acidic, environmental samples of the *in situ* microflora of natural and industrial, acidic environments should determine whether such strains are genuinely of limited distribution or simply do not appear readily in laboratory enrichment cultures that appear to favour isolation of a limited number of species.

7 References

Balashova, V.V., Vedenina I.Y., Markosyan, G.E. & Zavarzin, G.A. (1974). The autotrophic growth of *Leptospirillum ferrooxidans*. *Mikrobiologiya* (English translation) **43**, 491-494.

Barros, M.E.C., Rawlings, D.E. & Woods, D.R. (1984). Mixotrophic growth of a *Thiobacillus ferrooxidans* strain. *Applied and Environmental Microbiology* **47**, 593-595.

Beijerinck, M.W. (1904). Über die Bakterien, welche sich im Dunkeln mit Kohlensäure als Kohlenstoffquelle ernähren können. *Zentralblatt für Bakteriologie und Parasitenkunde, II Abteilung* **11**, 592-599.

Berthelot, D., Leduc, L.G. & Ferroni, G.D. (1997). Iron-oxidizing autotrophs and acidophilic heterotrophs from uranium mine environments. *Geomicrobiology Journal* **14**, 317-323.

Bhattacharyya, S., Chakrabarty, B.K., Das, A., Kunda, P.N. & Banerjee, P.C. (1991). *Acidiphilium symbioticum* sp. nov., and acidophilic heterotrophic bacterium from *Thiobacillus ferrooxidans* cultures isolated from Indian mines. *Canadian Journal of Microbiology* **37**, 78-85.

Blake, R.C., Shute, E.A., Greenwood, M.M., Spencer, G.H. & Ingledew, W.J. (1993). Enzymes of aerobic iron oxidation. *FEMS Microbiology Reviews* **11**, 9-18.

Brierley, C.L. (1997). Mining biotechnology: research to commercial development and beyond. In *Biomining: Theory, Microbes and Industrial Processes*, pp. 3-17. Edited by D.E. Rawlings. Berlin: Springer-Verlag.

Brierley, J.A. (1978). Thermophilic iron-oxidizing bacteria found in copper leaching dumps. *Applied and Environmental Microbiology* **36**, 523-525.

Brierley, J.A. & Brierley, C.L. (1986). Microbial mining using thermophilic microorganisms. In *Thermophiles: General, Molecular and Applied Microbiology*, pp. 279-305. Edited by T.D. Brock. New York: Wiley.

Bryant, R.D., McGroarty, K.M., Costerton, J.W. & Laishley, E.J. (1983). Isolation and characterisation of a new acidophilic *Thiobacillus* species (*T. albertis*). *Canadian Journal of Microbiology* **29**, 1159-1170.

Butler, B.J. & Kempton, A.G. (1987). Growth of *Thiobacillus ferrooxidans* on solid media containing heterotrophic bacteria. *Journal of Industrial Microbiology* **2**, 41-45.

Clark, D.A. & Norris, P.R. (1996). *Acidimicrobium ferrooxidans* gen. nov., sp. nov.: mixed-culture ferrous iron oxidation with *Sulfobacillus* species. *Microbiology* **142**, 785-790.

Colmer, A.R., Temple, K.L. & Hinkle, M.E. (1950). An iron-oxidizing bacterium from the acid mine drainage of some bituminous coal mines. *Journal of Bacteriology* **59**, 317-328.

Crundwell, F.K. (1997). Physical chemistry of bacterial leaching. In *Biomining: Theory, Microbes and Industrial Processes*, pp. 178-200. Edited by D.E. Rawlings. Berlin: Springer-Verlag.

Darland, G., Brock, T.D., Samsonoff, W. & Conti, S.F. (1970). A thermophilic acidophilic mycoplasm isolated from a coal refuse pile. *Science* **170**, 1416-1418.

Dufresne, S., Bousquet, J., Boissinot, M. & Guay, R. (1996). *Sulfobacillus disulfidooxidans* sp. nov., a new acidophilic, disulfide-oxidizing, Gram-positive, spore-forming bacterium. *International Journal of Systematic Bacteriology* **46**, 1056-1064.

Durand, P. (1996). Primary structure of the 16S rRNA gene of *Sulfobacillus thermosulfidooxidans* by direct sequencing of PCR amplified gene and its similarity with that of other moderately thermophilic chemolithotrophic bacteria. *Systematic and Applied Microbiology* **19**, 360-364.

Drobner, E., Huber, H., Rachel, R. & Stetter, K.O. (1992). *Thiobacillus plumbophilus* sp. nov., a novel galena and hydrogen oxidizer. *Archives of Microbiology* **157**, 213-217.

Edwards, K.J., Goebel, B.M., Rodgers, T.M., Schrenk, M.O., Gihring, T.M., Cardona, M.M., Hu, B., McGuire, M.M., Hamers, R.J., Pace, N.R. & Banfield, J.F. (1999). Geomicrobiology of pyrite (FeS2) dissolution: a case study at Iron Mountain, California. *Geomicrobiology Journal* **16**, 155-179.

Ehrich, S., Behrens, D., Lebedeva, E., Ludwig, W. & Bock, E. (1995). A new obligately chemolithotrophic, nitrate-oxidising bacterium, *Nitrospira moscoviensis* sp. nov. and its phylogenetic relationship. *Archives of Microbiology* **164**, 16-23.

Evangelou, V.P. & Zhang, Y.L. (1995). A review: pyrite oxidation mechanisms and acid mine drainage prevention. *Critical Reviews of Environmental Science and Technology* **25**, 141-199.

Fliermans, C.B. & Brock, T.D. (1972). Ecology of sulfur-oxidizing bacteria in hot, acid soils. *Journal of Bacteriology* **111**, 343-350.

Fuchs, T., Huber, H., Burggraf, S. & Stetter, K.O. (1996). 16S rDNA-based phylogeny of the archaeal order *Sulfolobales* and reclassification of *Desulfurolobus ambivalens* as *Acidianus ambivalens* comb. nov. *Systematic and Applied Microbiology* **19**, 56-60.

Gehrke, T., Hallmann, R. & Sand, W. (1995). Importance of exopolymers from *Thiobacillus ferrooxidans* and *Leptospirillum ferrooxidans* for bioleaching. In *Biohydrometallurgical Processing Vol. I*, pp. 1-11. Edited by C.A. Jerez, T. Vargas, H. Toledo & J.V. Wiertz. Santiago, University of Chile.

Goebel, B.M. & Stackebrandt, E. (1994a). The biotechnological importance of molecular biodiversity studies for metal bioleaching. In *Bacterial Diversity and Systematics*, pp. 259-273. Edited by F.G. Priest, A. Ramos-Cormenzana & B.J. Tindall. New York, Plenum Press.

Goebel, B.M. & Stackebrandt, E. (1994b). Cultural and phylogenetic analysis of mixed microbial populations found in natural and commercial bioleaching environments. *Applied and Environmental Microbiology* **60**, 1614-1621.

Goebel, B.M. & Stackebrandt, E. (1995). Molecular analysis of the microbial biodiversity in a natural acidic environment. In *Biohydrometallurgical Processing Vol II*, pp. 43-52. Edited by C.A. Jerez, T. Vargas, H. Toledo & J.V. Wiertz. Santiago, University of Chile.

Golovacheva, R.S. & Karavaiko, G.I. (1979). *Sulfobacillus* - a new genus of spore-forming thermophilic bacteria. *Mikrobiologiya* (Eng. trans.) **48**, 658-665.

Golovacheva, R.S., Golyshina, O.V., Karavaiko, K.I., Dorofeev, A.G., Pivovarova, T.O. & Chernykh, N.A. (1993). A new iron-oxidizing bacterium, *Leptospirillum thermoferrooxidans* sp. nov. *Mikrobiologiya*, **61**, 744-750.

Grogan, D.W. (1989). Phenotypic characterization of the archaebacterial genus *Sulfolobus*: comparison of five wild-type strains. *Journal of Bacteriology* **171**, 6710-6719.

Guay, R. & Silver, M. (1975). *Thiobacillus acidophilus* sp. nov: isolation and some physiological characteristics. *Canadian Journal of Microbiology* **21**, 281-288.

Hallberg, K.B. & Lindström, E.B. (1994). Characterization of *Thiobacillus caldus* sp. nov., a moderately thermophilic acidophile. *Microbiology* **140**, 3451-3456.

Hallman, R., Friedrich, A., Koops, H., Pommerening-Roser, A., Rohde, K., Zenneck, C. & Sand, W. (1992). Physiological characteristics of *Thiobacillus ferrooxidans* and *Leptospirillum ferrooxidans* and physicochemical factors influencing microbial metal leaching. *Geomicrobiology Journal* 10, 193-206.

Harrison Jr., A.P. (1982). Genomic and physiological diversity among strains of *Thiobacillus ferrooxidans* and genomic comparison with *Thiobacillus thiooxidans*. *Archives of Microbiology* 131, 68-76.

Harrison Jr., A.P. (1983). Genomic and physiological comparisons between heterotrophic thiobacilli *Acidiphilium cryptum, Thiobacillus versutus* sp. nov., and *Thiobacillus acidophilus* nom. rev. *International Journal of Systematic Bacteriology* 33, 211-217.

Harrison Jr., A.P. (1984). The acidophilic thiobacilli and other bacteria that share their habitat. *Annual Review of Microbiology* 38, 265-292.

Harrison Jr., A.P. (1986). Characteristics of *Thiobacillus ferrooxidans* and other iron-oxidizing bacteria, with emphasis on nucleic acid analysis. *Biotechnology and Applied Biochemistry* 8, 249-257.

Harrison Jr., A.P. & Norris, P.R. (1985). *Leptospirillum ferrooxidans* and similar bacteria: some characteristics and genomic diversity. *FEMS Microbiology Letters* 30, 99-102.

Harrison Jr., A.P., Jarvis, B.W. & Johnson, J.L. (1980). Heterotrophic bacteria from cultures of autotrophic *Thiobacillus ferrooxidans*: relationships as studie by means of deoxynucleic acid homology. *Journal of Bacteriology* 143, 448-454.

Helle, U. & Onken, U. (1988). Continuous microbial leaching of a pyritic concentrate by Leptospirillum-like bacteria. *Applied Microbiology and Biotechnology* 28, 553-558.

Hiraishi, A., Kishimoto, N., Kosako, Y., Wakao, N. & Tano, T. (1995). Phylogenetic position of the menaquinone-containing acidophilic chemo-organotroph *Acidobacterium capsulatum*. *FEMS Microbiology Letters* 132, 91-94.

Hiraishi, A., Nagashima, K.V.P., Matsuura, K., Shimada, K., Takaichi, S., Wakao, N. & Katayama, Y. (1998). Phylogeny and photosynthetic features of *Thiobacillus acidophilus* and related acidophilic bacteria: its transfer to the genus *Acidiphilium* as *Acidiphilium acidophilum* comb. nov. *International Journal of Systematic Bacteriology* 48, 1389-1398.

Hiskey, J.B. (1994). *In situ* leaching recovery of copper: what next?, pp. 43-65. In *Hydrometallurgy-94*. London; Chapman and Hill.

Huber, H. & Stetter, K.O. (1989). *Thiobacillus prosperus* sp. nov., represents a new group of halotolerant metal-mobilising bacteria isolated from a marine geothermal field. *Archives of Microbiology* 151, 479-485.

Huber, H. & Stetter, K.O. (1990). *Thiobacillus cuprinus* sp. nov., a novel facultatively organotrophic metal-mobilizing bacterium. *Applied and Environmental Microbiology* 56, 315-322.

Huber, G. & Stetter, K.O. (1991). *Sulfolobus metallicus*, sp. nov., a novel strictly chemolithotrophic thermophilic archaeal species of metal-mobilizers. *Systematic and Applied Microbiology* 14, 372-378.

Johnson, D.B. & Kelso, W.I. (1983). Detection of heterotrophic contaminants in cultures of *Thiobacillus ferrooxidans* and their elimination by subculturing in media containing copper sulphate. *Journal of General Microbiology* 123, 2969-2972.

Johnson, D.B. & McGinness, S. (1991). A new solid medium for the isolation and enumeration of *Thiobacillus ferrooxidans* and acidophilic heterotrophs. *Journal of Microbiological Methods* 13, 113-122.

Johnson, D.B. & Roberto, F.F. (1997). Heterotrophic acidophiles and their roles in the bioleaching of sulfide minerals. In *Biomining: Theory, Microbes and Industrial Processes*, pp. 259-279. Edited by D.E. Rawlings. Berlin: Springer-Verlag.

Johnson, D.B., Ghauri, M.A. & Said, M.F. (1992). Isolation and characterization of an acidophilic, bacterium capable of oxidizing ferrous iron. *Applied and Environmental Microbiology* 58, 1423-1428.

Johnson, D.B., Bacelar-Nicolau, P., Bruhn, D.F. & Roberto, F.F. (1995). Ubiquitous bacteria in leaching environments. In *Biohydrometallurgical Processing Vol II*, pp. 47-56. Edited by C.A. Jerez, T. Vargas, H. Toledo & J.V. Wiertz. Santiago, University of Chile.

Karavajko, G.I., Bulygina, E.S., Tsaplina, I.A., Bogdanova, T.I. & Chumakov, K.M. (1990). *Sulfobacillus thermosulfidooxidans*: a new lineage of bacterial evolution? *FEBS Letters* 261, 8-10.

Katayama, Y., Hiraishi, A. & Kuraishi, H. (1995). *Paracoccus thiocyanatus* sp. nov., a new species of thiocyanate-utilizing facultative chemolithotroph, and transfer of *Thiobacillus versutus* to the genus *Paracoccus* as *Paracoccus versutus* comb. nov. with emendation of the genus. *Microbiology* 141, 1469-1477.

Kelly, D.P., Norris, P.R. & Brierley, C.L. (1979). Microbiological methods for the extraction and recovery of metals. In *Microbial Technology: Current State, Future Prospects*, pp. 263-308. Edited by A.T. Bull. C. Ratledge & D.C. Ellwood. Cambridge: Cambridge University Press.

Kinsel, N.A. (1960). A new sulfur oxidizing iron bacterium: *Ferrobacillus sulfooxidans*. *Journal of Bacteriology* 80, 628-632.

Kishimoto, N. & Tano, T. (1987). Acidophilic heterotrophic bacteria isolated from acidic mine drainage, sewage and soils. *General and Applied Microbiology* 33, 11-25.

Kishimoto, N., Kosako, Y. & Tano, T. (1991). *Acidobacterium capsulatum* gen. nov., sp. nov.: An acidophilic chemoorganotrophic bacterium containing menaquinone from acidic mineral environment. *Current Microbiology* 22, 1-7.

Kishimoto, N., Kosako, Y. & Tano, T. (1993). *Acidiphilium aminolytica* sp. nov.: an acidophilic chemoorganotrophic bacterium containing menaquinone from acidic mineral environment. *Current Microbiology* 22, 1-7.

Kishimoto, N., Fukaya, F., Inagaki, K., Sugio, T., Tanaka, H. & Tano, T. (1995a). Distribution of bacteriochlorophyll *a* among aerobic acidophilic bacteria and light enhanced CO_2-incorporation in *Acidiphilium rubrum*. *FEMS Microbiology Ecology* 16, 291-296.

Kishimoto, N., Kosako, Y., Wakao, N., Tano, T. and Hiraishi, A. (1995b). Transfer of *Acidiphilium facilis* and *Acidiphilium aminolytica* to the genus *Acidocella* gen. nov., and emendation of the genus *Acidiphilium*. *Systematic and Applied Microbiology* 18, 85-91.

Kuenen, J.G., Robertson, L.A. & Tuovinen, O.H. (1992). The genera *Thiobacillus*, *Thiomicrospira* and *Thiosphaera*. In *The Procaryotes, Vol. 3*, pp. 2638-2657. Edited by A. Barlows, H.G. Trüper, M. Dwokin, W. Harder & K.-H. Schleifer. New York: Springer-Verlag.

Kurosawa, N. & Itoh, Y.H. (1993). Nucleotide sequence of the 16S rRNA gene from thermoacidophilic archaea *Sulfolobus acidocaldarius* ATCC 33909. *Nucleic Acids Research* 21, 357.

Lane, D.J., Stahl, D.A., Olsen, G.J., Heller, D.J. & Pace, N.R. (1985). Phylogenetic analysis of the genera *Thiobacillus* and *Thiomicrospira* by 5S rRNA sequences. *Journal of Bacteriology* 163, 75-81.

Lane, D.J., Harrison, A.P. Jr., Stahl, D.A., Pace, B., Giovannoni, S.J., Olsen, G.J. & Pace, N.R. (1992). Evolutionary relationships among sulfur- and iron-oxidizing eubacteria. *Journal of Bacteriology* 174, 269-278.

Lawson, E.N. (1997). The composition of mixed populations of leaching bacteria active in gold and nickel recovery from sulphide ores. In *IBS'97/Biomine'97*, pp. QP4.1-10. Glenside: Australian Mineral Foundation.

Leathen, W.W., Kinsel, N.A. & Braley, S.A. (1956). *Ferrobacillus ferrooxidans*: a chemosynthetic autotrophic bacterium. *Journal of Bacteriology* 72, 700-704.

Lizama, H.M. & Suzuki, I. (1987). Bacterial leaching of a sulfide ore by *Thiobacillus ferrooxidans* and *Thiobacillus thiooxidans*: I. Shake flask studies. *Biotechnology and Bioengineering* 32, 110-116.

Lobos, J.H., Chisolm, T.E., Bopp, L.H. & Holmes, D.S. (1986). *Acidiphilium organovorum* sp. nov., and acidophilic heterotroph isolated from *Thiobacillus ferrooxidans* culture. *International Journal of Systematic Bacteriology* 36, 139-144.

Maidak, B.L., Olsen, G.J., Larsen, N., Overbeek, R., McCaughey, M.J. & Woese, C.R. (1997). The RDP

(Ribosomal Database Project). *Nucleic Acids Research*. **25**, 109-110.

Markosyan, G.E. (1972). A new acidophilic iron bacterium *Leptospirillum ferrooxidans. Biologicheskii Zhurnal Armenii* **25**, 26-29.

Markosyan, G.E. (1973). A new mixotrophic sulfur bacterium developing in acid media, *Thiobacillus organoparus* sp. n. *Doklady Akademii Nauk USSR* **211**, 1205-1208.

Marsh, R.M., Norris, P.R. & Le Roux, N.W. (1983). Growth and mineral oxidation studies with *Sulfolobus*. In *Recent Progress in Biohydrometallurgy*, pp.71-81. Edited by G. Rossi and A.E. Torma. Iglesias: Associazione Mineraria Sarda.

Mason, J. & Kelly, D.P. (1988). Mixotrophic and autotrophic growth of *Thiobacillus acidophilus* on tetrathionate. *Archives of Microbiology* **149**, 317-323.

McDonald, I.R., Kelly, D.P., Murrell, J.C. & Wood, A.P. (1997). Taxonomic relationships of *Thiobacillus halophilus, T. aquaesulis*, and other species of *Thiobacillus*, as determined using 16S rDNA sequencing. *Archives of Microbiology* **166**, 394-398.

Merrettig, U., Wlotzka, P. & Onken, U. (1989). The removal of pyritic sulphur from coal by *Leptospirillum*-like bacteria. *Applied Microbiology and Biotechnology* **31**, 626-628.

Meulenberg, R., Pronk, J.T., Hazeu, W., Bos, P. & Kuenen, J.G. (1992). Oxidation of reduced sulphur compounds by intact cells of *Thiobacillus acidophilus. Archives of Microbiology* **157**, 161-168.

Moreira, D. & Amils, R. (1997). Phylogeny of *Thiobacillus cuprinus* and other mixotrophic thiobacilli: proposal for *Thiomonas* gen. nov. *International Journal of Systematic Bacteriology* **47**, 522-528.

Muyzer, G., DeBruyn, A.C., Schmedding, D.J.M., Bos, P., Westbroek, P. & Kuenen, G.J. (1987). A combined immunofluorescence-DNA-fluoresence staining technique for the enumeration of *Thiobacillus ferrooxidans* in a population of acidophilic bacteria. *Applied and Environmental Microbiology* **53**, 660-664.

Nordstrom, D.K. & Southam, G. (1997). Geomicrobiology of sulfide mineral oxidation. In *Geomicrobiology: Interactions Between Microbes and Minerals, Vol. 35*, pp. 361-382. Edited by J.F. Banfield & K.H. Nealson. Washington, DC: Mineralogical Society of America.

Norris, P.R. (1983). Iron and mineral oxidation with *Leptospirillum*-like bacteria. In *Recent Progress in Biohydrometallurgy*, pp. 83-96. Edited by G. Rossi and A.E. Torma. Iglesias: Associazione Mineraria Sarda,

Norris, P.R. (1997). Thermophiles and bioleaching. In *Biomining: Theory, Microbes and Industrial Processes*, pp. 248-258. Edited by D.E. Rawlings. Berlin: Springer-Verlag.

Norris, P.R. & Kelly, D.P. (1982). The use of mixed microbial cultures in metal recovery. In *Microbial Interactions and Communities*, pp. 443-474. Edited by A.T. Bull and J.H. Slater. London: Academic Press.

Norris, P.R. & Owen, J.P. (1993). Mineral sulfide oxidation by enrichment cultures of novel thermophilic bacteria. *FEMS Microbiology Reviews* **11**, 51-56.

Norris, P.R., Marsh, R.M. & Lindström, E.B. (1986). Growth of mesophilic and thermophilic acidophilic bacteria on sulfur and tetrathionate. *Biotechnology and Applied Biochemistry* **8**, 318-329.

Norris, P.R., Barr, D.W. & Hinson, D. (1988). Iron and mineral oxidation by acidophilic bacteria: affinities for iron and attachment to pyrite. In *Bioydrometallurgy, International Symposium Proceedings*, pp. 43-59. Edited by P.R. Norris and D.P. Kelly. Kew: Science and Technology Letters.

Norris, P.R., Clark, D.A., Owen, J.P. & Waterhouse, S. (1996). Characteristics of *Sulfobacillus acidophilus* sp. nov. and other moderately thermophilic mineral sulphide-oxidizing bacteria. *Microbiology* **142**, 775-783.

Pivovarova, T.E., Markosyan, G.E. & Karavaiko, G.I. (1981). Morphogenesis and fine structure of *Leptospirillum ferrooxidans. Mikrobiologiya* (Eng. Trans.) **50**, 339-344.

Pronk, J.T., Meesters, P.J.W., v. Dijken, J.P., Bos, P. & Kuenen, J.G. (1990). Heterotrophic growth of *Thiobacillus acidophilus* in batch and chemostat cultures. *Archives of Microbiology* **153**, 392-398.

Rawlings, D.E. (1995). Restriction enzyme analysis of 16S rRNA genes for the rapid identification of

Thiobacillus ferrooxidans, Thiobacillus thiooxidans and *Leptospirillum ferrooxidans* strains in leaching environments. In *Biohydrometallurgical Processing Vol. II*, pp. 9-17. Edited by C.A. Jerez, T. Vargas, H. Toledo, and J.V. Wiertz. Santiago: University of Chile.

Rawlings, D.E., (ed.) (1997). *Biomining: Theory, Microbes and Industrial Processes*. Berlin, Springer-Verlag.

Sand, W., Rohde, K., Sobotke, B. & Zenneck, C. (1992). Evaluation of *Leptospirillum ferrooxidans* for leaching. *Applied and Environmental Microbiology* 58, 85-92.

Sand, W., Gehrke, T., Hallmann, R. & Schippers, A. (1995). Sulfur chemistry, biofilms and the (in)direct attack mechanism - a critical evaluation of bacterial leaching. *Applied Microbiology and Biotechnology* 43, 961-966.

Schippers, A., Hallmann, R., Wentzien, S. & Sand, W. (1995). Microbial diversity in uranium mine waste heaps. *Applied and Environmental Microbiology* 61, 2930-2935.

Schippers, A., Jozsa, P.-G. & Sand, W. (1996). Sulfur chemistry in bacterial leaching of pyrite. *Applied and Environmental Microbiology* 62, 3424-3431.

Schrenk, M.O., Edwards, K.J., Goodman, R.M., Hamers, R.J. & Banfield, J.F. (1998). Distribution of *Thiobacillus ferrooxidans* and *Leptospirillum ferrooxidans*: implications for generation of acid mine drainage. *Science* 279, 1519-1522.

Segerer, A., Langworthy, T.A. & Stetter, K.O. (1988). *Thermoplasma acidophilum* and *Thermoplasma volcanium* sp. nov. from solfatara fields. *Systematic and Applied Microbiology* 10, 161-171.

Sievers, M., Ludwig, W. & Teuber, M. (1994). Phylogenetic positioning of *Acetobacter, Gluconobacter, Rhodopila* and *Acidiphilium* species as a branch of acidophilic bacteria in the alpha-subclass of Proteobacteria based on 16S ribosomal DNA sequences. *Systematic and Applied Microbiology* 17, 189-196.

Silverman, M.P. & Ehrlich, H.L. (1964). Microbial formation and degradation of minerals. *Advances in Applied Microbiology* 6, 153-206.

Skerman, V.B.D., McGowen, V. & Sneath, P.H.A. (1980). Approved lists of bacterial names. *International Journal of Systematic Bacteriology* 61, 225-420.

Stackebrandt, E., Rainey, F.A. & Ward-Rainey, N.L. (1997). Proposal for a new hierarchic classification system, *Actinobacteria* classis nov. *International Journal of Systematic Bacteriology* 47, 479-491.

Sugio, T., White, K.J., Shute, E., Choate, D. & Blake, R.C. (1992). Existence of a hydrogen sulfide:ferric ion oxidoreductase in iron-oxidizing bacteria. *Applied and Environmental Microbiology* 58, 431-433.

Sugio, T., Uemura, S., Makino, I., Iwahori, K., Tano, T. & Blake, R.C. (1994). Sensitivity of iron-oxidizing bacteria, *Thiobacillus ferrooxidans* and *Leptospirillum ferrooxidans*, to bisulfite ion. *Applied and Environmental Microbiology* 60, 722-725.

Temple, K.L. & Colmer, A.R. (1951). The autotrophic oxidation of iron by a new bacterium: *Thiobacillus ferrooxidans*. *Journal of Bacteriology* 62, 605-611.

Tourova, T.P., Poltoraus, A.B., Lebedeva, I.A., Tsaplina, A.L., Bogdanova, T.I. & Karavaiko, G.I. (1994). 16S ribosomal RNA (rDNA) sequence analysis and phylogentic position of *Sulfobacillus thermosulfidooxidans*. *Systematic and Applied Microbiology* 17, 509-512.

Tuovinen, O.H. & Kelly, D.P. (1972). Recommendation that the names *Ferrobacillus ferrooxidans* Leathen and Braley and *Ferrobacillus sulfooxidans* Kinsel be recognised as synonymns of *Thiobacillus ferrooxidans* Temple and Colmer. *International Journal of Systematic Bacteriology* 22, 170-172.

Tuttle, J.H. & Dugan, P.R. (1976). Inhibition of growth, iron and sulfur oxidation in *Thiobacillus ferrooxidans* by simple organic compounds. *Canadian Journal of Microbiology* 22, 719-730.

Tuttle, J.H., Dugan, P.R. & Apel, W.A. (1977). Leakage of cellular material from *Thiobacillus ferrooxidans* in the presence of organic acids. *Applied and Environmental Microbiology* 33, 459-469.

Ursing, J.B., Rosselló-Mora, R.A., García-Vald's, E. & Lalucat, J. (1995). Taxonomic note: a pragmatic approach to the nomenclature of phenotypically similar genomic groups. *International Journal of Systematic Bacteriology* **45**, 604.

Wakao, N., Shiba, T., Hiraishi, A., Ito, M. & Sakurai, Y. (1993). Distribution of bacteriochlorophyll *a* in species of the genus *Acidiphilium*. *Current Microbiology* **27**, 277-279.

Wakao, N., Nagasawa, N., Matsuura, T., Matsukura, H., Matsumoto, T., Hiraishi, A., Sakurai, Y. & Shiota, H. (1994). *Acidiphilium multivorum* sp. nov, an acidophilic chemoorganotrophic bacterium from pyritic acid mine drainage. *Journal of General and Applied Microbiology* **40**, 143-159.

Waksman, S.A. & Joffe, J.S. (1922). Microorganisms concerned with the oxidation of sulphur in soil II. *Thiobacillus thiooxidans*, a new sulfur-oxidizing bacterium isolated from soil. *Journal of Bacteriology* **7**, 239-256.

Wayne, L.G., Brenner, D.J., Colwell, R.R., Grimont, P.A.D., Kandler, O., Krichevsky, M.L., Moore, L.H., Moore, W.E.C., Murray, R.G.E., Stackebrandt, E., Starr, M.P. & Trüper, H.G. (1987). Report of the *ad hoc* committee on reconciliation of approaches to bacterial systematics. *International Journal of Systematic Bacteriology* **37**, 463-464.

Wichlacz, P.L., Unz, R.F. & Langworthy, T.A.. (1986). *Acidiphilium angustum* sp. nov., and *Acidiphilium rubrum* sp. nov.: acidophilic heterotrophic bacteria isolated from acid coal mine drainage. *International Journal of Systematic Bacteriology* **36**, 197-201.

Wisotzkey, J.D., Jurtshuk, J.R., Fox, G.E., Deinhard, G. & Poralla, K. (1992). Comparative sequence analyses on the 16S rRNA (rDNA) of *Bacillus acidocaldarius*, *Bacillus acidoterrestris*, and *Bacillus cycloheptanicus* and proposal for creation of new genus, *Alicyclobacillus* gen. nov. *International Journal of Systematic Bacteriology* **42**, 263-269.

Wulf-Durand, de P., Bryant, L.J. & Sly, L.I. (1997). PCR-mediated detection of acidophilic bioleaching-associated bacteria. *Applied and Environmental Microbiology* **63**, 2944-2948.

Zillig, W. (1993). Confusion in the assignments of *Sulfolobus* sequences to *Sulfolobus* species. *Nucleic Acids Research* **21**, 5273.

11   MICROBIAL COMMUNITIES IN OIL FIELDS

GERRIT VOORDOUW

Department of Biological Sciences, The University of Calgary,
2500 University Drive NW, Calgary (AB) T2N 1N4, Canada

1   The Formation and Production of Oil

It is generally thought that oil is formed from organic material of biogenic origin. A small fraction (<1%) of organic debris of plants, algae and microorganisms is incorporated in aquatic sediments. Burial and downward movement of these sediments over geological time causes temperatures to rise and leads to the formation of kerogen from humic and fulvic acids (Philp, 1986). Oil is formed from these precursors when temperatures of 100 to 200°C are reached during continued downward movement. Further temperature increases may lead to the demise of the newly formed oil by gasification. However, lateral and upward movement through cracks and fissures driven by high resident pressures may cause oil and gas in source rock to migrate and accumulate under sediment (rock) layers of low permeability (Fig. 1).

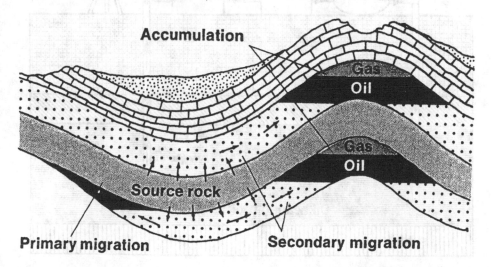

Figure 1. Formation of an oil reservoir. Oil moves from source rock into the carrier rock over short distances (primary migration). From the carrier it moves to the reservoir over longer distances (secondary migration). Adapted from Philp (1986).

F. G. Priest and M. Goodfellow (eds.), Applied Microbial Systematics, 315-332
© 2000 Kluwer Academic Publishers. Printed in the Netherlands.

When accessed by drilling a borehole the oil, which is held in rock pores, may initially flow spontaneously upward (primary production). However, once this pressure gradient has been exhausted continued (secondary) production requires water injection (Fig.2), yielding an oil-water mixture. The separated water is supplemented with make up water from a surface source and reinjected into the reservoir. Thus, an oil field in secondary production is a recirculating system that draws a certain amount of water from an external source. Oil-water mixtures emerge at producing wells and water is reinjected at injection wells. These, as well as other above ground units (e.g. oil storage tanks and connecting pipelines) can be used for sampling, using recommended procedures reviewed elsewhere (Herbert *et al.*, 1987; McInerney & Sublette, 1997). A wide variety of microorganisms can be isolated from these samples including sulfate-reducing bacteria (SRB) of the genera *Desulfobacter, Desulfobulbus, Desulfovibrio*, and sulfate-reducing archaea of the genus *Archaeoglobus*, sulfide-oxidizing bacteria from the genera *Arcobacter* and *Campylobacter* and a variety of heterotrophic and fermentative bacteria. The goal of oil field microbial systematics is to characterize the activities and numbers of individual species in this complex microbial community.

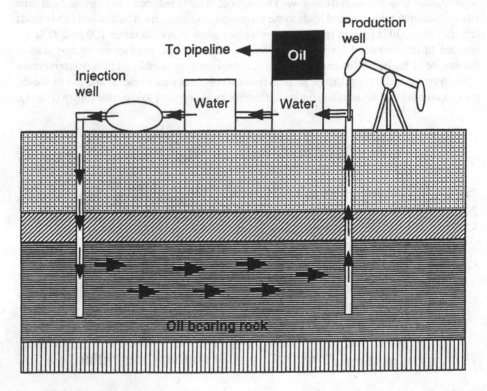

Figure 2. Diagram of a field from which oil is produced by water injection. Water is injected in the injection well and an oil-water mixture is produced in the production well. The oil is stored in tanks or transported through pipelines. The water is reinjected into the reservoir. Samples for microbiological analysis are obtained from any of these above ground installations.

2 The Origin of Microbial Life in Oil Fields

The view that oil bearing rock is sterile and that microorganisms are introduced by the drilling and subsequent production process, especially water injection, has long been held. An example of the controversies this can lead to is represented by several recent papers on the origin of a thermophilic microbial community present in deep North Sea and Alaskan North Slope oil reservoirs. Stetter and coworkers (Huber *et al.*, 1990) were the first to show that hyperthermophilic archaea (optimal growth temperatures 80 to 110°C) were present in the active zone of an erupting submarine volcano. Members of the genera *Archaeoglobus*, *Pyrococcus*, *Pyrodictium* and *Thermococcus* were demonstrated. Studies with production waters from deep North Sea and Alaskan oil reservoirs indicated the presence of similar archaea as judged by cross-hybridization of chromosomal DNAs (Stetter *et al.*, 1993). Because the hyperthermophiles can survive in sea water in low concentrations and since the fields in question were flooded with 10^5 m^3 of sea water per day, it was hypothesized that the oil field community originated from distant hydrothermal vents or erupting volcanoes. This idea was disputed by L'Haridon *et al.* (1995), who found hyperthermophilic archaea in a land-locked, deep oil field near Paris that had never been waterflooded. The question whether North Sea oil field archaea are merely similar or indeed identical to those found in distant high temperature environments could easily have been settled if the authors (Huber *et al.*, 1990; Stetter *et al.*, 1993) had used more definitive molecular systematic techniques than chromosomal DNA hybridization, e.g. restriction fingerprinting or sequencing of a large stretch of DNA, to establish strain identity. Differences would not be detected by any of these methods if these microorganisms had spread from hydrothermal sites in the last decennia. The issue is one of practical importance. If the bacteria are exclusively introduced through the injection process then sterilization treatment of waters prior to injection can be considered in principle as a method for preventing their establisment. However, recent work has indicated that the subsurface is not sterile so the idea that injection water sterilization might be beneficial would be based on a wrong premise.

Progress in subsurface microbiology has shown that microorganisms are present in rocks (Fredrickson & Phelps, 1997), where they sustain themselves through chemolithotrophic metabolism (Stevens & McKinley, 1995). The localized presence of sulfate-reducing bacteria in rock cores was demonstrated through incubation of sections with ^{35}S-sulfate and use of oxidized silver foil as a ^{35}S-sulfide trap (Krumholtz *et al.*, 1997). It was found that microbial activity was particularly high at sandstone-shale interfaces, possibly because free energy-liberating chemical reactions take place there or because these are subject to increased influx of metabolites through water transport. A system devoid of all diffusion and convection fluxes that a microorganism might exploit (e.g. oil-bearing rock trapped under a low permeability sedimentary layer in a geologically stable part of the subsurface) cannot sustain life for prolonged periods of time. All subsurface regions experience heat flow from the Earth's core to the surface, but it is unlikely that this flux can be exploited by microorganisms. However, when fluxes are induced by drilling holes and subsequent water injection the conditions required to sustain life vastly improve and microbes from neighbouring, more supportive subsurface regions become established.

In conclusion, there is a distinct possibility that microbes in oil fields are either already present or are transported from neighbouring subsurface regions. Bacteria living at the surface may be only a minor component of the oil field community. The halophilic and thermophilic properties of oil field bacteria are often lacking among bacteria in surface waters and soils.

3   Sulfate-Reducing Bacteria

3.1   TYPES OF SULFATE-REDUCING BACTERIA PRESENT IN OIL FIELDS

Sulfate Reducing Bacteria (SRB) derive energy for growth from the use of sulfate as the terminal electron acceptor in the oxidation of a variety of substrates (electron donors). Hydrogen and organic acids (acetate, lactate, propionate, pyruvate) are commonly used as electron donors for sulfate-reduction. SRB capable of using oil field hydrocarbons (e.g. hexadecane, toluene or more generally n-alkanes or alkylbenzenes) have been isolated and characterized by Widdels's group (Aeckersberg *et al.*, 1991; Rabus *et al.*, 1993, 1996; Rueter *et al.*, 1994). Many different genera of SRB are distinguished based on nutritional requirements, morphology and 16S rRNA sequences. The phylogenetic tree shown for mesophilic, nonsporeforming SRB (Fig. 3), indicates the relatedness of the Gram-negative genera *Desulfoarculus, Desulfobacter, Desulfobacterium, Desulfobotulus, Desulfobulbus, Desulfococcus, Desulfomicrobium, Desulfosarcina,* and *Desulfovibrio* and the Gram-positive genus *Desulfotomaculum.* Widdel & Bak (1992) proposed two families, the *Desulfovibrionaceae* (comprising the genera *Desulfovibrio* and *Desulfomicrobium*) and the *Desulfobacteriaceae* (comprising the other genera of Gram-negative bacteria indicated in Fig. 3) based on phylogenetic and physiological data. These families have also been referred to as the incomplete and complete oxidizers, respectively, indicating formation of CO_2 and acetate or of CO_2 only when organic compounds with more than 3 carbon atoms are used as electron donor in the reduction of sulfate. A detailed review of the growth and isolation of these different SRB has been given by Widdel & Bak (1992), only a limited number of their properties will be described here.

3.2   *DESULFOVIBRIONACEAE* AND OTHER INCOMPLETE OXIDIZERS

Members of the family *Desulfovibrionaceae* are most commonly cultivated using lactate as electron donor; the lactate is converted to acetate and CO_2. Other substrates used by these organisms are ethanol, glycerol, malate, pyruvate and in some instances sugars, like fructose. Hydrogen can also be used, provided acetate is available in addition to CO_2 as a carbon source for microbial growth. Media in general use for evaluating the presence of SRB in oil field fluids, e.g. Postgate's medium B, or API (American Petroleum Institute) medium, contain lactate and favour the growth of the *Desulfovibrionaceae.* The fact that these bacteria cannot use long chain fatty acids or hydrocarbons as electron donors for sulfate reduction raises the question as to whether these culture-based assays truly represent the numbers of SRB that thrive in an oil field environment.

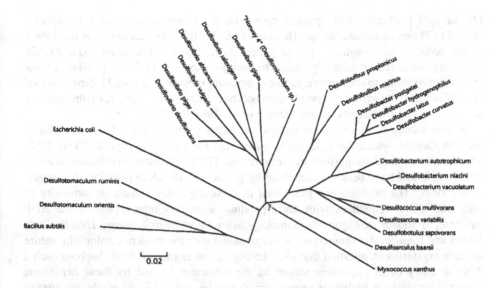

Figure 3. Phylogenetic relationships of mesophilic, sulfate-reducing bacteria from comparison of 16S rRNA sequences. Some non-sulfate reducing bacteria are also included in the tree (*Bacillus subtilis, Escherichia coli, Myxococcus xanthus*). The scale bar represents fixed nucleotide substitutions per sequence position. Reproduced from Widdel & Bak (1992).

A relatively recent example of the isolation and characterization of an oil field *Desulfovibrio* is the report by Magot *et al.* (1992) on the isolation of *Desulfovibrio longus* from an oil well in France. This organism prefers saline conditions (1 to 2%, w/v of NaCl) and can use formate, lactate or pyruvate as electron donor for sulfate reduction. Hydrogen can be used exclusively as electron donor provided acetate is provided as a carbon source. Sulfate, sulfite, thiosulfate, sulfur and fumarate, but not nitrate were also used as electron acceptors (Magot *et al.*, 1992). The enumeration and characterization of SRB in a wide variety of oil fields using lactate-based media was also recently reported by this group (Tardy-Jacquenod *et al.*, 1996). The results indicated that species similar to *Desulfovibrio desulfuricans* and *Desulfovibrio longus* are easily isolated from these environments at temperatures below 40°C. Although some nutritional versatility exists, none of these strains used hydrocarbons as electron donor for sulfate reduction. Members of the family *Desulfovibrionaceae* which tolerate high salinities (10%, w/v of NaCl or more) have been found in other environments (Caumette *et al.*, 1991) and are probably present in hypersaline oil reservoirs (6 to 9%, w/v of NaCl; Bhupathiraju *et al.*, 1993). An extremely halophilic SRB, capable of growth at 270 g/L of NaCl has been reported by Cord-Ruwisch *et al.* (1987). This organism was able to use ethanol, fatty acids and lactate, but not hydrogen as electron donor for sulfate reduction. Its phylogenetic status is unknown.

Oil field *Desulfovibrionaceae* can thus tolerate moderate to high salinities, but they are generally not isolated on lactate-sulfate media from higher temperature environments

(40 to 60°C) which yield spore forming *Desulfotomaculum* species or species related to *Thermodesulfobacterium* (Rosnes *et al.*, 1991; Tardy-Jacquenod *et al.*, 1996). Sulfate-reducing *Archaeoglobus* spp. are present at even higher temperatures (60 to 100°C). These organisms are nutritionally similar to *Desulfotomaculum* and *Desulfovibrionaceae* strains in their ability to use formate, lactate or pyruvate as electron donors. Hydrogen is used with thiosulfate, not sulfate, as the electron acceptor. Glucose and complex substrates are also used, but hydrocarbons are not (Stetter *et al.*, 1987, 1993).

A wide variety of *Desulfovibrionaceae* strains have been isolated from oil fields in western Canada, which are moderately saline and have temperatures of 25 to 30°C. These isolates were distinguished by chromosomal DNA hybridization (Voordouw *et al.*, 1991, 1992) and have been characterized by partial 16S rRNA sequencing (Voordouw *et al.*, 1996). The phylogenetic placement of these organisms, based on similarity of their 16S rRNA sequences with corresponding sequences represented in the RDP database, is shown in Figure 4. Homology with *Desulfovibrio gigas*, *Desulfovibrio longus* and *Desulfovibrio salexigens* is in agreement with the moderate halophilic nature of these organisms. A question that must be considered is why oil fields harbour such a wide diversity of *Desulfovibrionaceae* as the substrates needed by these organisms (lower chain alcohols, hydrogen, organic acids) may be scarce. Use of whole genomes to probe the occurrence of these different species indicated that selected *Desulfovibrionaceae* are highly enhanced in the biofilm populations covering metal surfaces thereby indicating a role in metal corrosion (Voordouw *et al.*, 1993).

3.3 *DESULFOBACTERIACEAE* AND OTHER COMPLETE OXIDIZERS

Evidence for the presence of completely oxidizing SRB in the oil field environment was first presented by Cord-Ruwisch *et al.* (1987) who used culturing methods to show that acetate-dependent *Desulfobacter* species (6 x 10^6 cfu/ml) outnumbered both benzoate-dependent SRB (1 x 10^6 cfu/ml) and culturable lactate-dependent *Desulfovibrio* species (1 x 10^5 cfu/ml). More recently, analysis of a mesophilic enrichment culture growing anaerobically on crude oil gave 6 x 10^6 to 8 x 10^6 cfu/ml, with benzoate and toluene as electron donors (Rabus *et al.*, 1996). A similar number was obtained with lactate. However, use of specific 16S rRNA-targeted oligonucleotide probes indicated that less than 5% of the cells in the enrichment culture belonged to the *Desulfovibrionaceae*, the remainder were all members of the family *Desulfobacteriaceae* (Rabus *et al.*, 1996). The presence of *Desulfobacteriaceae* in oil field fluids was also demonstrated by sequencing of 16S rDNA genes that were PCR amplified and then cloned from enrichment or purified cultures (Voordouw *et al.*, 1996). Use of acetate as electron donor in these cultures gave strains DBAC9 and DBAC14 with a *Desulfobacter* species as the closest RDP homologue (Fig. 4). Propionate gave enrichment of *Desulfobulbus* species, namely strains DBUL8, DBUL31 and DBUL35, as well as strain DCOC4, which is related to *Desulfococcus multivorans*.

The SRB able to use hydrocarbons as electron donors for sulfate reduction are all complete oxidizers which do not release intermediates (e.g. lower chain length fatty acids) that might be used for growth by *Desulfovibrionaceae* strains. These organisms have been isolated from a variety of environments. Strain Hxd3, a mesophile from anoxic marine sediment, grows on hexadecane, as well as on *n*-alkanes with 12, 14, 15, 17, 18, or

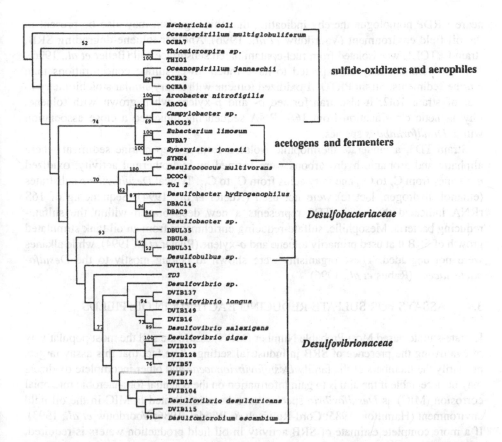

Figure 4. Phylogenetic placement of 16S rRNA sequences of bacteria in western Canadian oil fields (e.g. OCEA7 and THIO1) by comparison with sequences of RDP homologs (e.g. *Oceanospirillum multiglobuliferum*). The sequence of *Escherichia coli* served to root the tree. The numbers at the branches indicate the number of times the sequences group together in 100 bootstrap analyses. Reproduced from Voordouw *et al.* (1996).

19 carbon atoms (Aeckersberg *et al.*, 1991). Alcohols (1-hexadecanol and 2-hexadecanol) and fatty acids (butyrate to stearate) are also used. The phylogenetic position of this strain was not reported.

Strain Tol2, which was obtained from a similar environment used toluene as electron donor for sulfate reduction. Other aromatic compounds (e.g. benzaldehyde, benzoate, *p*-cresol) were also used. The stoichiometry for complete oxidation of toluene was reported as: $C_6H_5CH_3 + 4.5SO_4^{2-} + 2H^+ + 3H_2O \rightarrow 7HCO_3^- + 4.5H_2S$ (Rabus *et al.*, 1993). Sequencing of 16S rRNA indicated that strain Tol2 was related to *Desulfobacter* species. However, metabolically this strain showed greater similarity to some other members of the *Desulfobacteriaceae* (*Desulfobacterium, Desulfococcus, Desulfosarcina*); the organism was named *Desulfobacula toluolica*.

Analysis of 16S rDNA sequences from benzoate-sulfate enrichments of oil field production waters gave a clone (DBAC28) which had *Desulfobacula toluolica* as its nearest RDP homologue thereby indicating that this organism may also be present in the oil field environment (Voordouw *et al.*, 1996). Another toluene-degrading SRB, strain PRTOL1, was isolated from fuel-contaminated subsurface soil (Beller *et al.*, 1996). This environment may be expected to have much lower sulfate concentrations than marine sediments. Strain PRTOL1 oxidized toluene with a very similar stoichiometry to that of strain Tol2. It also transformed *o*- and *p*-xylene when grown with toluene. Phylogenetic classification from 16S rRNA sequencing indicated a close association with a *Desulforhabdus* species.

Strain TD3, a moderate thermophile isolated from a deep marine sediment where aliphatic and aromatic hydrocarbons were formed by hydrothermal activity, oxidized *n*-alkanes from C_6 to C_{16} and fatty acids from C_4 to C_{18}. Typical *Desulfovibrio* substrates (ethanol, hydrogen, lactate) were not used (Rueter *et al.*, 1994). Sequencing of 16S rRNA indicated that strain TD3 represents a new deep branch within the sulfate-reducing bacteria. Mesophilic, sulfate-reducing enrichments from an oil tank stimulated growth of SRB that used primarily toluene and *o*-xylene (Rueter *et al.* 1994), while alkanes were not degraded. These organisms were shown to belong mostly to the *Desulfo-bacteriaceae* (Rabus *et al.*, 1996).

3.4   ASSAYS FOR SULFATE-REDUCING BACTERIA IN OIL FIELDS

Lactate-sulfate based Most Probable Number (MPN) assays are still the most popular way of evaluating the presence of SRB in industrial settings. The fact that this assay targets primarily the members of the family *Desulfovibrionaceae* and other incomplete oxidizers may be acceptable if the aim is to gain information on the potential for anaerobic microbial corrosion (MIC) as *Desulfovibrio* species have been implicated in MIC in the oil field environment (Hamilton, 1985; Cord-Ruwisch & Widdel, 1986; Voordouw *et al.*, 1993). If a more complete estimate of SRB activity in oil field production waters is required, MPN assays with hydrocarbon substrates may be considered (Rabus *et al.*, 1996). These take more time and will remain technically difficult unless a premade medium is made commercially available for the cultivation of these organisms. These assays would catch the bulk of complete oxidizers present in production waters and may provide data on the potential or occurrence of reservoir souring.

The proportion of *Desulfobacteriaceae* and *Desulfovibrionaceae* can also be estimated by fluorescence microscopy using family-specific probes (Rabus *et al.*, 1996). Widespread adoption of such methods for evaluation of corrosion or souring potential may be hampered by the fact that these assays are technically complicated. This would of course not matter if the most economic way to exploit an oil field depended critically on obtaining this information.

Determination of the prevalence of genes or gene products that function in corrosion or souring has also been explored. The enzyme hydrogenase has been postulated to be important in MIC (Hamilton, 1985). The genes for three different hydrogenases in *Desulfovibrio* spp. have been cloned and sequenced. These cloned genes have been used to determine the diversity of these species in the oil field environment (Voordouw *et al.*, 1990). Also, a rapid colorimetric hydrogenase assay has been advocated as a means to obtain information on the prevalence of corrosive SRB at field sites (Bryant *et al.*, 1991).

Estimation of SRB concentrations by monitoring genes or gene products specific for the dissimilatory sulfate reduction pathway is another possibility.

Odom *et al.* (1991) have developed a rapid immunological assay for adenosine-5'-phosphosulfate reductase (APS reductase). This enzyme is sufficiently conserved between different species to allow correlation between the extent of the immunoreaction and the numbers of SRB present. The dissimilatory sulfite reductases of the thermophilic archaeum *Archaeoglobus fulgidus* and the mesophilic bacterium *Desulfovibrio vulgaris* have been shown to be highly conserved (Karkhoff-Schweizer *et al.*, 1995). Comparison of the two sequences allowed definition of primers that can be used to amplify the genes for this H_2S-producing enzyme from other dissimilatory sulfate-reducers or directly from environmental samples. Cloning of the product obtained from environmental samples may be expected to give information on the diversity of SRB present, while quantitative PCR or hybridization assays for the sulfite reductase genes could conceivably be used to estimate the potential for H_2S production activity.

4   Heterotrophic and Fermentative Bacteria

One might suppose that anaerobic fermentative bacteria that derive energy for growth from, for instance, the conversion of carbohydrates into lactate, will be scarce in oil fields. After all hydrocarbons cannot be fermented. Nevertheless, these organisms are readily isolated. This may be in part because carbohydrate polymers or other fermentable substrates are sometimes added to enhance oil recovery (EOR).

Microorganisms can have a positive role in EOR processes, e.g. subsurface fermentation can lead to
(i)   gas production, which may force oil from the pores in which it is held,
(ii) microbial production of surfactants may reduce the oil-rock interfacial tension to aid in oil displacement,
(iii) growth of microorganisms may block established and nonproductive paths of injection water to allow sweeping of regions that still hold considerable oil, or
(iv) acid produced in microbial fermentation may increase permeability by attack on carbonate rock. Processes in which microorganisms are thought to help in oil production are referred to as microbially enhanced oil recovery (MEOR).

Of course, the processes outlined in (i) to (iv) may also be achieved by chemical means. Instead of relying on bacteria to produce surfactants these may be injected directly. However, it must be realized that these compounds, which often comprise a hydrocarbon tail and a polar nitrogen- or phosphorus-containing headgroup, may be a much easier source of food for the resident bacteria than the oil hydrocarbons. It may thus be difficult to separate intended chemical EOR from enhanced production by microbial processes.

McInerney and coworkers have shown that a variety of aerobic and anaerobic heterotrophs and other bacteria that may be active in MEOR are present in petroleum reservoirs (Adkins *et al.*, 1992a; Bhupathiraju *et al.* 1993). They have also demonstrated that selected strains may help in oil recovery from model cores in laboratory studies (Raiders *et al.*, 1989; Adkins *et al.*, 1992b).

Grassia *et al.* (1996) isolated fermentative bacteria and archaea from 17 out of 36 petroleum reservoirs. Production waters from these fields came from depths of 400 to 3000 m with resident temperatures of 21 to 130°C. Fermentative microorganisms enriched by supplementing production waters with glucose and a rich medium component (e.g. yeast extract or peptone) were found to belong to three groups: the genera *Thermoanaerobacter* and *Thermoanaerobacterium* (Group 1), the order *Thermotogales* (Group 2), and archaea of the genus *Thermococcus* (Group 3). The latter had previously been reported by Stetter and coworkers (1993), who also identified archaea of the genus *Pyrococcus* and members of the sulfate-reducing genus *Archaeoglobus*. Stetter *et al.* (1993) stated that the predominant mode of nutrition in these high temperature reservoirs is currently unknown as none of these organisms use hydrocarbons for growth.

Grassia *et al.* (1996) considered that their data supported the view that fermentative organisms are widespread in subterranean oil fields, but they did not indicate how these organisms made their living. Hermann *et al.* (1992) obtained evidence for the presence of SRB, acetogens and methanogens in oil wells, and expressed the view that these bacteria formed a complete anaerobic community that was dependent on the injection of carbohydrate polymers for enhanced oil recovery. An interesting question is thus whether the failure of Grassia *et al.* (1996) to isolate fermentative bacteria and archaea from 19 of the 36 reservoirs examined, is related to the absence of introduced fermentable carbohydrate substrates in these reservoirs. Sequencing of 16S rRNA genes indicated the presence of acetogenic and/or fermentative bacteria in western Canadian oil fields (Voordouw *et al.*, 1996). Direct plating yielded a wide variety of heterotrophs, including members of the genera *Aeromonas* or *Shewanella*, *Bacillus*, *Oceanospirillum*, *Proteus* and *Vibrio*, from this environment (Telang *et al.*, 1997).

5   Sulfide-Oxidizing Bacteria

The growth of SRB in oil reservoirs in which oil is produced by water flooding can raise H_2S concentrations with time. Loss of production or revenue results if the souring process leads to H_2S concentrations exceeding maximum allowable limits, because this requires the introduction of H_2S scavenging equipment or the replacement of pipes and equipment to handle sour product streams. Microbiological sulfate reduction was confirmed as the main souring mechanism even in North Sea fields, which have a high resident temperature (>100°C) and are injected with sea water with a high sulfate concentration (Eden *et al.*, 1993). Souring is most prominent in a region referred to as the Thermal Viability Shell or TVS, which surrounds the injector and is cooled by sea water injection to below 85°C, through growth of thermophilic SRB and hyperthermophilic sulfate-reducing archaea.

Sulfide production can be reduced by injection of biocides (e.g. glutaraldehyde), but laboratory tests have shown that nitrate suppresses sulfide levels more effectively (Reinsel *et al.*, 1996). Jenneman *et al.* (1986) demonstrated that nitrate addition could inhibit biogenic sulfide production in sewage sludge, that had been amended with sulfate. Sublette *et al.* (1994) showed that *Thiobacillus denitrificans*, an obligate autotroph and facultative anaerobe, can oxidize H_2S to sulfate with reduction of nitrate to nitrogen. Efficient removal of H_2S from sour water or sandstone cores was achieved with a derivative (strain F) of this organism, which was more resistant to inhibition by high sulfide levels. These results suggested that sulfide might be removed from production waters by the addition of *Thiobacillus denitrificans* and sufficient nitrate (Sublette *et al.*, 1994). However, the isolation and characterization of SOB from oil fields suggest that the addition of exogenous bacteria may not be necessary.

Analysis of 16S rDNA sequences, following PCR amplification of total DNA from production waters or enrichment cultures, has indicated the presence of bacteria that are related to species of the genera *Arcobacter* and *Thiomicrospira*. Many of these organisms are able to oxidize sulfide at neutral pH (Voordouw *et al.*, 1996). Jenneman *et al.* (1997) have characterized strain CVO, a mesophilic bacterium isolated from a somewhat saline oil field in western Canada. In production waters this strain catalyzes the oxidation of sulfide by nitrate, according to equation 1:

$$5HS^- + 2NO_3^- + 7H^+ \rightarrow 5S_0 + N_2 + 6H_2O.$$

This reaction stoichiometry differs from that for sulfide oxidation by *Thiobacillus denitrificans*, which produces sulfate according to equation 2:

$$5HS^- + 8NO_3^- + 3H^+ \rightarrow 5SO_4^{2-} + 4N_2 + 4H_2O.$$

Removal of sulfide by strain CVO from oil field production waters may thus be achieved by nitrate addition and does not result in the generation of sulfate, the preferred electron acceptor of SRB. Strain CVO is a coccobacillus of diameter 0.5 μm. Sequencing of its 16S rRNA gene has indicated affiliation with the genus *Campylobacter* (Voordouw *et al.*, 1996). Strain CVO is widely distributed in production waters derived from western 'Canadian oil fields. This was demonstrated by hybridizing total community DNAs isolated from production waters before and after nitrate addition with a strain CVO-specific whole genome probe (Voordouw *et al.*, 1996).

Amendment of production waters with nitrate in an actual oil field also gave a significant reduction of sulfide levels (Telang *et al.*, 1997). It was shown that, even on this much larger field scale, strain CVO was specifically responsible for this effect (Figs 5 and 6). This indicates that the number of different bacteria capable of oxidizing sulfide with nitrate is very limited relative to the large diversity of SRB that is present in these fields (Fig. 4).

Injection of ammonium nitrate to a final concentration of 400 mg/l into a Saskatchewan oil field for fifty days (Fig. 5 from day 0 to day 50) led to a 40 to 90% reduction of sulfide levels in production and injection waters. The numbers of nitrate-reducing bacteria increased by 3 to 7 orders of magnitude immediately following nitrate addition, and decayed when this amendment was stopped. SRB decreased by 1 to 2 orders of magnitude (Fig. 5). The effects of nitrate addition on the oil field microbial community were monitored by reverse sample genome probing (RSGP), a technique that allows the tracking of multiple bacteria with a single hybridization assay. In RSGP the fractions of bacteria, calculated from the hybridization pattern, are plotted against species number in a bar diagram (Fig. 6). Average patterns for two injection wells and two production wells are shown. The data (Fig. 6) indicate that prior to nitrate injection strain CVO was a minor community component (f_x=1-5%), but during nitrate injection it became the dominant community component (f_x=60-80%). After injection was stopped strain CVO levels decreased immediately in the injector wells and more slowly at the production wells. The data strongly suggest that strain CVO (Fig. 6: species 32) is solely responsible for the reduction of sulfide levels upon nitrate injection.

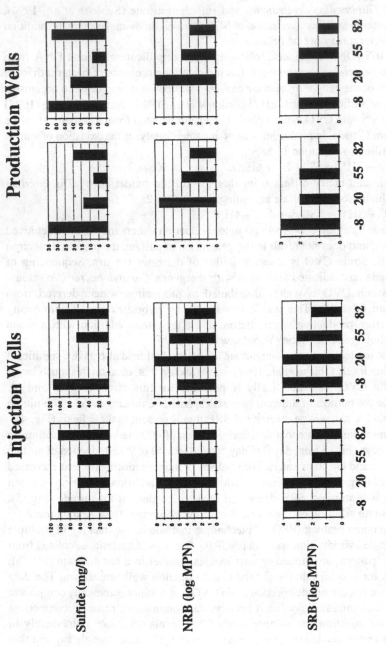

Figure 5. Effect of nitrate injection on sulfide concentrations (mg/l) and numbers of nitrate-reducing(NRB, logarithm of the MPN) and sulfate-reducing (SRB, logarithm of the MPN) bacteria. Data are shown for water samples collected from two injection and two production wells. Nitrate injection was from day 0 to day 50 and samples were collected before (−8 days), during (20 days) and after (55 and 82 days) nitrate injection. Adapted from data presented in Telang *et al.* (1997).

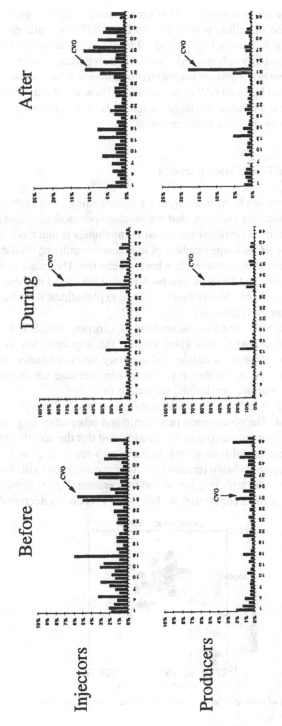

Figure 6. Effect of nitrate injection on the microbial population in an oil field. The vertical scale represents the calculated fractions (fx, %) at which each of 47 oil field bacteria (horizontal scale) are present. The fx values were calculated from quantitative reverse genomic hybridization. The diagrams are averages for two production and two injection wells. Adapted from data presented in Telang *et al.* (1997).

In the oil field community strain CVO is clearly the principal organism that derives energy from oxidation of sulfide with nitrate to form sulfur and nitrogen (equation 1). None of the SRB (Fig. 6: species 1 to 17, and 19 to 31) and heterotrophic or fermentative bacteria (species 18, and 33 to 47) showed a similarly strong increase upon nitrate injection. This indicates that these organisms are not able to use nitrate (e.g. for hydrocarbon oxidation) or the products released by strain CVO's metabolism. These observations are in agreement with the suggestion that the influx of a higher potential electron acceptor (nitrate) is targeted towards oxidation of sulfide, not hydrocarbon.

6   The Oil Field Food Chain Paradox

Considerable progress has been made in our understanding of microbial processes in oil fields, but an outstanding problem, that prevents proper modelling and application of the catalytic potential of the resident microbial communities is that the food chain is not defined in quantitative terms. Large numbers of hydrocarbon-utilizing SRB may be present, but these do not release substrates that other bacteria can use. The reasons for the diversity of *Desulfovibrio* species and fermentative bacteria that are found in these environments is therefore not understood. Nevertheless, some explanations may be found in the geochemical and geological literature.

Shock (1988) indicates that the maturation of kerogen, which is at the cradle of the formation of oil deposits, also gives rise to the accumulation of amino acids, carboxylic acids, cresols and phenols in sedimentary basin solutions. With respect to carboxylic acids the possibility that acetic acid and methane are in thermodynamic equilibrium under subsurface conditions, according to:

$$CH_3COOH \leftrightarrow CO_2 + CH_4 \qquad \text{(equation 3)}$$

has been investigated. The observation that acetic and other acids (e.g. propionic acid) are present in high concentrations (up to 6 g/L) indicated that the decarboxylation reaction is slow (even on a geological time scale). Perhaps as a result of slow temperature and pressure dependent decarboxylation reactions, an inverse correlation exists between the total concentration of carboxylic acids in solution and temperature (Fig. 7). It was proposed that the sudden lowering of these concentrations below 80°C is due to microbial metabolism.

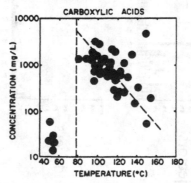

Figure 7. Concentrations of carboxylic acids in sedimentary basins as a function of temperature. Reproduced from Shock (1988).

The implication is that the process of diagenesis and catagenesis (Philp, 1986), which leads to formation of oil hydrocarbons, also leads to significant concentrations of organic acids in sediment layers immediately below the zone where active microbial metabolism may be expected. Like oil (Fig. 1), these organic acids may migrate upwards and be used as a carbon and energy source by locally booming subsurface populations of SRB, methanogens and fermentative bacteria. Subsurface flows, caused by oil production, may migrate these communities into the oil reservoir, where individual members may carve out a rich or meagre living depending on their metabolic potential and the influx of electron acceptors. Thus, the diversity of incompletely-oxidizing *Desulfovibrio* species that appear to be present in oil field production waters does not necessarily mean that these organisms make a living by oil degradation, but rather that the surrounding subsurface contains organic acid fed loci, where these organisms thrive.

7   Conclusions

The occurrence of a diversity of bacteria in oil fields is now widely accepted. Depending on the chemical and physical conditions in the oil-bearing subsurface, the resident microbial communities can be mesophilic or adapted to high salinities or high temperatures. Of the bacteria present, the SRB are best known. SRB are generally considered a nuisance, since they contribute to increased H_2S levels through souring, especially in fields where oil is produced by water injection. Souring or the direct action of SRB also contributes to metal corrosion of oil field equipment. Although the presence of easily culturable *Desulfovibrio* species can be demonstrated in most oil field production or tank bottom waters, bacteria of other genera (such as *Desulfobacter* or *Desulfobulbus*) may be more numerous. SRB capable of using oil hydrocarbons as electron donor for sulfate reduction are more closely related to members of these latter genera than to *Desulfovibrio* species. Thermophilic, Gram-positive SRB of the genus *Desulfotomaculum* and sulfate-reducing archaea of the genus *Archaeoglobus* have been found in high temperature oil fields.

Anaerobic fermentative and heterotrophic bacteria have also been reported. Thermophilic organisms are also present in these classes. Sulfide-oxidizing bacteria, that use a higher potential electron acceptor (e.g. ferric ions, nitrate, or oxygen) to reoxidize sulfide have also been found. Nitrate injection into an oil field can boost the fraction of SOB, while leaving other community members unaffected. These results have led to the generation of a model for the oil field microbial community in which oxidized forms of sulfur are the primary electron acceptors for hydrocarbon oxidation and in which influx of higher potential electron acceptors is primarily used to re-oxidize the sulfide formed, not the available hydrocarbons.

Because anaerobic oil field bacteria are hard to culture, surveys of multiple members of the oil field microbial community can best be done by molecular biological methods. These can help to define the changes in the microbial community in an oil field upon treatment aimed to reduce souring and corrosion or to increase production. Such information may aid in designing improved methods for oil production, based on exploiting the properties of the resident microbial community.

Acknowledgements

Research by the author on oil field microbial communities is funded by the Strategic Grants Program of the Natural Sciences and Engineering Research Council of Canada. This review was made possible through the efforts of, or collaborations with Sara Ebert, Julia Foght, Diane Gevertz, Gary Jenneman, Anita Telang and Don Westlake.

8 References

Adkins, J.P., Cornell, L.A. & Tanner, R.S. (1992a). Microbial composition of carbonate petroleum reservoir fluids. *Geomicrobiology Journal* 10, 87-97.

Adkins, J.P., Tanner, R.S., Udegbunam, E.O., McInerney, M.J. & Knapp, R.M. (1992b). Microbially enhanced oil recovery from unconsolidated limestone cores. *Geomicrobiology Journal* 10, 77-86.

Aeckersberg, F., Bak, F. & Widdel, F. (1991). Anaerobic oxidation of saturated hydrocarbons to CO_2 by a new type of sulfate-reducing bacterium. *Archives of Microbiology* 156, 5-14.

Beller, H.R., Spormann, A., Sharma, P.K., Cole, J.R. & Reinhard, R. (1996). Isolation and characterization of a novel toluene-degrading sulfate-reducing bacterium. *Applied and Environmental Microbiology* 62, 1188-1196.

Bhupathiraju, V.K., McInerney, M.J. & Knapp, R.M. (1993). Pretest studies for a microbially enhanced oil recovery field pilot in a hypersaline oil reservoir. *Geomicrobiological Journal* 11, 19-34.

Bryant, R.D., Jansen, W., Boivin, J., Laishley, E.J. & Costerton, J.W. (1991). Effect of hydrogenase and mixed sulfate-reducing bacterial populations on the corrosion of steel. *Applied and Environmental Microbiology* 57, 2804-2809.

Caumette, P., Cohen, Y. & Matheron, R. (1991). Isolation and characterization of *Desulfovibrio halophilus* sp. nov., a halophilic sulfate-reducing bacterium isolated from Solar Lake (Sinai). *Systematic and Applied Microbiology* 14, 33-38.

Cord-Ruwisch, R. & Widdel, F. (1986). Corroding iron as a hydrogen source for sulphate reduction in growing cultures of sulphate-reducing bacteria. *Applied Microbiology and Biotechnology* 25, 169-174.

Cord-Ruwisch, R., Kleinitz, W. & Widdel, F. (1987). Sulfate-reducing bacteria and their activities in oil production. *Journal of Petroleum Technology* 39, 97-106.

Eden, B., Laycock, P.J. & Fielder, M. (1993). Oilfield reservoir souring, *Health and Safety Executive-Offshore Technology Report*. ISBN 0717606376, pp. 1-85.

Fredrickson, J.K. & Phelps, T.J. (1997). Subsurface drilling and sampling. In *Manual of Environmental Microbiology*, 1st edn. pp. 523-540. Edited by C. J. Hurst, G. R. Knudsen, M. J. McInerney, L. D. Stetzenbach & M. V. Walter. ASM Press: Washington D.C.

Grassia, G.S., McLean, K.M., Glenat, P., Bauld, J. & Sheehy, A.J. (1996). A systematic survey for thermophilic fermentative bacteria in high temperature petroleum reservoirs. *FEMS Microbiology Ecology* 21, 47-58.

Hamilton, W.A. (1985). Sulphate-reducing bacteria and anaerobic corrosion. *Annual Review of Microbiology* 39, 195-217.

Herbert, B.N., Allison, P.W., Hardy, J.A., King, R.A., Sanders, P.F. & Stott, J. (1987). Review of current practices for monitoring bacterial growth in oilfield systems. In document No 001/87, *Corrosion Control Engineering Joint Venture and NACE* pp 1-16, Birmingham.

Hermann, M., Vandecasteele, J.-P. & Ballerini, D. (1992). Anaerobic microflora of oil reservoirs. Microbiological characterization of samples from some production wells. In *Bacterial Gas*, pp. 223-234. Edited by R. Viually. Paris: Editions Technip.

Huber, R., Stoffers, P., Cheminee, J.L., Richnow, H.H. & Stetter, K.O. (1990). Hyperthermophilic archaebacteria within the crater and open-sea plume of erupting Macdonald Seamount. *Nature* 345, 179-182.

Jenneman, G.E., McInerney, M.J. & Knapp, R.M. (1986). Effect of nitrate on biogenic sulfide production. *Applied and Environmental Microbiology* 51, 1205-1211.

Jenneman, G.E., Wright, M. & Gevertz, D. (1997). Sulfide bioscavenging of sour produced water by natural microbial populations. In *Proceedings of the 3rd International Petroleum Environmental Conference,* September 24 to 27, 1996, Albuquerque, NM.

Karkhoff-Schweizer, R.R., Huber, D.P.W. & Voordouw, G. (1995). Conservation of genes for dissimilatory sulfite reductase from *Desulfovibrio vulgaris* and *Archaeoglobus fulgidus* allows their detection by PCR. *Applied and Environmental Microbiology* 61, 290-296.

Krumholtz, L.R., McKinley, J.P., Ulrich, G.A. & Suflita, J.M. (1997). Confined subsurface microbial communities in Cretaceous rock. *Nature* 386, 64-66.

L'Haridon, S.L., Reysenbach, A.-L., Glenat, P., Prieur, D. & Jeanthon, C. (1995). Hot subterranean biosphere in a continental oil reservoir. *Nature* 377, 223-224.

Magot, M., Caumette, P., Desperrier, J.M., Matheron, R., Dauga, C., Grimont, F. & Carreau, L. (1992). *Desulfovibrio longus* sp. nov., a sulfate-reducing bacterium isolated from an oil-producing well. *International Journal of Systematic Bacteriology* 42, 398-402.

McInerney, M.J. & Sublette, K.L. (1997). Petroleum microbiology: Biofouling, souring, and improved oil recovery. In *Manual of Environmental Microbiology,* 1st edn., pp. 600-06. Edited by C.J. Hurst, G.R. Knudsen, M.J. McInerney, L.D. Stetzenbach & M.V. Walter. Washington D.C.: ASM Press.

Odom, J.M., Jessie, K., Knodel, E. & Emptage, M. (1991). Immunological cross-reactivities of adenosine-5'-phosphosulfate reductases from sulfate-reducing and sulfide-oxidizing bacteria. *Applied and Environmental Microbiology* 57, 727-733.

Philp, R.P. (1986). Geochemistry in search of oil. *Chemistry and Engineering News,* February 10, 28-43.

Rabus, R., Nordhaus, R., Ludwig, W. & Widdel, F. (1993). Complete oxidation of toluene under strictly anoxic conditions by a new sulfate-reducing bacterium. *Applied and Environmental Microbiology* 59, 1444-1451.

Rabus, R., Fukui, M., Wilkes, H. & Widdel, F. (1996). Degradative capacities and 16S rRNA-targeted whole cell hybridization of sulfate-reducing bacteria in an anaerobic enrichment culture utilizing alkylbenzenes from crude oil. *Applied and Environmental Microbiology* 62, 3605-3613.

Raiders, R.A., Knapp, R.M. & McInerney, M.J. (1989). Microbial selective plugging and enhanced oil recovery. *Journal of Industrial Microbiology* 4, 215-230.

Reinsel, M.A., Sears, J.T., Stewart, P.S. & McInerney, M.J. (1996). Control of microbiological souring by nitrate, nitrite or glutaraldehyde injection in a sandstone column. *Journal of Industrial Microbiology* 17, 128-136.

Rosnes, J.T., Torsvik, T. & Lien, T. (1991). Spore-forming thermophilic sulfate-reducing bacteria isolated from North Sea oil field waters. *Applied and Environmental Microbiology* 57, 2302-2307.

Rueter, R., Rabus, R., Wilkes, H., Aeckersberg, F., Rainey, F.A., Jannasch, H.W. & Widdel, F. (1994). Anaerobic oxidation of hydrocarbons in crude oil by new types of sulphate-reducing bacteria. *Nature* 372, 455-458.

Shock, E.L. (1988). Organic acid metastability in sedimentary basins. *Geology* 16, 886-890.

Stetter, K.O., Lauerer, G., Thomm, M. & Neuner, A. (1987). Isolation of extremely thermophilic sulfate reducers: evidence for a novel branch of archaebacteria. *Science* 236, 822-824.

Stetter, K.O., Huber, R., Blöchl, E., Kurr, M., Eden, R.D., Fielder, M., Cash, H. & Vance, I. (1993). Hyperthermophilic archaea are thriving in deep North Sea and Alaskan oil reservoirs. *Nature* 365, 743-745.

Stevens, T.O. & McKinley, J.P. (1995). Lithoautotrophic microbial ecosystems in deep basalt aquifers. *Science* 270, 450-454.

Sublette, K.L., McInerney, M.J., Montgomery, A.D. & Bhupathiraju, V. (1994). Microbial oxidation of sulfides by *Thiobacillus denitrificans* for treatment of sour water and sour gases. In *Environmental Geochemistry and Sulfide Oxidation*, pp. 68-78. Edited by C.N. Alpers and D.W. Blowes. Washington D.C.: American Chemical Society.

Tardy-Jacquenod, C., Caumette, P., Matheron, R., Lanau, C., Arnauld, O. & Magot, M. (1996). Characterization of sulfate-reducing bacteria isolated from oil-field waters. *Canadian Journal of Microbiology* 42, 259-266.

Telang, A.J., Ebert, S., Foght, J.M., Westlake, D.W.S., Jenneman, G.E., Gevertz, D. & Voordouw, G. (1997). Effect of nitrate on the microbial community in an oil field as monitored by reverse sample genome probing. *Applied and Environmental Microbiology* 63, 1785-1797.

Voordouw, G., Niviere, V., Ferris, F.G., Fedorak, P.M. & Westlake, D.W.S. (1990). The distribution of hydrogenase genes in *Desulfovibrio* and their use in identification of species from the oil field environment. *Applied and Environmental Microbiology* 56, 3748-3754.

Voordouw, G., Voordouw, J.K., Karkhoff-Schweizer, R.R., Fedorak, P.M. & Westlake, D.W.S. (1991). Reverse sample genome probing, a new technique for identification of bacteria in environmental samples by DNA hybridization, and its application to the identification of sulfate-reducing bacteria in oil field samples. *Applied and Environmental Microbiology* 57, 3070-3078.

Voordouw, G., Voordouw, J.K., Jack, T.R., Foght, J., Fedorak, P.M. & Westlake, D.W.S. (1992). Identification of distinct communities of sulfate-reducing bacteria in oil fields by reverse sample genome probing. *Applied and Environmental Microbiology* 58, 3542-3552.

Voordouw, G., Shen, Y., Harrington, C.S., Telang, A.J., Jack, T.R. & Westlake, D.W.S. (1993). Quantitative reverse sample genome probing of microbial communities and its application to oil field production waters. *Applied and Environmental Microbiology* 59, 4101-4114.

Voordouw, G., Armstrong, S.M., Reimer, M.F., Fouts, B., Telang, A.J., Shen, Y. & Gevertz, D. (1996). Characterization of 16S rRNA genes from oil field microbial communities indicates the presence of a variety of sulfate-reducing, fermentative, and sulfide-oxidizing bacteria. *Applied and Environmental Microbiology* 62, 1623-1629.

Widdel, F. & Bak, F. (1992). Gram-negative mesophilic sulfate-reducing bacteria. In *The Prokaryotes*, 2nd edn., pp. 3352-3378. Edited by A. Balows, H. G. Trüper, M. Dworkin, W. Harder & K. H. Schleifer. Springer-Verlag, New York.

12 SYSTEMATICS OF *SPHINGOMONAS* SPECIES THAT DEGRADE XENOBIOTIC POLLUTANTS

MARTINA M. EDERER *and* RONALD L. CRAWFORD

*Environmental Biotechnology Institute, University of Idaho,
Moscow ID 83844-1052, U.S.A.*

1 The Problem of Establishing a Meaningful Bacterial Phylogeny

Until recently, microbiologists were unable to establish a satisfactory prokaryotic phylogeny, a problem that some suggested was insolvable (Stanier & van Niel, 1962). Traditional methods for determining eukaryotic phylogenies according to cell morphology and physiological characteristics are not readily applicable to the classification of prokaryotes. Further, microbiologists have only been able to study those microorganisms that they could cultivate, and this strongly biased the perception of microbial diversity. Another obstacle that hampered the study of prokaryotic phylogeny is what Olsen *et al.* (1994) described as the "negative definition" of a prokaryote. Prokaryotes were defined as "lacking eukaryotic features," and such ideas as "if it is not a eukaryote, it is a prokaryote," dominated the field. With these underlying assumptions, microbiologists ignored the importance of phylogenetic relationships essential in understanding the nature of any organism, and reduced the study of prokaryotic diversity to the question, "How does *E. coli* differ from eukaryotes?" The characterization of rRNA nucleotide sequences in the early 1980s changed the paradigm.

Macromolecules have been used to determine the phylogenetic relationships of organisms since the 1950s with interest in this approach reaching a high point with the publication of "Molecules as Documents of Evolutionary History" by Zuckerkandl & Pauling (1965). However, the revolution in the study of prokaryotic phylogeny (Fox *et al.*, 1980) did not begin until the 1970s, when microbiologists first attempted to exploit rRNA sequences to help determine phylogenetic relationships. Since ribosomal sequences are generally considered to be well conserved and present in all genome-containing organisms, they lend themselves well to phylogenetic studies (Woese, 1991).

The prokaryotic ribosome consists of a 23S rRNA (~2900 nucleotides) and a 5S rRNA (~120 nucleotides) which are associated with about 30 ribosomal proteins to form the large ribosomal subunit, and a 16S rRNA associated with 20 ribosomal proteins to form the small subunit. Although the term 16S rRNA is commonly used, the actual size of the molecule may vary between 15S and 18S, or 1500 to 1900 nucleotides (Pace *et al.*, 1986). As of Sept. 12, 1999, more than 21,000 sequences are contained in the Ribosomal Database Project (RDP) SSU rRNA data files, with approximately one third of these sequences available in aligned form and placed on a phylogenetic tree (Maidak *et al.*, 1999).

F. G. Priest and M. Goodfellow (eds.), Applied Microbial Systematics, 333-366
© 2000 *Kluwer Academic Publishers. Printed in the Netherlands.*

Comparisons among the 16S rRNA nucleotide sequences have already had a profound effect on the structure of the phylogenetic tree of microorganisms. The narrow concept of *Escherichia coli* as the prokaryotic prototype was quickly abandoned. A whole new prokaryotic domain, the *Archaea*, was defined (Fig. 1). Woese and Fox (1977) described the archaea as more closely related to the eukaryotic domain than to the eubacteria. Forterre (1997), however, warned that until the universal tree has been rooted indisputably, such a notion may be explained by finding "human" features in archaea, a far more exciting endeavour than finding eubacterial features, but one that might bias phylogenetic analyses. There is also an ongoing discussion of whether the archaea represent a monophyletic group, a position supported by two pieces of evidence: all archaea identified thus far contain dextro-rotary ether-linked glycerolipids; and one of the large RNA polymerase subunits is split in two polypeptides (Forterre, 1997).

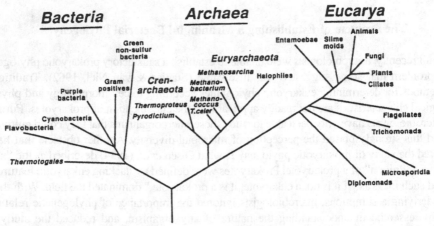

Figure 1. Universal tree of life (Woese, 1994). Xenobiotic-degrading bacteria had been classified as members of rather diverse genera within groups of the purple bacteria, flavobacteria, and Gram-positive bacteria. Recent classification studies, however, indicate that these bacteria seem to cluster in the new genus *Sphingomonas* within the α-subgroup of the purple bacteria.

The advent of the polymerase chain reaction (PCR) technology (Saiki *et al.*, 1985) further revolutionized the field of rRNA-based bacterial evolutionary phylogeny by allowing the rapid amplification, cloning, and nucleotide sequencing of 16S rRNA genes. Further developments in PCR technology have allowed microbial communities to be studied without cultivating the organisms (Amann *et al.*, 1995). Lee *et al.* (1996) reported the use of group- or species-specific PCR primers in quantitative PCR to determine the abundance and distribution of a particular bacterial species in a microbial community.

Since the introduction of 16S rRNA analyses, many improvements have been made to ensure that all members of a particular community are accurately represented (Reysenbach *et al.*, 1992). Caution must be taken so that the 16S rRNA sequences retrieved from such a community are not biased purely because of statistical reasons, leaving less abundant members of a community underrepresented in a 16S rRNA library.

This caution would also apply to members of bacterial species that are more resistant to lysis. Due to variation in the nucleotide sequence of the 16S rRNA genes, even in the very conserved regions used for primer design, some 16S rRNA genes may not be amplified efficiently under a given set of reaction conditions. Also, the possibility of creating chimeras poses a problem, leading to an exaggerated idea of bacterial diversity (Kopczynski *et al.*, 1994).

Many researchers caution that complete rRNA analysis cannot substitute for further characterization of a microorganism (Amann *et al.*, 1995). According to the rule of parsimony, one would expect that protein trees resembling rRNA trees would be considered the "good" ones (Forterre, 1997), and that the simplest explanation consistent with the data set should be chosen over more complex ones. The data are examined one character at a time, while alternative trees are being evaluated. The most parsimonious tree is determined to be the one requiring the fewest total events.

The definition of a bacterial species has not been unequivocally established. Stackebrandt & Goebel (1994) note that a 2.5% difference in 16S rRNA sequence would be sufficient for two sequences to be considered as derived from two species. However, two taxa may appear very similar when they are not, and, conversely, may appear very different when they are actually quite closely related (Pace *et al.*, 1986; Stewart, 1993). Other guidelines require 70% identity in DNA/DNA hybridization experiments for classification as one species (Wayne *et al.*, 1987). In addition, other relevant data can be obtained from fatty acid methyl ester analysis (FAME), which ideally should always accompany a 16S rRNA analysis (White, 1996).

Nonetheless, the study of phylogenetic relationships based on 16S rRNA analyses has not only broadened our view of microbial diversity, but has had, and will continue to have, a pronounced influence on the field of bacterial systematics. The results of 16S rRNA analyses have already led to the reclassification of many species and the establishment of numerous new genera. One such newly established genus is the genus *Sphingomonas* (Yabuuchi *et al.*, 1990).

2   The Genus *Sphingomonas*

As the recently proposed genus *Sphingomonas* (Yabuuchi *et al.*, 1990) becomes well established in bacterial taxonomy, many organisms previously classified as members of the Gram-negative genera *Beijerinckia*, *Flavobacterium*, and *Pseudomonas*, and even the Gram-positive genus *Arthrobacter*, are actually emerging as members of this new genus. Takeuchi *et al.* (1994) established the phylogeny of the genus *Sphingomonas* as a non-photosynthetic member of the α-4-subclass of the *Proteobacteria*, formerly "purple bacteria and their relatives" (Stackebrandt *et al.*, 1988).

The genus *Sphingomonas* groups Gram-negative, non-spore-forming, aerobic, yellow-pigmented, straight rods (Yabuuchi *et al.*, 1990). Its members can be distinguished from those of other genera by several unique characteristics. First, the yellow pigmentation in *Sphingomonas* is due to the carotenoid nostoxanthin, rather than to the more commonly occurring brominated arylpolyene xanthomonadin. The latter pigment is responsible for the yellow colour in members of the genus *Xanthomonas* (Jenkins *et al.*, 1979). Second, members of the new genus are characterized by the presence of sphingoglycolipids, which, although common in membranes of eukaryotic organisms, had not previously been found in prokaryotic cells. Cells also contain octadecanoic acid, 2-hydroxymyristic acid, cis-9-hexadecenoic acid, and hexadecanoic acid as major fatty acids (Yabuuchi et al., 1990). Other distinctive characteristics are the presence of the isoprenoid Q10, and DNA containing 62 to 67% GC. Finally, a comparison of the genus *Sphingomonas* and *Pseudomonas aeruginosa* shows that in the 16S rRNA sequence of the region between 1220 and 1376, nucleotide 1290 is deleted in the 16S rRNA of members of the genus *Sphingomonas*. *Sphingomonas paucimobilis* (NCTC 11030), a clinical isolate implicated in meningitis, septicemia, and wound infections (Holmes et al., 1977), was proposed as the type species. Other species identified were *S. adhaesiva*, *S. parapaucimobilis*, and *S. yanoikuyae* (a strain which was not yellow pigmented). In the same report it was determined that the non-motile "*Flavobacterium capsulatum*" (type strain ATCC 14666), isolated from distilled water (Leifson, 1962), should be reclassified *Sphingomonas capsulata*.

However, the organization of the genus has not been established unequivocally. In 1993, van Bruggen et al. found that according to 16S rRNA analyses *S. yanoikuyae* was more closely related to *Rhizomonas suberifaciens* than to other sphingomonads. Since *R. suberifaciens* does not seem to have sphingolipids, reexamination of this organism is in order. In the same report, it was recommended that *S. capsulata* should be placed into a separate genus, as-yet undetermined, on the basis of parsimony and bootstrapping analysis of 270 bases of the 16S rRNA gene. Takeuchi et al. (1994) supported this conclusion and suggested the division of the genus *Sphingomonas* into two subgroups. In addition, the reclassification of *S. yanoikuyae* into the genus *Rhizomonas* and the placement of *S. capsulata* into a separate genus were proposed by these workers.·

Another aspect of the genus *Sphingomonas* is the secretion of gellan-related polysaccharides (Pollock, 1993). These exopolysaccharides were first thought to be produced by a variety of different organisms, but Pollock (1993) found that they were mainly a characteristic of members of the genus *Sphingomonas*. The species he studied included environmental isolates from soil and water, as well as clinical specimens. It was postulated that the acidic heteropolysaccharides may play a role in pathogenesis (Decker et al., 1992).

In further studies, Pollock et al. (1998) found that the enzymes involved in the initial steps of exopolysaccharide biosynthesis are similar in different bacteria and determined that the glycosyl-(β1→4)-glucoronosyl transferase from *Sphingomonas* strain SS 88 can complement a *Rhizobium leguminosarum* mutant and *vice versa*. Generally, it is believed that exopolysaccharides, in addition to their importance in pathogenesis, play a role in tolerance to toxic chemicals. However, Richau et al. (1997) found that a non-gellan producing mutant of *Sphingomonas paucimobilis* R40 was less sensitive to Cu^{2+} ions. Clearly, much more research is required to elucidate the role and specificity of these exopolysaccharides.

Many *Sphingomonas* spp. can be isolated from a variety of oligotrophic environments, that is, distilled water, well water, and clinical environments. Even in the laboratory these organisms prefer oligotrophic media such as 1/10 strength tryptic soy agar (Ederer, unpublished observation). Eguchi et al. (1996) referred to *Sphingomonas* spp. as "model" oligotrophic bacteria on the basis of their finding that *Sphingomonas* strain RB2256, a marine isolate predominant in Resurrection Bay, Alaska, encodes for high-affinity nutrient uptake systems and grows only as very small (<0.08 μm) cells under nutrient-limiting conditions.

Further studies are needed to elucidate the mechanisms underlying the low nutrient requirements of this *Sphingomonas* isolate. The results of these studies may influence bioremediation efforts using sphingomonads, since these organisms are often exposed to oligotrophic environments. In summary, the genus *Sphingomonas* groups together an interesting collection of bacteria encoding potentially useful metabolic pathways. Therefore, a better understanding of the phylogeny of this genus, its members, and their respective catabolic pathways, along with the consideration of their particular biotope, will lead to advances in the field of bioremediation.

2.1 SPHINGOLIPIDS

Sphingoglycolipids, well-known molecules studied as important components of eukaryotic cell membranes, are implicated as playing a crucial role in cell surface antigenicity. Hannum & Bell (1989) discussed a possible role for sphingolipids in second messenger and other regulatory mechanisms in complex eukaryotic organisms, basing their argument on the notion that complex molecules such as sphingolipids would survive through the course of evolution only if they served an important role in cellular metabolism. Hakamori (1984) suggested a role for sphingolipids in cancer by stating that "alterations in sphingolipids may be implied to cause chaotic and undisciplined social interactions characteristic of cancer cells." Sphingolipids have also been found in the membranes of eukaryotic microorganisms, including fungi (Brennan *et al.*, 1974).

Sphingoglycolipids represent an important component of the eukaryotic cell membrane and have been implicated in cell surface antigenicity. The occurrence of sphingoglycolipids is rare in prokaryotic organisms. Yamamoto *et al.* (1978) were the first to isolate a novel sphingoglycolipid, glucuronosyl ceramide, which consists of the acid sugar glucuronic acid as the carbohydrate moiety and 2-hydroxymyristic acid as the fatty acid component from *Flavobacterium devorans* strain ATCC 10829. This unusual molecule had not been seen previously in either eukaryotic or prokaryotic systems and the 1978 report of Yamamoto *et al.* was the first describing a glycosphingolipid containing glucoronic acid as the sugar moiety. In 1990, Yabuuchi *et al.* postulated that the occurrence of glucuronosyl ceramide in the outer leaf of a bacterial membrane must play an important role in the determination of the taxonomic position of the new genus *Sphingomonas*. Thus, these glycosphingolipids, rare in the prokaryotic kingdom, might be of great significance in sphingomonad metabolism and antigenicity, and represent an important tool for establishing a meaningful phylogeny.

Kawahara *et al.* (1991) determined that *Sphingomonas paucimobilis* is devoid of lipopolysaccharide (LPS), which is an important component of the outer membrane of all other Gram-negative bacteria. Instead, the outer membrane of this organism contains other amphiphilic molecules replacing LPS in the membrane. These researchers purified and analysed the structure of two glycosphingolipids, replacing the LPS in the outer membrane of *Sphingomonas paucimobilis*. Kawasaki *et al.* (1994) confirmed these results and also suggested that the glycosphingolipid in the sphingomonad outer membrane plays the same roles as the LPS component in the outer membrane of the other prokaryotes. Wiese *et al.* (1996) found remarkable overlap in function of the two glycolipids in maintaining membrane fluidity, acting as a permeability barrier for bactericidal substances, and establishing the hydrophilic and antigenic character of the cell.

Since the genus *Sphingomonas* was established in 1990, many bacteria capable of degrading anthropogenic compounds have been (re)classified into this genus, indicating the possible importance of sphingolipids in the biodegradation of such compounds. Nohynek *et al.* (1995) hypothesized that the sphingolipids may act as a surfactant, improving the bioavailability of the anthropogenic compounds. In support of this hypothesis, they reported that four pentachlorophenol-degrading *Sphingomonas* strains were able to emulsify hexadecane. However, since the toxicity of pentachlorophenol (PCP) stems from its ability to uncouple ATP production (oxidative phosphorylation) from electron transport, controlling the bioavailability of PCP, which shows good solubility in both polar and non-polar solvents, may not affect its biodegradation. These hypotheses are speculative, and more research is needed to establish the roles of sphingolipids in the physiology of *Sphingomonas* spp. as well as their role, if any, in biodegradation processes.

3.   Biodegradation of Anthropogenic Compounds by *Sphingomonas* Species

Hundreds of bacterial species that degrade compounds of anthropogenic origin have been identified. The biodegradation of these compounds, especially xenobiotic ones alien to existing enzyme systems, has become a broad field of basic and applied research. Many of these catabolically diverse bacteria were previously classified into various Gram-negative genera, such as *Flavobacterium* and *Pseudomonas*, and even into the Gram-positive genus *Arthrobacter* (Fig. 1). With the advent of phylogenetic analysis based on 16S rRNA sequence data, many of these organisms were found to be very closely related to each other and were subsequently classified into one genus, *Sphingomonas*.

The diverse biodegradative activities mediated by these *Sphingomonas* strains will be considered here, and then their relatedness to each other and to members of other genera implicated in the biodegradation of xenobiotic compounds. The role played by *Sphingomonas* spp. in the degradation of the pesticide pentachlorophenol (PCP) will then be examined.

3.1   POLYETHYLENE GLYCOL

Polyethylene glycol (PEG; Fig. 2A) is a water-soluble lubricant which will not hydrolyse upon storage and is resistant to degradation by molds. Because of its low toxicity it has wide application in metal-forming operations, in food packaging, and in the cosmetic industry. Several bacteria have been reported to degrade PEG anaerobically, for example, *Bacteroides* spp. and *Desulfovibrio desulfuricans* (Dwyer & Tiedje, 1986). Aerobic PEG degradation has been demonstrated in *Pseudomonas aeruginosa* (Haines & Alexander, 1975) and *Pseudomonas stutzeri* (Obradors & Aguilar, 1991).

Biodegradation of PEG with a molecular weight of 4,000 has been reported for several different pure bacterial cultures which were initially classified as *Flavobacterium* spp. and as *Pseudomonas* spp. (Kawai *et al.*, 1977, 1984; Yamanaka & Kawai, 1989).

Kawai & Yamanaka (1986) also reported that a symbiotic mixed culture of two different bacterial strains, tentatively identified as a *Pseudomonas* sp. and *a Flavobacterium* sp., was required to degrade PEG 6,000 (Fig. 2B). These researchers identified three enzymes produced by the *Flavobacterium* component that were responsible for the initial degradation steps: PEG dehydrogenase, PEG aldehyde dehydrogenase, and PEG carboxylate dehydrogenase. The last enzyme was also present in the *Pseudomonas sp.* The *Pseudomonas* component of the obligate consortium degraded the by-product glyoxylate faster than the *Flavobacterium* sp. It was suggested that this intermediate, which strongly inhibited the enzyme PEG carboxylate dehydrogenase, represented the key compound necessitating the symbiosis.

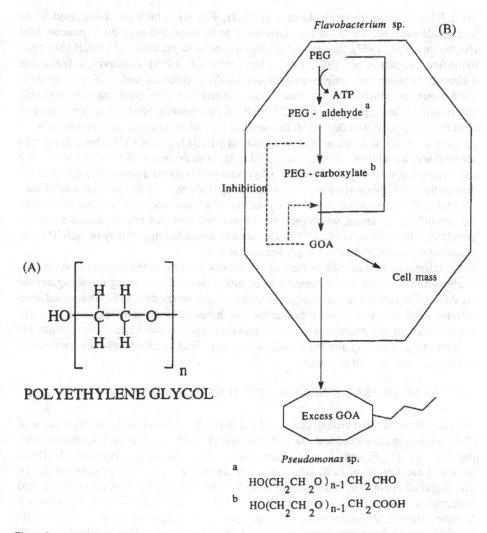

POLYETHYLENE GLYCOL

Figure 2. (A) Structure of polyethylene glycol (PEG) and (B) proposed symbiotic degradation pathway by *Flavobacterium* sp. and *Pseudomonas* sp. (adapted with permission from Kawai & Yamanaka, 1986).

Takeuchi *et al.* (1993) determined that six bacterial strains capable of PEG 4,000 degradation should be classified into the genus *Sphingomonas* and proposed the new species *Sphingomonas macrogoltabidus* sp. nov. They also found that the six strains implicated as the dominant components in the PEG 6,000 degradation consortium were members of the genus *Sphingomonas* and should be classified as *Sphingomonas terrae*. Both *Sphingomonas* strains were determined to be closely related to the genus *Rhizomonas* (Fig. 3). *Sphingomonas* and *Rhizomonas* are non-photosynthetic members of the α-4-subclass of the *Proteobacteria* (Takeuchi *et al.*, 1994).

3.2   γ-1,2,3,4,5,6-HEXACHLOROCYCLOHEXANE DEGRADATION

γ-1,2,3,4,5,6-Hexachlorocylcohexane (γ-HCH; Fig. 4), which has been used as an insecticide since the 1970s, is thus far known to be degraded only by *S. paucimobilis* (Nishiyama *et al.*, 1992). Several genes involved in the degradation of γ-HCH have been identified (Nagata *et al.*, 1994, 1997; Miyauchi *et al.*, 1998). However, a distinctive difference between the indigenous *S. paucimobilis* population and an *S. paucimobilis* SS86 population which was inoculated into the soil has been seen. Both strains were readily able to multiply in the presence of γ-HCH, but the *S. paucimobilis* SS86 populations vanished after the compound was degraded. In contrast, the indigenous strain is known to have survived in the soil since the annual application of γ-HCH began in 1973. This difference in survival was found to be directly correlated to the distribution of the bacteria within the soil fractions. It was suggested that the indigenous strains found a favourable microhabitat deep in the soil within soil micropores. A follow-up study showed that pretreatment of soils with γ-HCH before inoculation with *S. paucimobilis* strain SS86 ensured the survival of the non-indigenous strain, which probably became incorporated into the native microbial population (Nishiyama *et al.*, 1993). The aerobic degradation pathway of γ-HCH by *S. paucimobilis* strain SS86 has not yet been determined.

Ecotaxonomic studies such as these are extremely relevant to the design of systems for biodegradation of xenobiotic compounds by either naturally occurring or bioengineered organisms. The survival of the biodegradative agent must be ensured, and the bioremediation activity must be reliable and reproducible. A thorough understanding of the growth requirements of the organisms involved, and their specific tolerance or intolerance for both naturally occurring and man-made compounds, will accelerate the development of efficient bioremediation processes.

3.3   2,4-DICHLOROPHENOXYACETIC ACID DEGRADATION

The intensively studied biodegradation of the herbicide 2,4-dichlorophenoxyacetic acid (2,4-D) has been associated with the pJP4 plasmid of *Alcaligenes eutrophus*, which carries the *tfdB, -C, -D, -E,* and *-F* genes (Don *et al.*, 1985). Recently, Leveau *et al.* (1998) identified another gene, *tfdK*, which encodes an active transporter protein for 2,4-D. The degradation pathway is shown in Figure 5. Other bacteria have also been implicated in degrading 2,4-D; among these are members of *Achromobacter, Alcaligenes, Arthrobacter, Corynebacterium, Flavobacterium, Pseudomonas,* and *Streptomyces* spp. However, the classification of these organisms may have to be reexamined, since much of the work

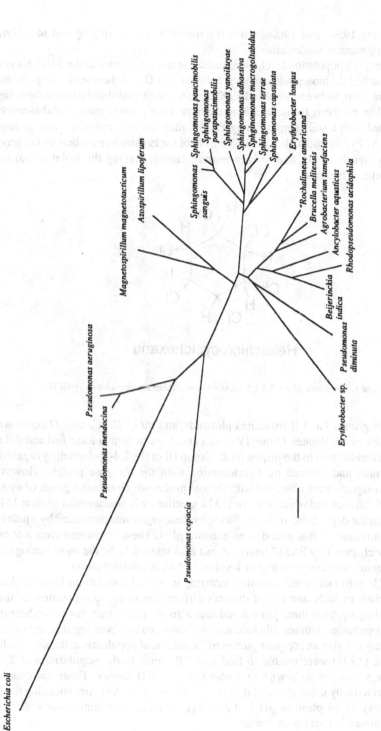

Figure 3. Unrooted phylogenetic tree displaying the relationship of different *Sphingomonas* species in comparison to several reference organisms. The topology of the tree was evaluated by bootstrap analysis. (With permission from Takeuchi *et al.*, 1994).

was done in the 1950s and 1960s, when the methods for classifying and identifying bacteria were generally inadequate.

Furthermore, at a population level, bioremediation is not very well understood. Ka *et al.* (1994a) approached the biodegradation of the herbicide 2,4-D with the question of population composition in mind. Between 1989 and 1992 these researchers isolated bacteria from eight agricultural plots, purifying the isolates from the most dilute positive most-probable-number tube inoculated with a soil sample. Using FAME analyses and repetitive extra genomic palindromic (REP) sequences, they identified 57% of the isolates as members of the genera *Alcaligenes*, *Pseudomonas*, and *Sphingomonas*, further placing the isolates into four different groups of bacteria.

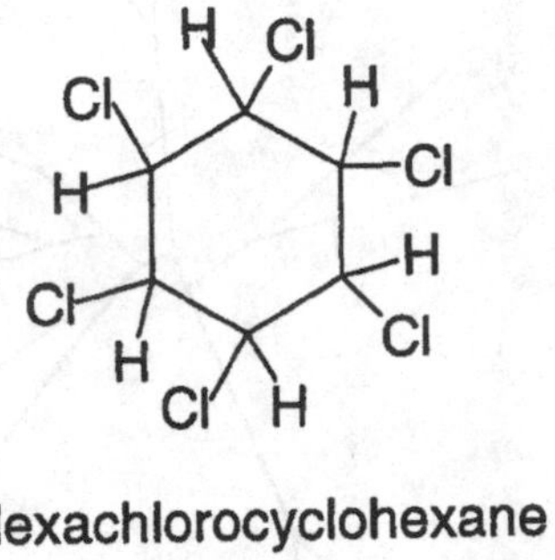

Hexachlorocyclohexane

Figure 4. Structure of γ-1,2,3,4,5,6-hexachlorocylcohexane degradation (γ-HCH).

Bacteria in groups I and II contained plasmids, and *tfdA, -B, -C*, and *-D* genes were located for 75% of the isolates. Group IV strains could not be further identified and did not exhibit any hybridization to the probes used. Group III of the 2,4-D-degrading organisms lacked plasmids and showed no hybridization with the *tfd* gene probes. However, hybridization signals were observed with the *spa* probe, which encodes genes of as-yet-undetermined function and which is known to be specific for *S. paucimobilis* (isolate 1443), suggesting that the degradation of the herbicide by these organisms proceeded by a pathway completely unrelated to that encoded on plasmid pJP4. These organisms were not only more closely clustered by FAME analyses, but also seemed to be the most ecologically successful group, tolerating the higher level of herbicide contamination.

Ka *et al.* (1994b) monitored naturally occurring microbial populations from Michigan and Saskatchewan soils and found distinct differences in the composition of these 2,4-D-degrading communities. They noted that a single probe may not be sufficient to monitor the populations with specific function, and cautioned that even together, the *tfdA* and *spa* probes may not give an adequate picture of the microbial population at the sites studied. Holben *et al.* (1992) were unable to find any difference in the populations of 2,4-D degrading organisms in soils with or without prior 2,4-D history. These communities could be detected only using *tfdA* and *tfdB* probes, but not with the genes encoding the rest of the pathway as on plasmid pJP4 of *Alcaligenes eutrophus*, indicating a different catabolic pathway in those communities.

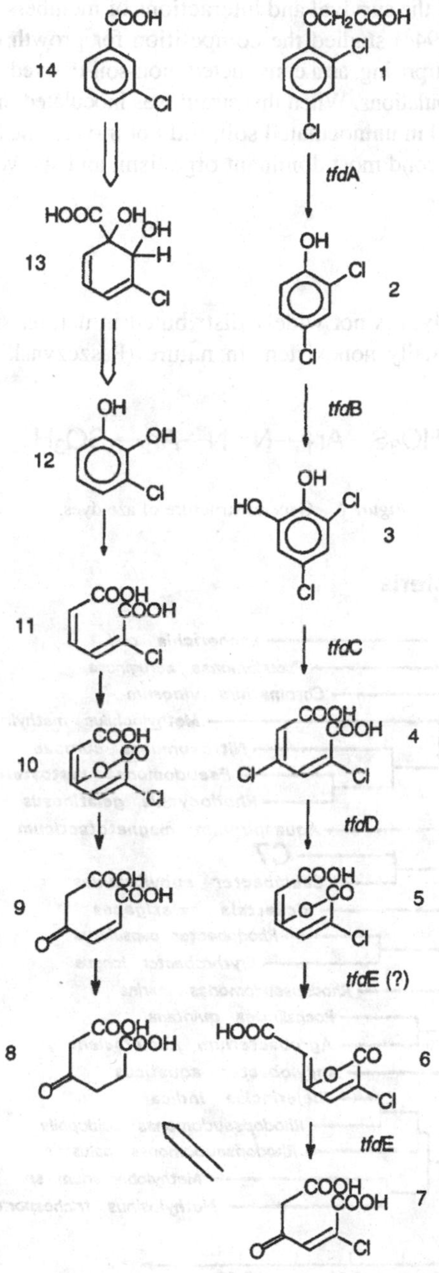

Figure 5. Proposed degradation pathways for the degradation of 3-chlorobenzoate and 2,4-D by *Alcaligenes eutrophus* strain JMP134. The genes involved are *tfdA*, 2,4-D monooxygenase; *tfdB*, 2,4 dichlorophenol hydroxylase; *tfdC*, chlorocatechol 1,2-dioxygenase; *tfdD*, chloromuconate cycloisomerase; *tfdE*, 4-ca boxymethylenebut-2-en-4-olide (diene-lactone)hydrolase. Trans-2-chlorodiene-lactone is probably transformed by *tfdF* encoding atrans-chlorodiene-lactone isomerase. (For more details see Don *et al.*, 1985).

M. M. Ederer and R. L. Crawford

To further understand the survival and interactions of members of different species in a biotope, Ka *et al.* (1994c) studied the competition for growth on 2,4-D in the field. The results were rather surprising, as a constructed, non-soil-derived strain was able to grow faster than the native populations. When this strain was inoculated into soil, the indigenous strains, which dominated in uninoculated soil, did not thrive. The *S. paucimobilis* population represented the second most dominant organism, and showed an intermediate lag time in all experiments.

3.4   AZO DYES

Biodegradation of azo dyes is not widely distributed in nature, since sulfonic and azo groups (Fig. 6) are virtually nonexistent in nature (Paszczynski & Crawford, 1995).

$$HO_3S—Ar_1—N{=}N—Ar_2—SO_3H$$

Figure 6.   General structure of azo dyes.

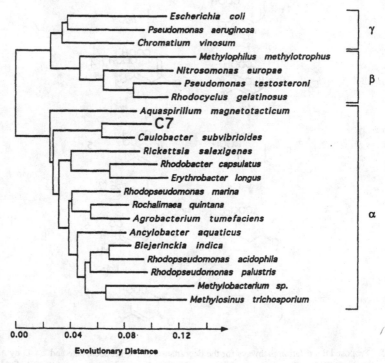

Proteobacteria

Figure 7.   Phylogenetic relationship of selected proteobacteria from 16S rRNA sequences. The tree was constructed using a distance matrix method (Olsen, 1987) and rooted with *Clostridium perfringens* (taken with permission from Godvindaswami *et al.*, 1993).

Govindaswami *et al.* (1993) conducted a phylogenetic study of a bacterial isolate capable of aerobic azo dye degradation. Initially this isolate was identified as *Pseudomonas paucimobilis,* but API NFT test strips, the Biolog Identification System, and fatty acid profile analysis revealed distinctive differences between isolate C7 and *P. paucimobilis.* Phylogenetic analysis of the 16S rRNA sequence identified strain C7 as a previously undescribed strain within the α-subgroup of the *Proteobacteria,* closely related to *Caulobacter subvibrioides* (Fig. 7) (Stahl *et al.,* 1992). Ederer *et al.* (1997) confirmed this observation and also placed both of these strains close to the genus *Sphingomonas* (Fig. 8).

Keck *et al.* (1997) studied azo dye degradation by *Sphingomonas* sp. strain BN6. These researchers found that the anaerobic reduction of the sulfonated azo dye amaranth was stimulated by the aerobic conversion of 2-naphthalenesulfonate to salicylate. This reaction generates a redox mediator which is responsible for the reduction of the azo bond. The reductase activity could be measured in the cytoplasm as well as in the membrane fraction of *Sphingomonas* sp. strain BN6 (Kudlich *et al.,* 1997).

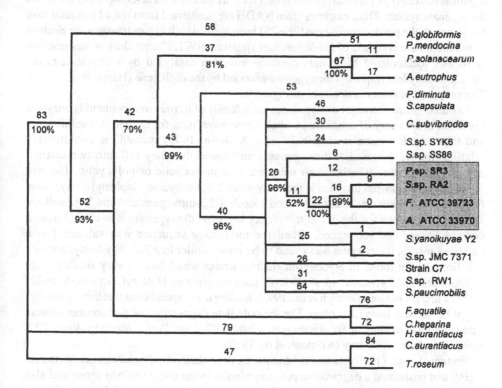

Figure 8. Phylogenetic tree resulting from 23 taxa of the four PCP-degrading bacteria (shaded), different *Sphingomonas* spp., and members of the genera *Arthrobacter, Flavobacterium,* and *Pseudomonas.* The tree was obtained from a bootstrap analysis of the variable positions of the 23 taxa. The most parsimonious tree had a length of 1368 steps (Swofford, 1991). The g1 statistic of 10 000 randomly generated trees was –0.781. The numbers about the branches describe the branch length; numbers below indicate the percentages of detection of each clade in 1000 bootstrap iterations (taken from Ederer *et al.,* 1997, with permission).

3.5   DIBENZO-P-DIOXIN AND DIBENZOFURAN

Several groups have studied dibenzo-p-dioxin and dibenzofuran degradation by
Sphingomonas spp. (Happe *et al.*, 1993; Wittich *et al.*, 1992; Fortnagel *et al.*, 1990).
Wittich *et al.* (1992) isolated strain RW1 from the Elbe River by enrichment with
dibenzo-p-dioxin; the purified strain grew with a doubling time of 8 h on dibenzo-p-dioxin
and 5 h on dibenzofuran. This Gram-negative, strict aerobic rod was identified as a
Sphingomonas sp. on the basis of fatty acid analysis and the presence of Q10. Subsequently,
Moore *et al.* (1993) confirmed the identification by 16S rRNA analysis, which revealed
94% sequence identity between *Sphingomonas* sp. strain RW1 and *S. paucimobilis.*

Bünz & Cook (1993) studied the first enzyme involved in dibenzofuran degradation
by *Sphingomonas* sp. strain RW1. They purified the four different components of
the dibenzofuran 4,4a-dioxygenase system (DBFDOS), catalysing the reaction leading from
dibenzofuran to 2,2'3-trihydroxybiphenyl. Armengaud and Timmis (1997) characterized
a putidaredoxin-type [2Fe-2S] ferredoxin, Fdx1, as part of a three-component class IIA
dioxygenase system. Thus, electrons from NADH are transferred from one of two reductase
monomers to the putidaredoxin-type [2Fe-2S] ferredoxin, which in turn transfers the electron
to the dioxin dioxygenase of *Sphingomonas* sp. strain RW1. The product of this reaction
is further metabolized by a *meta*-cleavage reaction catalysed by a single enzyme,
the 2,2',3-trihydroxybiphenyl dioxygenase encoded by the *dbfB* gene (Happe *et al.*, 1993).
The proposed pathway is depicted in Figure 9.

Extradiol (*meta* cleavage) dioxygenases are a family of ferrous-iron-containing enzymes
involved in a variety of aromatic catabolic pathways, e.g., for biphenyl, naphthalene,
and toluene (Harayama *et al.*, 1992). Eltis & Bolin (1996) studied the evolutionary
relationship of 35 extradiol dioxygenases and found that they fell into two distinct
groups according to whether their substrate was monocyclic or polycyclic. The two
Sphingomonas enzymes included in the study were 2,2',3-trihydroxy biphenyl dioxygenase
from *Sphingomonas* sp. strain RW1 and catechol 2,3-dioxygenase from *S. yanoikuae*
strain B1. The gene for the 2,2',3-trihydroxy biphenyl dioxygenase was cloned from a
cosmid library and sequenced. When the nucleotide sequence was subjected to a
database search, the enzyme was found to be most similar to a 2,3-dihydroxybenzoate
1,2 dioxygenase found in biphenyl-degrading strains which have a very similar range
of substrates. *Sphingomonas* sp. strain RW1 does not grow on biphenyl, but adapts readily
within a few generations (Wittich *et al.*, 1992), leading to the speculation that these pathways
might be derived from each other. For the complete degradation of 4-chlorobenzofuran
a consortium consisting of *Sphingomonas* sp. strain RW1 and *Burkholderia* sp. strain JWS
was shown to be necessary (Arfmann *et al.* 1997).

Fortnagel *et al.* (1990) studied dibenzofuran degradation by *Pseudomonas* sp. isolate
HH69 and postulated a degradative pathway similar to the one described above and also
involving a trihydroxylated biphenyl. This isolate initially grew on biphenyl and other
aromatic compounds, but lost that ability after four months of cultivation with dibenzofuran
as sole source of carbon.

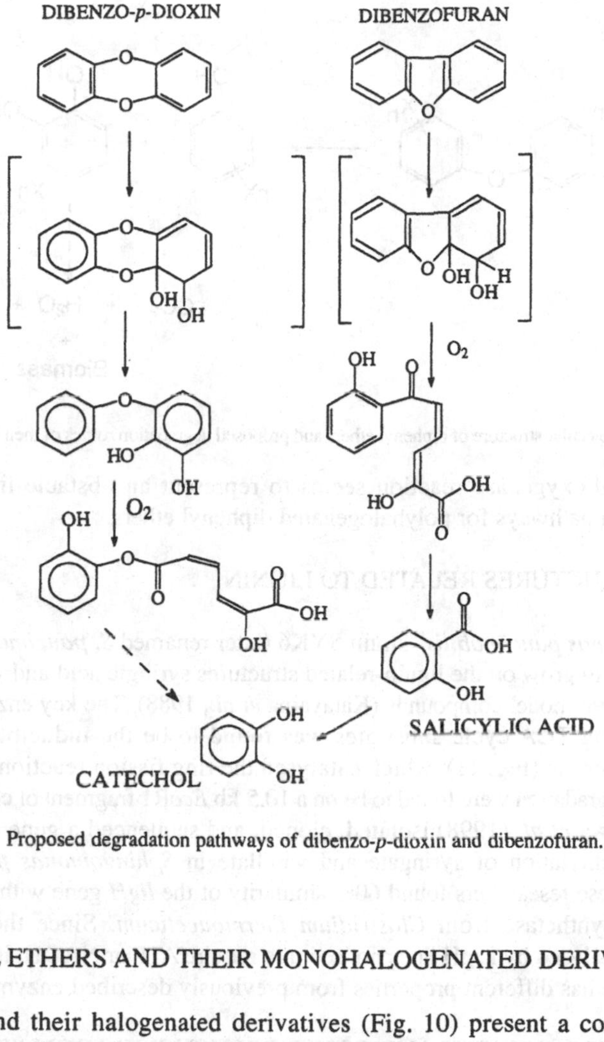

Figure 9. Proposed degradation pathways of dibenzo-*p*-dioxin and dibenzofuran.

3.6   DIPHENYL ETHERS AND THEIR MONOHALOGENATED DERIVATIVES

Diphenyl ethers and their halogenated derivatives (Fig. 10) present a considerable environmental hazard. Schmidt *et al.* (1992) isolated *Sphingomonas* sp. strain SS3 from enrichment cultures supplemented with 4-fluoro diphenyl ether. This organism was shown to co-metabolize isomeric monohalogenated diphenyl ethers. The proposed degradation pathway is similar to the pathway implicated for the degradation of the diaryl ethers dibenzofuran (Fig. 10, Fortnagel *et al.*, 1990) and dibenzo-*p*-dioxin for *Sphingomonas* sp. strain RW1 (Wittich *et al.*, 1992; Wilkes *et al.*, 1996). Phenol and catechol were found as intermediates, but did not seem to induce oxidation activity of diphenyl ethers.

Bacterial cultures capable of degrading dihalogenated diphenyl ethers could not be isolated directly. After several weeks of adaptation on 4,4'-difluoro diphenyl ether, *Sphingomonas* sp. strain SS33 was isolated from the parent culture, *Sphingomonas* sp. strain SS3 (Schmidt *et al.*, 1993). The conversion of dihalogenated diphenyl ethers seems to depend on steric effects elicited by the bulkiness of the halo substituent; therefore,

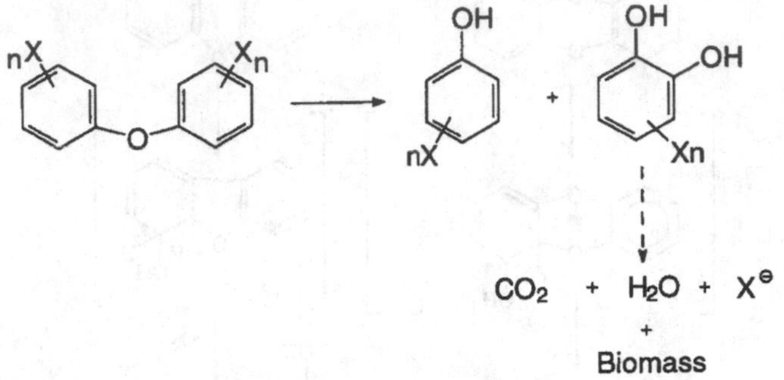

Figure 10. Molecular structure of diphenyl ethers and proposed degradation routes of their halogenated derivatives.

the initial dioxygenase reaction seems to represent an obstacle in the evolution of degradation pathways for polyhalogenated diphenyl ethers.

3.7   STRUCTURES RELATED TO LIGNIN

"*Pseudomonas paucimobilis*" strain SYK6 (later renamed *S. paucimobilis* strain SYK6) was shown to grow on the lignin-related structures syringic acid and vanillate, and other lignin dimeric model compounds (Katayama *et al.*, 1988). The key enzyme in the process of producing TCA cycle substrates was found to be the inducible protocatechuate 4,5-dioxygenase (Fig. 11), which catalysed the ring fission reaction. The enzymes for vanillate degradation were found to be on a 10.5-kb *Eco*R1 fragment of chromosomal origin.

Nishikawa *et al.* (1998) isolated, cloned, and sequenced a gene (*ligH*) involved in the *o*-demethylation of syringate and vanillate in *Sphingomonas paucimobilis* strain SYK 6. These researchers found 60% similarity of the *ligH* gene with the formyltetrahydrofolate synthetase from *Clostridium thermoaceticum*. Since the *o*-demethylation reaction catalysed by LigH does not seem to require ATP or NADH, it would appear that the enzyme has different properties from previously described enzymes.

3.8   FLUORANTHENE AND OTHER POLYCYCLIC AROMATIC
         HYDROCARBONS

The soil bacterium *S. paucimobilis* was shown to use fluoranthene, a high-molecular-weight polynuclear aromatic compound, as well as a variety of other aromatic compounds, as sole sources of carbon and energy (Mueller *et al.*, 1990). In another study, Stringfellow & Aitken (1994) isolated two dissimilar pseudomonads capable of fluoranthrene degradation. These organisms were identified by FAME analysis as *Pseudomonas stutzeri*, belonging to the rRNA group I and the γ-subdivision of the purple bacteria, and *Pseudomonas saccharophila*, a member of the rRNA group III and β-subdivision of the purple bacteria and therefore distinct from the *Sphinogmonas* isolate that is a member of the α-subgroup of the purple bacteria. It was determined that these two organisms

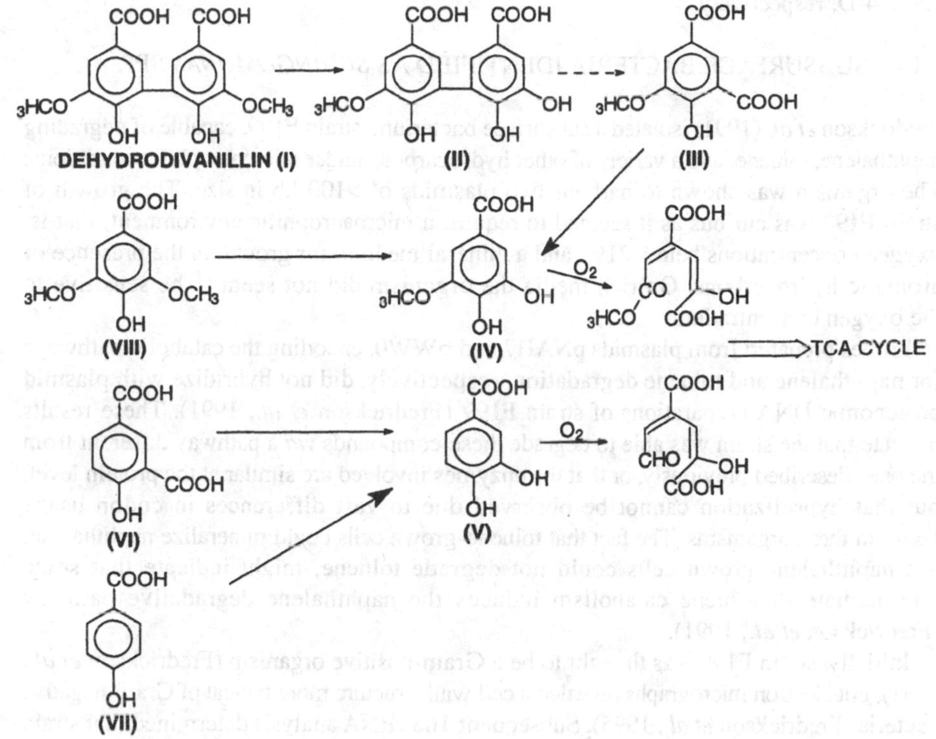

Figure 11. Proposed biodegradation pathway for lignin-model compounds.

degraded the xenobiotic compound by differing pathways with different induction requirements. Kästner and coworkers (1994) identified *S. paucimobilis* strain BA2, which is capable of degrading anthracene.

3.9   CARBOFURAN

Carbofuran (Fig. 12A) has been used extensively as an insecticide on crops including corn, potato, and strawberry. Since the compound is relatively toxic to mammals, the biodegradation of carbofuran is of considerable interest. Feng *et al.* (1997) found that *Sphingomonas* strain CF06 was able to use carbofuran as a sole carbon and nitrogen source. The organism contained five plasmids, one of which encodes the carbofuran degradation pathway, as shown by conjugation experiments with both *Pseudomonas fluorescens* strain M480R and carbofuran-degradation-deficient *Sphingomonas* strain CF06. The latter strain was found to be closely related to other xenobiotic-degrading sphingomonads (Fig.12B), but was distinct from *Sphingomonas* sp. strain RW1 (Wittich *et al.*, 1992) and *S. paucimobilis* isolate 1443 (Ka *et al.*, 1994a) as it could not degrade benzofuran and dibenzofuran, and 2,4-D, respectively.

3.10   SUBSURFACE BACTERIA IDENTIFIED AS *SPHINGOMONAS* SPP.

Fredrickson *et al.* (1991) isolated a subsurface bacterium, strain F199, capable of degrading naphthalene, toluene, and a variety of other hydrocarbons under microaerophilic conditions. The organism was shown to harbour two plasmids of >100 kb in size. The growth of strain F199 was curious as it seemed to require a microaerophilic environment, that is, oxygen concentrations below 21%, and a mineral medium for growth in the presence of aromatic hydrocarbons. On rich media the organism did not seem to be sensitive to the oxygen concentration.

Probes prepared from plasmids pNAH7 and pWW0, encoding the catabolic pathways for naphthalene and toluene degradation, respectively, did not hybridize with plasmid or genomic DNA preparations of strain F199 (Fredrickson *et al.*, 1991). These results indicate that the strain was able to degrade these compounds *via* a pathway different from the ones described previously, or that the enzymes involved are similar at the protein level, but that hybridization cannot be observed due to vast differences in codon usage between these organisms. The fact that toluene-grown cells could mineralize naphthalene, but naphthalene-grown cells could not degrade toluene, might indicate that some intermediate of toluene catabolism induces the naphthalene degradative pathway (Fredrickson *et al.*, 1991).

Initially, strain F199 was thought to be a Gram-positive organism (Fredrickson *et al.*, 1991), but electron micrographs revealed a cell wall structure more typical of Gram-negative bacteria (Fredrickson *et al.*, 1995). Subsequent 16S rRNA analysis determined that strain F199 was most closely related to *Sphingomonas capsulata*, a result supported by fatty acid analyses, which revealed the presence of sphingolipids, a major characteristic of the genus *Sphingomonas*. Although the subsurface isolates seemed to be most closely related to *S. capsulata*, these researchers pointed out that there were distinctive differences between the two, indicating that the subsurface isolates possibly represented a separate species in the genus. Balkwill *et al.* (1997) studied six different aromatic-degrading, deep-terrestrial subsurface bacteria and identified strain F199 as *Sphingomonas aromaticivorans* (a new species in the genus *Sphingomonas*) and described two additional new *Sphingomonas* species, *S. stygia and S. subterranea*. Further, strain F199 was found to contain a large plasmid of >100 kb, pNL1. Stillwell *et al.* (1995) determined that catechol 2,3-dioxygenase activity was linked to two distinct regions on the 180-kb pNL1 plasmid, indicating that at least part of the aromatic catabolism activity of this *Sphingomonas* strain is plasmid encoded.

Kim *et al.* (1996) investigated the differences between hydrocarbon-degrading *Sphingomonas* strains isolated from the surface and the deep subsurface. Using molecular probes for biphenyl (*bphC*) and xylene degradation (*xylE*), they found similarity between the aromatic hydrocarbon-degrading *Sphingomonas* strains and the cloned region from *S. yanoikuyae* strain B1 encoding for biphenyl and *m*-xylene degradation. Four of the five deep subsurface isolates contained a plasmid which hybridized to the molecular probe, and one isolate contained two different plasmids of 180 and 800 kb with sequence

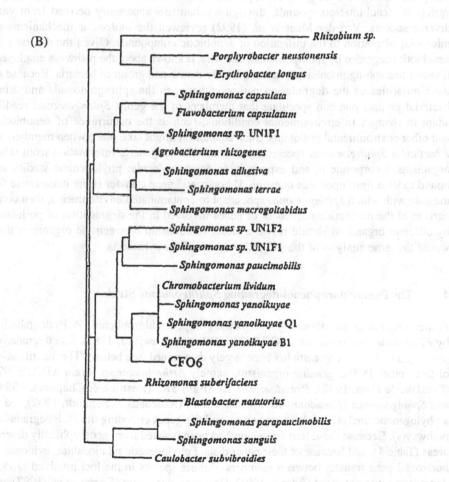

Figure 12. (A) Molecular structure of carbofuran. (B) Consensus tree for the carbofuran-degrading CFO6 and 22 selected eubacterial strains. The 16S rRNA sequence alignments were analysed using the distance matrix method from the PHYLIP package of computer programs (taken from Feng *et al.*, 1997, with permission).

similarity to the *xyl* and *bph* genes from *S. yanoikuyae* strain B1. While the surface isolates *Sphingomonas paucimobilis* strain Q1 and *Sphingomonas yanoikuyae* strain B1 also contained large plasmids, the genes hybridizing to the *xyl* and *bph* probes were found to be located on the chromosome, clearly distinguishing the surface from the subsurface *Sphingomonas* isolates. This result may indicate the migration of these genes from a plasmid to the chromosome, or *vice versa*, at some point in evolution. The type strain *S. capsulata* strain ATCC 14666 degraded cresol and salicylate, whereas *S. paucimobilis* strain ATCC 29837 degraded only salicylate (Fredrickson *et al.*, 1995).

3.11   OVERVIEW

The genus *Sphingomonas* groups a large number of soil bacteria capable of degrading a variety of xenobiotic compounds, through mechanisms apparently derived from very diverse sources. Van der Meer *et al.* (1992) reviewed the molecular mechanisms of microbial adaptation to the utilization of xenobiotic compounds. Given the diversity of xenobiotics degraded by sphingomonads and what is known about the pathways employed, it seems that sphingomonads represent a widely distributed group of bacteria. Because of the similarities of the degradation pathways between the sphingomonads and other bacterial groups, one can speculate that members of the genus *Sphingomonas* readily adapt to changes in environmental conditions, such as the occurrence of xenobiotics and other environmental pollutants. Such adaptation might take place when members of a particular *Sphingomonas* species are able to take up genetic information from other organisms, incorporate it, and express it. In-depth molecular phylogenetic studies are bound to shed light upon this question. If horizontal gene transfer is the main cause for the ease with which *Sphingomonas* spp. adapt to contaminated environments, then comparison of the nucleotide sequences of genes involved in the degradation of pollutants by different organisms should reveal a closer relationship between the organisms than would the same analysis of the 16S rRNA genes of these bacteria.

4   The Pentachlorophenol-degrading *Sphingomonas* Strains

Earlier research in our laboratory was focused on pentachlorophenol (PCP) degradation by *Flavobacterium* sp. strain ATCC 39723 (Saber & Crawford, 1985). The degradative pathway used by this organism has been largely determined (see below). The identification of three other PCP-degrading organisms, namely *Arthrobacter* sp. strain ATCC 33970 (Stanlake & Finn, 1982), *Pseudomonas* sp. strain SR3 (Resnick & Chapman, 1994), and *Sphingomonas* (*Pseudomonas*) sp. strain RA2 (Radehaus & Schmidt, 1992), led to a phylogenetic analysis of these organisms and the genes encoding the PCP degradation pathway(s). Because these four organisms had been isolated from geographically diverse areas (Table 1), and because of their rather distant phylogenetic relationships, evidence of horizontal gene transfer between members of these species in the loci involved in PCP degradation was expected (Ederer, 1994). However, our group (Ederer *et al.*, 1997) and others (Karlson *et al.*, 1995; Nohynek *et al.*, 1995) found that the PCP-degrading organisms were not only very closely related but might even represent a single species (see below).

4.1 BACKGROUND TO PENTACHLOROPHENOL

Because of its general toxicity, which affects animals, bacteria, fungi, and plants alike by uncoupling oxidative phosphorylation (Steiert & Crawford, 1988), the aromatic chlorinated compound PCP has been used extensively as a biocide. The compound is unusually acidic, with a pK_a of 4.70 in water, making it soluble in water and organic solvents (Crosby, 1981). In humans, PCP is readily taken up by absorption through the skin, by inhalation, and by consumption of contaminated foods. The compound is most widely employed in the wood-preserving industry, where it is used to prevent bluestaining and sapstaining caused by fungi. In 1985, the world production of PCP was about 100,000 tons (Wild *et al.*, 1993).

So far, bacteria have been primarily implicated in the degradation of PCP, although a number of wood-deteriorating fungi were found to at least partially degrade PCP, representing a threat to PCP-treated wood products. The discovery of PCP-mineralizing bacteria has led to their application in bioremediation processes at PCP-contaminated sites. In the past, wood products such as fencing, poles, and railway ties were treated by dipping into open treatment basins, leading to extensive contamination of soil and water. PCP contamination has also been detected in smoke from sawmills; in leather products, paint, paper, and rope; and in wool and other textiles (U.S. Environmental Protection Agency, 1984).

Because PCP is toxic to fish (Rao, 1978), its use was banned in Japanese paddy and upland rice fields in 1971; PCP production became illegal in Finland in 1988 (McAllister *et al.*, 1996). However, the forests of the world are rapidly diminishing, so the use of PCP and other wood preservatives will probably continue, posing further environmental and biological risks. The U.S. National Priorities List alone mentions 1400 sites seriously contaminated with PCP, justifying research aimed at finding inexpensive and efficient alternatives to traditional cleanup methods such as charcoal filtration, incineration, and long-term storage. Bioremediation is a promising approach, as several studies have shown that bacteria can be successfully used to decontaminate PCP-laden soil (Edgehill & Finn, 1982; Crawford & Mohn, 1985; Colores *et al.*, 1995; Schmidt *et al.*, 1995). In a recent study, Mietling & Karlson (1996) studied the survival and mineralization capacity of *Mycobacterium chlorophenolicum* strain PCP1 and *Sphingomonas chlorophenolica* strain RA2 in actual PCP-contaminated soils. One reason *Sphingomonas chlorophenolica* strain RA-2 can grow on relatively high concentrations of PCP without being inhibited by its degradation products seems to be the fast and efficient degradation process. McCarthy *et al.* (1997) found that degradation intermediates do not accumulate in the cell at levels toxic to the cells. Work on the use of bioreactors for the remediation of PCP contamination has been initiated (Mueller *et al.*, 1993) and should lead to more economical environmental cleanup efforts for PCP.

4.2 PENTACHLOROPHENOL DEGRADATION BY *SPHINGOMONAS* (*FLAVOBACTERIUM*) SP. STRAIN ATCC 39723

Bacterial PCP degradation is thought to be mediated by two different pathways. The first was described by Juha *et al.* (1986) and Häggblom *et al.* (1988) in two "*Rhodococcus*" chlorophenolicus isolates (PCP-1 and CP-1) which were shown to produce 2,3,5,6-tetrachloro-*p*-

hydroquinone as an intermediate. Uotila *et al.* (1992) found that the first degradation step, the *para* hydroxylation of PCP, was catalysed by a cytochrome P450 enzyme.

The second pathway was first studied in *Flavobacterium* sp. strain ATCC 39723, isolated from a PCP contaminated soil in Minnesota (Saber & Crawford, 1985). This organism was found to utilize high concentrations of PCP (100-200 ppm) as its sole carbon source. The first degradation step is catalysed by a PCP-4-monooxygenase encoded by the *pcpB* gene (Orser *et al.*, 1993). This enzyme has a wide substrate range (Xun *et al.*, 1992a) and considerable similarity at both protein and nucleotide sequence levels with other monooxygenases (Orser & Lange, 1994). The second and third degradation steps were found to be catalyzed by a reductive dehalogenation by PcpC with glutathione as a cofactor (Xun *et al.*, 1992b). The next step in the PCP degradation pathway is catalysing by the PcpA. The expression of the *pcpA* gene is induced by PCP (Lee & Xun, 1997) and *pcpA* 2,6-dichlorohydroquinone, and its oxidized products (Chanama, 1996). Initially, it was proposed that *pcpA* encodes a chlorohydrolase catalysing the removal of the fourth chlorine from the aromatic ring (Lee & Xun, 1997). However, Xu *et al.* (1999) recently presented evidence that *pcpA* encodes a dichloroquinone dioxygenase. The fission product of this reaction is still to be determined.

Downstream of *pcpB* reading in the same direction, Lange (1994) found two further open reading frames possibly involved in PCP-degradation, *pcpD* and *pcpR*. The open reading frame (ORF) of *pcpD* showed high similarity to the ORFs of different oxygenase reductases and may transfer electrons from NADPH through a redox centre to the flavine adenine dinucleotide group of the PCP-4-monooxygenase. The second ORF, *pcpR*, showed significant similarity with the nucleotide sequences encoding a group of regulatory proteins called the LysR type positive regulators (Lange, 1994). These similarities have been studied only by nucleotide and amino acid alignments; the actual activities remain to be determined. Figure 13 summarizes the PCP degradation pathway known thus far. McCarthy *et al.* (1997) found that *Sphingomonas chlorophenolica* strain RA2 degraded PCP using a degradation pathway identical to that identified in *Flavobacterium* (*Sphingomonas*) sp. strain ATCC 39723. A detailed review of microbial PCP degradation has been published (McAllister *et al.*, 1996).

4.3   CLASSIFICATION OF FOUR PCP-DEGRADING *SPHINGOMONAS* STRAINS

The ability to degrade PCP has been considered a trait widely distributed in the bacterial domain. Three independent studies recently addressed this assumption (Karlson *et al.*, 1995; Nohynek *et al.*, 1995; Ederer *et al.*, 1997). The authors of all three studies concluded that four PCP-degrading organisms previously identified as *Arthrobacter*, *Flavobacterium*, *Pseudomonas*, and *Sphingomonas* should be considered members of the genus *Sphingomonas*, and may possibly even constitute one species within the genus. The four strains seemed to be more closely related to each other than to the other members of the genus (Fig. 8). The results of database searches indicating *Caulobacter subvibrioides* as a close relative to the PCP-degrading bacteria was rather unexpected. However, Stahl *et al.* (1992) found that *C. subvibrioides*, which is also associated with the α-subgroup of the class *Proteobacteria*, did not fit into the *Caulobacter* assemblage, a fresh-water

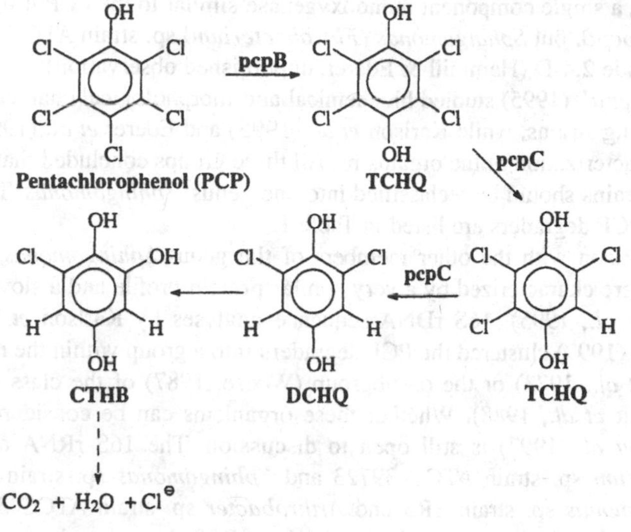

Figure 13. Proposed PCP degradation pathway for *Sphingomonas* spp.

and a salt-water grouping, thus most likely requiring reclassification. The 16S rRNA gene sequence of this organism identified *Brevundimonas (Pseudomonas) diminuta* as the closest relative (Sly *et al.*, 1997) and found that 2-hydroxymyristic acid, which is characteristic of *Sphingomonas* spp., was absent.

The nucleotide sequence analysis of the *pcpB* gene, encoding PCP-4-monooxygenase, in all four of these organisms revealed that this gene was identical in three of the isolates, and showed only a slightly different nucleotide sequence in the fourth (*Arthrobacter* sp. strain ATCC 33970) (Ederer *et al.*, 1997). Similar studies on the *pcpC* gene, which encodes 2,3,5,6-tetrachloro-*p*-hydroquinone (TeCH) dehalogenase, revealed a lower degree of genetic conservation between the four PCP-degrading species (Ederer & Orser, unpublished observation). This observation is consistent with a more general detoxification role of the *pcpC* gene product, which was shown to share similarities with a variety of gluthathione-S-transferases, and also with the fact that transcription of the gene is constitutive and is not affected by the presence or absence of PCP in the medium (Orser & Lange, 1994).

Many pathways for degradation of xenobiotic compounds, e.g., 2,4-D, methyl phenols, naphthalene, and toluene, are encoded on extrachromosomal entities, but no one has been able to determine whether this is true for the PCP degradation pathway of the sphingomonads. The sequence similarity of the *pcpB* genes would strongly support the possibility of a plasmid location for the gene, and perhaps even for the whole pathway. Ka *et al.* (1994a) found the 2,4-D pathway gene *tfdB* on a plasmid in a *Pseudomonas* sp., but did not find any hybridization of *tfdB* with the isolates identified as *S. paucimobilis*. However, a *tfdB* probe from *Alcaligenes eutrophus* plasmid pJMC134 hybridized strongly with the *pcpB* genes of the PCP degraders. The *tfdB* gene encodes dichlorophenol hydroxylase, a single component monooxygenase similar to the PCP-4-monooxygenase encoded by *pcpB,* but *Sphingomonas (Flavobacterium)* sp. strain ATCC 39723 was not able to degrade 2,4-D (Hammill & Ederer, unpublished observation).

Nohynek *et al.* (1995) studied biochemical and morphological characteristics of four PCP-degrading strains, while Karlson *et al.* (1995) and Ederer *et al.* (1997) focused on genetic characterization of the organisms. All three groups concluded that the four PCP-degrading strains should be reclassified into one genus, *Sphingomonas*. The phenotypes of the four PCP degraders are listed in Table 1.

In comparison with the other members of the genus *Sphingomonas,* the four PCP degraders were characterized by a very similar protein profile and a slower growth rate (Nohynek *et al.*, 1995). 16S rDNA sequence analyses by Karlson *et al.* (1995) and Ederer *et al.* (1997) clustered the PCP degraders into a group within the rRNA group IV (Palleroni *et al.*, 1973) or the α-subgroup (Woese, 1987) of the class *Proteobacteria* (Stackebrandt *et al.*, 1988). Whether these organisms can be considered one species (McCarthy *et al.*, 1997) is still open to discussion. The 16S rRNA analysis groups *Flavobacterium* sp. strain ATCC 39723 and *Sphingomonas* sp. strain RA2 together, and *Pseudomonas* sp. strain SR3 and *Arthrobacter* sp. strain ATCC 33970 together. Nohynek *et al.* (1995) suggested that the four PCP degraders be grouped into one single species, *Sphingomonas chlorophenolica* sp. nov. (Karlson *et al.*, 1995).

There are no strict rules about how divergent the nucleotide sequence of two organisms can be while still being regarded as members of one species. Some researchers base speciation on 70% sequence identity according to DNA:DNA hybridization data (Wayne *et al.*, 1987) and require at least 97% identity between rRNA genes (Stackebrandt & Goebel, 1994). Both Noyhynek's group (Noyhynek *et al.*, 1995) and ours (Ederer *et al.*, 1997) found only very little rRNA sequence variation, supporting the classification of these four strains into one species. However, there are significant differences between the four bacterial strains. First, *Arthrobacter* sp. strain ATCC 33970 shows an approximately 10% difference in nucleotide sequence of the *pcpB* gene in comparison to the other three organisms. Second, total DNA *Eco*R1 digests from *Arthrobacter* sp. ATCC 33970 probed with the *Flavobacterium* sp. *pcpC* gene resulted in a weak hybridization signal, also indicating a substantial amount of divergence between the two strains. Finally, *Pseudomonas* sp. strain SR3 degrades 2,4 dichlorophenol (Resnick & Chapman, 1994) and seems to exhibit a generally larger substrate range. Further investigations and better definition of bacterial species will be needed to unambiguously classify these and other organisms.

5   Practical Implications of Phylogenetic Analyses

Recently, the gene for a sphingomonad glutathione *S*-transferase (GST) was proposed as a potential probe to detect bacteria involved in polycyclic aromatic hydrocarbon degradation (Lloyd-Jones & Lau, 1997). The designed primer set allowed the amplification of GSTs from a number of aromatic-hydrocarbon-degrading sphingomonads of diverse origins. Using nucleotide sequences of common genes to establish a phylogeny and comparing these results with a 16SrRNA based phylogeny will certainly lead to a better understanding of bacterial systematics.

TABLE 1. Phenotypes of PCP-degrading strains. Listed are the highest PCP concentrations (PCP mg/l) which can be degraded, the presence or absence of ubiquinone Q10 in the respiratory chain, %GC of the chromosomal DNA, and presence of the yellow pigment nostoxanthin. Δ1290 represents a deletion of nucleotide 1290 in the 16S rRNA gene (numbering with respect to the E. coli gene). The presence of sphingolipids and octadecenoate is indicative of the genus Sphingomonas.

Strain	Origin (USA)	Shape	Gram stain	PCP mg/l	Pigment	Motility	Fimbriae	Sphingo-lipids	Octa-decenoate	Q10	Δ1290	%GC
ATCC 33790	NY	rod	negative	100-200	+	-	nd	+	62	+	+	66±1
ATCC 39723	MN	rod	negative	100-200	+	+	+	+	57	+	+	66±1
SR3	FL	rod	negative	175	+	+	nd	+	59	+	+	64.2
RA2	CO	rod	negative	200	+	+	nd	+	62	+	+	64±1

+, present; -, absent. nd, not done.

The bacteria that degrade pentachlorophenol have great potential for use in restoration of environments contaminated by chlorinated phenols. The commercial use of these organisms as bioremediation agents is, in fact, already a reality. To be profitable, most commercial technologies must be proprietary to the companies that market them. This is as true for a bioremediation technology as for any other. Thus, PCP degraders and their genes have been patented by their discoverers so that they might be licensed to industry for profitable application in commerce. For example, *Flavobacterium* sp. strain ATCC 39723 was patented for use as a bioremediation agent by the University of Minnesota in 1987 (Crawford, 1987). The PCP pathway genes *pcpB* (PCP-4-hydroxylase) and *pcpC* (chlorohydroquinone dehalogenase) from this organism, and the proteins they encode, were patented by the Idaho Research Foundation on behalf of the inventors at the University of Idaho (Orser *et al.*, 1994; 1996). The phylogenetic analyses of PCP-degrading bacteria are likely to strongly influence the ability to practice as proprietary the technologies described by these patents. Patents of microbial strains generally reflect phenotypic characteristics of the patented organism, protecting the use of other organisms similar to the patented strains.

Phylogenetic analyses based on characteristics such as 16S RNA sequences and lipid compositions offer firm evidence of the similarity or dissimilarity of any two strains. Phylogenetic evidence indicates that most of the PCP-degrading strains examined to date are either of the same genus (*Sphingomonas*) or the same species of that genus (see discussion above). Thus, the original patent is likely to protect commercial use of all of these strains. In the case of the genes encoding enzymes of the PCP pathway, it is the sequence of the open reading frame that is patented. In the case of the *pcpB* gene, the sequences of the gene from the different strains of *Sphingomonas* are identical. Thus, the patents that have been issued regarding *pcpB* should protect the use of the alleles from all the strains if these are used to construct commercially valuable genetically engineered organisms. Thus, it is clear that modern phylogenetic analyses such as those discussed here for the genus *Sphingomonas* have implications well beyond the science of phylogeny.

6   References

Amann, R.I., Ludwig, W. & Schleifer, K.-H. (1995). Phylogenetic identification and *in situ* detection of individual microbial cells without cultivation. *Microbiological Reviews* 59, 143-169.

Arfmann, H.-A., Timmis, K.N. & Wittich, R.-M. (1997). Mineralization of 4-chlorodibenzofuran by a consortium consisting of *Sphingomonas* sp. strain RW1 and *Burkholderia* sp. strain JWS. *Applied and Environmental Microbiology* 63, 3458-3462.

Armengaud, J. & Timmis, K. (1997). Molecular characterization of Fdx1, a putidaredoxin-type [2Fe-2S] ferredoxin able to transfer electrons to the dioxin dioxygenase of *Sphingomonas* sp. RW1. *European Journal of Biochemistry* 247, 833-842.

Balkwill, D. L., Drake, G. R., Reeves, R. H., Fredrickson, J.K., White, D.C., Ringelberg, D.B., Chandler, D.P., Romine, M.F., Kennedy, D.W. & Spadoni, C. M. (1997). Taxonomic study of aromatic-degrading bacteria from deep-terrestrial-subsurface sediments and description of *Sphingomonas aromaticivorans* sp. nov., *Sphingomonas subterranea*, sp. nov., and *Sphingomonas stygia* sp. nov. *International Journal of Systematic Bacteriology* 47, 191-201.

Brennan, P.J., Griffin, P.F.S., Loesel, D.M. & Tyrrell, D. (1974). The lipids of fungi. *Progress in the Chemistry of Fats and other Lipids* 14, 49-89.

Bünz, P.V. & Cook, A.M. (1993). Dibenzofuran 4,4a-dioxygenase from *Sphingomonas* sp. strain RW1: Angular dioxygenation by a three-component enzyme system. *Journal of Bacteriology* 175, 6467-6475.

Chanama, S. (1996). Molecular Characterization of the Catabolic Pathway of Pentachlorophenol Degradation in *Flavobacterium* sp. strain ATCC 39723. Ph.D. Dissertation. University of Idaho, USA.

Colores, G. M., Radehaus, P.M. & Schmidt, S.K. (1995). Use of a pentachlorophenol degrading bacterium to bioremediate highly contaminated soil. *Applied Biochemistry and Biotechnology* 54, 271-275.

Crawford, R.L. (1987). Biodegradation of pentachlorophenol. U.S. patent 4,713,340.

Crawford, R.L. & Mohn, W.W. (1985). Microbiological removal of pentachlorophenol from soil using a *Flavobacterium*. *Enzyme and Microbiological Technology* 7, 617-620.

Crosby, D.G. (1981). Environmental chemistry of pentachlorophenol. *Pure and Applied Chemistry* 53, 1052-1080.

Decker, C.F., Hawkins, R.E., Simon, G.L. (1992). Infections with *Pseudomonas paucimobilis*. *Clinical Infectious Diseases* 14, 783-784.

Don, R.H., Weightman, A.J., Knackmuss, H.J. & Timmis, K.N. (1985). Transposon mutagenesis and cloning analysis of the pathways for degradation of 2,4-dichlorophenoxyacetic acid and 3-chlorobenzoate in *Alcaligenes eutrophus* JMP134(pJP4). *Journal of Bacteriology* 161, 85-90.

Dwyer, D.F. & Tiedje, J.M. (1986). Metabolism of polyethylene glycol by two anaerobic bacteria, *Desulfovibrio desulfuricans* and a *Bacteroides* sp. *Applied and Environmental Microbiology* 52, 852-856.

Ederer, M.M. (1994). PCP Degradation: An Evolutionary Study. Ph.D. Dissertation. University of Idaho, USA.

Ederer, M.M., Crawford, R.L., Herwig, R.P. & Orser, C.S. (1997). PCP degradation is mediated by closely related strains of the genus *Sphingomonas*. *Molecular Ecology* 6, 39-49.

Edgehill, R.U. & Finn, R.K. (1983). Microbial treatment of soil to remove pentachlorophenol. *Applied and Environmental Microbiology* 45, 1122-1125.

Eguchi, M., Nishikawa, T., MacDonald, K., Cavicchioli, R., Gottschal, J.C. & Kjelleberg, S. (1996). Responses to stress and nutrient availability by the marine ultramicrobacterium *Sphingomonas* sp. strain RB2256. *Applied and Environmental Microbiology* 62, 1287-1294.

Eltis, L.D. & Bolin, J.T. (1996). Evolutionary relationships among extradiol dioxygenases. *Journal of Bacteriology* 178, 5930-5937.

Feng, X., Ou, L.-T. & Ogram, A. (1997). Plasmid-mediated mineralization of carbofuran by *Sphingomonas* sp. strain CF06. *Applied and Environmental Microbiology* 63, 1332-1337.

Forterre. P. (1997). Protein versus rRNA: Problems in rooting the universal tree of life. *ASM News* 63, 89-95.

Fortnagel, P., Harms, H., Wittich, R.-M., Krohn, S., Meyer, H., Sinnwell, V., Wilkes, H. & Francke, W. (1990). Metabolism of dibenzofuran by *Pseudomonas* sp. strain HH69 and the mixed culture HH27. *Applied and Environmental Microbiology* 56, 1148-1156.

Fox, G.E., Stackebrandt, E., Hespell, R.B., Gibson, J., Maniloff, J., Dyer, T.A., Wolfe, R.S., Balch, W.E., Tanner, R., Magrum, L., Zablen, L.B., Blakemore, R., Gupta, R., Bonen, L., Lewis, B.J., Stahl, D.A., Luehrsen, K.R., Chen, K.N. & Woese, C.R. (1980). The phylogeny of prokaryotes. *Science* 209, 457-463.

Fredrickson, J.K., Brockman, F.J., Workman, D.J., Li, S.W. & Stevens, T.O. (1991). Isolation and characterization of a subsurface bacterium capable of growth on toluene, naphthalene and other aromatic compounds. *Applied and Environmental Microbiology* 57, 796-803.

Fredrickson, J.K., Balkwill, D.L., Drake, G.R., Romine, M.F., Ringelberg, D.B. & White, D.C. (1995). Aromatic-degrading *Sphingomonas* isolates from the deep subsurface. *Applied and Environmental Microbiology* 61, 1917-1922.

Govindaswami, M., Schmidt, T.M., White, D.C. & Loper, J.C. (1993). Phylogenetic analysis of a bacterial aerobic degrader of azo dyes. *Journal of Bacteriology* 175, 6062-6066.

Häggblom, M.M., Apajalahti, J.H.A. & Salkinoja-Salonen, M.S. (1988). O-methylation of chlorinated *para*-hydroxyquinones by *Rhodococcus chlorophenolicus*. *Applied and Environmental Microbiology* 54, 1818-1824.

Haines, J.R. & Alexander, M. (1975). Microbial degradation of polyethylene glycols. *Applied Microbiology* 29, 621-615.

Hakamori, S. (1984). Glycosphingolipids. *Scientific American* 245, 44-53.

Hannum, Y.A. & Bell, R.M. (1989). Function of sphingolipids and sphingolipid breakdown products in cellular regulation. *Science* 243, 500-507.

Happe, B., Eltis, L.D., Poth, H., Hedderich, R. & Timmis, K.N. (1993). Characterization of 2,2',3-trihydroxy-biphenyl dioxygenase, an extradiol dioxygenase from the dibenzofuran- and dibenzo-*p*-dioxin-degrading bacterium *Sphingomonas* sp. strain RW1. *Journal of Bacteriology* 175, 7313-7320.

Harayama, S., Kok, M. & Neidle, E.L. (1992). Functional and evolutionary relationship among diverse dioxygenases. *Annual Review in Microbiology* 46, 565-601.

Holben, W.E., Schroeter, B.M., Calabrese, V.G.M., Olsen, R.H., Kukor, J.K., Biederbeck, B.O., Smith A.E. & Tiedje, J.M. (1992). Gene probe analysis of soil microbial populations selected by amendment with 2,4-dichlorophenoxyacetic acid. *Applied and Environmental Microbiology* 58, 3941-3948.

Holmes, B., Owen, R.J., Evans, J., Malnick, H. & Wilcox, W.R. (1977). *Pseudomonas paucimobilis*, a new species isolated from human clinical specimens, the hospital environment, and other sources. *International Journal of Systematic Bacteriology* 27, 133-146.

Jenkins, C.L., Andrews, A.G., McQuade, T.J. & Starr, M.P. (1979). The pigment of *Pseudomonas pauci-mobilis* is a carotenoid (nostoxanthin), rather than a brominated aryl-polyene (xanthomonadin). *Current Microbiology* 3, 1-4.

Juha, H.A.P., Karpanoja, P. & Salkinoja-Salonen, M.S. (1986). *Rhodococcus chlorophenolicus* sp. nov., a chlorophenol-mineralizing actinomycete. *International Journal of Systematic Bacteriology* 36, 246-251.

Ka, J.O., Holben, W.E. & Tiedje, J.M. (1994a). Genetic and phenotypic diversity of 2,4-dichlorophenoxyacetic acid (2,4-D)-degrading bacteria isolated from 2,4-D-treated field soils. *Applied and Environmental Microbiology* 60, 1106-1115.

Ka, J.O., Holben, W.E. & Tiedje, J.M. (1994b). Use of gene probes to aid in recovery and identification of functionally dominant 2,4-dichlorophenoxyacetic acid-degrading populations in soil. *Applied and Environmental Microbiology* 60, 1116-1120.

Ka, J.O., Holben, W.E. & Tiedje, J.M. (1994c). Analysis of competition in soil among 2,4-dichlorophe-noxyacetic acid-degrading bacteria. *Applied and Environmental Microbiology* 60, 1121-1128.

Karlson, U., Rojo, F., van Elsas, J.D. & Moore, E. (1995). Genetic and serological evidence for the recognition of four pentachlorophenol-degrading bacterial strains as a species of the genus *Sphingomonas*. *Systematic and Applied Microbiology* 18, 539-548.

Kästner, M., Breuer-Jammali, M. & Mahro, B. (1994). Enumeration and characterization of the soil microflora from hydrocarbon-contaminated soil sites able to mineralize polycyclic aromatic hydrocarbons (PAH). *Applied Microbiological Biotechnology* 41, 267-273.

Katayama, Y., Nishikawa, S., Murayama, A., Yamasaki, M., Morohoshi, N. & Haraguchi, T. (1988). The metabolism of biphenyl structures in lignin by the soil bacterium (*Pseudomonas paucimobilis* SYK-6). *FEBS Letters* 233, 129-133.

Kawahara, K., Seydel, U., Matsuura, M., Danbara, H., Rietschel, E.T. & Zähringer, U. (1991). Chemical structure of glycosphingolipids isolated from *Sphingomonas paucimobilis*. *FEBS Letters* 292, 107-110.

Kawai, F. & Yamanaka, H. (1986). Biodegradation of polyethylene glycol by symbiotic mixed culture (obligate mutualism). *Archives of Microbiology* **146**, 125-129.

Kawai, F., Fukaya, M., Tani, Y. & Ogata, K. (1977). Identification of polyethylene glycols (PEG)-assimilable bacteria and culture characteristics of PEG 6,000. *Journal of Fermentation Technology* **55**, 429-434.

Kawai, F., Kimura, T., Tani, Y. & Yamada, H. (1984). Involvement of polyethylene glycol (PEG) oxidizing enzyme in the bacterial metabolism of PEG. *Agricultural and Biological Chemistry* **48**, 1349-1351.

Kawasaki, S., Moriguchi, R., Sekiya, K., Nakai, T., Ono, E., Kume, K. & Kawahara, K. (1994). The cell envelope structure of the lipopolysaccharide-lacking gram-negative bacterium *Sphingomonas paucimobilis*. *Journal of Bacteriology* **176**, 284-290.

Keck, A., Klein, J., Judlich, M., Stoltz, A., Knackmuss, H.-J. & Mattes, R. (1997). Reduction of azo dyes by redox mediators originating in the naphthalenesulfonic acid degradation pathway of *Sphingomonas* sp. strain BN6. *Applied and Environmental Microbiology* **63**, 3684-3690.

Kim, E., Aversano, P.J., Romine, M.F., Schneider, R.P. & Zylstra, G.J. (1996). Homology between genes for aromatic hydrocarbon degradation in surface and deep-surface *Sphingomonas* strains. *Journal of Environmental Microbiology* **62**, 1467-1470.

Kopczynski, E.D., Bateson, M.M. & Ward, D.M. (1994). Recognition of chimeric small-subunit ribosomal DNAs composed of genes from uncultured microorganisms. *Applied and Environmental Microbiology* **60**, 746-748.

Kudlich, M., Keck, A., Klein, J. & Stolz, A. (1997). Localization of the Enzyme Systems Involved in Anaerobic Reductions of Azo Dyes by *Sphingomonas* sp. Strain BN6 and Effect of Artificial Redox Mediators on the Rate of Azo Dyes Reduction. *Applied and Environmental Microbiology* **63**, 3691-3694.

Lange, C.C. (1994). Molecular Analysis of PCP Degradation by *Flavobacterium* sp. Strain ATCC 39723. Ph.D. Dissertation. University of Idaho, USA.

Lee, J.-Y. & Xun, L. (1997). Purification and characterization of 2,6-dichloro-*p*-hydroxyquinone chlorohydrolase from *Flavobacterium* sp. strain ATCC 39723. *Journal of Bacteriology* **179**, 1521-1524.

Lee, S.-Y., Bollinger, J., Bezdicek, D. & Ogram, A. (1996). Estimation of the abundance of an uncultured soil bacterial strain by a competitive quantitative PCR method. *Applied and Environmental Microbiology* **62**, 3787-3793.

Leifson, E. (1962). The bacterial flora of distilled and stored water. I. General observations, techniques and ecology. *International Bulletin of Bacteriological Nomenclature and Taxonomy* **12**, 133-153.

Leveau, J.H.J., Zehnder, A.J.B. & van der Meer, J.R. (1998). The *tfdK* gene product facilitates uptake of 2,4-dichlorophenoxyacetate by *Ralstonia eutropha* JMP134(pJP4). *Journal of Bacteriology* **180**, 2237-2243.

Lloyd-Jones, G. & Lau, P.C.K. (1997). Glutathione *S*-transferase-encoding gene as a potential probe for environmental bacterial isolates capable of degrading polycyclic aromatic hydrocarbons. *Applied and Environmental Microbiology* **63**, 3286-3290.

Maidak, B.L., Cole J.R., Parker C.T. Jr., Garrity G.M., Larsen, N., Li B., Lilburn T.G., McCaughey, M.J., Olsen, G.J., Overbeek, R., Pramanik S., Schmidt T.M., Tiedje J.M. & Woese, C.R. (1999). A new version of the RDP (Ribosomal Database Project). *Nucleic Acids Research* **27**, 171-173.

McAllister, K., Lee, H. & Trevors, J.T. (1996). Microbial degradation of pentachlorophenol. *Biodegradation* **7**, 1-40.

McCarthy, D.L., Claude, A.A. & Copley, S.D. (1997). *In vivo* levels of chlorinated hydroquinones in a pentachlorophenol-degrading bacterium. *Applied and Environmental Microbiology* **63**, 1883-1888.

Mietling, R. & Karlson, U. (1996). Accelerated mineralization of pentachlorophenol in soil upon inoculation with *Mycobacterium chlorophenolicum* PCP1 and *Sphingomonas chlorophenolica* RA2. *Applied and Environmental Microbiology* **62**, 4361-4366.

Miyauchi, K., Suh, S.-K., Nagata, Y. & Takagi, M. (1998). Cloning and sequencing of a 2,5-dichlorohydro-quinone reductive dehaolgenase gene whose product is involved in degradation of γ-hexachlorocyclohexane by *Sphingomonas paucimobilis*. *Journal of Bacteriology* 180, 1354-1359.

Moore, E.R.B., Wittich, R.-M., Fortnagel, P. & Timmis, K.N. (1993). 16S ribosomal RNA gene sequence characterization and phylogenetic analysis of a dibenzo-*p*-dioxin-degrading isolate within the new genus *Sphingomonas*. *Letters in Applied Microbiology* 17, 117-118.

Mueller, J.G., Chapman, P.J., Blattmann, B.O. & Pritchard, P.H. (1990). Isolation and characterization of a fluoranthene-utilizing strain of *Pseudomonas paucimobilis*. *Applied and Environmental Microbiology* 56, 1079-1086.

Mueller, J.G., Lantz, S.E., Ross, D., Colvin, R.J., Middaugh, D.P. & Pritchard, P.H. (1993). Strategy using bioreactors and specially selected microorganisms for bioremediation of groundwater contaminated with creosote and pentachlorophenol. *Environmental Science and Technology* 27, 691-698.

Nagata, Y., Ohtomo, R., Miyauchi, K., Fukuda, M., Yano, K. & Takagi, M. (1994). Cloning and sequencing of a 2,5-dichloro-2,5-cyclohexadiene-1,4-diol dehydrogenase gene involved in degradation of γ-hex-achlorocyclohexane in *Pseudomonas paucimobilis*. *Journal of Bacteriology* 176, 3117-3125.

Nagata, Y., Miyauchi, K., Damborsky, J., Manova, K., Ansorgova, A. & Takagi, M. (1997). Purification and characterization of a haloalkane dehalogenase of a new substrate class from a γ-hexachlorocyclohexane-degrading bacterium, *Sphingomonas paucimobilis* UT26. *Applied and Environmental Microbiology* 63, 3707-3710.

Nishikawa, S., Sonoki, T., Kasahara, T., Obi, T., Kubota, S., Kawai, S., Morohoshi, N. & Katayama, Y. (1998). Cloning and sequencing of the *Sphingomonas (Pseudomonas) paucimobilis* gene essential for the *o*-demethylation of vanillate and syringate. *Applied and Environmental Microbiology* 64, 836-842.

Nishiyama, M., Senoo, K., Wada, H. & Matsumoto, S. (1992). Identification of soil micro-habitats for growth, death and survival of a bacterium, gamma-1,2,3,4,5,6-hexachlorocyclohexane-assimilating *Sphingomonas paucimobilis*, by fractionation of soil. *FEMS Microbiology Ecology* 101, 145-150.

Nishiyama, M., Senoo, K. & Matsumoto, S. (1993). Establishment of gamma-1,2,3,4,5,6-hexachlorocyclo-hexane-assimilating bacterium, *Sphingomonas paucimobilis* strain SS86, in soil. *Soil Biological Biochemistry* 25, 769-774.

Nohynek, L.J., Suhonen, E.L., Nurmiaho-Lassila, E.L., Hantula J. & Salkinoja-Salonen M. (1995). Description of four pentachlorophenol-degrading bacterial strains as *Sphingomonas chlorophenolica* sp. nov. *Systematic and Applied Microbiology* 18, 527-538.

Obradors, U. & Aguilar, J. (1991). Efficient biodegradation of high molecular-weight polyethylene glycols by pure cultures of *Pseudomonas stutzeri*. *Applied and Environmental Microbiology* 57, 2383-2388.

Olsen G.J. (1987). The earliest phylogenetic branchings: comparing rRNA-based evolutionary trees inferred with various techniques. *Cold Spring Harbor Symposia on Quantitative Biology*, 52, 825-838.

Olsen, G.J., Woese, C.R. & Overbeek, R. (1994). The winds of (evolutionary). change: Breathing new life into Microbiology. *Journal of Bacteriology* 176, 1-6.

Orser, C.S. & Lange, C.C. (1994). Molecular analysis of pentachlorophenol degradation. *Biodegradation* 5, 277-288.

Orser, C.S., Lange, C.C., Xun, L., Zahrt, T.C. & Schneider, B.J. (1993). Cloning, sequence analysis, and expression of the *Flavobacterium* pentachlorophenol 4-monooxygenase gene in *Escherichia coli*. *Journal of Bacteriology* 175, 411-416.

Orser, C.S., Xun, L. & Lange, C.C. (1994). Genes and enzymes involved in the microbial degradation of pentachlorophenol and used in bioremediation applications. U.S. patent 5,363,787.

Orser, C.S., Xun, L. & Lange, C.C. (1996). Genes and enzymes involved in the microbial degradation of pentachlorophenol and used in bioremediation applications. U.S. patent 5,512,478.

Pace, N.R., Stahl, D.A., Lane, D.J. & Olsen, G.J. (1986). The analysis of natural microbial populations by ribosomal RNA sequences. *Advances in Microbial Ecology* 11-55.

Palleroni N.J., Kunisawa R., Contopoulou R. & Doudoroff M. (1973). Nucleic acid homologies in the genus *Pseudomonas*. *International Journal of Systematic Bacteriology* 23, 333-339.

Paszczynski, A. & Crawford, R.L. (1995). Potential for bioremediation of xenobiotic compounds by the white-rot fungus *Phanerochaete chrysosporium*. *Biotechnological Progress* 11, 368-379.

Pollock, T.J. (1993). Gellan-related polysaccharides and the genus *Sphingomonas*. *Journal of General Microbiology* 139, 1939-1945.

Pollock, T.J., van Workum, W.A.T., Thorne, L., Mikolajczak, M.J., Yamazaki, M., Kijne, J.W. & Armentrout, R.W. (1998). Assignment of biochemical functions to glycosyl transferase genes which are essential for biosynthesis of exopolysaccharides in *Sphingomonas* strain S88 and *Rhizobium leguminosarum*. *Journal of Bacteriology* 180, 586-593.

Radehaus, P. & Schmidt, S.K. (1992). Characterization of a novel *Pseudomonas* sp. that mineralizes high concentrations of pentachlorophenol. *Applied and Environmental Microbiology* 58, 2879-2885.

Rao, K.R. (1978). *Pentachlorophenol: Chemistry, Pharmacology and Environmental Toxicology.* New York: Plenum Press.

Resnick, S.M. & Chapman, P.J. (1994). Physiological properties and substrate specificity of a pentachlorophenol-degrading *Pseudomonas* species. *Biodegradation* 5, 47-54.

Reysenbach, A.L., Giver, L.J., Wickham, G.S. & Pace, N.R. (1992). Differential amplification of rRNA genes by polymerase chain reaction. *Applied and Environmental Microbiology* 58, 3417-3418.

Richau, J.A., Choquenet, D., Fialho, A.M., Moreira, L.M. & Sá-Correia, I. (1997). The biosynthesis of the exopolysaccharide gellan results in the decrease of *Sphingomonas paucimobilis* tolerance to copper. *Enzyme and Microbial Technology* 20, 510-515.

Saber, D.L. & Crawford, R.L. (1985). Isolation and characterization of *Flavobacterium* strains that degrade pentachlorophenol. *Applied and Environmental Microbiology* 50, 1512-1518.

Saiki, R.K., Scharf, S., Faloona, F., Mullis, K.B., Horn G.T., Erlich H.A. & Arnheim N. (1985). Enzymatic amplification of ß-globin genomic sequences and restriction site analysis for diagnosis of sickle cell anemia. *Science* 230, 1350-1354.

Schmidt S., Wittich R.-M., Erdmann D., Wilkes H., Francke W. & Fortnagel P. (1992). Biodegradation of diphenyl ether and its monohalogenated derivatives by *Sphingomonas* sp. strain SS3. *Applied and Environmental Microbiology* 58, 2744-2750.

Schmidt, S., Fortnagel, P. & Wittich, R.M. (1993). Biodegradation and transformation of 4,4'- and 2,4-dihalophenyl ethers by *Sphingomonas* sp. strain SS33. *Applied and Environmental Microbiology* 59, 3931-3933.

Schmidt, S.K., Colores, G.M., Hess, T.F. & Radehaus, P.M. (1995). A simple method for quantifying activity and survival of microorganisms involved in bioremediation processes. *Applied Biochemistry and Biotechnology* 54, 259-270.

Sly, L.I., Cahill, M.M., Majeed, K. & Jones, G. (1997). Reassessment of the phylogenetic position of *Caulobacter subvibrioides*. *International Journal of Systematic Bacteriology* 47, 211-213.

Stackebrandt, E. & Goebel, B.M. (1994). Taxonomic note: A place for DNA-DNA reassociation and 16S rRNA sequence analysis in the present species definition in bacteriology. *International Journal of Systematic Bacteriology* 44, 846-849.

Stackebrandt, E., Murray, R.G.E. & Trüper, H.G. (1988). *Proteobacteria* classis nov., a name for the phylogenetic taxon that includes the "Purple bacteria and their relatives". *International Journal of*

Systematic Bacteriology **38**, 321-325.

Stahl, D.A., Key, R., Flesher, B. & Smit, J. (1992). The phylogeny of marine and freshwater caulobacters reflects their habitat. *Journal of Bacteriology* **174**, 2193-2198.

Stanlake, G.J. & Finn, R.K. (1982). Isolation and characterization of a pentachlorophenol-degrading bacterium. *Applied and Environmental Microbiology* **44**, 1421-1427.

Stanier, R.Y. & van Niel, C.B. (1962). The concept of a bacterium. *Archives of Microbiology* **42**, 17-35.

Steiert, J.G. & Crawford, R.L. (1988). Catabolism of pentachlorophenol by a *Flavobacterium* sp. *Biochemical and Biophysical Research Communications* **141**, 825-830.

Stewart, C.-B. (1993). The powers and pitfalls of parsimony. *Nature* **361**, 603-607.

Stillwell, L.C., Thurston, S.J., Schneider, R.P., Romine, M.F., Fredrickson, J.K. & Saffer, J.D. (1995). Physical mapping and characterization of a catabolic plasmid from the deep-subsurface bacterium *Sphingomonas* sp. strain F199. *Journal of Bacteriology* **177**, 4537-4539.

Stringfellow, W.T. & Aitken, M. (1994). Comparative physiology of phenanthrene degradation by two dissimilar pseudomonads isolated from a creosote-contaminated soil. *Canadian Journal of Microbiology* **40**, 432-438.

Swofford, D.L. (1991). PAUP: *Phylogenetic Analysis Using Parsimony*, 3.0s. Illinois Natural History Survey, USA: Champaign, Illinois.

Takeuchi, M., Kawai, F., Shimada, Y. & Yokota, A. (1993). Taxonomic study of polyethylene glycol-utilizing bacteria: Emended description of the genus *Sphingomonas* and new descriptions of *Sphingomonas macrogoltabidus* sp. nov., *Sphingomonas sanguis* sp. nov. and *Sphingomonas terrae* sp. nov. *Systematic and Applied Microbiology* **16**, 227-238.

Takeuchi, M., Sawada, H., Oyaizu, H. & Yokota, A. (1994). Phylogenetic evidence for *Sphingomonas* and *Rhizomonas* as non-photosynthetic members of the alpha-subgroup of the *Proteobacteria*. *International Journal of Systematic Bacteriology* **44**, 308-314.

United States Environmental Protection Agency (1984). Wood preservative pesticides: creosote, pentachlorophenol, and the inorganic arsenicals. EPA Office of Pesticide Programs, Registration Division. Washington D.C.

Uotila, J.S., Kitunen, V.H., Apajalahti, J.H.A. & Salkinoja-Salonen, M.S. (1992). Environment-dependent mechanism of dehalogenation by *Rhodococcus chlorophenolicus* PCP-1. *Applied Microbiological Biotechnology* **38**, 408-412.

Van Bruggen, A.H.C, Jochimsen, K.N., Steinberger, E.M., Segers, P. & Gillis, M. (1993). Classification of *Rhizomonas suberifaciens*, an unnamed *Rhizomonas* species, and *Sphingomonas* spp. In rRNA Superfamily IV. *International Journal of Systematic Bacteriology* **43**, 1-7.

Van der Meer, J.R., de Vos, W.M., Harayama, S. & Zehnder, A.J.B. (1992). Molecular mechanisms of genetic adaptation to xenobiotic compounds. *Microbiological Reviews* **56**, 677-694.

Wayne, L.G., Brenner, D.J., Colwell, R.R., Grimont, P.A.D., Kandler, O., Kichevsky, M.I., Moore, L.H., Moore, W.E.C., Murray, R.G.E., Stackebrandt, E., Starr, M.P. & Trüper, H.G. (1987). Report of the *ad hoc* committee on reconciliation of approaches to bacterial systematics. *International Journal of Systematic Bacteriology* **37**, 463-364.

White, D.C. (1996). Lipid biomarker analysis for *in situ* microbial community ecology. ASM General Meeting, New Orleans, Session 65; Application of Molecular Techniques for Environmental Problems.

Wild, S.R., Harrad, S.J. & Jones, K.C. (1993). Chlorophenols in digested U.K. sewage sludges. *Water Research* **27**, 1527-1534.

Wiese, A., Reiners, J.O., Brandenburg, K., Kawahara, K., Zähringer, U. & Seydel, U. (1996). Planar asymmetric lipid bilayers of glycosphingolipid or lipopolysaccharide on one side and phospholipids on the other: membrane potential, porin function, and complement activation. *Biophysical Journal* **70**, 321-329.

Wilkes, H., Wittich, R.-M., Timmis, K.N., Fortnagel, P. & Franke, W. (1996). Degradation of chlorinated dibenzofurans and dibenzo-*p*-dioxins by *Sphingomonas* sp. strain RW1. *Applied and Environmental Microbiology* **62**, 367-371.

Wittich, R.M., Wilkes, H., Sinnwell, V., Francke, W. & Fortnagel, P. (1992). Metabolism of dibenzo-*p*-dioxin by *Sphingomonas* sp. strain RW1. *Applied and Environmental Microbiology* **58**, 1005-1010.

Woese, C.R. (1987). Bacterial evolution. *Microbiological Reviews* **51**, 221-271.

Woese, C.R. (1991). The use of ribosomal RNA in reconstructing evolutionary relationships among bacteria. In: Evolution at the Molecular Level, pp 1-24. Edited by R.K. Selander, A. G. Clark & T. S. Whittam. Sinauer Associates, Inc., Sunderland, Mass.

Woese, C.R. (1994) There must be a prokaryote somewhere: Microbiology's search for itself. *Microbiological Reviews* **58**, 1-9.

Woese, C.R. & Fox, G.E. (1977). Phylogenetic structure of the prokaryotic domain: the primary kingdoms. *Proceedings of the National Academy of Sciences USA* **74**, 5088-5090.

Xu L., Resing, K., Lawson, S.L., Babbit P.C. & Copley, S.D. (1999). Eviddence that *pcpA* encodes 2,6-dichlorohydroquinone dioxygenase, the ring cleavage enzyme required for pentachlorophenol degradation in *Sphingomonas chlorophenolica* strain ATCC 39723. *Biochemistry* **38**, 7659-7669.

Xun, L. & Orser, C.S. (1991). Purification of a *Flavobacterium* pentachlorophenol-induced periplasmic protein (PcpA). and nucleotide sequence of the corresponding gene (*pcpA*). *Journal of Bacteriology* **173**, 2920-2926.

Xun, L., Topp, E. & Orser, C.S. (1992a). Diverse substrate range of a *Flavobacterium* pentachlorophenol hydrolase and reaction stoichiometries. *Journal of Bacteriology* **174**, 2898-2902.

Xun, L., Topp, E. & Orser, C.S. (1992b). Purification and characterization of a tetrachloro-*p*-hydroquinone reductive dehalogenase from a *Flavobacterium* sp. *Journal of Bacteriology* **174**, 8003-8007.

Yabuuchi, E., Yano, I., Oyaizu, H., Hashimoto, Y., Ezaki, T. & Yamamoto, H. (1990). Proposal of *Sphingomonas paucimobilis* gen. nov. and comb. nov., *Sphingomonas parapaucimobilis* sp. nov., *Sphingomonas yanoikuyae* sp. nov., *Sphingomonas adhaesiva* sp. nov., *Sphingomonas capsulata* comb. nov., and two genospecies of the genus *Sphingomonas*. *Microbiological Immunology* **34**, 99-119.

Yamanaka, K. & Kawai, F. (1989). Purification and characterization of constitutive polyethylene glycol (PEG). dehydrogenase of PEG 4,000-utilizing *Flavobacterium* sp. no.203. *Journal of Fermentation Technology* **67**, 324-330.

Yamamoto, A., Yano, I., Masui, M. & Yabbuchi, M. (1978). Isolation of a novel sphingolipid containing glucuronic acid and 2-hydroxy fatty acid from *Flavobacterium devorans* ATCC 10829. *Journal of Biochemistry* **83**, 1213-1216.

Zuckerkandl, E. & Pauling, L. (1965). Molecules as documents of evolutionary history. *Journal of Theoretical Biology* **8**, 357-366.

V   FOOD AND MEDICINE

13   LACTIC ACID BACTERIA

LARS AXELSSON [1] *and* SIV AHRNÉ [2]

[1] *MATFORSK, Norwegian Food Research Institute, Osloveien 1,*
   N-1430 ÅS, Norway
[2] *Department of Food Technology, University of Lund, P.O. Box 124,*
   S-221 00 LUND, Sweden

1   Introduction

In ancient times, the ways of preserving raw agricultural materials were essentially restricted to salting, drying and fermentation. In many parts of the world, this is also the case today. Fermentation in this sense has a meaning other than the scientific, that is, the non-respiratory metabolism of organic substrates by microorganisms. According to this definition, a food is fermented if it "has been subject to the action of microorganisms or enzymes so that desirable biochemical changes cause significant modification of the food" (Campbell-Platt, 1987). The importance of fermented foods in the human diet has been immense, perishable raw material was preserved and microbial growth often enriched the food with vitamins. Since the turn of the century, the production of fermented foods has become a significant part of the food processing industry. Along with the results of typical yeast-fermentations, such as beer, wine and leavened bread, fermentations involving lactic acid production are the most important.

The acidification of food by lactic acid (and acetic acid) inhibits many pathogenic organisms. This is probably the single most important factor that has resulted in the large variety of fermented foods based on lactic fermentations that exist around the world (Table 1) (see Steinkraus, 1996). Traditionally, these fermentations were empirical processes and relied on a certain way of handling the raw material and the activities of the natural flora. During the course of time, man learned how to promote the desired changes in the product and also discovered ways to improve the processes. One such improvement was to include a small portion from a successful production at the start of the next batch. Such "back-slopping" techniques were one of the essential factors for the industrial development of fermented foods, and this process is still in use even in industrialized countries. It is now known that back-slopping is an inoculum of the dominating microorganisms in the finished product to the new batch. In lactic fermentations, these microorganisms are bacteria which produce

F. G. Priest and M. Goodfellow (eds.), Applied Microbial Systematics, 367-388

lactic acid as the main end-product. This rather diverse group of bacteria is often collectively called the *lactic acid bacteria* (LAB). This term is frequently used, but, as will be shown, it has a certain degree of ambiguity.

TABLE 1.   Fermented foods worldwide for which lactic acid production plays a major role

Group	Examples	Major Region/ Country of production
Milk	Cheese	All
	Yoghurt	All
	Kefir (goat, sheep or cow's milk)	Europe, USA, Russia
	Fil, viili (ropy, sour milk)	Sweden, Finland
Cereals	Sourdough bread	All
	Idli (rice)	India
	Ogi (maize, sorghum)	Nigeria
	Enjera (tef, barley, wheat)	Ethiopia
Starch crops	Gari (cassava)	West Africa
Legumes	Soy sauce	East Asia
Vegetables	Pickles	All
	Sauerkraut	Europe, USA
	Kimchi (cabbage, radish)	Korea
Fish/seafood	Fish sauce	South-East Asia
	Rakfisk (trout)	Norway
	Balao balao (rice-shrimp)	Philippines
Meat	Dry fermented sausage	Europe, USA
	Nham (pork)	Thailand

The investigation of lactic fermentations has its roots in Pasteur's germ theory of fermentative changes and therefore belongs to the era that marks the dawn of microbiology as a science. In 1873 Joseph Lister by chance isolated the first bacterial pure culture, following up Pasteur's work on the souring of milk. He named the organism "*Bacterium lactis*"; it was later renamed "*Streptococcus lactis*" and is now known as *Lactococcus lactis*.

Work on LAB around the turn of the century was heavily biased towards milk and the dairy industry. At that time, more or less defined starter cultures of "mesophilic streptococci" were already being used. The LAB group name was actually created for bacteria which produced milk acid from milk sugar (Weigmann, 1899). Other scientists, however, noted that similar bacteria could be found in other habitats (Henneberg, 1904; Löhnis, 1907) and also drew attention to the newly described genus *Lactobacillus*. These early workers on bacterial classification struggled with the problem of what characters should be of importance in defining families, genera and species of bacteria. For the lactic group, this shifted somewhat after 1910 from being focused on the reactions of the bacteria with milk towards a more general view of their activities, such as type of metabolism, sugar fermentation capacity, and the capacity to grow at various temperatures.

A synthesis of the views outlined above was made by Orla-Jensen (1919) in his classic monograph. In the summary he wrote: "The true lactic acid bacteria form a great natural group of immotile, sporeless, Gram-positive cocci and rods, which in fermenting sugar form chiefly lactic acid". He recognized seven genera: *"Betabacterium"*, *"Betacoccus"*, *Microbacterium*, *"Streptobacterium"*, *Streptococcus*, *"Tetracoccus"* and *"Thermobacterium"*. Of these, only *Streptococcus* remains a valid name today (*Microbacterium* is also a valid name, but Orla-Jensen's strains were later renamed *Brochothrix* and are generally not included in the LAB group). Nevertheless, Orla-Jensen's work has had an enormous influence on the classification of the LAB.

The division of the LAB into seven genera was based on morphology (cocci or rods, tetrads), end-product formation from glucose (homo- or heterofermentation) and growth at certain temperatures (15°C and/or 45°C). After some taxonomic revisions, where essentially previously described genera were recognized as being identical to the genera described by Orla-Jensen, the LAB group consisted of *Lactobacillus* (with Orla-Jensen's designations *"Betabacterium"*, *"Streptobacterium"* and *"Thermobacterium"* included as subgenera), *Leuconostoc* (*"Betacoccus"*), *Pediococcus* (*"Tetracoccus"*) and *Streptococcus*. These genera had a firm standing in the systematics of lactic acid bacteria until the advent of molecular taxonomy in the 1970s and 1980s. From 1984 onwards, major taxonomic revisions have resulted in a further division of these genera. In addition, new genera, related to the lactic group, have been described (for thorough reviews on the current taxonomy of LAB, see Pot *et al.*, 1994; Wood & Holzapfel, 1995).

It is appropriate at this point to discuss the exact meaning of the term *lactic acid bacteria*. In the scientific community, especially among food microbiologists, there is a kind of consensus to the meaning of the term, but looking at it more closely, there is no strict definition. There are two ways of using the term: the first is of historical/practical use where LAB means those bacteria that are intimitely associated with food and feed fermentations and related bacteria normally associated with the (healthy) mucosal surfaces of humans and animals. The latter was once thought to be one of the sources of inoculum for fermentations (Orla-Jensen, 1919). The second way is the physiological/phylogenetic use which include the above group, but also others that are related by their physiological characters and phylogenetic position.

This ambiguity in the use of the LAB term was evident also at the time of Orla-Jensen's pioneering work. For example, the genus *Streptococcus* has always been included as one of the "core" genera (Ingram, 1975). It is interesting to note, however, that pathogenic streptococci are very rarely referred to as being "lactic acid bacteria". This is the only case when the LAB group should be described with " objective" definitions (e.g. see Wood & Holzapfel, 1995). Similarly, conferences and meetings with Lactic Acid Bacteria as the heading (e.g. the series of symposia in the Netherlands from 1983) hardly ever deal with streptococci other than *S. thermophilus*, currently the only streptococcal species associated with food fermentation.

The situation is similar with enterococci. Serious problems concerning antibiotic resistance among enterococci are never associated with these organisms being "lactic acid bacteria". However, when enterococci are found to possibly be important in food fermentations (e.g. artisanal cheeses), they suddenly are! Thus, it seems that in practice the term LAB has a " hidden code", that is, it means something like " food", "edible", "health", "beneficial" or "fermentation". The physiological/phylogenetic view of the term might be more objective, but we believe that the "hidden code" will always be there in practice. There will probably never be a scientific communication starting with: " *Streptococcus pneumoniae*, a member of the lactic acid group of organisms....", although this would be as correct as if the species under study was *Lactococcus lactis*. This chapter will mostly deal with LAB in the historical/ practical sence (with the "hidden code"!), although there will inevitably be some reference to the physiology and phylogeny of the group and therefore to " all" LAB.

The physiological/morphological description of LAB is that they are Gram-positive, catalase-negative rods or cocci which produce lactic acid as the sole, major or important end-product from sugar fermentation with a non-respiring metabolism (Axelsson, 1998). It is also often added that they are non-sporing, generally non-motile and that they do not reduce nitrate under standard conditions. One important character that should be mentioned is their inability to synthesize porphyrinoid groups, the actual background for the catalase-negative reaction and the non-respiring (cytochrome-negative) mode of metabolism (Axelsson, 1998). Some LAB may change these characters when porphyrins (e.g. heme) are provided in the growth medium.

After the recent taxonomic revisions, based on chemotaxonomic, genetic and phylogenetic data, the following genera are included in the broad sense of the term LAB: *Aerococcus, Alloiococcus, Carnobacterium, Dolosigranulum, Enterococcus, Lactobacillus, Lactococcus, Lactosphaera, Leuconostoc, Melissococcus, Oenococcus, Pediococcus, Streptococcus, Tetragenococcus, Vagococcus* and *Weissella* (Schleifer & Ludwig, 1995b). Phylogenetically, these genera all belong to a super-cluster within the low-G+C subdivision of Gram-positive bacteria, which also includes aerobic and facultatively anaerobic bacteria such as *Bacillus, Listeria* and *Staphylococcus*. Species of the genus *Bifidobacterium* are often included in this group, mainly because they fulfill the basic physiological criteria and share almost the same ecological niche as some lactobacilli (the gastrointestinal tract). However, bifidobacteria belong phylogenetically to the high-G+C subdivision of Gram-positive bacteria (Schleifer & Ludwig, 1995a) and the metabolic pathway leading to lactic acid (and acetic acid) formation is quite distinct from the so-called "genuine" LAB (Sgorbati *et al.*, 1995).

2   Methods Used in Classifying and Identifying Lactic Acid Bacteria

2.1   PHENOTYPIC METHODS

Traditionally, morphology has been a major distinguishing character for classifying LAB. For example, lactobacilli were distinguished from streptococci, pediococci and leuconostocs and pediococci were unique in forming tetrads. Morphology is still important, but is now regarded as a poor indicator of relatedness. This is evident from some of the newly described genera of LAB, e.g. *Weissella*, which encompasses strains previously regarded as leuconostocs (cocci) or lactobacilli (rods) (Collins *et al.*, 1993).

Simple physiological tests, such as growth at different temperatures, acid, alkaline and salt tolerance and gas production are still useful, but it is increasingly difficult to use these for genus identification (Axelsson, 1998). Gas (CO_2) production from glucose, for example, is used for distinguishing between the two major modes of fermentation occurring within LAB, namely homolactic and heterolactic fermentation. Under standard conditions, homolactic fermentation of glucose results in the formation of almost entirely lactic acid as the end product, whereas with heterolactic fermentation equimolar amounts of lactic acid, CO_2 and ethanol/acetic acid are formed (for reviews on the metabolism of LAB, see Kandler (1983) and Axelsson (1998)). The largest group of LAB possesses both pathways, they appear homofermentative on hexoses and heterofermentative on pentoses and gluconate and are therefore designated facultatively heterofermentative. The different fermentative types: obligately homofermentative, obligately heterofermentative and facultatively heterofermentative, are used to divide lactobacilli into three groups (Hammes & Vogel, 1995), a tradition with its roots in the concepts of Orla-Jensen (see above).

Carbohydrate fermentation patterns are used in standard phenotypic tests to differentiate species. Although very useful, one should be aware of the limitations of this method, notably the large degree of variation within species, interlaboratory variation and poor reproducability (Pot *et al.*, 1994), and pronounced differences in the results depending on the method used (Axelsson & Lindgren, 1987). However, databases prepared from results using standardized, commercially available systems (e.g. API 50 CH; API System, la Balme les Grottes, Montalieu-Vercieu, France) should be valuable since the reliability of these kinds of methods will increase with standardization and the number of strains tested.

Numerical taxonomy has also been used to clarify LAB. This method has been proven useful in connection with LAB in association with meat (see below), but has also posed some difficulties, e.g the inability of the numerical analyses to create sound clusters (Priest & Barbour, 1985). Some lactobacilli seem to have an unusual ability for spontaneous phenotype alterations (Ahrné *et al.*, 1992) which could affect the outcome of numerical analyses.

Other phenotypic methods include more chemical/biochemical approaches, such as the type of interbridging in the peptidoglycan, electrophoretic mobility of lactic acid dehydrogenases, extraction and analyses of fatty acids and SDS-PAGE of whole cell proteins. The latter method has become very useful due to a standardization of the procedure and the preparation of a database of digitized and normalized protein patterns of the known species of LAB (Pot *et al.*, 1994; Vandamme *et al.*, 1996). The results of this method correlate well with similarity data obtained with DNA:DNA hybridization which means that the method is valuable for species recognition (see below).

2.2 GENOTYPIC METHODS

Determination of DNA base composition and percentage of similarity after DNA:DNA hybridization were the first genotypic methods used in bacterial systematics. DNA base composition is of limited value in identification schemes, but is still important in the description of species. DNA:DNA hybridization studies were applied extensively for LAB in the 1970s and 1980s and clarified much of the confusion in the systematics prevailing at that time. DNA homology values are still the basis for the definition of species as the basic unit of bacterial taxonomy (Vandamme *et al.*, 1996).

Comparative analysis of ribosomal RNA (rRNA) sequences in relation to bacterial systematics, first by DNA:rRNA hybridization and 16S rRNA cataloguing techniques, now replaced by direct sequencing, is regarded as the optimal measure of phylogenetic relationships (Woese, 1987; Vandamme et al., 1996). The method has been used extensively for LAB and has clarified the phylogeny of the group (for reviews, see e.g. Schleifer & Ludwig, 1995a,b).

The direct sequencing of the 16S rRNA molecule by PCR technology is now relatively simple (Pettersson, 1997). Together with the availability of a considerable number of sequences in databases, this makes the method one of the most powerful in the classification of an unknown strain in one single step. However, there are some pitfalls which have been discussed by Vandamme et al. (1996), e.g. some clearly different species may have the same 16S rRNA sequence (Pettersson, 1997) and the reliability of some sequences in the databases can be questioned. Also, it is still not clear to what extent there exist interoperon sequence variation (within the same clone) and/or strain variation within species.

The availability of 16S rRNA (and to some extent 23S rRNA) sequences has also made it possible to design species specific oligonucleotide probes for recognition of many species of LAB (Schleifer et al., 1995). These are very useful in confirmatory tests, but generally the isolate requires a certain degree of preidentification. A very useful variation of the use of these probes is the reverse dot-blot hybridization method (Ehrmann et al., 1994). In conjunction with a single PCR amplification step directly on a food sample, this method can be used to determine the dominant species in a food fermentation. Oligonucleotide probes are very important in ecological studies of LAB. With colony hybridization techniques, large number of strains can be screened for particular species or groups of LAB. For example, specific probes for the subspecies lactis and cremoris of Lactococcus lactis have been developed (Klijn et al., 1991; Salama et al., 1991) and used to identify new strains from traditional milk products from third world countries (Salama et al., 1993, 1995) and non-dairy environments (Klijn et al., 1995). These studies represent very important measures for the development of new dairy products and for combating problems like phage attacks in starter cultures of dairy foods. An interesting outcome of these investigations was that the present dairy starter strains appear to have lost important traits for survival outside the dairies since the "new" strains isolated differed phenotypically in some key characteristics (Klijn et al., 1995).

Other genotypic methods used to identify and/or classify LAB, even below species level, are those based on DNA restriction fragment analysis or fragment patterns obtained from PCR using arbitrary primers (randomly amplified polymorphic DNA, RAPD; Welsh & McClelland, 1990; Williams et al., 1990). The former can involve either an analysis of digested total DNA using image analysis and multivariate statistics (restriction endonuclease analysis, REA; Ståhl et al., 1990) or restriction fragment length polymorphism (RFLP) of certain genes. When used with a universal rRNA probe, the method is generally referred to as ribotyping (Grimont & Grimont, 1992). It is at present difficult to generalize about the taxonomic level the different methods can resolve. This could probably differ between species, as illustrated by a comparison between Lactobacillus reuteri and Lacotobacillus plantarum using ribotyping (Ståhl et al., 1994; Johansson et al., 1995a). For L. plantarum, ribotyping can be used for species recognition, RAPD for group classification and REA for strain identification (Johansson, 1995) (see also below). Reliable strain typing methods will become increasingly important in the study of the performance of LAB starter cultures and cultures used as additives in functional food type products (see below).

3 Lactic Acid Bacteria and Applied Systematics

To begin this section, it is appropriate to cite Orla-Jensen (1919) in his introduction where he points out the importance of bacterial systematics to a very applied science, namely dairy technology:

" We are still, however, far from having arrived at a complete elucidation of all the questions involved. It is particularly difficult to understand how various sorts of hard cheese, apparently containing the same microflora, should each have its own characteristic taste and smell. There can hardly be any doubt that these sorts of cheese in reality contain different species of bacteria, only we are unable to distinguish them by the methods hitherto employed. The object of the present work is primarily to meet this want by describing the useful bacteria of the dairy industry, so thoroughly that it may be possible in the future to identify the strains encountered."

Today, almost 80 years later, essentially the same issues are of interest, and now we do have the methods to distinguish the bacteria. In particular, the development of rapid and reliable methods where large number of strains can be processed with accuracy is beginning to get us where Orla-Jensen wanted to be, but, seen in retrospect, perhaps never was. However, his work should be greatly acknowledged and it is remarkable how good his classfication system was considering the means he had available.

Orla-Jensen was focused on dairy fermentations. These are of course also very important today, but it is perhaps time to acknowledge the large number of other lactic fermentations and other applications of LAB which are or can be very important for a healthy global food supply. To fully explore the potential of LAB in optimizing the production of fermented products (e.g. control of the process, accelerating the process, improved hygienic quality, improved taste) by using selected strains as starter cultures and for other applications it is necessary to obtain knowledge about

(i) which genera and species of LAB tend to dominate a particular fermentation or niche,
(ii) the diversity of strains in these niches,
(iii) the ecology of LAB inhabiting these environments, and
(iv) the physiology and genetics of important genera and species.

Methods for the accurate classification and identification of LAB are essential in these investigations. Most of this research is aiming at applying selected strains for certain purposes (e.g. starters, protective cultures, probiotics). Methods for distinguishing bacteria at the strain level (i.e. strain typing) must also be developed to be able to follow the fate of these strains in particular applications.

It is very difficult to cover this vast area, that is, ways in which taxonomic concepts have been, are or will be of value for understanding and exploiting LAB in various applications, but a few examples will be given to illustrate the point with reference to meat/meat fermentation and LAB in the gastrointestinal tract. These areas are chosen because they illustrate some interesting points and they represent topics which have an old tradition in LAB research, but are perhaps less known than, for example dairy research, and where the thorough use of modern systematics has changed or may change some previously held views. In addition, some aspects of lactic fermentations are also considered.

3.1   GENERAL ASPECTS OF LACTIC FERMENTATIONS

Lactic acid bacteria are generally fastidious on artificial media and require a number of growth factors (e.g. amino acids, vitamins) and a fermentable sugar for growth. They are thus able to grow only in environments which provide these nutrients. Since most agricultural raw materials (cereals, meat, milk, plants) contain the necessary nutrients (especially after release of intracellular material due to cutting, grinding or soaking), LAB generally have the potential to grow in such products.

Some LAB are very competitive if provided with the right conditions, such as somewhat reduced oxygen tension, mesophilic temperatures and reduced water activity. Although the initial load of microbes plays a role, these competitive LAB often outgrow the competitors despite lower initial numbers and initiate a lactic fermentation. Once pH starts to fall due to lactic acid production, conditions become even more selective for LAB and a genuine lactic fermentation of the material will occur with the most acid tolerant LAB dominating at the end. In general terms, this sequence of events is true for all spontaneous lactic fermentations, although the product's intrinsic properties (e.g. buffering capacity, pH, sugar content), the physical containment (e.g. packing, submerging in brine, aeration, temperature) and initial microbial composition will influence the course of fermentation and the types of LAB that will be able to grow.

The manufacture of sauerkraut is the prototype fermentation of the kind outlined above and has been studied as a model system for spontaneous, mixed culture fermentations since 1930 (cited in Steinkraus, 1983). A generalized diagram for the appearance ("succession") of different types of LAB in a spontaneous fermentation is shown in Figure 1. There are no numerical data included in the diagram as the numbers of LAB may differ considerably depending on the product. However, in a "genuine" lactic fermentation, the number of LAB may reach 10^8-10^9 CFU/g.

The type A LAB are characterized by having a short lag phase, a high growth rate, generally quite oxygen tolerant and often produce CO_2, but are usually not very acid tolerant; this factor leads to a high death rate later in the fermentation. Type B LAB have relatively high growth rates and are also acid tolerant. The most acid tolerant species, type C LAB, tend to have a comparatively slow growth rate. The type A LAB are thus very important in "setting the stage", they are, for instance, responsible for the initial lowering of pH and sometimes for creating more anaerobic conditions (by producing CO_2). In sauerkraut and other fermentations of plant origin, the type A LAB are often different *Leuconostoc* species, while types B and C are represented by *Lactobacillus brevis* and *Lactobacillus plantarum*, respectively (Steinkraus, 1983). In meat fermentations, type A could likewise be leuconostocs, but also species of *Carnobacterium* or *Weissella*, while B and C are often *Lactobacillus curvatus* or *Lactobacillus sakei* (previously *L. sake*). In milk fermentations (e.g. cheese or yoghurt), *Lactococcus lactis* or *Streptococcus thermphilus* may represent LAB type A and *Lactobacillus delbrückii* subsp. *bulgaricus*, *Lactobacillus casei* or *Lactobacillus helveticus* types B or C (Davis, 1975).

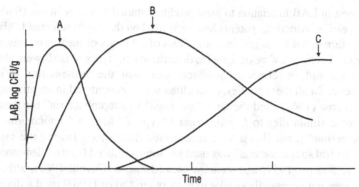

Figure 1. Generalized diagram representing numbers of lactic acid bacteria in a typical spontaneous lactic fermentation. A, B and C denote different types of lactic acid bacteria. See text for details.

To be able to study and ultimately understand these fermentations in detail, the classification of important groups and species is pertinent. Sauerkraut fermentation can again be taken to illustrate this point. The initial heterofermentation by leuconostocs is essential for the taste and aroma of the product. It is, therefore, not advantageous to add the dominating flora (e.g. *L. plantarum*) as a starter (Pederson & Albury, 1969). However, with the knowledge of the sequential appearance of different LAB, mixed starters of suitable strains can be developed to obtained more controlled fermentations (Harris *et al.*, 1992).

It should be noted that there is a fine line between desirable lactic fermentations, as in properly prepared fermented foods, and spoilage of certain products where a lactic fermentation is undesirable. The factors governing these reactions are essentially the same, encouraged in one instance and discouraged in the other. Meat and meat products provide some good examples of this dual character of lactic fermentations, this is discussed below with the emphasis placed on the importance of systematics.

3.2 MEAT AND MEAT FERMENTATIONS

For meat products, the term fermented food applies to most types of raw sausages and raw hams. The latter are mainly preserved by salting and drying with little activity of microorganisms. However, nitrate-reducing bacteria and/or LAB may be used to improve colour and flavour and to accelerate the ripening process (Leistner & Lücke, 1989). The major application of LAB is, however, in the area of fermented sausages where their activity is essential (Hammes *et al.*, 1990).

Fresh meat is a perfect substrate for microbial growth and aerobically Gram-negative rods will dominate. The handling of meat for preparing fermented sausages (grinding, salting and stuffing) eventually selects for LAB (Kröckel, 1995). The use of vacuum-packing and modified atmosphere packing (MAP) to prolong shelf-life of fresh meat or cooked products has introduced a "new" biotope which also selects for LAB. In such instances, LAB may act as spoilage flora, but addition of selected LAB strains could also enhance the shelf-life (Stiles, 1996). In this application, the added bacteria should not change the appearance of the product and it is therefore not a fermentation in the normal sense. The term "protective cultures" has been suggested to describe these strains (Stiles, 1996).

The interest in LAB in relation to meat products started seriously in the 1950s although some years earlier, American patents had been filed on the use of selected LAB as starter cultures for fermented sausages. Investigations of all kinds of meat products, especially vacuum-packed versions, were designed to determine what types of LAB were able to grow, cause spoilage and, in fermented products, dominate the fermentation to obtain the desired product. In all these studies, difficulties were encountered in identifying many of the LAB. Sharpe (1962) used the term " unclassified streptobacteria" for these LAB as they had some similarities to *L. plantarum* (" typical" for the *Lactobacillus* subgenus *"Streptobacterium"*), but also some characteristic differences. During the same period, the term " atypical streptobacteria" was used for strains isolated from chilled-stored chicken (Thornley & Sharpe, 1959) and vacuum-packed bacon (Cavett, 1963), although these organisms were not necessarily similar to each other. Cavett (1963) used a diagnostic key to differentiate LAB, among them both " atypical lactobacilli" (which were motile) and " atypical streptobacteria". It is remarkable how these terms became almost species- or group-specific epithets during the following years. Reuter (1975) summarized the criteria used to differentiate " atypical streptobacteria" from the " typical streptobacteria" in investigations of a variety of meat products (Table 2). Reuter's atypical streptobacteria were found not only in fresh and cured, vacuum-packed whole meat, but also as the dominant LAB in fermented sausages with no starter culture added. They were, therefore, probably responsible for the essential lactic fermentation in these products. This group of bacteria was finally identified as belonging to the species *L. curvatus* and *L. sakei* (Kagermeier, 1981; Egan, 1983; Shaw & Harding, 1984; Champomier *et al.*, 1987), two known species of streptobacteria, which had not previously been associated with meat.

TABLE 2.   Differentiation of "typical" and "atypical" lactobacilli of the sub-group "streptobacteria" isolated from meat (Reuter, 1975)

	Streptobacteria	Atypical streptobacteria
Growth on meat extract tryptone agar:		
Growth rate	Reduced	Not markedly reduced
Colony size	Pin point	Small
Cell shape	Rods	Coccoid
Growth on lactobacilli-media:		
Growth rate	Not reduced	Not reduced
Colony size	Diameter, 1-2 mm	Diameter, 0.5-1 mm
Cell shape	Rods	Short rods
Growth in liquid media:		
Minimum temperature for growth	4-8°C	2-4°C
Maximum temperature for growth	42-45°C	40-42°C
Final pH	3.7-3.8	3.9-4.1
Growth at pH 3.9	Yes	No

In retrospect, it is interesting to note that it took almost 25 years before it was concluded that the two *Lactobacillus* species were involved in the fermentation of the meat products. It seems that classification and identification of bacteria are always "biased" by previous knowledge. Since the investigation of meat was a relatively new research topic, the methods and the conclusions were heavily influenced by investigations of other habitats containing similar bacteria, in this case most notably dairy products. It almost appears that certain species were expected to be found, that is, *L. casei* and *L. plantarum*. When the encountered bacteria showed unusual properties compared to these species they were classified as atypical. With a more "non-biased" approach, strains of known species like *L. curvatus* and *L. sakei* should have been included as reference strains in the investigations. The example outlined above clearly shows the need for objective classification criteria and identification methods.

One objective approach is numerical taxonomy, which was used to examine many LAB isolated from vacuum-packed meat (Shaw & Harding, 1984). One of the phenetic clusters identified conformed to the atypical streptobacteria and was tentatively identified as *L. sakei*. Another cluster was allotted to the genus *Leuconostoc*. It was not possible to identify strains assigned to a third cluster, though bacteria with similar characteristics, most notably the inability to grow on acetate agar (and hence separated from lactobacilli in general), had been detected in previous studies. Attention was drawn to a study by Holzapfel & Gerber (1983) where a new species, *Lactobacillus divergens*, was designated as an unusual heterofermentative LAB unable to grow on acetate agar. Shaw & Harding (1985) subsequently allotted their strains to this taxon or to a new albeit related species, *L. carnis*. The taxonomic studies of these strains culminated with the description of the new genus *Carnobacterium* to accommodate these "atypical lactobacilli" (Collins *et al.*, 1987). This work was ingenious in that it included strains isolated by Thornley & Sharpe (1959) (see above) and in a way, therefore, embraced all the studies describing these unusual LAB since that time. Interestingly, carnobacteria, which form a coherent phylogenetic unit, are more related to enterococci than to lactobacilli (Wallbanks *et al.*, 1990). Carnobacteria are typically found on meat, especially at low temperatures, but can be isolated from other habitats, e.g. cheese (Millière *et al.*, 1994) and fish (Collins *et al.*, 1987).

Lactobacillus curvatus and *L. sakei* and are also very competitive in meat products, but seem to have a broader range of possible habitats. This is especially true for *L. sakei*, which can be characterized as being quite resistant to adverse conditions, e.g. low temperatures, high salt concentration, smoke, ethanol, low water activity and radiation, but not as acid tolerant as, for example *L. plantarum*. *Lactobacillus sakei* has been shown to be a significant part of the fermenting flora in rice wine (the original source for the description of the species; Kandler & Weiss, 1986), sauerkraut (Klein *et al.*, 1996), silage (Gundersen, 1994), and a traditional, Norwegian fermented fish product (Axelsson, unpublished). The third group of LAB that is frequently isolated from meat are the so-called leuconostoc-like organisms. It has recently been shown that these bacteria really belong to two genera, *Leuconostoc sensu stricto* and *Weissella*, a genus proposed to encompass *Leuconostoc paramesenteroides* and related heterofermentative LAB, such as *Lactobacillus confusus* (Collins *et al.*, 1993).

The improved classification of LAB makes it possible to study the diversity, ecology and distribution on meat products in a more accurate way. Several identification schemes have been published, for example, to distinguish between carnobacteria and meat lactobacilli (Montel et al., 1991). 23S rRNA targeted oligonucleotide probes have also been developed to detect typical meat lactobacilli (Hertel et al., 1991). Some interesting work has also been done with genus-specific oligonucleotide probes developed for carnobacteria and leuconostocs (Nissen et al., 1994, 1996). These probes were used to monitor the LAB flora of the same starting material packed under different atmospheres and stored at different temperatures. It was shown that at certain temperatures, differences in atmosphere will select for either carnobacteria (N_2) or leuconostocs (CO_2) early in the storage period. Longer storage periods lead to a "succession" where ultimately L. sakei/curvatus tend to dominate. This succession is analogous to that already noted for general lactic fermentations (see above). It is also interesting that the major groups of LAB that have been isolated from meat products were represented in the same material, the dominating group reflecting the external (environmental) conditions (atmosphere, temperature, pH). In another experiment, the type of meat (pork or lamb) had a decisive role in determining the dominant flora under the same storage conditions. Information such as that described above is essential for the selection of appropriate strains of LAB for application as protective cultures (Nissen et al., 1994; Stiles, 1996).

For fermented sausage, novel starter cultures are appearing on the market bearing in mind their competitiveness in meat (Hammes et al., 1990). The first generation of starter cultures were selected merely for acid production capability (and probably ease of handling, e.g. lyophilization and distribution), such as strains of L. plantarum and Pediococcus acidilactici. However, the second generation of starter cultures are based on L. curvatus or L. sakei which, due to their competitiveness, control the fermentation better (Hammes et al., 1990), and influence the aroma development and ripening process to a larger extent (Kröckel, 1995). This latter property can be positive, but is also the reason why strains of these species may act as spoilage organisms in certain instances (Björkroth & Korkeala, 1996b). Strain typing methods are being developed for these species and this will facilitate studies where the fate of starter cultures and undesirable contaminants is to be investigated (Björkroth & Korkeala, 1996a; Björkroth et al., 1996). Important fundamental knowledge about the genetics and physiology of members of these species is beginning to accumulate (Wijtzes et al., 1995; Hammes & Hertel, 1996).

3.3 INTESTINAL LACTOBACILLI

The role of lactic acid bacteria, mainly lactobacilli, in the gastrointestinal (GI) tract has been a subject of interest since the beginning of this century. Elie Metchnikoff launched his idea of their positive effect, counteracting the negative part of the intestinal flora, as early as 1907 (Metchnikoff, 1907). He stated the beneficial effects of changing the intestinal wild-flora into a tamer one. This is also the reasoning when it comes to the concept of probiotics. Fuller (1989) defined probiotics as " A live microbial feed supplement which beneficially affects the host animal by improving its intestinal microbial balance". One could perceive the role of lactobacilli in the intestines being similar to their role in lactic fermentations in general. By growing and producing organic acids, lowering the pH and sometimes producing antimicrobial substances they compete with other bacteria in the habitat. However, recent studies indicate that their mode of action could be somewhat more sophisticated.

Reports of changing the immunoresponse of mucosal cells (Gaskins *et al.*, 1996) and finding specific receptor-binding at the same site (Adlerberth *et al.*, 1996) points in this direction.

Probiotic bacteria have to survive and thrive in the GI tract so the most obvious place to find such strains is in this habitat. Lactobacilli are found throughout the entire GI tract (Moore & Holdeman, 1974; Finegold *et al.*, 1983; Lidbeck *et al.*, 1987; Mikelsaar & Mändar, 1993; Molin *et al.*, 1993; Ahrné *et al.*, 1998) where they more or less contribute to the fermentation depending on the location. In the small intestine these bacteria play a somewhat dominant role, while further down in the large bowel they are only a minor part of the flora (Lidbeck, 1991; Lichtenstein & Goldin, 1993).

The composition of the *Lactobacillus* flora in the GI tract has, for practical reasons, mostly been studied in faecal samples and occasionally by looking at intestinal content (Reuter, 1965). Fecal composition might reflect the situation of the large intestine but is a poor indicator of flora in the small intestine (Savage, 1977; Sharpe, 1981). Many of the validly described *Lactobacillus* spp. have been isolated from human faeces. *Lactobacillus acidophilus, L. fermentum* and *L. salivarius* are frequently reported as members of the faecal flora. *Lactobacillus casei, L. crispatus, L. gasseri, L. plantarum* and *L. reuteri* are also thought to be typical representatives (Moore & Holdeman, 1974; Sharpe, 1981; Hentges, 1983; Axelsson, 1990). Strain identification has been quite difficult, but improved methods are being introduced. One example is the so-called "*L. acidophilus* group", which consists of six genera that cannot be separated by traditional phenotypic methods, but can now be distinguished on the basis of SDS-PAGE protein (Pot *et al.*, 1993) or RAPD patterns (Du Plessis & Dicks, 1995).

Investigations of the composition of the LAB flora other than in faecal material are rare, but one survey of the *Lactobacillus* species residing at intestinal mucosa has been undertaken (Molin *et al.*, 1993). Seventy-five persons were sampled by biopsies from both the small and the large intestine, the result showed that the flora was heterogeneous and differed widely between different individuals (Table 3). Often, members of one or two species dominated within one person though no general differences were found in the type of dominating *Lactobacillus* flora between mucosa from different parts of the intestine (Johansson, 1995).

Following this investigation an administration study (intake of live *Lactobacillus* strains as " food additives") was undertaken in order to find strains capable of colonising the mucosa. Colonising ability was scored as positive if a strain was detected as one of the dominant strains at the mucosa 11 days after the administration stopped (Johansson *et al.*, 1993). The strains administered were chosen to represent the different clusters found at the mucosa by Molin *et al.* (1993). Similar studies, but performed on faecal samples, have been carried out with selected *Lactobacillus* strains. Saxelin *et al.* (1991) found that *Lactobacillus rhamnosus* GG could be recovered in faeces 7 days after stopping the administration, whereas for *L. acidophilus* NCDO 1748, Lidbeck *et al.* (1987) noted that the strain was only detected as long as the participants were given the preparation. Unfortunately, these workers did not use methods likely to discriminate between bacteria at the strain level. Discrimination at this level is highly important in such ecological studies, just as it is in epidemiological studies where particular strains are target organisms. The detection of a specific strain in its natural habitat (read a probiotic strain in the GI tract) demands a means of identification that is very hard to obtain. In the administration study mentioned above, restriction endonuclease analysis (REA) of the total chromosomal DNA was used for this purpose (Johansson *et al.*, 1993). This approach turned out to be very powerful.

REA has been used successfully for classification of *L. reuteri* strains (Ståhl & Molin, 1994). Following the administration study mentioned above, this method was also used to classify *L. plantarum* (Johansson *et al.*, 1995b), since two strains of this species were shown to be the best colonisers. The REA-classification of *L. plantarum* placed the colonising strains, *L. plantarum* 299 and *L. plantarum* 299v, in a specific sub-cluster of the species; this taxon encompassed all of the intestinal derived strains together with some strains from other habitats (Johansson *et al.*, 1995b). In the meantime, the same set of *L. plantarum* strains, approximately twenty strains of various origin, where classified by RAPD (Johansson *et al.*, 1995c) and by ribotyping (Johansson *et al.*, 1995a). It can be concluded from these studies that ribotyping was of value as discriminating at the species level (Johansson *et al.*, 1995a), RAPD was sufficiently sensitive to detect the subgroups of *L. plantarum* (Johansson *et al.*, 1995c) whereas REA was the only method capable of discrimination at the strain level (Johansson *et al.*, 1995b).

The significance of studying intestinal lactobacilli from a taxonomic point of view proved to be enlightning following subsequent findings. Members of the " intestinal subgroup" of *L. plantarum* turned out to have a specific trait as they were shown, by Adlerberth *et al.* (1996), to have a mannose-specific adhesin that mediated adherence to human colonic cells. Furthermore, in a recent investigation of the lactobacilli of the intestinal mucosa of perfectly healthy persons, *L. plantarum* strains were found to dominate the rectal flora of as much as 26% of the participating individuals (Table 3; Ahrné *et al.*, 1998). Again, the majority of these strains showed mannose-sensitive adherence to colonic cells (Ahrné *et al.*, 1998). Comparing the frequency of *L. plantarum* in the previously mentioned study (Molin *et al.*, 1993), which was 5% (Table 3), with the current data suggests that persons seeking medical care may not be appropriate subjects for assessing the normal intestinal flora of healthy individuals (Ahrné *et al.*, 1998).

The oral cavity, as reflected by samples taken from the back of the tongue, is dominated by the same set of lactobacilli as is the GI tract (Tables 3 and 4; Ahrné *et al.*, 1998). Similarly, *L. plantarum*, belonging to the the " intestinal sub-group" (determined by RAPD-typing) and showing mannose-sensitive adherence are very common on the tongue. Up to 43% of the individuals in this study turned out to be dominated by *L. plantarum* (Ahrné *et al.*, 1998). It has been suggested for *Escherichia coli* strains which possess mannose-sensitive binding trough the type 1 fimbriaes (Bloch *et al.*, 1992) that " Colonization of the throat would provide a constant source of bacteria entering the stomach and might thus increase the chances for an incoming strain to colonize the intestine" (Salyers & Whitt, 1994). This may well be the strategy adopted by lactobacilli.

It can be seen from Tables 3 and 4 that the dominating *Lactobacillus* flora of human intestinal mucosa, as well as in that of the oral cavity, were identified as *L. casei* subsp. *pseudoplantarum, L. plantarum* and *L. rhamnosus*. Members of these species have to be regarded as the least fastidious lactobacilli (Morishita *et al.*, 1981); they can be isolated at many different locations in nature (Kandler & Weiss, 1986); they also tend to dominate in traditionally prepared lactic fermentations of food (Oyewole & Odunfa, 1990; Fernández Gonzalez *et al.*, 1993; Johansson *et al.*, 1995d; Figueroa *et al.*, 1997). Man has, for thousands of years, been able to use such fermentations to preserve food, and in many

TABLE 3.   Occurrence of the dominant identified lactobacilli found on human intestinal mucosa of: (i) persons seeking medical advice for gastrointestinal complaints (Molin *et al.*, 1993), and (ii) completely healthy individuals (Ahrné *et al.*, 1998). Clusters were based on phenotypic data (Molin *et al.*, 1993) or on phenotypic testing followed by confirmation by DNA:DNA hybridization for the species *Lactobacillus casei* subsp. *pseudoplantarum, L. plantarum* and *L. rhamnosus* (Ahrné *et al.*, 1998). The phenotypic based clustering was defined at 76% similarity (Sj, UPGMA analysis; Molin *et al.*, 1993) and 79% similarity (Sj, UPGMA analysis; Ahrné *et al.*, 1998)

Species	Occurrence (%)[a]	
	Patients[b]	Healthy persons[c]
L. acidophilus/gasseri/jenseni/crispatus[d]	15	0
L. casei subsp. *pseudoplantarum*	5	7
L. plantarum	5	24
L. reuteri	5	2
L. rhamnosus	16	12
L. salivarius	9	2
L. vaginalis	ND[e]	2

[a] Percentage of the total number of persons harbouring a member of the actual species as dominant member of the *Lactobacillus* flora; [b] according to Molin *et al.*(1993); [c] according to Ahrné *et al.* (1998); [d] these species could not be distinguished by using phenotypic data; [e] the type strain of *L. vaginalis* was not included in this study.

TABLE 4   Occurrence of the dominant identified lactobacilli found in the oral cavity (from the back of the tongue) of perfectly healthy persons. Clusters were based on phenotypic data followed by confirmation by DNA:DNA hybridization for the species *Lactobacillus casei* subsp. *pseudoplantarum, L. plantarum* and *L. rhamnosus* (Ahrné *et al.*, 1998). The phenotypic based clustering was defined at 76% similarity (Sj, UPGMA analysis; Ahrné *et al.*, 1998)

Species	Occurence (%)[a]
L. acidophilus/gasseri/jenseni/crispatus[b]	7
L. animalis	2
L. casei subsp. *pseudoplantarum*	12
L. fermentum	2
L. oris	2
L. plantarum	43
L. rhamnosus	21
L. salivarius	10

[a] Percentage of the total number of persons harbouring a member of the actual species as the dominant member of the *Lactobacillus* flora;
[b] these species could not be distinguished by using phenotypic data.

African cultures this is still the most important technique for this purpose (Salovaara, 1993). The inoculation of such foods by oral or faecal lactobacilli seems to be a plausible scenario. The intake of huge numbers of living lactobacilli has been the state of art, and the adaption of, for instance *L. plantarum*, by developing specific adherence mechanisms towards an intestinal niche seems a logical development. However, recently, in this perspective, the habit of eating live lactobacilli was unwittingly stopped by the invention of new food preservation methods, mainly chilling. This practice means that the route of passing intestinal lactobacilli back to the intestine by fermented foods was disrupted. And, the prospect of probiotic lactobacilli can be regarded as an attempt to turn back to a " normal" situation.

4   Conclusions and Future Perspectives

Acid fermentation of food, where lactic acid bacteria play the most important role, is of great importance for a wholesome global food supply. In developed countries, production has been industrialized and demands fundamental knowledge of the processes to effectively produce high quality products. The number and variety of products produced in developing countries are great and represent an enormous source for gaining new knowledge and for developing new products. Technology transfer to these countries is important for the progress of their own food industries, which supply traditional products. At a time where global strategies are needed to meet shortages in both energy and food supply, it is timely to acknowledge that food fermentations are one of several alternatives for low-cost and low-energy-demanding means for food preservation. If this alternative also means more healthy and nutritious food, the benefit is even greater.

The new food technology depends upon a detailed understanding of the microorganisms, namely the *lactic acid bacteria*. At the beginning of the century, Orla-Jensen recognized that it was necessary to know what types and activities of LAB were involved in a certain process in order to be able to predict, control and improve it. Although the industrial use of these bacteria resembles the use of any additive, be it salt or spices, there is a fundamental difference since bacterial additives are live organisms which may change or adapt their behaviour depending on the prevaling conditions. Microbial systematics will play a crucial part in the understanding of the role of LAB in different processes and ecological niches, since it is the basis for a historical record on the abilities, benefits and perhaps peculiarities of the species involved.

5   References

Adlerberth, I., Ahrné, S., Johansson, M.-L., Molin, G., Hanson, L.A. & Wold, A.E. (1996). A mannose-specific adherence mechanism in *Lactobacillus plantarum* conferring binding to the human colonic cell line HT-29. *Applied and Environmental Microbiology* 62, 2244-2251.

Ahrné, S., Casas, I., Lindgren, S.E., Molin, G. & Dobrogosz, W.J. (1992). Spontaneous and SDS-induced phenotype and plasmid alterations in starter cultures of *Lactobacillus plantarum. Systematic and Applied Microbiology* 15, 285-288.

Ahrné, S., Nobaek, S., Jeppsson, B., Adlerberth, I., Wold, A. & Molin, G. (1998). The normal *Lactobacillus* flora of healthy human oral and rectal mucosa. *Journal of Applied Microbiology* **85**, 88-94.

Axelsson, L. (1990). *Lactobacillus reuteri*, a member of the gut bacterial flora. Studies on antagonism, metabolism and genetics. PhD thesis, Swedish University of Agricultural Sciences, Uppsala, Sweden.

Axelsson, L. (1998). Lactic acid bacteria: classification and physiology. In *Lactic Acid Bacteria: Microbiology and Functional Aspects, 2nd Edition, Revised and Expanded*, pp. 1-72. Edited by S. Salminen & A. von Wright. New York, Marcel Dekker, Inc.

Axelsson, L. & Lindgren, S. (1987). Characterization and DNA homology of *Lactobacillus* strains isolated from pig intestine. *Journal of Applied Bacteriology* **62**, 433-438.

Björkroth, K.J. & Korkeala, H.J. (1996a). rRNA gene restriction patterns as a characterization tool for *Lactobacillus sakei* strains producing ropy slime. *International Journal of Food Microbiology* **30**, 293-302.

Björkroth, K.J. & Korkeala, H.J. (1996b). Evaluation of *Lactobacillus sakei* contamination in vacuum-packaged sliced cooked meat products by ribotyping. *Journal of Food Protection* **59**, 398-401.

Björkroth, J., Ridell, J. & Korkeala, H. (1996). Characterization of *Lactobacillus sakei* strains associating with production of ropy slime by randomly amplified polymorphic DNA (RAPD) and pulsed-field gel electrophoresis (PFGE) patterns. *International Journal of Food Microbiology* **31**, 59-68.

Bloch, C., Stocker, B. & Orndorff, P. (1992). A key role for type 1 pili in enterobacterial communicability. *Molecular Microbiology* **6**, 697-701.

Campbell-Platt, G. (ed.) (1987). *Fermented Foods of the World*. London: Butterworths.

Cavett, J.J. (1963). A diagnostic key for identifying the lactic acid bacteria out of vacuum packed bacon. *Journal of Applied Bacteriology* **26**, 453-470.

Champomier, M.-C., Montel, M.-C., Grimont, F. & Grimont, P.A.D. (1987). Genomic identification of meat lactobacilli as *Lactobacillus sakei. Annales Institut Pasteur (Paris)* **138**, 751-758.

Collins, M.D., Farrow, J.A.E., Phillips, B.A., Ferusu, S. & Jones, D. (1987). Classification of *Lactobacillus divergens, Lactobacillus piscicola* and some catalase-negative, asporogenous, rod-shaped bacteria from poultry in a new genus, *Carnobacterium. International Journal of Systematic Bacteriology* **37**, 310-316.

Collins, M.D., Samelis, J., Metaxopoulos, J. & Wallbanks, S. (1993). Taxonomic studies on some leuconostoc-like organisms from fermented sausages: description of a new genus *Weissella* for the *Leuconostoc paramesenteroides* group of species. *Journal of Applied Bacteriology* **75**, 595-603.

Davis, J.G. (1975). The microbiology of yoghurt. In *Lactic Acid Bacteria in Beverages and Food*, pp. 245-263. Edited by J.G. Carr, C.V. Cutting & G.C. Whiting. London: Academic Press.

Du Plessis, E.M. & Dicks, L.M.T. (1995). Evaluation of random amplified polymorphic DNA (RAPD)-PCR as a method to differentiate *Lactobacillus acidophilus, Lactobacillus crispatus, Lactobacillus amylovorus, Lactobacillus gallinarum, Lactobacillus gasseri*, and *Lactobacillus johnsonii. Current Microbiology* **31**, 114-118.

Egan, A.F. (1983). Lactic acid bacteria of meat and meat products. *Antonie van Leeuwenhoek* **49**, 327-336.

Ehrmann, M., Ludwig, W. & Schleifer, K.H. (1994). Reverse dot blot hybridization: a useful method for the direct identification of lactic acid bacteria in fermented food. *FEMS Microbiology Letters* **117**, 143-150.

Fernández Gonzalez, M.J., García García, P., Garrido Fernández, A. & Durán Quintana, M.C. (1993). Microflora of the aerobic preservation of directly brined green olives from Hojiblanca cultivar. *Journal of Applied Bacteriology* **75**, 226-233.

Figueroa, C., Davila, A.M. & Pourquie, J. (1997). Original properties of ropy strains of *Lactobacillus plantarum* isolated from the sour cassava starch fermentation. *Journal of Applied Microbiology* **82**, 68-72.

Finegold, S.M., Sutter, V.L. & Mathisen, G.E. (1983). Normal indigenous intestinal flora. In *Human Intestinal Microflora in Health and Disease*, pp. 3-31. Edited by D.J. Hentges. London: Academic Press.

Fuller, R. (1989). Probiotics in man and animals. *Journal of Applied Bacteriology* **66**, 365-378.

Gaskins, H.R., McCracken, V.J., Baldeon, M.E., Finlay, B.B. & Mackie, R.J. (1996). Adherent commensal and pathogenic bacteria differentially modulate inflammatory cytokine gene expression by colonic epithelial cells. XXIst International Congress on Microbial Ecology and Disease. October 28-30. Institut Pasteur, Paris, France.

Grimont, F. & Grimont, P.A.D. (1992). Identification and typing by rRNA gene restriction patterns. In *Proceedings of the Conference on Taxonomy and Automated Identification of Bacteria*, pp. 15-18. Edited by J. Schindler. Prague: Czechoslovak Society for Microbiology.

Gundersen, A. (1994). Lactic acid bacteria in grass silage. PhD thesis, Agricultural University of Norway, Ås.

Hammes, W.P. & Hertel, C. (1996). Selection and improvement of lactic acid bacteria used in meat and sausage fermentation. *Lait* **76**, 159-168.

Hammes, W.P. & Vogel, R.F. (1995). The genus *Lactobacillus*. In *The Genera of Lactic Acid Bacteria*, pp. 19-54. Edited by B.J.B. Wood & W.H. Holzapfel. London: Chapman & Hall.

Hammes, W.P., Bantleon, A. & Min, S. (1990). Lactic acid bacteria in meat fermentation. *FEMS Microbiology Reviews* **87**, 165-173.

Harris, L.J., Fleming, H.P. & Klaenhammer, T.R. (1992). Novel paired starter culture system for sauerkraut, consisting of a nisin-resistant *Leuconostoc mesenteroides* strain and a nisin-producing *Lactococcus lactis* strain. *Applied and Environmental Microbiology* **58**, 1484-1489.

Henneberg, W. (1904). Zur Kenntnis der Milchsäurebakterien der Brennerei-Maische, der Milch, des Bieres, der Presshefe, der Melasse, des Sauerkohls, der säuren Gurken und des Sauerteigs; sowie einige Bemerkungen über die Milchsäurebakterien des menschlishen Magens. *Zentralblatt für Bakteriologie, Parasitenkunde, Infektionskrankheiten und Hygiene, Abteilung II* **11**, 154-170.

Hentges, D.J. (ed.) (1983). *Human Intestinal Microflora in Health and Disease.* New York, USA: Academic Press.

Hertel, C., Ludwig, W., Obst, M., Vogel, R.F., Hammes, W.P. & Schleifer, K.H. (1991). 23S rRNA-targeted oligonucleotide probes for the rapid identification of meat lactobacilli. *Systematic and Applied Microbiology* **14**, 173-177.

Holzapfel, W.H. & Gerber, E.S. (1983). *Lactobacillus divergens* sp. nov., a new heterofermentative *Lactobacillus* species producing L(+)-lactate. *Systematic and Applied Microbiology* **4**, 522-534.

Ingram, M. (1975). The lactic acid bacteria - a broad view. In *Lactic Acid Bacteria in Beverages and Food*, pp. 1-13. Edited by J.G. Carr, C.V. Cutting & G.C. Whiting. London: Academic Press.

Johansson, M.-L. (1995). Systematics and starter culture selection of *Lactobacillus* for human intestine and Nigerian ogi, with special reference to *Lactobacillus plantarum*. PhD thesis, University of Lund, Lund, Sweden.

Johansson, M.-L., Molin, G., Jeppsson, B., Nobaek, S., Ahrné, S. & Bengmark, S. (1993). Administration of different *Lactobacillus* strains in fermented oatmeal soup: *in vivo* colonization of human intestinal mucosa and effect on the indigenous flora. *Applied and Environmental Microbiology* **59**, 15-20.

Johansson, M.-L., Molin, G., Pettersson, B., Uhlén, M. & Ahrné, S. (1995a). Characterization and species recognition of *Lactobacillus plantarum* strains by restriction fragment length polymorphism (RFLP) of the 16S rRNA gene. *Journal of Applied Bacteriology* **79**, 536-541.

Johansson, M.-L., Quednau, M., Ahrné, S. & Molin, G. (1995b). Classification of *Lactobacillus plantarum* by restriction endonuclease analysis of total chromosomal DNA using conventional agarose gel electrophoresis. *International Journal of Systematic Bacteriology* **45**, 670-675.

Johansson, M.-L., Quednau, M., Molin, G. & Ahrné, S. (1995c). Randomly amplified polymorphic DNA (RAPD) for rapid typing of *Lactobacillus plantarum* strains. *Letters in Applied Microbiology* **21**, 155-159.

Johansson, M.-L., Sanni, A., Lönner, C. & Molin, G. (1995d). Phenotypically based taxonomy using API 50CH of lactobacilli from Nigerian ogi, and the occurrence of starch fermenting strains. *International Journal* *of Food Microbiology* **25**, 159-168.

Kagermeier, A. (1981). Taxonomie und Vorkommen von Milchsäurebakterien in Fleischprodukten. PhD thesis, Ludwig-Maximilian University, Munich.

Kandler, O. (1983). Carbohydrate metabolism in lactic acid bacteria. *Antonie van Leeuwenhoek* **49**, 209-224.

Kandler, O. & Weiss, N. (1986). Regular, non-sporing gram-positive rods. In *Bergey's Manual of Systematic Bacteriology Vol. 2*, pp. 1208-1234. Edited by P.H.A. Sneath, N.S. Mair, M.E. Sharpe & J.G. Holt. Baltimore: Williams and Wilkins Co.

Klein, G., Dicks, L.M.T., Pack, A., Hack, B., Zimmermann, K., Dellaglio, F. & Reuter, G. (1996). Emended descriptions of *Lactobacillus sakei* (Katagiri, Kitahara, and Fukami) and *Lactobacillus curvatus* (Abo-Elnaga and Kandler): numerical classification revealed by protein fingerprinting and identification based on biochemical patterns and DNA-DNA hybridizations. *International Journal of Systematic Bacteriology* **46**, 367-376.

Klijn, N., Weerkamp, A.H. & de Vos, W.M. (1991). Identification of mesophilic lactic acid bacteria by using polymerase chain reaction-amplified variable regions of 16S rRNA and specific DNA probes. *Applied and Environmental Microbiology* **57**, 3390-3393.

Klijn, N., Weerkamp, A.H. & de Vos, W.M. (1995). Detection and characterization of lactose-utilizing *Lactococcus* spp. in natural ecosystems. *Applied and Environmental Microbiology* **61**, 788-792.

Kröckel, L. (1995). Bacterial fermentation of meats. In *Fermented Meats*, pp. 69-109. Edited by G. Campbell-Platt & P.E. Cook. London: Blackie A & P.

Leistner, L. & Lücke, F.-K. (1989). Bioprocessing of meat. In *Biotechnology, Vol 5*, pp. 273-286. Edited by S.-D. Kung, D.D. Bills & R. Quatrano. Boston: Butterworths.

Lichtensteln, A.H. & Goldin, B.R. (1993). Lactic acid bacteria and intestinal drug and cholesterol metabolism. In *Lactic Acid Bacteria*, pp. 227-235. Edited by S. Salminen & A. von Wright. New York, Marcel Dekker, Inc.

Lidbeck, A. (1991). Studies on the impact of *Lactobacillus acidophilus* on human microflora and some cancer-related intestinal ecological variables. PhD thesis, Karolinska Institutet, Huddinge University Hospital, Huddinge, Sweden.

Lidbeck, A., Gustafsson, J.-Å. & Nord, C.E. (1987). Impact of *Lactobacillus acidophilus* supplement on the human oropharyngeal and intestinal microflora. *Scandinavian Journal of Infectious Disease* **19**, 531-537.

Löhnis, F. (1907). Versuch einer Gruppierung der Milchsäurebakterien. *Zentralblatt für Bakteriologie, Parasitenkunde, Infektionskrankheiten und Hygiene, Abteilung II* **18**, 97-149.

Metchnikoff, E. (1907). *The Prolongation of Life*. London: Heinemann.

Mikelsaar, M. & Mändar, R. (1993). Development of individual lactic acid microflora in the human microbial ecosystem. In *Lactic Acid Bacteria*, pp. 237-293. Edited by S. Salminen & A. von Wright. New York: Marcel Dekker, Inc.

Millière, J.B., Michel, M., Mathieu, F. & Lefebvre, G. (1994). Presence of *Carnobacterium* spp. in French surface mould-ripened soft-cheese. *Journal of Applied Bacteriology* **76**, 264-269.

Molin, G., Jeppsson, B., Johansson, M.L., Ahrné, S., Nobaek, S., Ståhl, M. & Bengmark, S. (1993). Numerical taxonomy of *Lactobacillus* spp. associated with healthy and diseased mucosa of the human intestines. *Journal of Applied Bacteriology* **74**, 314-323.

Montel, M.-C., Talon, R., Fournoud, J. & Champomier, M.-C. (1991). A simplified key for identifying homofermentative *Lactobacillus* and *Carnobacterium* spp. from meat. *Journal of Applied Bacteriology* **70**, 469-472.

Moore, W.E.C. & Holdeman, L.V. (1974). Human fecal flora of 20 Japanese-Hawaiians. *Applied Microbiology* **27**, 961-979.

Morishita, T., Deguchi, Y., Yajima, M., Sakurai, T. & Yura, T. (1981). Multiple nutritional requirements of lactobacilli: genetic lesions affecting amino acid biosynthetic pathways. *Journal of Bacteriology* **148**, 64-71.

Nissen, H., Holck, A. & Dainty, R.H. (1994). Identification of *Carnobacterium* spp. and *Leuconostoc* spp. in meat by genus-specific 16S rRNA probes. *Letters in Applied Microbiology* **19**, 165-168.

Nissen, H., Sørheim, O. & Dainty, R. (1996). Effects on vacuum, modified atmospheres and storage temperature on the microbial flora of packaged meat. *Food Microbiology* **13**, 183-191.

Orla-Jensen, S. (1919). *The Lactic Acid Bacteria.* Copenhagen: Host and Son.

Oyewole, O.B. & Odunfa, S.A. (1990). Characterization and distribution of lactic acid bacteria in cassava fermentation during fufu production. *Journal of Applied Bacteriology* **68**, 145-152.

Pederson, C.S. & Albury, M.N. (1969). The sauerkraut fermentation. *New York State Agricultural Experiment Station, Technical Bulletin 824*, Geneva, New York State Agricultural Experiment Station.

Pettersson, B. (1997). Direct solid-phase 16S rDNA sequencing: a tool in bacterial phylogeny. PhD thesis, Royal Institute of Technology, Stockholm, Sweden.

Pot, B., Hertel, C., Ludwig, W., Descheemaeker, P., Kersters, K. & Schleifer, K.-H. (1993). Identification and classification of *Lactobacillus acidophilus*, *L. gasseri* and *L. johnsonii* strains by SDS-PAGE and rRNA-targeted oligonucleotide probe hybridization. *Journal of General Microbiology* **139**, 513-517.

Pot, B., Ludwig, W., Kersters, K. & Schleifer, K.-H. (1994). Taxonomy of lactic acid bacteria. In *Bacteriocins of Lactic Acid Bacteria*, pp. 13-90. Edited by L. De Vuyst & E.J. Vandamme. London: Chapman & Hall.

Priest, F.G. & Barbour, E.A. (1985). Numerical taxonomy of lactic acid bacteria and some related taxa. In *Computer-assisted Bacterial Systematics.*, pp. 137-163. Edited by M. Goodfellow, D. Jones & F.G. Priest. London, Academic Press.

Reuter, G. (1965). Das Vorkommen von Laktobazillen in Lebensmitteln und ihr Verhalten im Menschlichen Intestinaltrakt. *Zentralblatt für Bakteriologie, Parasitenkunde, Infektionskrankheiten und Hygiene, Abteilung I B Originale* **197** S, 468-487.

Reuter, G. (1975). Classification problems, ecology and some biochemical activities of lactobacilli of meat products. In *Lactic Acid Bacteria in Beverages and Food*, pp. 221-229. Edited by J.G. Carr, C.V. Cutting & G.C. Whiting. London, Academic Press.

Salama, M.S., Sandine, W.E. & Giovannoni, S.J. (1991). Development and application of oligonucleotide probes for identification of *Lactococcus lactis* subsp. *cremoris*. *Applied and Environmental Microbiology* **57**, 1313-1318.

Salama, M.S., Sandine, W.E. & Giovannoni, S.J. (1993). Isolation of *Lactococcus lactis* subsp. *cremoris* from nature by colony hybridization with rRNA probes. *Applied and Environmental Microbiology* **59**, 3941-3945.

Salama, M.S., Musafija-Jeknic, T., Sandine, W.E. & Giovannoni, S.J. (1995). An ecological study of lactic acid bacteria: isolation of new strains of *Lactococcus* including *Lactococcus lactis* subspecies *cremoris*. *Journal of Dairy Science* **78**, 1004-1017.

Salovaara, H. (1993). Lactic acid bacteria in cereal-based products. In *Lactic Acid Bacteria*, pp. 111-126. Edited by S. Salminen & A. von Wright. New York: Marcel Dekker, Inc.

Salyers, A.A. & Whitt, D.D. (1994). *Escherichia coli* gastrointestinal infections. In *Bacterial Pathogenesis: a Molecular Approach*, pp. 190-204. (Edited by A.A. Salyers & D.D. Whitt). Washington DC: ASM Press.

Savage, D.C. (1977). Microbial ecology of the gastrointestinal tract. *Annual Review of Microbiology* **31**, 107-133.

Saxelin, M., Elo, S., Salminen, S. & Vapaatalo, H. (1991). Dose response colonisation of faeces after oral administration of *Lactobacillus* GG. *Microbial Ecology in Health and Disease* **4**, 209-214.

Schleifer, K.H. & Ludwig, W. (1995a). Phylogenetic relationships of lactic acid bacteria. In *The Genera of Lactic Acid Bacteria*, pp. 7-18. Edited by B.J.B. Wood & W.H. Holzapfel. London: Chapman & Hall.

Schleifer, K.H. & Ludwig, W. (1995b). Phylogeny of the genus *Lactobacillus* and related genera. *Systematic and Applied Microbiology* **18**, 461-467.

Schleifer, K.H., Ehrmann, M., Beimfohr, C., Brockmann, E., Ludwig, W. & Amann, R. (1995). Application of molecular methods for the classification and identification of lactic acid bacteria. *International Dairy Journal* **5**, 1081-1094.

Sgorbati, B., Biavati, B. & Palenzona, D. (1995). The genus *Bifidobacterium*. In *The Genera of Lactic Acid Bacteria*, pp. 279-306. Edited by B.J.B. Wood & W.H. Holzapfel. London: Chapman & Hall.

Sharpe, M.E. (1962). Lactobacilli in meat products. *Food Manufacture* **37**, 582-589.

Sharpe, M.E. (1981). The genus *Lactobacillus*. In *The Procaryotes. A Handbook on Habitats, Isolation and Identification of Bacteria*, pp. 1653-1674. Edited by M.P. Starr, H. Stolp, H.G. Trüper, A. Balows & H.G. Schlegel. Berlin: Springer-Verlag.

Shaw, B.G. & Harding, C.D. (1984). A numerical taxonomic study of lactic acid bacteria from vacuum-packed beef, pork, lamb and bacon. *Journal of Applied Bacteriology* **56**, 25-40.

Shaw, B.G. & Harding, C.D. (1985). Atypical lactobacilli from vacuum-packaged meats: comparison by DNA hybridization, cell composition and biochemical tests with a description of *Lactobacillus carnis* sp. nov. *Systematic and Applied Microbiology* **6**, 291-297.

Steinkraus, K.H. (1983). Lactic acid fermentation in the production of foods from vegetables, cereals and legumes. *Antonie van Leeuwenhoek* **49**, 337-348.

Steinkraus, K.H. (ed.) (1996). *Handbook of Indigenous Fermented Food, 2nd edition.* New York: Marcel Dekker, Inc.

Stiles, M.E. (1996). Biopreservation by lactic acid bacteria. *Antonie van Leeuwenhoek* **70**, 331-345.

Ståhl, M. & Molin, G. (1994). Classification of *Lactobacillus reuteri* by restriction endonuclease analysis of chromosomal DNA. *International Journal of Systematic Bacteriology* **44**, 9-14.

Ståhl, M., Molin, G., Persson, A., Ahrné, S. & Ståhl, S. (1990). Restriction endonuclease patterns and multivariate analysis as a classification tool for *Lactobacillus* spp. *International Journal of Systematic Bacteriology* **40**, 189-193.

Ståhl, M., Pettersson, B., Molin, G., Uhlén, M. & Ahrné, S. (1994). Restriction fragment length polymorphism of *Lactobacillus reuteri* and *Lactobacillus fermentum*, originating from intestinal mucosa, based on 16S rRNA genes. *Systematic and Applied Microbiology* **17**, 108-115.

Thornley, M.J. & Sharpe, M.E. (1959). Microorganisms from chicken meat related to both lactobacilli and aerobic sporeformers. *Journal of Applied Bacteriology* **22**, 368-376.

Vandamme, P., Pot, B., Gillis, M., de Vos, P., Kersters, K. & Swings, J. (1996). Polyphasic taxonomy, a consensus approach to bacterial systematics. *Microbiological Reviews* **60**, 407-438.

Wallbanks, S., Martinez-Murcia, A.J., Fryer, J.L., Phillips, B.A. & Collins, M.D. (1990). 16S rRNA sequence determination for members of the genus *Carnobacterium* and related lactic acid bacteria and description of *Vagococcus salmonarium* sp. nov. *International Journal of Systematic Bacteriology* **40**, 224-230.

Weigmann, H. (1899). Versuch einer Einteilung der Milchsäurebakterien des Molkereigewerbes. *Zentralblatt für Bakteriologie, Parasitenkunde, Infektionskrankheiten und Hygiene, Abteilung II* **5**, 825-831.

Welsh, J. & McClelland, M. (1990). Fingerprinting genomes using PCR with arbitraty primers. *Nucleic Acids Research* **18**, 7213-7218.

Wijtzes, T., de Wit, J.C., Huis in't Veld, J.H.J., van't Riet, K. & Zwietering, M.H. (1995). Modelling bacterial growth of *Lactobacillus curvatus* as a function of acidity and temperature. *Applied and Environmental Microbiology* **61**, 2533-2539.

Williams, J.G., Kubelik, A.R., Livak, K.J., Rafalski, J.A. & Tingey, S.V. (1990). DNA polymorphisms amplified by arbitrary primers are useful as genetic markers. *Nucleic Acids Research* **18**, 6531-6535.

L. T. Axelsson and S. Ahrné

Woese, C.R. (1987). Bacterial evolution. *Microbiological Reviews* **51**, 221-271.
Wood, B.J.B. & Holzapfel, W.H. (eds.) (1995). *The Genera of Lactic Acid Bacteria.* London: Chapman & Hall.

14   A SLOW RAMBLE IN THE ACID-FAST LANE

THE COMING OF AGE OF MYCOBACTERIAL TAXONOMY

LAWRENCE G. WAYNE

*Tuberculosis Research Laboratory, Veterans Administration
Medical Center, Long Beach, California 90822, U.S.A.*

1   Introduction

This chapter does not present a comprehensive description of the currently recognized species in the genus *Mycobacterium*, nor does it include tables of features, phenetic dendrograms or phylogenetic matrices. It is rather a history of the development of a modern mycobacterial taxonomy written to illustrate the mutual influences upon one another of an increasingly urgent practical need for a consistent taxonomy of this genus and the gradual emergence of sophisticated tools for taxonomic analysis. This, then, is an essentially anecdotal history of the evolution of mycobacterial taxonomy from the time around the late 1940's when Ruth Gordon was introducing polythetic principles to the study of the rapid growers while the taxonomic interest in slow growers was largely expressed in the search for virulence tests to distinguish among " varieties" of tubercle bacilli, and extending to the present, when nucleic acid sequences and probes are making their contributions to the polyphasic systematic mosaic of this important genus.

Before proceeding to the subject itself, I want to pay tribute to the memory of Ernest H. Runyon (1904-1994), whose seminal observations on the so-called " atypical mycobacteria" and whose unique personality played such an important role in bringing together investigators from all over the world to participate in multi-decade cooperative ventures that helped to bring order to the systematics of the slowly growing mycobacteria. Uncle Ernest was at heart an old fashioned naturalist and it was a joy to tramp the woods and seashores with him as he stopped to inspect, touch, smell and identify all the examples of flora that were unique to each environment. He would have preferred to study mycobacteria that way too, but, of course, that could not be done, so he accommodated to the need for less direct observation, such as biochemical and serological testing, and eventually even to such an impersonal approach as the use of computers to fine tune the classification of the " Runyon Groups" of mycobacteria. In fact, though, it was his personal touch that had such a great impact on the evolution of some very productive cooperative efforts; Ernest did not tolerate formality very well, and his open friendly manner and personal effervescent enthusiasm for the task melded a constellation of diverse investigators into an extended family of mycobacteriological devotees.

F. G. Priest and M. Goodfellow (eds.), Applied Microbial Systematics, 389-419
© 2000 *Kluwer Academic Publishers. Printed in the Netherlands.*

The sixth edition of Bergey's Manual of Determinative Bacteriology, which was published in 1948 (Breed *et al.* 1948), listed only 13 named species of mycobacteria. Six of them, *M. avium, M. leprae, M. lepraemurium, M. marinum* (erroneously listed with the "rapid growers"), *M. paratuberculosis,* and *M. tuberculosis* occupy positions on the "slowly growing" branch of the genus, and all are still accepted as valid taxa today with some redistribution as to species and subspecies status. On the other hand, of the seven named species from the "rapidly growing" branch, only one, *M. phlei,* has kept its standing to the present time. Thus, only seven of the species of mycobacteria that were described in the span of over 60 years between the first description of *M. tuberculosis* by Robert Koch (1882) and the beginning of the era of chemotherapy in the late 1940s, have stood the test of time. Furthermore, two of these seven, *M. leprae* and *M. lepraemurium,* were recognized only by the pathology they produced in man or experimental animals, respectively, as they were not considered to be cultivable, and a third, *M. paratuberculosis,* was difficult to culture and required a factor extracted from cells of other mycobacteria to grow at all. The other four were largely characterized on the basis of their ability to grow on various culture media and at different temperatures, of the morphology of that growth, and of their ability to produce disease in selected hosts. Some limited serological information was also available to help distinguish between the species. By and large, the classification of mycobacteria at this point was based on dichotomous keys, thus placing inordinate weight on single properties, that is, it was essentially a monothetic taxonomy.

As early as 1935, Max Pinner had recognized a number of strains of mycobacteria that looked very different in culture from *M. tuberculosis* and which did not produce progressive disease in guinea pigs, but which did appear to be implicated in human tuberculosis-like disease (Pinner 1935a, 1935b). The very limited array of properties that were available at that time for characterizing mycobacteria made it impossible to develop a consistent taxonomy of these strains and no significant progress was made in the ensuing decade.

2   Early Stirrings

There are two main elements in the motivation to carry out studies on bacterial systematics, and, while they are not mutually exclusive, the dominant element will strongly influence the path that the studies take; that is, the studies may be predominantly curiosity-driven or utility-driven. The curiosity-driven aspects emphasize classification in the broadest sense, and utility-driven aspects lead to emphasis on identification of individual strains that may belong to selected categories of interest. The risk in the latter approach is that too much emphasis is placed on a small number of diagnostic features, as in the use of dichotomous keys, where a single deviation in one selected property may lead to a vast gulf of separation between two strains that actually are very closely related.

In the period immediately following the publication of the sixth edition of Bergey's Manual of Determinative Bacteriology, the most influential and productive practitioner of curiosity-driven mycobacterial systematics was Ruth Gordon. Her efforts were directed toward the study of the so-called rapid growers, that is, mycobacteria that yield visible colonies from dilute inocula in a few days. (I recently reviewed some correspondence from the year 1958 in which Dr. Gordon responded to my comments that I was having trouble recovering some " rhodochrous" strains she had sent me by writing " If you were here I should laugh at you and make a crack about medical microbiologists who believe that all microorganisms grow well on blood agar at 37°C". In reply I had to confess that " I am not one of those microbiologists who believes that all microorganisms grow on blood agar at 37°C; worse yet, I am one of those who automatically inoculates everything to Lowenstein-Jensen medium at 37°C!").

In 1953 Gordon published an exhaustive survey of 170 strains, 70 of which she received with labels ascribing them to 17 different species, with the remainder simply designated as " *Mycobacterium* sp." (Gordon & Smith, 1953). The cultures were characterized in terms of twenty biochemical and physiological properties in addition to their ability to grow on different media at various temperatures. Based on overall similarity to one another, Gordon was able to ascribe to the species *M. phlei*, 21 strains that were received with 3 different labels, and to *M. smegmatis*, 56 strains that were received with 14 different labels. The reduction of these strains to synonymy with one or the other of these two species resulted in the loss of standing of the names of almost all of the rapid growers listed in the sixth edition of Bergey's Manual and represented a major housecleaning achievement. This was followed by another paper in 1955 characterizing a third species of rapid grower, *M. fortuitum* (Gordon & Smith, 1955). Addition of even larger numbers of tests and strains to her studies (Gordon & Mihm, 1959) confirmed that a number of strains in each species could be expected to deviate from the majority pattern for one or more tests and that if a monothetic dichotomous key had been used to classify them, serious fragmentation of each species would have resulted. Her decision to take the polythetic approach and base the classification on similarity of strains to overall patterns, was a major step in demonstrating that the systematic study of mycobacteria could be approached rationally.

After the major consolidation of species efffected by Gordon, there was little taxonomic activity along the " rapid grower" branch of the mycobacterial tree for about a decade; the first additional species was not described until the recognition by Bojalil in 1962 of *M. flavescens* in the first application of modern numerical taxonomy to the genus *Mycobacterium* (Bojalil *et al.*, 1962).

In 1949 a slowly growing acid-fast organism which caused skin lesions in humans, and which yielded little or no growth at temperatures above 33°C was named *M. ulcerans* (MacCallum *et al.*, 1948). Three years later, a different organism which was associated with human skin lesions (swimming pool granuloma) and also intolerant of high temperatures was described (Norden & Linell, 1951) and later named " *M. balnei*" (Linell & Norden 1963). This organism attracted considerable attention in 1961 when a rash of tuberculin conversions, which were at first considered to be an indication of an outbreak of tuberculosis, was eventually ascribed to immunologic response in a large number of cases of the relatively benign and self limiting swimming pool granuloma acquired in a pool in Colorado that was contaminated with " *M. balnei*" (Schaefer & Davis, 1961). This organism is now recognized to be the same as the legitimate species *M. marinum*, a pathogen of fish which had been described and named by Aronson (1926).

3   The Growing Need

In the late 1940s several chemotherapeutic agents became available that were effective
in the specific treatment of tuberculosis. Prior to that time treatment of this devastating
chronic disease had been nonspecific, directed toward alleviating the clinical condition,
which could not always be distinguished from some other chronic granulomatous diseases,
but not toward the causative agent itself. Of the first three drugs that were applied on a
large clinical scale, streptomycin, p-aminosalicylic acid, and isoniazid, the latter two were
highly specific for *M. tuberculosis* and *M. bovis,* and were essentially without effect on
other species of mycobacteria known at that time or on members of other bacterial genera.
Hitherto, the main reason for the bacteriologic examination of specimens from tuberculosis
patients was to help in establishing the initial diagnosis, determining the infectivity of
the patient for contact individuals, and in following the course of remission in cases where
the palliative treatment, which usually involved bedrest and/or various surgical interventions
to effect collapse of a lung and closure of cavities, might have caused sputum conversion.
Furthermore, much of the bacteriologic examination of sputum involved simple concentration and microscopic examination of specimens treated with one of the acid-fast stains,
since that was much more economical to do and yielded immediate results, as compared
to culturing or inoculation of guinea pigs.

The introduction of specific chemotherapy for the treatment of tuberculosis had several
consequences and these, in turn, had a profound effect on the course of systematic study
of the slowly growing members of the genus *Mycobacterium.* The cultural confirmation of
a diagnosis became more important as was the availability of cultures that could be tested
for the emergence of drug resistance. This led to a marked increase in the numbers of
specimens that were examined from each patient and the repeated and consistent isolation
from some patients of mycobacteria that did not have the recognizable attributes of
M. tuberculosis, especially its ability to produce rapidly progressive and fatal disease in
guinea pigs, which was then considered the "gold standard" for identifying tubercle bacilli.
At that time, considerable effort was being applied to a search for "virulence" tests for
two reasons. One reason was that recognition of properties that distinguished virulent
from nonvirulent strains of mycobacteria might lead to insights into the mechanisms of
pathogenicity itself and the other reason was to provide a faster and more economical
way to distinguish *M. tuberculosis* from all other mycobacteria, which had been widely
considered to be nonpathogenic contaminants when recovered from sputum. Two such
properties that were described in the laboratory of Rene Dubos were the tendency of
virulent tubercle bacilli to form tightly "corded" colonies of parallel cells of bacilli
(Middlebrook *et al.*, 1947) and their ability to bind neutral red dye in its red salt form under
mildly alkaline conditions (Dubos & Middlebrook, 1948). Other features considered to
be potentially useful for grading the virulence of mycobacteria included their ability to
reduce selected redox indicator dyes (Clark & Forrest, 1953) and the minimal inocula
needed to yield ulcerating lesions after intradermal injection into guinea pigs (Richmond &
Cummings, 1950). Ultimately, although not designed as a virulence test *per se*, the niacin test,
which detects the accumulation of niacin in the medium surrounding colonies of
M. tuberculosis, turned out to be the simplest and most definitive test for distinguishing
colonies of that species from other species of mycobacteria, including *M. bovis* (Konno, 1956)

and became a routine test in the diagnostic laboratory. At about the same time the guinea pig was deposed from its golden pedestal of authority when Gardner Middlebrook and colleagues observed that strains of *M. tuberculosis* that developed resistance to high concentrations of isoniazid lost the ability to produce progressive disease in that animal model but were still capable of persisting in lesions of human patients (Middlebrook & Cohn, 1953; Oestreicher *et al.*, 1955). These highly resistant organisms were subsequently reported to have lost also the ability to produce the enzyme catalase (Middlebrook, 1954).

The increased frequency of isolation of colonies that were not at all typical of classical *M. tuberculosis* in texture, pigmentation, and/or animal pathogenicity had already stimulated interest in these " atypical" mycobacteria. As early as 1949, Cuttino and McCabe isolated from tissues of a child who died of disseminated infection, an acid fast organism which did not produce progressive disease in guinea pigs. They showed slides of the organism, which they believed to be a *Nocardia,* to two experts on fungi and mycobacteria, Norman Conant and David Smith, who noted that the microscopic appearance was compatible with that of a *Nocardia* (personal communication). In the absence at the time of any satisfactory methods for definitively characterizing these mycobacteria, Cuttino and McCabe (1949) erroneously named their new organism " *Nocardia intracellularis*"). It was not until 1965 that its true identity as a *Mycobacterium* was recognized formally when Runyon (1965) established it as a member of the " Battey" group of mycobacteria (many of the early isolates of this organism were from patients at the Battey State Hospital in Georgia) and designated it *M. intracellularis* (sic) (now *M. intracellulare*). Even then, as will be discussed later, the relationship between *M. intracellulare* and *M. avium* remained a matter of some controversy.

Another consequence of the introduction of chemotherapy into the regimen for treatment of tuberculosis was the fairly frequent use of resectional surgery to remove pulmonary lesions that did not appear to be resolving at a satisfactory rate. As noted above, the earlier use of surgery was largely limited to collapse therapy to immobilize a lung and enhance healing of open cavities. Because of the risks of spreading bacilli to unaffected regions of the lung, surgeons had been reluctant to risk cutting directly into the diseased tissue. Once the efficacy of chemotherapy was established in managing tuberculosis, surgical intervention under cover of chemotherapy became acceptable and was widely practiced throughout the 1950s. In ensuing years, as more drugs and improved regimens became available, the need for resectional surgery diminished, and it is rarely used today in the treatment of tuberculosis. By 1953 a slowly growing variably pigmented *Mycobacterium* (the " yellow bacillus") that was isolated from a resected lung specimen from one patient and from autopsy specimens of another had been described in terms of animal pathogenicity and colonial characteristics (Buhler & Pollak, 1953; Pollak & Buhler, 1955); this organism was named *M. kansasii* by Hauduroy (1955). In 1956 a slowly growing bright yellow *Mycobacterium* that had been isolated from diseased human tissues, primarily pus aspirated from cervical lesions of children, was described in a similarly limited way and named *M. scrofulaceum* (Prissick & Masson, 1956, 1957).

4   The Runyon Revolution

Up to this time the study of the systematics of members of the " slow grower" branch of the mycobacterial tree was anything but systematic! Descriptions of the new species often came out of clinical laboratories and were based largely on local isolates. Very little effort had been expended on the development of biochemical or physiologic tests, other than the in vitro virulence tests mentioned above, that could be used reliably for classifying these organisms. Furthermore, although the increase in isolation and recognition of unidentifiable mycobacteria could be explained simply enough on the basis of the increasing numbers of culture studies that were performed on patients, that abrupt increase in their numbers caused some investigators to consider them to be drug induced mutants of *M. tuberculosis*, that is, " atypical" tubercle bacilli, especially since they were almost invariably resistant to specific antituberculosis drugs. Thus Tarshis, after examining a collection of 26 strains of chromogenic mycobacteria received from a number of sources throughout the United States and arriving at no clearcut definition of species (Tarshis & Frisch, 1952), attempted to convert a strain of *M. tuberculosis* to a chromogen by blind serial passage in streptomycin- and isoniazid-containing broth. He considered the non-acid fast bacteria isolated from these cultures to be atypical variants of *M. tuberculosis* (Tarshis, 1958). This work has not been confirmed; the long term serial blind passage offered ample opportunity for contamination. As late as 1979, Tsukamura and his colleagues described experiments in which a culture of the scotochromogenic *M. gordonae* was subjected to repeated UV irradiation and subculture of selected colonies that had certain desired characteristics; cultures were produced that were phenotypically similar to *M. triviale* (Tsukamura et al., 1979). It was not claimed that the *M. gordonae* culture had become *M. triviale*, only that it could not be distinguished from the latter by the techniques available to the authors.

The year 1954 is generally recognized as a turning point in the systematic approach to the genus *Mycobacterium* with the publication by Alice Timpe and Ernest Runyon of the first large scale study on " atypical" mycobacteria isolated from patients with pulmonary disease, and the introduction of the " Runyon Grouping" system (Timpe & Runyon, 1954). This preliminary report established three Groups of slowly growing mycobacteria on the basis of colonial morphological and pigment characteristics and their abilities to produce lesions in experimental animals. There was a fair degree of correlation between membership in one of the three original Groups and the probable etiologic role of the strain in the source patient's disease. The actual descriptions of the Groups in this report were rather vague, but in the ensuing years an informal understanding of the Group characteristics evolved among mycobacteriologists. In 1959 Runyon published a more complete set of criteria for membership in what had by now become four Runyon Groups, representing isolates from over 400 patients (Runyon, 1959), and this definition of the Groups became the standard in the field.

Rigorous criteria were established for implicating an organism in one of these Groups as the etiologic agent of a patient's disease. The strain had to be isolated repeatedly from specimens of a patient with disease compatible with tuberculosis but *M. tuberculosis* must not have been found in that patient; most convincing was the isolation of the organism from lesions in surgically resected tissues, again in the absence of colonies of *M. tuberculosis*.

Groups I, II and III were made up of slowly growing strains of mycobacteria, and the newly introduced Group IV consisted of rapid growers. Group I was comprised of photochromogens, that is, strains whose colonies became yellow only after exposure to light, and which were capable of causing lesions after intravenous inoculation to mice. Beyond that, the descriptions of morphological and growth characteristics were rather loose. At the time, only *M. kansasii* was recognized as belonging to Group I and this organism was considered to be the etiologic agent of the disease in all 120 patients from whom it had been isolated.

Group II strains produced yellow to orange pigment even when grown in the dark, that is, were scotochromogenic, and only rarely produced lesions in mice. Convincing implication of members of Group II in human disease was quite rare, with most isolates believed to represent contaminants from the upper respiratory tract. No mention was made in this publication of the fact that the original description of *M. scrofulaceum*, which was regularly associated with lesions of pediatric cervical adenitis (Prissick & Masson, 1956, 1957), was consistent with membership in Group II. The majority of the strains of Group II were undoubtedly members of a category generally referred to at that time as " tap water scotochromogens", and which are now represented by the nonpathogenic species, *M. gordonae* (Wayne, 1970).

Group III strains produced little or no pigment and varied widely in their ability to cause lesions in experimental animals. Furthermore, only about 75% of the 143 Group III strains reported in this study met the criteria for etiologic significance in the source patients, with the remaining 25% considered to represent contaminants. Many of the strains were from Battey State hospital in Georgia, but at this time Runyon did not mention the similarity of Cuttino and McCabe's " *Nocardia intracellularis*" (*M. intracellulare*) (Runyon, 1965), to members of Group III. The newly introduced Group IV was a catchall category comprised of rapid growers, with no defining characteristics other than the growth rate. The only pathogen in Group IV that was recognized at that time was *M. fortuitum*.

5   In Search Of Tools

The Runyon Group convention was quickly adopted by clinical laboratories as the most useful means of reporting isolates of mycobacteria other than *M. tuberculosis*; it represented the only method of giving even approximate interpretation of probable significance of an isolate. However, a number of investigators recognized that the Groups were not homogeneous, and that further subdivision should yield useful information for both the curiosity-driven and the utility-driven systematicist. The problem with all taxonomic descriptions of slowly growing mycobacteria to that time is that they depended on such a small number of strain characteristics. Animal pathogenicity studies were very expensive to perform and yielded very limited information. There were just so many colony types to describe, and these were not consistent, and only three pigment patterns, consisting of those used to define the first three Runyon Groups, could be recognized. Many more features were needed before it would be possible to expand the taxonomy of these mycobacteria in a rational and consistent way. Conventional taxonomic tests, such as those for fermentation or utilization of various carbon sources, while useful in study of rapidly growing mycobacteria (Gordon & Mihm, 1959), did not yield satisfactory results with the slow growers and it became necessary both to develop new tests that were specifically useful for mycobacterial taxonomy or to adapt some existing tests to make them applicable to very slowly growing bacteria.

Between the years 1957 and 1964 only three new species of slowly growing mycobacteria, *M. gordonae, M. microti,* and *M. xenopi* were named. *M. microti* is really a low virulence variant of *M. tuberculosis,* while *M. gordonae* is a name that was applied to the previously recognized group of tap-water scotochromogens to replace a discredited name of ambiguous provenance, " *M. aquae*". This *de facto* eight year moratorium on the recognition of new species was probably related to the fact that many investigators were now concentrating their efforts on developing or evaluating tests that could apply to the problem. Several approaches were being brought to bear on the accumulation of a battery of properties that could be used to classify mycobacteria.

In 1964, I discussed the deterrent effect of a " mycobacterial mystique" on the development of a consistent mycobacterial taxonomy (Wayne, 1964). Part of the problem was the myopic tendency to place *M. tuberculosis* in the centre of the genus rather than at the periphery, and to judge all other mycobacteria in terms of their relationship to that species. Another reason why mycobacterial systematics lagged behind that of other genera that were of equal clinical interest and urgency was that an extra dimension, time, is so very influential in the study of mycobacteria.

One of the problems in testing biochemical reactions of mycobacteria was the wide range of growth rates observed between species, and the need to determine how long a test should be carried out before calling the result positive or negative. However, the status of a feature can be accorded to a selected level of reaction, based on relative intensity or time to reach a selected intensity, thus allowing several features to be derived from the different levels. Indeed, actual plots of strain frequency against intensities or rates of reactions in several tests yielded bi- or polymodal distributions in several tests that were applied to mycobacteria, thus providing up to four character states for a single feature (Wayne, 1967). Furthermore, and most importantly, these distributions did not necessarily reflect growth rates, and thus represented novel rather than redundant information. In at least one case, that of catalase, the quantitative differences in response to the test corresponds to the presence of at least two different kinds of catalase that may be produced singly or in combination in different species of mycobacteria (Wayne & Diaz, 1982, 1986).

Virulence tests that were originally designed to separate *M. tuberculosis* from all other mycobacteria were reevaluated to see if they could contribute to the systematics of the more recently recognized slow growers. Whitehead and colleagues had originally described an aryl sulfatase test that was based on hydrolysis of phenolphthalein disulfate and was negative for strains of *M. tuberculosis* and positive for most other mycobacteria (Whitehead *et al.,* 1953). Later studies demonstrated that most, if not all mycobacteria produce aryl sulfatase, but that there are quantitative differences in the rate at which the enzyme is expressed. On this basis a prolonged incubation (six week) test for aryl sulfatase demonstrated that " M. rhodochrous" differed from all other mycobacteria in its complete lack of that enzyme (Wayne *et al.,* 1958); this species has since been transferred to the genus *Rhodococcus* (Skerman *et al.,* 1980). A ten-day version of the test provided some basis for distinguishing between *M. avium* and *M. intracellulare* (Meissner *et al.,* 1974) and a three day version separated members of the potentially pathogenic *M. fortuitum* complex from other rapid growers (Wayne, 1961). The virulence tests based on colonial morphology, especially cording (Middlebrook *et al.*, 1947), and on neutral red binding reaction (Dubos & Middlebrook, 1948) were combined into a microcolonial test in which colonies of mycobacteria grown on membrane filters were subjected to a modified neutral red staining procedure, and proved useful in supporting the divisions within Runyon Groups (Wayne *et al.*, 1957), as did a quantitative modification of the neutral red test (Wayne, 1959). Although the original interest in catalase activity of mycobacteria was directed toward the associated loss of virulence and of the ability to produce this enzyme in strains of *M. tuberculosis* that had acquired a high level of resistance to isoniazid, subsequent studies demonstrated a bimodal quantitative distribution of results that proved valuable in classifying other mycobacteria (Wayne, 1967).

Some biochemical tests that were useful in classifying members of other bacterial genera were also evaluated to see if they could be adapted for application to mycobacteria. For example, a modified ß-galactosidase test was first applied to distinguishing between rapidly growing mycobacteria and members of other related genera (Tsukamura, 1974) and later provided a sharp distinction of opportunistic pathogens from some essentially saprophytic species within Runyon's Group III (Meissner *et al.*, 1974). Virtanen (1960) adapted the nitrate reduction test to use for mycobacterial identification; subsequent studies demonstrated a polymodal distribution of reactions of mycobacteria that permitted further resolution of mycobacteria into four sharply demarcated quantitative categories of response (Wayne, 1967). As noted above, conventional bacterial tests for fermentation and specific substrate utilization were being applied to mycobacteria, but their value was largely limited to classification of the rapidly growing members of this genus (Gordon & Mihm, 1959; Gordon & Smith, 1953, 1955).

During this same time period novel biochemical, chemical and immunologically based tests, as well as phage typing techniques, were developed specifically to apply to the classification of mycobacteria. Among the most powerful of the biochemical features were the results of amidase tests performed on ten substrates included in Rudolf Bönicke's "amidreihe" (Bönicke 1960a, 1960b, 1962). All ten of the aliphatic or aromatic acid amides were useful in the establishment of characteristic reaction patterns among rapidly growing mycobacteria. Among the slow growers only three substrates, urea, nicotinamide and pyrazinamide, were useful. The test was based on detection of ammonia released from very thoroughly washed cells after incubation with substrate and was cumbersome to perform. Consequently, only a few laboratories routinely performed these tests, but in experienced hands the tests gave very striking and consistent patterns for respective subgroups of bacilli.

Another useful test evolved serendipitously in our laboratory as we were exploring the possibility that the potential pathogenicity of the newly recognized mycobacteria was in part a reflection of their tolerance to harsh physical conditions. To test this we grew different strains in broth containing high concentrations of the detergent Tween 80. On examination of a culture of *M. kansasii* in a 5% Tween 80 medium we noted very high optical absorbance readings that suggested a remarkable degree of stimulation of growth by the detergent. However, by relying on our eyes rather than on the spectrophotometer alone, we noted that the culture appeared more opalescent than truly turbid. Subculture and chemical analysis of the spent medium proved that the high readings were not due to growth but to release of oleic acid into the medium by hydrolysis of the Tween 80. In searching for a simple way to evaluate the ability of mycobacteria to hydrolyse Tween 80, we called on our experience with the neutral red test. We knew that neutral red was yellow in its free base form, and red in the ionic state. Reasoning that a high concentration of Tween 80 in neutral buffer might absorb the dye into its micelles in the nonionized yellow form, we added the dye to a series of dilutions of the detergent and noted that the colour was amber in the highest concentrations and red in the lowest. On this basis we developed the Tween hydrolysis test, in which mycobacteria were suspended in a mixture of the detergent and dye; the suspension was incubated for a week or more and observed for development of a red colour in the supernate, which reflected destruction of the Tween 80. The test provided sharp resolution among slow growers, and contributed to the definition of species (Wayne, 1962).

A serologic technique that was especially attractive to some mycobacterial systematists was the immunodiffusion test, which permitted analysis of complex antigenic mixtures and recognition of shared components of the mixtures. Generally this technique was used to establish individual antigens or antigen patterns that were characteristic of different species (Lind & Norlin, 1963; Weiszfeiler et al., 1968; Norlin et al., 1969; Stanford & Beck, 1969; Kwapinski et al., 1970; Stanford & Gunthorpe, 1971; Meissner et al., 1974). Lines of identity between strain pairs were recognized as fusions, whereas partial identities were seen as deflections or spurs (Chaparas et al., 1983). Subsequent studies, to be discussed in more detail later in this chapter, demonstrated that isofunctional proteins that occur in many species of mycobacteria are highly conserved, but have regions where the epitopes are specific (Wayne & Diaz, 1979, 1982; Tasaka et al., 1983; Tasaka & Matsuo, 1984; Tasaka et al., 1985; Shivannavar et al., 1996). Such proteins do cross react serologically, but exhibit quantitative differences in their affinities, which explains the complexity of the cross reactive patterns in immunodiffusion tests.

A seroagglutination test pioneered by Werner Schaefer was directed toward detection of highly specific antigens on the surface of mycobacterial cells (Schaefer, 1965, 1967, 1968). Whole sets of serovars could be ascribed to certain species, although monovar species were also recognized, and a serovar numbering scheme was eventually proposed to account for and group the recognized serovars by putative species or species complexes (Wolinsky & Schaefer, 1973). Eventually Patrick Brennan isolated and characterized the small epitopes on the common glycopeptidolipid and glycolipid cores of the surface antigens that were responsible for the exquisitely specific reactivity of these antigens (Brennan, 1984; Brennan et al., 1981).

Mogens Magnusson pioneered a third immunologically based system for classifying mycobacteria that was based on comparative skin test reactions in immunized guinea pigs, to crude cell extract "sensitins" of homologous and heterologous provenance (Magnusson, 1962). While guinea pigs that were immunized with sensitins from one species yielded positive skin reactions when tested with heterologous sensitins, the difference in size between a homologous and heterologous reaction could be quantified and expressed as a specificity difference (SPD); the SPD reflected the net antigenic similarities between the components of the paired complex mixtures that comprised the sensitins. While this method did permit quantitative expression of the relatedness between strains and contributed to the definition of species (Baess & Magnusson, 1982; McIntyre et al., 1986), the cross testing of large numbers of strains by skin testing of guinea pigs required an enormous number of animals and expenditure of effort and was therefore limited in practice to a very few laboratories.

The extraordinary richness and variety of lipids produced by mycobacteria have made them attractive targets for chemotaxonomic studies. In some cases specific lipids belonging to certain classes such as mycolic acids and their esters (Minnikin *et al.*, 1982), isoprenoid quinones (Minnikin *et al.*, 1978) and glycolipid and glycopeptidolipid surface antigens (Brennan *et al.*, 1981; Brennan, 1984) have been identified. In other cases lipid patterns that were seen on thin layer (Szulga *et al.*, 1966; Marks *et al.*, 1969a, 1969b; Jenkins & Marks 1972) or high performance liquid chromatography (HPLC) (Butler *et al.*, 1991; Butler & Kilburn, 1990) could be ascribed to different taxa. Multiple patterns were usually seen within any given species, suggesting utility of these techniques for infrasubspecific rather than specific categorization; they have proven most useful for identification of individual strains rather than classification at the species level.

Very early in the evolution of modern mycobacterial taxonomy bacteriophages were isolated and applied to the typing of isolates (Penso & Ortali, 1949a, 1949b; Will *et al.*, 1957; Froman & Scammon, 1964). Of special interest in this regard was the report by Seymour Froman in 1954 of the isolation of a phage that attacked *M. tuberculosis* (Froman *et al.*, 1954). Within individual species a variety of phage susceptibility patterns were seen (Hobby *et al.*, 1967; Wayne *et al.*, 1979). Here again the technique is most useful for typing of individual strains rather than classification at the species level. An international effort has been made to standardize these procedures, which is critical for reproducibility of phage typing (Rado *et al.*, 1975). Typing of members of the *M. avium* and *M. tuberculosis* complexes with selected batteries of mycobacteriophages has proven to be a useful epidemiologic tool (Bates & Mitchison, 1969; Crawford *et al.*, 1981; Crawford & Bates, 1985).

6   A Blueprint For The Tower Of Babel

The introduction of chemotherapy into the treatment of tuberculosis and the consequent resurgence of interest in the bacteriology of this disease thus resulted in the introduction of a great number and variety of tests that could be used for the systematic study of members of the genus *Mycobacterium*. The individual efforts in many laboratories yielded techniques that were initially directed toward identification of individual strain isolates, but not necessarily toward the larger goal of developing a consistent and logical taxonomy of the genus. This reflected the orientation of most of the test developers, people with diagnostic responsibilities. The apparently paradoxical reversal of the appropriate functional positions of cart and horse seemed to make sense at first, when a few properties of the members of the small number of Runyon Groups appeared to correlate well with their clinical significance. However, as more tests were applied and larger numbers of strains examined, it became clear that each of these Groups contained strains that differed from the majority of their neighbours, and that a far greater number of species probably existed than had been anticipated.

A risk associated with the adoption of different sets of favoured features in different laboratories around the world was the haphazard erection of a nomenclatural Tower of Babel, upon which communication was severely compromised. In 1959 I attended an informal round table discussion on the "atypical mycobacteria", during which a number of comments were made about *M. marinum* and "*M. marianum*", which at that time represented two distinct species. As the discussion progressed it became apparent that those referring to the skin pathogen *M. marinum* did not realize that others were referring to the agent of cervical adenitis, "*M. marianum*" and vice versa. Some in the room did not even know those were two different species. This problem was later resolved by rejection of the more recent of these two names, "*M. marianum*" and validation of its later synonym, *M. scrofulaceum*, for this species (Wayne, 1968, 1975). Other problems were not so easily resolved. I recall a conversation I had with Rudolph Bönicke in 1963, in which I commented that I had seen strains of Runyon's Group III that did not show the characteristic nicotinamidase and pyrazinamidase reaction pattern attributed to that Group. "Ah" he replied "then they are not Group III strains". Since Group III was originally based on growth rate and pigment patterns, this position constituted an implicit redefinition of the Group, but that was not automatically understood by other workers in the field. Some time later I was discussing some criteria for definition of mycobacterial species with Werner Schaefer, and he agreed these might be useful, as long as they agreed with what was "the real basis for classification, the serotype".

7   The Numbers Game

While the development of a wide variety of tests for classifying mycobacteria offered hope of progress from the primitive monothetic taxonomy that had existed up to 1948, the introduction of this large resource presented some novel problems. One of them was the selective application of favoured tests in individual laboratories. Many of the tests were novel and unfamiliar and laboratory directors tended to favour the ones with which they were most comfortable. This led in part to the situations described above. Secondly, even if a large number of tests could be performed in a given laboratory, how were the masses of data so generated to be analysed in a systematic way?

Fortunately the emerging clinical need to identify the newly appreciated opportunistic mycobacterial pathogens meshed with the concurrent development by Peter Sneath of the techniques of numerical taxonomy (NT) (Sneath & Cowan, 1958; Sneath, 1962; Sneath & Sokal, 1973). Sneath recognized that the recent appearance of computers made it possible for the first time to contemplate a practical application of the 200 year old Adansonian concept of taxonomy (Adanson, 1763), that is, one based on the greatest number of features possible, with assignment of equal weight to all features, and establishment of overall similarity between strain pairs as a function of the proportion of shared features (Sneath, 1962). As noted above, Ruth Gordon had brought this philosophical approach to her vision of bacterial species (Gordon & Smith, 1953, 1955; Gordon & Mihm, 1959), but had not formalized it in a way that could be exploited by the use of the emerging electronic computers. Sneath, on the other hand, developed the mathematical concepts and methods for actually bringing the power of the computer to bear on the Adansonian approach.

In 1962 Luis Bojalil *et al.* published a paper entitled "Adansonian classification of mycobacteria", in which 32 attributes were used to sort 229 strains of mycobacteria into groups (Bojalil *et al.*, 1962). Twenty three of these properties were based on carbo-

fermentation or carbon source utilization tests but these properties were almost universally negative among the slow growers. On review of the data in this paper, it is evident that the algorithm used for performing the NT was one in which negative matches were excluded from the calculations of similarity coefficients. As a consequence, the apparently sharp resolution among species of slow growers is based on only a few features, and is essentially a reiteration of the Runyon Groups, i.e. it is based on growth rates and nature of pigmentation. Nevertheless, the recognition of *M. gordonae* as a new species that was distinct from the other scotochromogenic species "*M. marianum*" (*M. scrofulaceum*) has stood the test of time.

A number of other-NT based studies of mycobacteria followed the seminal paper of Bojalil in short order and these incorporated many of the tests described above which were either specifically developed for or adapted to use with slowly growing mycobacteria. In 1966 Wayne applied a battery of these specialized tests that yielded 25 features for characterizing a number of strains that met the criteria for Runyon's Group III. At that time we did not have the computer resources to fully exploit the methods of NT. However Liston *et al.* (1963) had just described the use of Hypothetical Median Strain (HMS) patterns for analysis of NT data. We were able to apply that approach to the analysis of our data to compare individual strains to HMS patterns that we defined and polished as the interpretation progressed. In this study we defined and named two new clinically innocuous species of Group III mycobacteria, that is *M. gastri* and *M. terrae*, that were very sharply distinguished from the opportunistic pathogens *M. avium* and "Battey bacilli" (*M. intracellulare*) (Wayne, 1966). A subsequent study conducted along the same lines provided clear resolution between the opportunistic pathogen *M. scrofulaceum* and the tap water scotochromogen "*M. aquae*" in Runyon Group II (Wayne *et al.*, 1967), although it was not until 1970 that it was fully recognized that "*M. aquae*" was synonymous with the *M. gordonae* of Bojalil (Wayne, 1970).

About the same time George Kubica's group at the Centers for Disease Control reported their studies on the use of the HMS strategy for NT analysis of slowly growing mycobacteria (Kestle *et al.*, 1967). These investigators used many of the new tests that were tailored to mycobacteria, and also introduced resistance to a number of inhibitory agents into the battery of features for use in mycobacterial taxonomy. Another of the important contributions in this paper was an evaluation of the relative clinical significance of members of the different new subgroups and species (Kestle *et al.*, 1967). They corroborated the status of the new species *M. gastri* and *M. terrae* noted above, and also recognized a new subgroup related to *M. nonchromogenicum* and to *M. terrae*, which they designated the "V" group (Kestle *et al.*, 1967), later named *M. triviale* (Kubica *et al.*, 1970).

In 1966 Michio Tsukamura published an Adansonian classification of slowly and rapidly growing mycobacteria based on a battery of tests yielding 94 features; some represented tests that were among the newer ones designed specifically for mycobacteria but most were based on growth on a variety of carbon and nitrogen sources. In this paper Tsukamura (1966) proposed two new species of rapid growers, *M. aurum* and *M. thermoresistibile*, and one new species, *M. terrae*, that fell into Runyon's Group III. The latter could have represented a conflict with *M. terrae* Wayne, cited above, except that Tsukamura (1965) had already assigned the name *M. nonchromogenicum* to these organisms and was now trying to rename them as *M. terrae*. Since this was not allowed under the Bacteriological Code of Nomenclature, the original name had to apply. *Mycobacterium nonchromogenicum* Tsukamura and *M. terrae* Wayne are phenotypically similar to one another but both are recognized as distinct and valid species. Tsukamura has written voluminously on mycobacterial taxonomy, and his contributions are especially important to our understanding of the taxonomy of the rapidly growing mycobacteria and of their relationship to such genera as *Nocardia* and *Rhodococcus*.

As noted earlier, Gordon brought a polythetic, if not an actual NT orientation to the study of rapidly growing mycobacteria and to other genera which appeared to be closely related to them. In 1959 she noted that there were problems in ascribing some organisms to one or another of these genera; as she noted "rhodochrous" was a "good species in search of a genus", and her assignment of the name "*Mycobacterium rhodochrous*" to the organism was temporary, since she could not find another existing genus to which it could comfortably be ascribed (Gordon & Mihm, 1955). After extensive comparison of the rhodochrous taxon with many strains of rapidly growing mycobacteria by Goodfellow *et al.* (1972), this taxon was elevated to genus status with the name *Rhodococcus rhodochrous;* nine other species were also assigned to this genus (Goodfellow & Alderson, 1977).

After the "housecleaning" successes of Gordon in the 1950s there was a lengthy lacuna in the table of new valid species of rapid growers. Then, from 1962 through 1967 six new species of rapid growers were described that are still valid; four of them were named by Tsukamura and his colleagues. After another four year recess, five more presently valid species were named, and then there was another hiatus from 1974 to 1979. Presumably the inactivity in the sphere of the rapid growers was related to the failure to detect many pathogens among them. Ever since then, there has been a spate of new species among the rapid growers, with 27 names validated from 1979 through 1996. Some of them were revivals of names that had been lost due to omission from the Approved Lists of Bacterial Names (Skerman *et al.*, 1980). However, the renewal in activity may also be ascribed in large degree both to developments in taxonomic technology and to new practical reasons for reexamining them.

8   Conversations In Babel

The specific development of tests that were suitable for classifying slowly growing mycobacteria and the application of computer supported numerical methods to the analysis of the data derived from them led to the recognition and naming of twelve new species of slow growers in the period from 1965 through 1978. The year 1965 also marked the beginning of the informal gatherings of mycobacteriologists that eventually led to the organization that is now known as the International Working Group on Mycobacterial Taxonomy (IWGMT). In the beginning Ernest Runyon had been in the habit of collaring everyone he knew who was interested in mycobacterial taxonomy and who was in attendance at one or another international meeting, finding an empty room, and getting us to sit around and talk about the state of the art. After one such meeting, in Munich in 1965, Ernest Runyon asked me to initiate a newsletter (The Forum Mycobacteriorum) to maintain contact among the participants in these "salons".

By 1967 we started identifying ourselves as members of the IWGMT, even though we had no formal structure, organizational affiliation, nor specific funding, and began to discuss the possibility of establishing cooperative studies. In 1969 we initiated our biennial meetings of the IWGMT at various venues around the world in order to plan and review our studies and the general state of the art. The increasing numbers and relative complexities of some of the new tests had made it unlikely that all laboratories would use the same tests, or if they did, would interpret them in the same way. It had already been suggested that it would be desirable to establish cooperative studies in which individual investigators would pool their data in the hopes that a comprehensive and consistent taxonomy would evolve (Wayne, 1964). Since many of us were beginning to believe our tests and strategies represented the path to the "true" taxonomy of mycobacteria, it took some courage to agree to the pooling of our data into one study where this "truth" could be tested against consensus.

A number of ground rules were established for the conduct of the initial permissive cooperative studies that would minimize bias while allowing complete freedom for each member to select the tests and specific procedures he or she would use (Wayne, 1981). All investigators in a given study received subcultures of the same set of coded strains and submitted their test results to the study coordinator. Consensus results for tests that were done in more than one laboratory were used in performing NT analyses. Only after all data were submitted was the code broken, and no new data were accepted after that time.

The first six studies of the IWGMT, which were published between 1971 and 1979, were designed to give a broad baseline phenotypic description of already recognized species within the genus *Mycobacterium*. In order to permit examination of a fairly large number of strains within each taxon, the six studies, comprising an average of about 75 strains per study, were principally organized according to Runyon Groups, with one study each devoted to the slow growers representing Runyon's Group I (Wayne *et al.*, 1979), Group II (Wayne *et al.*, 1971), and Group III (Meissner *et al.*, 1974), two studies to the rapidly growing members of Group IV (Kubica *et al.*, 1972; Saito *et al.*, 1977), and one study to the shadowy borderline between rapidly growing mycobacteria and members of other selected genera (Goodfellow *et al.*, 1974).

One of the valuable products of these initial studies was the compilation of feature frequency tables for each recognized taxon, reflecting results of all of the tests that were performed on these strains. On analysis of the combined results, it became evident that the most definitive descriptions of species were derived from consensus biochemical and growth tests. Other tests based on phage or immunology-based methods tended to yield subdivisions of species at the infrasubspecific level. In as much as biochemical and growth tests can be performed in any laboratory by simple acquisition of standard chemicals and reagents, whereas the phage and immunologic tests required a source of specific reference materials that were not universally available, the next two studies conducted by the IWGMT were directed toward precise description and evaluation of the biochemical and growth tests that seemed most definitive at the species level. The original data derived from a given test were examined separately for each laboratory that had performed it to establish which laboratory's results were in the best agreement with consensus results for that test. The individual who was responsible for production of those "best" results was then asked to provide a detailed protocol for performance of that procedure in the ensuing two studies on reproducibility of techniques. The results of these two studies were published in 1974 and 1976, and yielded precisely described reproducible procedures for the generation of over a dozen features that were most reproducible and definitive for characterizing slowly growing species of mycobacteria (Wayne *et al.*, 1974, 1976). These tests were among those later selected for inclusion in a large taxonomic probability matrix for use in recognizing possible candidates for new species status (Wayne *et al.*, 1980).

Among the species of slowly growing mycobacteria named during the period from 1965 through 1982, many were represented by very few strains and/or were not thoroughly characterized. The next project initiated by the IWGMT was designated as an "open ended" study, in which strains of slow growers that did not conform to any of the species that had been adequately characterized in the prior Group studies were accessed and were distributed in small lots over about a decade. The pooled data were analysed at intervals. As sufficient numbers of strains of some of the rare or poorly characterized named species were accumulated and characterized, their descriptions were published, and they were deleted from subsequent NT analyses and reports (Wayne *et al.*, 1981, 1983, 1989, 1991). In some cases, only single isolated strains of a given potential taxon were recognized over the course of the study, and in others well circumscribed phenetic clusters emerged which were considered candidates for new species status. However, even as this work was progressing, methods and concepts for defining species on a phylogenetic basis were emerging and the IWGMT avoided naming the new potential phenetic species until the new methods could be applied.

The 14-year period from 1979 through 1992 was marked by the validation of only one new name of a slowly growing species of *Mycobacterium*. That was the revival of *M. shimoidei* (Tsukamura, 1982), which had originally been represented by a single strain and had been omitted from the 1980 Approved Lists of Bacterial Names (Skerman *et al.*, 1980), but which clustered with additional strains in the first report of the IWGMT open-ended study (Wayne *et al.*, 1981). This hiatus in the recognition of new species reflected, in part at least, an exhaustion of the powers of phenetic NT to provide acceptable resolution among species that are very similar to one another. Another reason was simply that the numbers of nonclassifiable mycobacteria that were encountered in diagnostic laboratories was small and clinical interest in them declined accordingly. That was to change drastically as the decade of the 1980s dawned, bringing with it the specter of the acquired immune deficiency syndrome (AIDS), a disease marked by the failure of the cell mediated defense mechanism that is largely responsible for defense against mycobacterial disease.

9   Semantides Rule! ...Or Do They?

In 1962, Floodgate (1962) had argued that it was important to recognize that classifications are motivated by specific needs and reflect our phenotypic descriptive resources, but cannot be considered to reflect actual phylogenetic relationships; that argument was true to the extent that resources for establishing bacterial phylogenetic relatedness were almost nonexistent at that time. Sneath (1962) noted that the affinities determined by the Adansonian approach are treated as independent of phylogeny. As far back as 1965, Zuckerkandl and Pauling (1965), had applied the term "semantides" to the large primary, secondary, and tertiary (DNA, RNA and protein, respectively) classes of information bearing molecules, referring to these "molecules as documents of evolutionary history". Soon thereafter

DNA/DNA hybridization methods were being applied to mycobacteria to try to establish the extent to which the phenetic classification reflected evolutionary relationships. Within the limited range of interpretable hybridization values achievable by these techniques, the evidence confirmed that the phenetic taxonomy of mycobacteria bore the same general relationship to phylogeny as did that of other bacterial genera (Tewfik & Bradley, 1967; Wayne & Gross, 1968; Gross & Wayne, 1970; Bradley, 1972, 1973; Baess, 1979; Baess & Magnusson, 1982). However, DNA/DNA hybridization was largely limited to confirming two strains as phylogenetically "same" or "not-same" at the species level (Wayne et al., 1987), with limited further quantitative inferences possible only among closely related "not-same species" strains (Gross & Wayne, 1970; Wayne, 1978).

An interesting example is the way that DNA/DNA hybridization provided the resolution to a problem that had been troubling mycobacterial taxonomists for several decades. Originally Werner Schaefer assigned members of Runyon's Group III that were pathogenic to chickens to his M. avium serovars 1, 2, and 3 (Schaefer, 1968; Marks et al., 1969b). Other serovars observed among strains of Group III "Battey" bacilli (M. intracellulare) were assigned names corresponding to the patient from whom the original strain of the type was derived. Subsequently Wolinsky & Schaefer (1973) reorganized the nomenclature of these agglutinating serovars, substituting numbers for names and assigning serovar numbers 1, 2 and 3 to M. avium, serovars 4 through 20 to M. intracellulare, and serovars 41, 42, and 43 to M. scrofulaceum. Based on independently conducted phenetic NT analyses, both Wayne (1966) and Kubica et al. (1973) argued that strains of M. intracellulare should be considered to be biovars or pathovars of M. avium, and not recognized as belonging to a separate species. On the other hand, Gertrud Meissner and her colleagues had been conducting chicken virulence tests along with skin tests, using selected sensitins, on guinea pigs that were infected with members of a variety of serovars (Anz & Meissner, 1969; Anz et al., 1970). Their results led them to the conclusion that members of serovars 1, 2 and 3 were classical M. avium, serovars 7, 12, 14, 16, 18, 19 and 20 represented M. intracellulare, and that members of serovars 4, 5, 6, and 8 belong to an intermediate group most closely related to M. avium. Even the completion of an IWGMT cooperative study devoted to members of Runyon's Group III did not resolve the problem (Meissner et al., 1974). Although two phenetic clusters, putatively assigned to M. avium and M. intracellulare, were resolved at the 85% level of similarity, almost complete fusion of these two clusters occurred at 80%. Furthermore, there was some correlation between serovar designation and the phenetic sequence of strains in the M. intracellulare cluster, even though there was no clear division of that cluster. The majority of the 22 authors of this paper concluded that M. intracellulare should be reduced to synonymy with M. avium. A minority statement by six of the authors, including the senior author, arguing for continued separation of the two species was incorporated into the text, to the subsequent amusement, I am told, of some members of the British scientific press. I can recall from meetings around this period, that Gertrud Meissner would repeatedly return to this question, asking me "but what are these intermediate strains, really?". I would reply, "they are not anything 'really', they are just what we choose to call them, depending on where we decide to draw the species lines." In retrospect, I conclude that my reply was much too glib. It was not long before Inga Baess, after a major commitment of time and effort, resolved the problem by establishing the relationship of strains of different serovars in the "M. avium - M. intracellulare complex" through extensive DNA/DNA hybridization studies. She established that members of serovars 1, 2 and 3, the most pathogenic strains, along with those serovars that were included in Meissner's group of strains of intermediate virulence, all hybridized as *M. avium*, and the remaining serovars represented *M. intracellulare* (Baess, 1983). Baess's hybridization studies also demonstrated that *M. paratuberculosis* should be considered a subspecies of *M. avium* (Saxegaard & Baess, 1988). These results have been corroborated in subsequent molecular studies to be discussed below.

It was not long after the quantitative limitations of DNA/DNA hybridization for study of more distant relationships were fully appreciated before methods became available for providing quantitative measurements of evolutionary distances over broader ranges of divergence using the secondary and tertiary semantides. Titration of specific proteins with heterologous and homologous antisera permitted immunological distances to be calculated, and these reflected structural divergence of selected isofunctional proteins from different species. This method has been applied to the generation of evolutionary trees based on immunological distances of two distinct classes of mycobacterial catalases (Wayne & Diaz, 1976, 1979, 1986; Katoch *et al.*, 1982) and of superoxide dismutase (Shivannavar *et al.*, 1996).

In the meantime, evolutionary distances derived from analysis of the divergence of sequences of 16S rRNA were being reported for actinomycetes, including a *Mycobacterium* (Stackebrandt & Woese, 1981). Pitulle and colleagues (1992), working with data on rapidly growing mycobacteria, as well as Wayne (1991), analysing published data on slow growers, concluded that there is a very poor correlation between phenetic distances and 16S rRNA sequence divergence within the genus *Mycobacterium*. We had anticipated these discrepancies on the basis of the nature of phenetic data that are used to generate cluster diagrams (Wayne, 1978). We reasoned that if two strains share almost all phenotypic features it is unlikely that they would have evolved such a complete set of comparable characters *via* different evolutionary pathways. However, if two strains exhibit markedly less phenotypic similarity to one another, it is reasonable to expect that even some of the apparently shared characters reflect different coding, that is, that they are not the same characters genetically speaking. The necessarily large battery of fairly crude tests used in an NT analysis of mycobacteria may yield observable reactions that appear to be the same, but are actually based on different mechanisms. From this reasoning one may expect that two strains that exhibit a very high order of phenotypic similarity do belong to the same genotaxon, but that as the phenotypic similarity between two strains diminishes, the apparent phenetic distance is only erratically related to the actual phylogenetic divergence between them, i.e. the correlations between phenetic and phylogenetic measures decay as phenetic distances increase.

On the other hand, the immunologic distances between T-catalases derived from seven species of slowly growing mycobacteria other than *M. bovis* or *M. tuberculosis* exhibited a high correlation (r=0.85) with reported distances between the same strain pairs based on 16S rRNA sequences. When values for *M. bovis* and *M. tuberculosis* were used for the calculations, the correlation disappeared (r=0.56), suggesting these essentially obligate mammalian pathogens underwent rapid phenotypic evolution in adapting to a highly restricted niche in nature that may reflect the late emergence of their selected mammalian hosts (Wayne, 1991).

After the long hiatus in the establishment of new species of slow growers between 1979 and 1992, a burst of taxonomic activity resulted in the naming of six species over the next three years. A similar acceleration in the naming of new species of rapid growers also occurred during this period. This renewal of interest was probably the result of the interaction of at least two motivating factors. As noted above, the arrival of the AIDS epidemic on the clinical scene had triggered a major revival in the previously declining interest in the so-called opportunistic mycobacteria and in *M. tuberculosis* itself. Initial interest was triggered by the recognition that many AIDS patients developed disseminated infections, including severe bacteremia, with members of the *M. avium - M. intracellulare* complex (Greene *et al.*, 1982; Zakowski *et al.*, 1982; Gottlieb *et al.*, 1983), as well as infections with classical *M. tuberculosis* (Pitchenik *et al.*, 1984). After a period in which interest was focused on the predominant *M. avium - M. intracellulare* complex infections, it became evident that almost any opportunistic mycobacteria, whether fast or slow growing (including some that could not be identified) could invade AIDS patients (Wayne & Sramek, 1992). Another reason for renewed interest in mycobacteria was their unique ability to metabolize hydrocarbons, with the concomitant possibility that some of the nonpathogenic rapid growers could be used for bioremediation of environmental contamination with various petroleum products or toxic halogenated hydrocarbons (Hägglblom *et al.*, 1994; Kleespies *et al.*, 1996).

Concurrent with these events was the development of technology that permitted rapid characterization of mycobacterial isolates, either through application of highly specific DNA probes tailored to signature regions of 16S rRNA (Drake *et al.*, 1987; Saito *et al.*, 1987; Body *et al.*, 1990) or through amplification of 16S rRNA genes by reverse transcriptase PCR and automated sequencing of the product and computerized comparison to sequences of known species (Böttger, 1989; Rogall *et al.*, 1990; Stahl & Urbance, 1990; Böttger *et al.*, 1993; Meier *et al.*, 1993; Springer *et al.*, 1993). Another attractive feature of these techniques was that they could be applied to organisms that were hard to grow, or even, for a time, uncultivable. Particularly striking in early phases of these studies, was the confirmation by 16S rRNA sequence comparisons that the phenotypically defined division of the genus *Mycobacterium* into "slow" and "rapid" growers represented a genuine major genetically based branching within the genus (Stahl & Urbance, 1990).

The alacrity and enthusiasm with which practitioners of the new molecular taxonomy discovered and named new species of mycobacteria, sometimes with limited phenotypic characterization, elicited a sense of deja vu among old time practioners of mycobacterial taxonomy. The new technology, based on the sequences of selected regions of the powerful semantide, 16S rRNA, was proclaimed to be the key to the "true" taxonomy, just as had its predecessors, virulence tests, amidase tests, phage typing, serology, sensitin testing, phenetic numerical taxonomy, and lipid chemistry. (Several years ago an RNA advocate stood up at an IWGMT meeting and announced that the work that the rest of us were doing was now irrelevant). However, as is developed in a detailed and very well reasoned discussion by Stackebrandt and Goebel (1994), because of differences in ages of various evolutionary branches, it is not possible to define a precise numerical value for the difference between two 16S rRNA sequences that can unequivocally evoke the designation of a new species. The International Committee on Systematic Bacteriology early recognized that there was a need for a reconciliation among different approaches to bacterial systematics and organized a work shop in 1987 to examine the problem. Based on discontinuities seen in the distribution of percent hybridization values of paired cultures from many species and genera of bacteria, with a null value seen between 60 and 70% hybridization, this multidisciplinary task group concluded that the pragmatic gold standard for defining bacterial species should be DNA/DNA hybridization (Wayne *et al.*, 1987. This position was reiterated and supported in 1994 by Stackebrandt and Goebel (1994).

10   Escape From Babel; The Road To Consensus

In order to avoid chaos there must be some consensus among the various makers and users of bacterial taxonomy and nomenclature. The more broadly based the taxonomy is the more likely it is to be universally accepted and used. As proposed by Rita Colwell (1970a, 1970b) and reiterated and expanded upon twenty six years later by Vandamme and colleagues (1996), a consensus based polyphasic approach to bacterial systematics, using data from many different disciplines and representing a range of interests and applications, will be necessary if this goal is to be approached. The long held dream of reconciling phenetic and phylogenetic bacterial taxonomy now seems to be an approachable goal.

Recognizing these principles as well as the limits we had reached in taxonomic resolutions that were based on exclusive reliance on phenetic, predominantly biochemically based NT, the IWGMT has undergone a gradual evolution into polyphasic approaches to mycobacterial systematics. Two recent studies illustrate how we integrated serologic, chemotaxonomic and semantide based disciplines with the earlier phenetic NT methodologies. One of these studies was initially intended to evaluate interlaboratory reproducibility of agglutination serotyping of members of the species *M. avium, M. intracellulare,* and *M. scrofulaceum,* but was expanded to include limited semantide based and chemotaxonomic characterizations of the 62 strains included in the study (Wayne *et al.*, 1993). The serovars were determined by agglutination and confirmed by cross-absorption and/or thin layer chromatography (TLC) of the extracted surface antigens. One of the semantide tests was based on use of DNA probes directed against specific regions of the 16S rRNA (Gen Probe, Inc., San Diego) and the other was based on a dot blot test for antibody capture of intrinsic enzyme that was directed toward specific epitopes on mycobacterial catalase (Wayne & Diaz, 1987). The chemotaxonomic analyses were based on interpretation of selected HPLC peak patterns of cellular lipids (Butler & Kilburn, 1990; Butler *et al.*, 1991;). Comparison of the consensus serovar results to both the DNA probe and intrinsic catalase capture results confirmed prior conclusions about the distribution of serovars within the species *M. avium* (serovars 1 to 6 and 8 to 11) and *M. intracellulare* (serovars 7 and 12 to 20). Furthermore, serovar 21, previously unascribed, now fell into *M. avium,* and serovars 23 and 24 were ascribed to *M. intracellulare.* The HPLC ascriptions were in agreement with the semantide results for all strains of *M. intracellulare* and *M. scrofulaceum,* but 5 of 20 strains of *M. avium* were assigned to *M. intracellulare* by HPLC. (Subsequent applications of HPLC to identification of strains has used an expanded pattern recognition program that is more definitive). Only four of the 62 strains examined in this study could not be assigned to a species by consensus results of this polyphasic study.

The second polyphasic IWGMT study (Wayne *et al.*, 1996) used a much broader range of semantide and chemotaxonomic features to assist in the classification of 66 isolated strains or members of novel clusters from the open-ended phenetic NT study of unclassified slowly growing mycobacteria (Wayne *et al.*, 1991). In this new study 19 type strains, representing almost every known species of slow grower were also included as coded unknowns to provide reference information and confirmation of reliability of results. Among the classes of information included in the study were DNA/DNA hybridization values, DNA RFLP patterns, 16S rRNA sequence groups, responses to specific 16S rRNA probes, and responses to specific cross-absorbed antibodies to isofunctional protein antigens. Among the chemotaxonomic information were mycolate HPLC patterns, lipid TLC and gas chromatography patterns, and multilocus enzyme electrophoresis patterns.

The first analysis conducted with the data from the semantide and chemotaxonomy based IWGMT cooperative study was an examination of the extent of agreement between the location of strains in the phenetic NT diagram and in a phylogenetic dendrogram based on the extent of divergence of their 16S rRNA sequences. One large phenetic cluster consisted of the atypical strains that resembled *M. intracellulare* but exhibited significant disagreements in key diagnostic phenotypic or probe responses. All of these strains fell on the same branch of the 16S rRNA sequence tree, with between one and eight nucleotide differences among them.

All bound specifically to the *M. intracellulare* DNA/DNA hybridization probes but most did not react with the *M. intracellulare* DNA/RNA Gen-Probe that was considered to be specific to the 16S rRNA signature region for that species, nor to the corresponding catalase antibody probe. The overall results confirmed previous impressions that *M. intracellulare* is far less homogeneous in its 16S rRNA nucleotide composition than other related species such as *M. avium* (McFadden *et al.*, 1987). It must, however, be reiterated that the variations described here occurred in strains that were specially selected for this study because they had some characteristics that were not typical of the species, and that most strains of *M. intracellulare* do react in a predictable way with these probes (Wayne *et al.*, 1993).

Another major cluster of strains, designated cluster 7, had been recognized in the fourth report of the IWGMT phenetic open-ended study (Wayne *et al.*, 1991). It was considered to be an important cluster because its members were associated with human disease, but diagnostic features used in most clinical laboratories could not distinguish them from *M. gordonae*, a nonpathogenic chromogen of Runyon's Group II. In fact, even though the phenetic NT analysis clearly separated members of cluster 7 from *M. gordonae*, we had been unable to identify any single phenotypic feature that could be used routinely and reliably to separate these clusters. Although we suggested that this cluster 7 probably represented a novel species, we declined to name it until such time as molecular studies could confirm its status (Wayne *et al.*, 1991). In the meantime, Springer and colleagues (1993) recovered two cultures of an unidentifiable *Mycobacterium* from a child with chronic lymphadenitis. They considered it to resemble *M. scrofulaceum* phenotypically, but recognized a unique 16S rRNA sequence and named it *M. interjectum*. In the course of the IWGMT semantide study *M. interjectum* was found to correspond to the novel phenetic cluster 7.

A number of isolated strains and of strains that fell into small novel phenetic clusters or which exhibited novel 16S rRNA sequences that closely resembled those of other species, such as *M. interjectum* or *M. simiae*, could not be assigned to new species with confidence. Inasmuch as an intraspecies range of up to 8 or more nucleotide divergences was seen, it has been concluded that a precise cutoff value in 16S rRNA sequences between closely related organisms cannot be established as the basis for unequivocal definition of new mycobacterial species (Stackebrandt & Goebel, 1994; Wayne *et al.*, 1996). DNA/DNA hybridization appears to be destined to remain a key player in the polyphasic study of mycobacterial systematics. However, because so few laboratories now maintain the capability for performing these hybridizations, there remains a need for a simplified and reliable method for taxonomic (rather than diagnostic) performance of these hybridizations. As suggested in the discussion of the IWGMT semantide study (Wayne *et al.*, 1996), some minor modifications in the checkerboard plate method (Kusunoki *et al.*, 1991) may meet this need. Another need is for a central reference bank to be maintained of DNA from every named species of *Mycobacterium* to permit block hybridizations to be done when candidate taxa are being considered for designation as new species.

Regardless of the general availability of practical methods for routine application of these procedures, we continue to need a balance between molecular and phenetic approaches to mycobacterial systematics to meet the needs of all users of the system.

11   And, In Conclusion....

Overall, I have tried in this chapter to show how taxonomic methodologies and philosophies that were evolving over the past 50 years, and the specific applied diagnostic needs that were also evolving, came together to shape the direction of mycobacterial systematics. This confluence of interests during that period contributed significantly to the transformation of the dry bones of mycobacterial taxonomy to a living and exciting subject, whose complete story is still unfolding.

12   References

Adanson, M. (1763). Familles des plantes. Volume 1. Paris: Vincent.

Anz, W. & Meissner, G. (1969). Serotypen von Stammen der aviaren Mykobakteriengruppe, isolierte von Mensch und Tier. Ihre epidemiologischen Beziehungen. *Praxis der Pneumonologie* 23, 221-230.

Anz, W., Lauterbach, D., Meissner, G. & Willers, I. (1970). Vergleich von Sensitin-testen an Meerschweinchen mit Serotyp und Huhnervirulenz bei M. avium und M. intracelllulare Stammen. *Zentralblatt Bakteriologie* 215, 536-549.

Aronson, J.D. (1926). Spontaneous tuberculosis in salt water fish. *Journal of Infectious Diseases* 39, 314-320.

Baess, I. (1979). Deoxyribonucleic acid relatedness among species of slowly-growing mycobacteria. *Acta Pathologia et Microbiologia Scandinavica* 87, 221-226.

Baess, I. (1983). Deoxyribonucleic acid relationships between different serovars of *Mycobacterium avium*, *Mycobacterium intracellulare* and *Mycobacterium scrofulaceum*. *Acta Pathologia, Microbiologia et Immunologia Scandinavica* 91, 201-203.

Baess, I. & Magnusson, M. (1982). Classification of *Mycobacterium simiae* by means of comparative reciprocal intradermal sensitin testing on guinea pigs and deoxyribonucleic acid hybridization.

Acta Pathologia, Microbiologia et Immunologia Scandinavica **90**, 101-107.

Bates, J.H. & Mitchison, D.A. (1969). Geographic distribution of bacteriophage types of *Mycobacterium tuberculosis. American Review of Respiratory Disease* **100**, 189-193.

Body, B.A., Warren, N.G., Spicer, A., Henderson, D. & Chery, M. (1990). Use of Gen-Probe and Bactec for rapid isolation and identification of mycobacteria. *American Journal of Clinical Pathology* **93**, 415-420.

Bojalil, L.F., Cerbon, J. & Trujillo, A. (1962). Adansonian classification of mycobacteria. *Journal of General Microbiology* **28**, 333-346.

Bönicke, R. (1960a). The classification of atypical mycobacteria by their power to metabolize amides. *Tuberculosearzt* **14**, 209-211.

Bönicke, R. (1960b). Über das Verkommen von Acylamidasen in Mycobacterium. *Zentralblatt Bakteriologie und Hygiene* **178**, 209-222.

Bönicke, R. (1962). Identification of mycobacteria by biochemical methods. *Bulletin of the International Union Against Tuberculosis & Lung Disease* **32**, 13-68.

Böttger, E.C. (1989). Rapid determination of bacterial ribosomal RNA sequences by direct sequencing of enzymatically amplified DNA. *FEMS Microbiology Letters* **65**, 171-176.

Böttger, E.C., Hirschel, B. & Coyle, M.B. (1993). *Mycobacterium genavense* sp. nov. *International Journal of Systematic Bacteriology* **43**, 841-843.

Bradley, S.G. (1972). Reassociation of deoxyribonucleic acid from selected mycobacteria with that from *Mycobacterium bovis* and *Mycobacterium farcinica. American Review of Respiratory Disease* **106**, 122-124.

Bradley, S.G. (1973). Relationship among mycobacteria and nocardiae based upon deoxyribonucleic acid reassociation. *Journal of Bacteriology* **113**, 645-651.

Breed, R.S., Murray, E.G.D. & Hitchens, A.P. (1948). *Bergey's Manual of Determinative Bacteriology.* Baltimore: The Williams and Wilkins Co.

Brennan, P.J. (1984). Antigenic peptidoglycolipids, phospholipids and glycolipids. In *The Mycobacteria: a Sourcebook*, pp. 467-490. Edited by G.P. Kubica & L.G. Wayne. New York: Marcel Dekker, Inc.

Brennan, P.J., Mayer, H., Aspinall, G.O. & Nam Shin, J.E. (1981). Structures of the glycopeptidolipid antigens from serovars in the *Mycobacterium avium/Mycobacterium intracellulare/ Mycobacterium scrofulaceum serocomplex. European Journal of Biochemistry* **115**, 7-15.

Buhler, V.B. & Pollak, A. (1953). Human infection with atypical acid-fast organisms. Report of two cases with pathologic findings. *American Journal of Clinical Pathology* **23**, 363-374.

Butler, W.R. & Kilburn, J.O. (1990). High-performance liquid chromatography patterns of mycolic acids as criteria for identification of *Mycobacterium chelonae, Mycobacterium fortuitum*, and *Mycobacterium smegmatis. Journal of Clinical Microbiology* **28**, 2094-2098.

Butler, W.R., Jost, K.C. Jr. & Kilburn, J.O. (1991). Identification of mycobacteria by high-performance liquid chromatography. *Journal of Clinical Microbiology* **29**, 2468-2472.

Chaparas, S.D., Lind, A., Ouchterlony, O. & Ridell, M. (1983). Terminology guidelines for serotaxonomic studies using immunodiffusion and immunoelectrophoresis. *International Journal of Systematic Bacteriology* **33**, 414-416.

Clark, M.E. & Forrest, E.S. (1953). Results with oxidation-reduction dyes in the determination of virulence of mycobacteria. *American Review of Tuberculosis* **68**, 786-787.

Colwell, R.R. (1970a). Polyphasic taxonomy of the genus *Vibrio*: Numerical taxonomy of *Vibrio cholerae, Vibrio parahaemolyticus*, and related *Vibrio species. Journal of Bacteriology* **104**, 410-433.

Colwell, R.R. (1970b). Polyphasic taxonomy of bacteria. In *Culture Collections of Microorganisms.* Proceedings of the *International Conference on Culture Collections.* October 1968, Tokyo, pp. 421-436. Edited by H. Iizuka and T. Hasegawa. Tokyo: University of Tokyo Press:

Crawford, J.T. & Bates, J.H. (1985). Phage typing of the *Mycobacterium avium-intracellulare-scrofulaceum* complex. *American Review of Respiratory Disease* 132, 386-389.

Crawford, J.T., Fitzhugh, J.K. & Bates, J.H. (1981). Phage typing of the *Mycobacterium avium-intracellulare-scrofulaceum* complex. *American Review of Respiratory Disease* 124, 559-562.

Cuttino, J.L. & McCabe, A.M. (1949). Pure granulomatous nocardiosis: a new fungus disease distinguished by intracellular parasitism. A description of a new disease in man due to a hitherto undescribed organism, *Nocardia intracellularis*, n.sp., including a study of the biologic and pathogenic properties of this species. *American Journal of Pathology* 25, 1-34.

Drake, T.A., Hindler, J.A., Berlin, O.G.W. & Bruckner, D.A. (1987). Rapid identification of *Mycobacterium avium* complex in culture using DNA probes. *Journal of Clinical Microbiology* 25, 1442-1445.

Dubos, R.J. & Middlebrook, G. (1948). Cytochemical reaction of virulent tubercle bacilli. *American Review of Tuberculosis* 58, 698-699.

Floodgate, G.D. (1962). Some remarks on the theoretical aspects of bacterial taxonomy. *Bacteriological Reviews* 26, 277-291.

Froman, S. & Scammon, L. (1964). Effect of temperature on the bacteriophage susceptibility of strains of *Mycobacterium avium* isolated from fowl. *American Review of Respiratory Disease* 89, 236-239.

Froman, S., Will, D.W. & Bogen, E. (1954). Bacteriophage active against virulent *Mycobacterium tuberculosis*. I. Isolation and activity. *American Journal of Public Health* 44, 1326-1333.

Goodfellow, M. & Alderson, G. (1977). The actinomycete-genus *Rhodococcus*: a home for the "rhodochrous" complex. *Journal of General Microbiology* 100, 99-122.

Goodfellow, M., Fleming, A. & Sackin, M.J. (1972). Numerical taxonomy of "Mycobacterium" rhodochrous and Runyon's Group IV mycobacteria. *International Journal of Systematic Bacteriology* 22, 81-98.

Goodfellow, M., Lind, A., Mordarska, H., Pattyn, S. & Tsukamura, M. (1974). A co-operative numerical analysis of cultures considered to belong to the "rhodochrous" taxon. *Journal of General Microbiology* 85, 291-302.

Gordon, R.E. & Mihm, J.M. (1959). A comparison of four species of mycobacteria. *Journal of General Microbiology* 21, 736-748.

Gordon, R.E. & Smith, M.M. (1953). Rapidly growing, acid fast bacteria. I. Species descriptions of *Mycobacterium phlei* Lehmann and Neumann and *Mycobacterium smegmatis* (Trevison) Lehmann and Neumann. *Journal of Bacteriology* 66, 41-48.

Gordon, R.E. & Smith, M.M. (1955). Rapidly growing, acid fast bacteria. II. Species description of *Mycobacterium fortuitum* Cruz. *Journal of Bacteriology* 69, 502-507.

Gottlieb, M.S., Groopman, J.E., Weinstein, W.M., Fahey, J.L. & Detels, R. (1983). The acquired immunodeficiency syndrome. *Annals of Internal Medicine* 99, 208-220.

Greene, J.B., Sidhu, G.S., Lewin, S., Levine, J.F., Masur, H., Simberkoff, M.S., Nicholas, P., Good, R.C., Zolla-Pazner, S.B., Pollock, A.A., Tapper, M.L. & Holzman, R.S. (1982). *Mycobacterium avium-intracellulare* (sic): A cause of disseminated life-threatening infection in homosexuals and drug abusers. *Annals of Internal Medicine* 97, 539-546.

Gross, W.M. & Wayne, L.G. (1970). Nucleic acid homology in the genus *Mycobacterium*. *Journal of Bacteriology* 104, 630-634.

Hauduroy, P. (1955). *Derniers aspects du monde des mycobactéries*. Paris: Masson et Cie.

Hägglblom, M.M., Nohynek, L.J., Palleroni, N.J., Kronqvist, K., Nurmiaho-Lassila, E.-L., Salkinoja-Salonen, M.S., Klatte, S. & R. Kroppenstedt, M. (1994). Transfer of polychlorophenol-degrading *Rhodococcus chlorophenolicus* (Apajalahti *et al.* 1986) to the genus *Mycobacterium* as *Mycobacterium chlorophenolicum* comb. nov. *International Journal of Systematic Bacteriology* 44, 485-493.

Hobby, G.L., Redmond, W.B., Runyon, E.H., Schaefer, W.B., Wayne, L.G. & Wichelhausen, R.H. (1967). A study on pulmonary disease associated with mycobacteria other than *Mycobacterium tuberculosis*: identification and characterization of the mycobacteria. *American Review of Respiratory Disease* **95**, 954-971.

Jenkins, P.A. & Marks, J. (1972). Thin-layer chromatography of mycobacterial lipids as an aid to classification: The scotochromogenic mycobacteria, including *M. scrofulaceum, M. xenopi, M. aquae, M. gordonae, M. flavescens. Tubercle* **53**, 118-127.

Katoch, V.M., Wayne, L.G. & Diaz, G.A. (1982). Characterization of catalase by micro-immunoprecipitation in tissue-derived cells of *M. lepraemurium* TMC 1701. *International Journal of Systematic Bacteriology* **32**, 414-418.

Kestle, D.G., Abbott, V.D. & Kubica, G.P. (1967). Differential identification of mycobacteria. II. Subgroups of Groups II and III (Runyon) with different clinical significance. *American Review of Respiratory Disease* **95**, 1041-1052.

Kleespies, M., Kroppenstedt, R.M., Rainey, F.A., Webb, L.E. & Stackebrandt, E. (1996). *Mycobacterium hodleri* sp nov; a new member of the fast-growing mycobacteria capable of degrading polycyclic aromatic hydrocarbons. *International Journal of Systematic Bacteriology* **46**, 683-687.

Koch, R. (1882). Die aetiologie der tuberculose. *Klinische Wochenschrift* **19**, 221.

Konno, K. (1956). New chemical method to differentiate human-type tubercle bacilli from other mycobacteria. *Science* **124**, 985.

Kubica, G.P., Baess, I., Gordon, R.E., Jenkins, P.A., Kwapinski, J.B.G., McDurmont, C., Pattyn, S.R., Saito, H., Silcox, V.A., Stanford, J.L., Takeya, K. & Tsukamura, M. (1972). A cooperative numerical analysis of rapidly growing mycobacteria. *Journal of General Microbiology* **73**, 55-70.

Kubica, G.P., Silcox, V.A. & Hall, E. (1973). Numerical taxonomy of selected slowly growing mycobacteria. *Journal of General Microbiology* **74**, 159-167.

Kubica, G.P., Silcox, V.A., Kilburn, J.O., Smithwick, R.W., Beam, R.E., Jones, W.D. Jr. & Stottmeier, K.D. (1970). Differential identification of mycobacteria. VI. *Mycobacterium triviale. International Journal of Systematic Bacteriology* **20**, 161-174.

Kusunoki, S., Ezaki, T., Tamesada, M., Hatanaka, Y., Asano, K., Hashimoto, Y. & Yabuuchi, E. (1991). Application of colorimetric microdilution plate hybridization for rapid genetic identification of 22 *Mycobacterium* species. *Journal of Clinical Microbiology* **29**, 1596-1603.

Kwapinski, J.B.G., Alcasia, A. & Palser, H. (1970). Serological relationships of endoplasm antigens of saprophytic mycobacteria. *Canadian Journal of Microbiology* **16**, 871-876.

Lind, A. & Norlin, M. (1963). A comparative serological study of *M. avium, M. ulcerans, M. balnei* and *M. marinum* by means of double diffusion-in-gel methods. A preliminary investigation. *Scandinavian Journal of Clinical Laboratory Investigation* **15**, 152-163.

Linell, F. & Norden, A. (1954). *Mycobacterium balnei*: A new acid-fast bacillus occurring in swimming pools and capable of producing skin lesions in humans. *Acta Tuberculosea Scandinavica, Supplement* **33**, 1.

Liston, J., Wiebe, W. & Colwell, R.R. (1963). Quantitative approach to the study of bacterial species. *Journal of Bacteriology* **85**, 1061-1070.

MacCallum, P., Tolhurst, J.C., Buckle, G. & Sissons, H.A. (1948). A new mycobacterial infection in man. *Journal of Pathology and Bacteriology* **60**, 93-122.

Magnusson, M. (1962). Specificity of sensitins. III. Further studies in guinea pigs with sensitins of various species of *Mycobacterium* and *Nocardia. American Review of Respiratory Disease* **86**, 395-404.

Marks, J., Jenkins, P.A. & Schaefer, W.B. (1969a). Identification and incidence of a third type of *Mycobacterium avium. Tubercle* **50**, 394-395.

Marks, J., Jenkins, P.A. & Schaefer, W.B. (1969b). Thin layer chromatography of mycobacterial lipids as an acid to classification. *Tubercle* 52, 219-225.

McFadden, J.J., Butcher, P.D., Thompson, J., Chiodini, R. & Hermon-Taylor, J. (1987). The use of DNA probes identifying restriction-fragment-length polymorphisms to examine the *Mycobacterium avium* complex. *Molecular Microbiology* 1, 283-291.

McIntyre, G., Belsey, E. & Stanford, J.L. (1986). Taxonomic differences between *Mycobacterium avium* and *Mycobacterium intracellulare* elucidated in man by skin tests with three new tuberculins. *European Journal of Respiratory Disease* 69, 146-152.

Meier, A., Kirschner, P., Schröder, K.-H., Wolters, J., Kroppenstedt, R.M. & Böttger, R.M. (1993). *Mycobacterium intermedium* sp. nov. *International Journal of Systematic Bacteriology* 43, 204-209.

Meissner, G., Schröder, K.H., Amadio, G.E., Anz, W., Chaparas, S., Engel, H.W.B., Jenkins, P.A., Käppler, W., Kleeberg, H.H., Kubala, E., Kubin, M., Lauterbach, E., Lind, A., Magnusson, M., Mikova, ZD., Pattyn, S.R., Schaefer, W.B., Stanford, J.L., Tsukamura, M., Wayne, L.G., Willers, I. & Wolinsky, E. (1974). A cooperative numerical analysis of non-scoto and nonphotochromogenic slowly growing mycobacteria. *Journal of General Microbiology* 83, 207-235.

Middlebrook, G. (1954). Isoniazid-resistance and catalase activity of tubercle bacilli. *American Review of Tuberculosis* 69, 471-472.

Middlebrook, G. & Cohn, M.L. (1953). Some observations on the pathogenicity of isoniazid-resistant variants of tubercle bacillis. *Science* 118, 297-299.

Middlebrook, G., Dubos, R.J. & Pierce, C. (1947). Virulence and morphological characteristics of mammalian tubercle bacilli. *Journal of Experimental Medicine* 86, 175-184.

Minnikin, D.E., Collins, M.D. & Goodfellow, M. (1978). Menaquinone patterns in the classification of nocardioform and related bacteria. *Zentalblatt für Bakteriologie und Hygiene*. Abteilung 1, *Supplement* 6, 85-90.

Minnikin, D.E., Minnikin, S.M. & Goodfellow, M. (1982). The oxygenated mycolic acids of *Mycobacterium fortuitum*, *M. farcinogenes* and *M. senegalense*. *Biochimica et Biophysica Acta* 712, 616-620.

Norden, A. & Linell, F. (1951). A new type of pathogenic mycobacterium. *Nature* 168, 826.

Norlin, M., Lind, A. & Ouchterlony, O. (1969). A serologically based taxonomic study of *M. gastri*. *Zeitschrift für Immunologie, Allergie und Klinische Immunologie* 137, 241-248.

Oestreicher, R., Dressler, S.H., Russell, W.F.Jr., Grow, J.B. & Middlebrook, G. (1955). Observations on the pathogenicity of isoniazid-resistant mutants of tubercle bacilli for tuberculous patients. *American Review of Tuberculosis and Pulmonary Disease* 71, 390-400.

Penso, G. & Ortali, V. (1949a). Studi e recerche sui micobatteri. Nota II. I fagi dei micobatteri. *Rendiconti dell' Istituto Superiore di Sanita* 12, 903-918.

Penso, G. & Ortali, V. (1949b). Isolamento e studio di fagi attivi sui micobatteri. *Rendiconti dell'Accademia Nazionale Lincei* 6, 109-115.

Pinner, M. (1935a). Atypical acid-fast microorganisms. IV. Smooth-growing tubercle bacilli. *American Review of Tuberculosis* 32, 440-445.

Pinner, M. (1935b). Atypical acid-fast microorganisms. *American Review of Tuberculosis* 32, 424-439.

Pitchenik, A.E., Cole, C., Russell, B.W., Fischl, M.A., Spira, T.J. & Snider, D.E.Jr. (1984). Tuberculosis, atypical mycobacteriosis, and the acquired immunodeficiency syndrome among Haitian and Non-Haitian patients in South Florida. *Annals of Internal Medicine* 101, 641-645.

Pitulle, C., Dorsch, M., Kazda, J., Wolters, J. & Stackebrandt, E. (1992). Phylogeny of rapidly growing members of the genus *Mycobacterium*. *International Journal of Systematic Bacteriology* 42, 337-343.

Pollak, A. & Buhler, V.B. (1955). The cultural characteristics and animal pathogenicity of an atypical acid-fast organism which causes human disease. *American Review of Tuberculosis* **71**, 74-87.

Prissick, F.H. & Masson, A.M. (1956). Cervical lymphadenitis in children caused by chromogenic mycobacteria. *Canadian Medical Association Journal* **75**, 798-803.

Prissick, F.H. & Masson, A.M. (1957). Yellow-pigmented pathogenic mycobacteria from cervical lymphadenitis. *Canadian Journal of Microbiology* **3**, 91-100.

Rado, T.A., Bates, J.H., Engel, H.W.B., Mankiewicz, E., Murohashi, T., Mizuguchi, Y. & Sula, L. (1975). World Health Organization studies on bacteriophage typing of mycobacteria - subdivision of the species *Mycobacterium tuberculosis*. *American Review of Respiratory Disease* **111**, 459-468.

Richmond, L. & Cummings, M.M. (1950). An evaluation of methods of testing the virulence of acid-fast bacilli. *American Review of Respiratory Disease* **62**, 632-637.

Rogall, T., Wolters, J., Flohr, T. & Böttger, E.C. (1990). Towards a phylogeny and definition of species at the molecular level within the genus *Mycobacterium*. *International Journal of Systematic Bacteriology* **40**, 323-330.

Runyon, E.H. (1959). Anonymous mycobacteria in pulmonary disease. *Medical Clinics of North America* **43**, 273-290.

Runyon, E.H. (1965). Pathogenic mycobacteria. *Advances in Tuberculosis Research* **14**, 235-287.

Saito, H., Gordon, R.E., Juhlin, I., Käppler, W., Kwapinski, J.B.J., McDurmont, C., Pattyn, S.R., Runyon, E.H., Stanford, J.L., Tarnok, I., Tasaka, H., Tsukamura, M. & Weiszfeiler, J. (1977). Cooperative numerical analysis of rapidly growing mycobacteria. *International Journal of Systematic Bacteriology* **27**, 75-85.

Saito, H., Tomioka, H., Sato, K., Asano, K. & Kusunoki, S. (1987). Usefulness of Gen-Probe for identification and classification of *Mycobacterium avium* complex. *Kekkaku* **63**, 47.

Saxegaard, F. & Baess, I. (1988). Relationship between *Mycobacterium avium*, *Mycobacterium paratuberculosis* and "wood pigeon mycobacteria". *Acta Pathologia, Microbiologia et Immunologia Scandinavica* **96**, 37-42.

Schaefer, W.B. (1965). Serologic identification and classification of atypical mycobacteria by their agglutination. *American Review of Respiratory Disease* (Supplement) **92**, 85-93.

Schaefer, W.B. (1967). Serologic identification of the atypical mycobacteria and its value in epidemiologic studies. *American Review of Respiratory Disease* **96**, 115-118.

Schaefer, W.B. (1968). Incidence of the serotypes of *Mycobacterium avium* and atypical mycobacteria in human and animal diseases. *American Review of Respiratory Disease* **97**, 18-23.

Schaefer, W.B. & Davis, C.L. (1961). A bacteriologic and histopathologic study of skin granulomas due to *Mycobacterium balnei*. *American Review of Respiratory Disease* **84**, 837-844.

Shivannavar, C.T., Katoch, V.M., Sharma, V.D., Patil, M.A., Katoch, K., Bharadwaj, V.P. & Agrawal, B.M. (1996). Development of a SOD ELISA to determine the immunological relatedness among mycobacteria. *International Journal of Leprosy and Other Mycobacterial Diseases* **64**, 58-65.

Skerman, V.B.D., McGowan, V. & Sneath, P.H.A. (1980). Approved lists of bacterial names. *International Journal of Systematic Bacrteriology* **30**, 255-420.

Sneath, P.H.A. (1962). The construction of taxonomic groups. In *Microbial Classification*. Edited by G.C. Ainsworth & P.H.A.Sneath, pp.289-332. Cambridge: Cambridge University Press.

Sneath, P.H.A. & Cowan, S.T. (1958). An electro-taxonomic survey of bacteria. *Journal of General Microbiology* **19**, 551-565.

Sneath, P.H.A. & Sokal, R.R. (1973). *Numerical Taxonomy*. San Francisco: W.H. Freeman and Co.

Springer, B., Kirschner, P., Rost-Meyer, G., Schröder, K.-H., Kroppenstedt, R.M. & Böttger, E.C. (1993). *Mycobacterium interjectum*, a new species isolated from a patient with chronic lymphadenitis. *Journal of Clinical Microbiology* **31**, 3083-3089.

Stackebrandt, E. & Goebel, B.M. (1994). Taxonomic Note: A place for DNA-DNA reassociation and 16S rRNA sequence analysis in the present species definition in bacteriology. *International Journal of Systematic Bacteriology* **44**, 846-849.

Stackebrandt, E. & Woese, C.R. (1981). Towards a phylogeny of the actinomycetes and related organisms. *Current Microbiology* **5**, 197-202.

Stahl, D.A. & Urbance, J.W. (1990). The division between fast- and slow-growing species corresponds to natural relationships among the mycobacteria. *Journal of Bacteriology* **172**, 116-124.

Stanford, J.L. & Beck, B. (1969). Bacteriological and serological studies of fast growing mycobacteria identified as *Mycobacterium friedmannii*. *Journal of General Microbiology* **58**, 99-106.

Stanford, J.L. & Gunthorpe, J. (1971). A study of some fast-growing scotochromogenic mycobacteria including species descriptions of *Mycobacterium gilvum* (new species) and *Mycobacterium duvalii* (new species). *British Journal of Experimental Pathology* **52**, 627-637.

Szulga, T., Jenkins, P.A. & Marks, J. (1966). Thin-layer chromatography of mycobacterial lipids as an aid to classification; *Mycobacterium kansasii* and *Mycobacterium marinum* (balnei). *Tubercle* **47**, 130-136.

Tarshis, M.S. (1958). Preliminary observations on the development of atypical (chromogenic) variants of the H37Rv strain of *M. tuberculosis* under the influence of streptomycin and isoniazid in vitro. *American Review of Tuberculosis and Pulmonary Disease* **78**, 921-926.

Tarshis, M.S. & Frisch, A.W. (1952). Chromogenic acid-fast bacilli from human sources.Cultural studies, pathology and hypersensitivity. *American Review of Tuberculosis* **65**, 278-315.

Tasaka, H. & Matsuo, Y. (1984). Specificity and distribution of alpha antigens of *Mycobacterium kansasii* and *Mycobacterium marinum*. *American Review of Respiratory Disease* **130**, 647-649.

Tasaka, H., Kiyotani, K. & Matsuo, Y. (1983). Purification and antigenic specificity of alpha protein (Yoneda and Fukui) from *Mycobacterium tuberculosis* and *Mycobacterium intracellulare*. *Hiroshima Journal of Medical Science* **32**, 1-8.

Tasaka, H., Nomura, T. & Matsuo, Y. (1985). Specificity and distribution of alpha antigens of *Mycobacterium avium-intracellulare, Mycobacterium scrofulaceum*, and related species of mycobacteria. *American Review of Respiratory Disease* **132**, 173-174.

Tewfik, E.M. & Bradley, S.G. (1967). Characterization of deoxyribonucleic acids from streptomycetes and nocardiae. *Journal of Bacteriology* **94**, 1994-2000.

Timpe, A. & Runyon, E.H. (1954). The relationship of "atypical" acid-fast bacteria to human disease. *Journal of Laboratory and Clinical Medicine* **44**, 202-209.

Tsukamura, M. (1965). A group of mycobacteria from soil sources resembling nonphotochromogens (Group 3). A description of *Mycobacterium nonchromogenicum*. *Medicine and Biology* **71**, 110-113.

Tsukamura, M. (1966). Adansonian classification of mycobacteria. *Journal of General Microbiology* **45**, 253-272.

Tsukamura, M. (1974). Differentiation of the "Mycobacterium" rhodochrous group from nocardiae by ß-galactosidase activity. *Journal of General Microbiology* **80**, 553-555.

Tsukamura, M. (1982). *Mycobacterium shimoidei* sp. nov., nom. rev., a lung pathogen. *International Journal of Systematic Bacteriology* **32**, 67-69.

Tsukamura, M., Mizuno, S. & Tsukamura, S. (1979). Artificial induction of an organism similar to *Mycobacterium triviale* from *Mycobacterium gordonae* by successive ultraviolet irradiations. *International Journal of Systematic Bacteriology* **29**, 32-37.

Vandamme, P., Pot, B., Gillis, M., De Vos, P., Kersters, K. & Swings, J. (1996). Polyphasic taxonomy, a consensus approach to bacterial systematics. *Microbiological Reviews* **60**, 407-438.

Virtanen, S. (1960). A study of nitrate reduction by mycobacteria. *Acta Tuberculosea Scandinavica, Supplement* **48**, 1-119.

Wayne, L.G. (1959). Quantitative aspects of neutral red reactions of typical and "atypical" mycobacteria. *American Review of Tuberculosis and Pulmonary Disease* 79, 526-530.

Wayne, L.G. (1961). Recognition of *Mycobacterium fortuitum* by means of a three-day phenolphthalein sulfatase test. *American Journal of Clinical Pathology* 36, 185-187.

Wayne, L.G. (1962). Differentiation of mycobacteria by their effect on Tween 80. *American Review of Respiratory Disease* 86, 579-581.

Wayne, L.G. (1964). (Editorial) The mycobacterial mystique: deterrent to taxonomy. *American Review of Respiratory Disease* 90, 255-257.

Wayne, L.G. (1966). Classification and identification of mycobacteria. III. Species within Group III. *American Review of Respiratory Disease* 93, 919-928.

Wayne, L.G. (1967). Selection of characters for the Adansonian analysis of mycobacterial taxonomy. *Journal of Bacteriology* 93, 1382-1391.

Wayne, L.G. (1968). *Mycobacterium marinum* and *M. marianum* (Correspondence). *American Review of Respiratory Disease* 98, 317.

Wayne, L.G. (1970). On the identity of *Mycobacterium gordonae* Bojalil and the so-called tap water scotochromogens. *International Journal of Systematic Bacteriology* 20, 149-153.

Wayne, L.G. (1975). Proposal to reject the specific epithet marianum in the name *Mycobacterium marianum* Penso 1953 and to conserve the specific epithet scrofulaceum in the name *Mycobacterium scrofulaceum* Prissick and Masson 1956 - Request for an Opinion. *International Journal of Systematic Bacteriology* 25, 230-231.

Wayne, L.G. (1978). Mycobacterial taxonomy: a search for discontinuities. *Annales de Microbiologie (Institut Pasteur)* 129, 13-27.

Wayne, L.G. (1981). Numerical taxonomy and cooperative studies: roles and limits. *Reviews of Infectious Disease* 3, 822-828.

Wayne, L.G. (1991). The mycobacteria: a leisurely climb in the family tree. *United States Federation of Culture Collections Newsletter* 21, 1-8.

Wayne, L.G. & Diaz, G.A. (1976). Immunoprecipitation studies of mycobacterial catalase. *International Journal of Systematic Bacteriology* 26, 38-44.

Wayne, L.G. & Diaz, G.A. (1979). Reciprocal immunological distances of catalase derived from strains of *Mycobacterium avium*, *Mycobacterium tuberculosis*, and closely related species. *International Journal of Systematic Bacteriology* 29, 19-24.

Wayne, L.G. & Diaz, G.A. (1982). Serological, taxonomic and kinetic studies of the T and M classes of mycobacterial catalase. *International Journal of Systematic Bacteriology* 32, 296-304.

Wayne, L.G. & Diaz, G.A. (1986). Differentiation between T-catalases derived from *Mycobacterium avium* and *M. intracellulare* by a solid-phase immunosorbent assay. *International Journal of Systematic Bacteriology* 36, 363-367.

Wayne, L.G. & Diaz, G.A. (1987). An intrinsic catalase dot-blot immunoassay for identification of *Mycobacterium tuberculosis*, *Mycobacterium avium* and *Mycobacterium intracellulare*. *Journal of Clinical Microbiology* 25, 1687-1690.

Wayne, L.G. & Diaz, G.A. (1988). Detection of a novel catalase in extracts of *Mycobacterium avium* and *Mycobacterium intracellulare*. *Infection and Immunity* 56, 936-941.

Wayne, L.G. and Gross, W.M. (1968). Base composition of deoxyribonucleic acid isolated from mycobacteria. *Journal of Bacteriology* 96, 1915-1919.

Wayne, L.G. & Sramek, H.A. (1992). Agents of newly recognized or infrequently encountered mycobacterial diseases. *Clinical Microbiology Reviews* 5, 1-25.

Wayne, L.G., Krasnow, I. & Huppert, M. (1957). Characterization of atypical mycobacteria and of nocardia species isolated from microcolonial test. *American Review of Tuberculosis and Pulmonary Disease* 76, 451-467.

Wayne, L.G., Juarez, W.J. & Nichols, E.G. (1958). Aryl sulfatase activity of aerobic actinomycetales. *Journal of Bacteriology* 75, 367-368.

Wayne, L.G., Doubek, J.R. & Diaz, G.A. (1967). Classification and identification of mycobacteria. IV. Some important scotochromogens. *American Review of Respiratory Disease* 96, 88-95.

Wayne, L.G., Andrade, L., Froman, S., Käppler, W., Kubala, E., Meissner, G. & Tsukamura, M. (1979). A co-operative numerical analysis of *Mycobacterium gastri*, *Mycobacterium kansasii*, and *Mycobacterium marinum*. *Journal of General Microbiology* 109, 319-327.

Wayne, L.G., Dietz, T.M., Gernez-Rieux, C., Jenkins, P.A., Käppler, W., Kubica, G.P., Kwapinski, J.B.G., Meissner, G., Pattyn, S.R., Runyon, E.H., Schröder, K.H., Silcox, V.A., Tsukamura, M. & Wolinsky, E. (1971). A cooperative numerical taxonomic analysis of scotochromogenic slowly growing mycobacteria. *Journal of General Microbiology* 66, 255-271.

Wayne, L.G., Engbaek, H.C., Engel, H.W.B., Froman, S., Gross, W., Hawkins, J., Käppler, W., Karlson, A.G., Kleeberg, H.H., Krasnow, I., Kubica, G.P., McDurmont, C., Nel, E.E., Pattyn, S.R., Schröder, K.H., Showalter, S., Tarnok, I., Tsukamura, M., Vergmann, B. & Wolinsky, E. (1974). Highly reproducible techniques for use in systematic bacteriology in the genus *Mycobacterium*: tests for pigment, urease, resistance to sodium chloride, hydrolysis of Tween 80, and β-galactosidase. *International Journal of Systematic Bacteriology* 24, 412-419.

Wayne, L.G., Engel, H.W.B., Grassi, C., Gross, W., Hawkins, J., Jenkins, P.A., Käppler, W., Kleeberg, H.H., Krasnow, I., Nel, E.E., Pattyn, S.R., Richards, P.A., Showalter, S., Slosarek, M., Szabo, I., Tarnok, I., Tsukamura, M., Vergmann, B. & Wolinsky, E. (1976). Highly reproducible techniques for use in systematic bacteriology in the genus *Mycobacterium*. II. Tests for niacin and catalase and for resistance to isoniazid, thiophene 2-carboxylic acid hydrazide, hydroxylamine and p-nitrobenzoate. *International Journal of Systematic Bacteriology* 26, 311-318.

Wayne, L.G., Krichevsky, E.J., Love, L.L., Johnson, R. & Krichevsky, M.I. (1980). Taxonomic probability matrix for use with slowly-growing mycobacteria. *International Journal of Systematic Bacteriology* 30, 528-538.

Wayne, L.G., Good, R.C., Krichevsky, M.I., Beam, R.E., Blacklock, Z., Chaparas, S.D., Dawson, D., Froman, S., Gross, W., Hawkins, J., Jenkins, P.A., Juhlin, I., Käppler, W., Kleeberg, H.H., Krasnow, I., Lefford, M.J., Mankiewicz, E., McDurmont, C., Meissner, G., Nel, E.E., Pattyn, S.R., Portaels, F., Richards, P.A., Rüsch, S., Schröder, K.H., Szabo, I., Tsukamura, M. & Vergmann, B. (1981). First report of the cooperative, open-ended study of slowly growing mycobacteria by the International Working Group on Mycobacterial Taxonomy. *International Journal of Systematic Bacteriology* 31, 1-20.

Wayne, L.G., Good, R.C., Krichevsky, M.I., Beam, R.E., Blacklock, Z., David, H.L., Dawson, D., Gross, W., Hawkins, J., Jenkins, P.A., Juhlin, I., Käppler, W., Kleeberg, H.H., Krasnow, I., Lefford, M.J., Mankiewicz, E., McDurmont, C., Nel, E.E., Portaels, F., Richards, P.A., Rüsch, S., Schröder, K.H., Silcox, V.A., Szabo, I., Tsukamura, M., Van Den Breen, L. & Vergmann, B. (1983). Second report of the cooperative, open ended study of slowly growing mycobacteria by the International Working Group on Mycobacterial Taxonomy. *International Journal of Systematic Bacteriology* 33, 265-274.

Wayne, L.G., Brenner, D.J., Colwell, R.R., Grimont, P.A.D., Kandler, O., Krichevsky, M.I., Moore, L.H., Moore, W.E.C., Murray, R.G.E., Stackebrandt, E., Starr, M.P. & Trüper, H.G. (1987). Report of the ad hoc committee on reconciliation of approaches to bacterial systematics. *International Journal of Systematic Bacteriology* 37, 463-464.

Wayne, L.G., Good, R.C., Krichevsky, M.I., Blacklock, Z., David, H.L., Dawson, D., Gross, W., Hawkins, J., Jenkins, P.A., Juhlin, I., Käppler, W., Kleeberg, H.H., Levy-Frebault, V., McDurmont, C., Nel, E.E., Portaels, F., Rüsch-Gerdes, S., Schröder, K.H., Silcox, V.A., Szabo, I., Tsukamura, M., Van Den Breen, L., Vergmann, B. & Yakrus, M.A. (1989). Third report of the cooperative, open-ended study of slowly growing mycobacteria by the International Working Group on Mycobacterial Taxonomy. *International Journal of Systematic Bacteriology* 39, 267-278.

Wayne, L.G., Good, R.C., Krichevsky, M.I., Blacklock, Z., David, H.L., Dawson, D., Gross, W., Hawkins, J., Levy-Frebault, V.V., McManus, C., Portaels, F., Rüsch-Gerdes, S., Schröder, K.H., Silcox, V.A., Tsukamura, M., Van Den Breen, L. & Yakrus, M.A. (1991). Fourth report of the cooperative open-ended study of slowly growing mycobacteria of the International Working Group on Mycobacterial Taxonomy. *International Journal of Systematic Bacteriology* 41, 463-472.

Wayne, L.G., Good, R.C., Tsang, A., Butler, R., Dawson, D., Groothuis, D., Gross, W., Hawkins, J., Kilburn, J., Kubin, M., Schröder, K.H., Silcox, V.A., Smith, C., Thorel, M.-F., Woodley, C. & Yakrus, M.A. (1993). Serovar determination and molecular taxonomic correlation in *Mycobacterium avium*, *Mycobacterium intracellulare*, and *Mycobacterium scrofulaceum*: a cooperative study of the International Working Group on Mycobacterial Taxonomy. *International Journal of Systematic Bacteriology* 43, 482-489.

Wayne, L.G., Good, R.C., Böttger, E.C., Butler, R., Dorsch, M., Ezaki, T., Gross, W., Jonas, V., Kilburn, J., Kirschner, P., Krichevsky, M.I., Ridell, M., Shinnick, T.M., Springer, B., Stackebrandt, E., Tarnok, I., Tarnok, Z., Tasaka, H., Vincent, V., Warren, N.G., Knott, C.A. & Johnson, R. (1996). Semantide- and chemotaxonomy-based analyses of some problematic phenotypic clusters of slowly growing mycobacteria, a cooperative study of the International Working Group on Mycobacterial Taxonomy. *International Journal of Systematic Bacteriology* 46, 280-297.

Weiszfeiler, J., Jokay, I., Carczag, E., Almassy, K. & Somos, P. (1968). Taxonomic studies of mycobacteria on the basis of their antigenic structure. *Acta Microbiologica Academiae Scientiarum Hungaricae* 15, 69-76.

Whitehead, J.E.M., Wildy, P. & Engbaek, H.C. (1953). Arylsulfatase activity of mycobacteria. *Journal of Pathology and Bacteriology* 65, 451-460.

Will, D.W., Froman, S., Krasnow, I. & Bogen, E. (1957). Cultural characteristics and drug and bacteriophage resistance of avian tubercle bacilli. *American Review of Tuberculosis and Pulmonary Disease* 76, 435-450.

Wolinsky, E. & Schaefer, W.B. (1973). Proposed numbering scheme for mycobacterial serotypes. *International Journal of Systematic Bacteriology* 23, 182-183.

Zakowski, P., Fligiel, S., Berlin, O.G.W. & Johnson, B.L. (1982). Disseminated *Mycobacterium avium-intracellulare* (sic) infection in homosexual men dying of acquired immunodeficiency. *Journal of the American Medical Association* 248, 2980-2982.

Zuckerkandl, E. & Pauling, L. (1965). Molecules as documents of evolutionary history. *Journal of Theoretical Biology* 8, 357-366.

VI   REGULATORY ASPECTS

15   MICROBIAL RESOURCE CENTRES AND EX-SITU CONSERVATION

VANDERLEI P. CANHOS [1]   and   GILSON P. MANFIO [2]

[1] Faculdade de Engenharia de Alimentos (FEA) - Universidade Estadual de Campinas; Rua Professor Zeferino Vaz, s/n; Cidade Universitária, Barão Geraldo; CEP 13081-970; Campinas (SP), Brazil
[2] Coleção de Culturas Tropical (CCT) - Fundação André Tosello; Rua Latino Coelho, 1301; Parque Taquaral; CEP 13087-010; Campinas (SP), Brazil

1   Extent and Knowledge of Microbial Diversity

Biodiversity encompasses 'genetic diversity', the diversity of genomes within and between populations of organisms, 'species diversity', the number of species in a site or habitat (commonly used as a synonym of 'species richness) and 'ecosystem diversity', the quantitative assessment of diversity at the ecosystem, habitat or community level, taking into account species richness, abundance, size classes, trophic, functional and taxonomic groups (Norse et al., 1986; World Conservation Monitoring Centre, 1992; Harper & Hawksworth, 1994).

The ecological importance of biodiversity, and especially of its microbial component (archaea, bacteria, fungi, microalgae and protozoa), is poorly understood. Microbial diversity has been linked to the sustainability of life support systems on Earth (Lovelock, 1988; Stolz et al., 1989; Hawksworth et al., 1991a; Trüper, 1992; Freckman et al., 1997; Staley et al., 1997). Indeed, the apparent redundancy in the diversity of species, genomes and physiological groups of microorganisms may be a resilience mechanism that provides stability to the primary ecological processes involving the microbiota, notably oxygenic photosynthesis, turnover of organic matter and biogeochemical cycles. The role of microorganisms in ecosystem functioning of soils and sediments is poorly understood but extremely important in nutrient cycling and soil structure (Stolz et al., 1989; Hawksworth, 1991b; Trüper, 1992; Brussaard et al., 1997; Freckman et al., 1997).

Despite their major importance, the number of microbial taxa known and described (species diversity) represents only a very small fraction of the global estimated microbial diversity (Table 1). Studies based on the direct analysis of bacterial diversity in environmental samples by using molecular methods (mainly 16S rRNA sequence analysis)

F. G. Priest and M. Goodfellow (eds.), Applied Microbial Systematics, 421-446 have revealed a new scenario of novel and environmentally important diversity yet to be studied (Pace, 1997; Hugenholtz *et al.*, 1998a,b). Current data suggest the existence of at least 36 major lines of descent (Divisions) in the Domain Bacteria alone (Hugenholtz *et al.*, 1998a). This number is at least three times greater than the number known a decade ago, based mainly on the analysis of cultivated bacteria (Woese, 1987).

TABLE 1.   Number of known and estimated microbial species in relation to material held in service culture collections[a]

Group[a]	Number of species		Material held in culture collections		
	Known	Estimated	Total number	Known species %	Estimated species %
Algae	40 000	60 000	1 600	4.0	2.6
Fungi	69 000	1 500 000	11 500	16.7	0.8
Prokaryotes[b]	4 204	40 000 to 3 500 000	2 300	54.7	0.07 to 5.8
Viruses	5 000	130 000	2 200	44.0	1.7

[a] Group names are used in the colloquial not in the formal taxonomic sense. Data compiled from Stork (1988), Wilson (1988), Hawksworth (1991b), Nisbet and Fox (1991), Bull *et al.* (1992) and World Conservation Monitoring Centre (1992).
[b] Including *Archaea* and *Bacteria*. Upper limit for predictions include uncultivated bacteria.

Results from these and other studies indicate that some groups of the Domain *Bacteria* have cosmopolitan distribution whereas others appear to be restricted to particular habitats (Schlegel & Jannasch, 1992; Hugenholtz *et al.*, 1998a). Some of the phylogenetic groups with cosmopolitan distribution are well known from traditional isolation and cultivation studies, such as the representatives of the Cytophagales, low and high G+C Gram-positive bacteria and the *Proteobacteria*. In contrast, molecular analyses of environmental DNA have revealed substantial phylogenetic diversity with little or no representation among organisms previously studied. Because of their abundance and wide distribution some of the organisms represented by 16S rRNA sequences are likely to contribute significantly to the global biogeochemical cycles. Relatively abundant groups of bacteria are still poorly studied and many have not yet been detected by culture methods. Examples of these are the cosmopolitan organisms from the Division *Acidobacterium* and *Verrucomicrobia* (Hugenholtz *et al.*, 1998a). Similar comprehensive data are still not available for many microbial taxa, such as the organisms allocated in the Domains *Archaea* (archaeobacteria) and *Eucarya* (filamentous fungi, yeasts, microalgae and protozoa).

Factors which have contributed to the lack of knowledge on microbial diversity include the limitations of traditional methods for isolation and cultivation of a variety of microorganisms in the laboratory (Palleroni, 1996). It is reasonable to argue that the description of microbial communities cannot be based solely on the use of techniques that involve isolation and cultivation (Pace *et al.*, 1985). Data from comparative studies indicate that only a small fraction of microorganisms in nature (between <0.1 to 1%, depending on the habitat and nature of sample) can be cultivated by using conventional microbiological techniques (Amann *et al.*, 1995).

Lack of expertise and financial resources have limited the use of modern molecular techniques in routine activities of culture collections. A large portion of the known deposited bacterial type cultures remains to be studied by 16S rDNA sequencing (<50% sequenced to date; Stackebrandt, 1997; Garrity *et al.,* 1998).

Predictions of the loss of microbial diversity are unreliable given the meager knowledge base available at this time. Destruction, conversion and degradation of ecosystems represent serious threats to biodiversity. On a daily basis, common environmental mismanagement, including chemical pollution, over-exploitation of natural renewable and non-renewable resources, large-scale physical interventions in the environment related to land- and water-use changes, many of these derived from human population growth, contribute to loss of diversity (Bull *et al.,* 1992; Harvey & Pimentel, 1996).

There are several well documented reports about the loss of animal and plant species (Wilson & Peter, 1988; McNeely *et al.,* 1990; Wilson, 1992; World Conservation Monitoring Centre, 1992) though estimates of species loss are also speculative, suggesting that from one quarter to one half of the earth's species will become extinct in the next 30 years unless current trends are reversed (Myers, 1979; Ehrlich & Ehrlich, 1981; World Conservation Monitoring Centre, 1992). Reported cases of microbial loss are very rare and evidence for them is often indirect. For example, Rifai (1989) attributed the loss of *Penicilliopsis clavariaeformis* from the Bogor Botanical Gardens to the demise of the host tree species. Large-scale decline of European fungal species has also been reported (Jaenike, 1991). It is possible that some groups of microorganisms may be adaptable to environmental changes but endemic or host-specific associations may not resist habitat perturbation (Bull *et al.,* 1992; Wilson, 1992; Lovejoy, 1994; Freckman *et al.,* 1997).

Loss of biodiversity, with the consequent loss of valuable genetic resources, leads to the depletion of the gene pool on a global scale. The long-term effects of this process may have a profound impact on the mechanisms of resilience and sustainability of life on the planet. The conservation issue is extremely complex, involving parameters from the biological, political, and particularly the socio-economical aspects (Marggraf & Birner, 1998). Nevertheless, comprehensive and workable guidelines are available for the selection of sites to be conserved (National Science Foundation USA, 1991; Bull *et al.,* 1992).

Despite their enormous economic and ecological importance, the relevance of *ex-situ* preservation of microorganisms in culture collections has not yet been perceived on a global scale and the majority of the existing service collections are not properly supported on a long-term basis (Bull *et al.,* 1992; Hawksworth & Colwell, 1992; Zedan, 1993; Colwell, 1997; Hunter-Cevera, 1996, 1998).

2   The Debate on *Ex-Situ* Conservation

The problems experienced in compiling inventories of microbial species surpass in complexity those encountered with animals and plants. It is often pointed out that less than 10% of bacteria in soil, as determined by direct counts with microscopy techniques, and less than 0.1% of bacteria in marine environments can be cultured (Bull *et al.,* 1992; O'Donnell *et al.,* 1994). A few estimates of the extent of microbial diversity have been published (Hawksworth *et al.,* 1991b; Harper & Hawksworth, 1994), calculated by using several different methods and experimental models. These figures are the subject of controversy and debate, given the assumptions and limitations of the models used to estimate them. Although not precise, they serve to illustrate the extent of microbial diversity still to be unravelled.

Ex-situ conservation of microorganisms, i.e., conservation of components of biological diversity outside their natural habitats, by maintaining them in *ex-situ* culture collections, is currently the best way to guarantee long-term access and to facilitate distribution. Traditionally, only cultured microorganisms have been maintained in culture collections. Recent studies have uncovered a wealth of as yet uncultured novel microbial taxa, the ecological importance of which remains to be studied (Pace, 1997; Hugenholtz *et al.*, 1998a,b). The lack of knowledge on these poorly studied taxa reinforces the need for *ex-situ* as well as *in-situ* preservation (Wilson, 1992).

The conservation of the full range of microbial diversity may not be feasible. As data from environmental studies using molecular approaches accumulate, it becomes clear that only a very small fraction of microbial genetic diversity is, or will ever be, preserved in culture collections. The *ex-situ* conservation of microbial genetic resources in culture collections requires scientific expertise and incurs high maintenance costs. Data on the distribution of microbial resource centers show that both the major and majority of culture collections (considering holdings and scientific expertise) are in developed countries in the northern hemisphere (Stackebrandt, 1997). As stated by Glowka *et al.* (1994) "*... In general, the countries richest in species are the ones where scientific knowledge on individual species is least*". The *ex-situ* conservation of microbial genetic resources in the countries of origin, as advocated by the Convention on Biological Diversity, is currently compromised by the lack of expertise and funds for development in many countries, a situation that needs to be solved in the long term if countries are to benefit from the exploitation of their biological resources.

Several technical and political developments must be tackled by culture collections to improve on *ex-situ* preservation of microbial diversity, some of which came out as conclusions from the 1995 Workshop "Priorities for Microbial Biodiversity Research: Summary and Recommendations" (www.cme.msu.edu/homepage.html/PUBLICA-TIONS/BIODIVERSITY/biodiversity.report.html). There include:

- novel approaches for the preservation of mixed communities, such as microbial consortia and natural communities from environmental samples;
- improving culture preservation strategies, such as miniaturization and optimized regimes for difficult-to-grow microbial groups (fastidious microorganisms, extremophiles, strict anaerobes etc.);
- support to maintain a worldwide coordinated network of microbial resource centers;
- concentrated efforts for the preservation of organisms from habitats with rare or threatened species;
- development of integrated databases for microbial diversity and bioinformatics; and,
- expand training of new scientists in microbial systematics and physiology.

Taxonomic training and development of modern approaches in microbial systematics were considered at the Workshop "Priorities in Systematic Research and Training" held in 1995 at the Linnean Society of London (www.nhm.ac.uk/hosted_sites/uksf/priorit.htm) and in the recommendations of the 1996 Montreal SBSTTA Meeting report "Practical

Approaches for Capacity Building for Taxonomy", which outlined a framework for a Global Taxonomy Initiative (www.biodiv.org/sbstta2/sb205.html). Several sites on the Internet, many of which are organized by microbial culture collections, have online information on systematics, taxonomy and nomenclature of microorganisms (Table 2).

TABLE 2.   List of Internet sites relevant to nomenclature, taxonomy and systematics

Taxonomy and Systematics

BIOSIS	Taxonomy and nomenclature information from BIOSIS, UK.	www.york.biosis.org/zrdocs/ tax_nom.htm
Dictionary of the Fungi	Systematic Arrangement of Genera from the 8th Edition of the Dictionary of the Fungi. Searchable index for generic names of fungi and systematic position in the accepted family/order/class hierarchy, along with an indication of synonymy. Organized by CAB International/IMI, UK.	www.cabi.org/institut/imi/taxon.htm
ICTVdB	The Universal Virus Database, Australia. Searchable database compiled from the ICTV (International Committee on Taxonomy of Viruses) list of approved virus names.	www.ncbi.nlm.nih.gov/ICTVdb/Ictv/ fr-index.htm (also at life.anu.edu.au:80/viruses/Ictv/ fr-index.htm)
Index Nominum Genericorum (ING)	Searchable index of the compilation of generic names published for all organisms covered by the International Code of Botanical Nomenclature, including fungi. Organized by the International Association for Plant Taxonomy (IAPT) and the Smithsonian Institution (USA).	www.nmnh.si.edu/ing/
Gomales (INVAM)	Webpage on the classification of Gomales (mycorrhizal) from the International Culture Collection of Arbuscular and Vesicular-Arbuscular Mycorrhizal Fungi (IVAM).	invam.caf.wvu.edu/classification.htm
NCBI Taxonomy	Taxonomic database maintained by NCBI/ GenBank on the classification of archaea, bacteria, eukaryotes and viruses.	www.ncbi.nlm.nih.gov/Taxonomy tax.html
Protist Information Server	Information database on taxonomy and classification of protists (in Japanese).	202.250.194.136/WWW/ Protist_menuE.html
Quinone Database	Database of quinone systems in Gram-negative bacteria and ubiquinones systems in filamentous fungi.	wdcm.nig.ac.jp/wdcm/Quinone.html

TABLE 2.   continued

Taxonomy and Systematics

RDP   The Ribosomal Database Project, USA.   www.cme.msu.edu/RDP/
                    Database of 16S and 23S rDNA sequences
                    of archaea, bacteria, filamentous fungi and
                    yeasts with their respective phylogenetic
                    classification schemes and analytical tools.

Species2000   Database project aimed at listing all known   www.sp2000.org/
                    species of plants, animals, fungi and
                    microbes on Earth as the baseline dataset
                    for studies of global biodiversity.

The   Interactive database with the phylogenetic   www.ucmp.berkeley.edu/alllife/
Phylogeny   classification of organisms in the three   threedomains.html
of Life   Domains of life.

Tree of life   Distributed Internet project containing infor-   phylogeny.arizona.edu/tree/
                    mation about phylogeny and biodiversity.   phylogeny.html

Nomenclature

Bacterial   Searchable index of valid bacterial names   www.dsmz.de/bactnom/bactname.htm
Nomenclature   organized by the DSMZ (Germany)
up-to-date   from data published in the IJSB.

Chron Bact   Chronicle of Bacterial Names-RIKEN,   www.jcm.riken.go.jp/chronbact/
                    DSMZ. Searchable index of valid bacterial
                    names, including search options by type
                    strain number and author. Developed by
                    The Institute of Physical and Chemical
                    Research (RIKEN), Japan (updated until
                    IJSB 1995).

Gomales   Webpage on the names and authorities
(INVAM)   of fungi classified in the Glomales (mycorrhizal) invam.caf.wvu.edu/myc_info/taxonomy/
                    from the International Culture Collection   authors/authors.htm
                    of Arbuscular and Vesicular-Arbuscular
                    Mycorrhizal Fungi (INVAM).

Index Virum   A list of ICTV (International Committee on   www.ncbi.nlm.nih.gov/ICTVdb/Ictv/fr-
                    Taxonomy of Viruses) approved virus names,   index.htm (also at life.anu.edu.au:80/
                    including virus identification numbers used in   viruses/Ictv/index.html)
                    the ICTVdB.

Latin Names   Practical guidelines for naming organisms in   www.bdt.org.br/bdt/about/truper/
                    Latin, organized by Prof. H.G. Trüper (Germany).

TABLE 2.   continued

Nomenclature

List of Bacterial Names with Standing in Nomenclature	List of valid bacterial names compiled by J.P. Euzéby (ENVT, France) from data published in the *International Journal of Systematic Bacteriology*. (IJSB).	www-sv.cict.fr/bacterio/
NPPB	International Society for Plant Pathology (ISPP, UK) list of "Names of Plant Pathogenic Bacteria, 1864-1995".	www.bspp.org.uk/ispp/nppb.html
TRITON	Taxonomy Resource & Index To Organism Names (BIOSIS, UK). Includes data from BIOSIS Register of Bacterial Nomenclature (BRBN) and other databases for fungi.	www.york.biosis.org/triton/nameind.htm

3   Microbial Resource Centres from a Historical Perspective

Microbial culture collections throughout the world are critical infrastructural resources responsible for the acquisition, conservation, preservation, characterization, and documentation of the world's microbiota. As scientific resources, microbial culture collections serve diverse functions, such as:

- reference centres for the distribution of certified material for scientific research and applications in industry, agriculture and health;
- centers of expertise in polyphasic taxonomy and systematics;
- providers of genetic materials for development of predictive phylogenetic tools;
- foundation for collaborative efforts in microbial research;
- training centers for systematics, preservation techniques and bioinformatics;
- education and public awareness in microbial diversity, ecological roles and applications of microorganisms.

The first collection of cultures available to the local microbiology community was established in 1890 by Frantisek Král, in Prague, Czechoslovakia. The centenary of this collection was celebrated at a meeting in Osaka, Japan (Sly *et al.*, 1990). In the decades that followed several collections were established around the world.

In the Netherlands, the Centraalbureau voor Schimmelcultures (CBS) was founded in 1903 following a proposal of the "Association Internationale des Botanistes" to establish a central collection of fungi. In 1906, the first CBS list of 80 available cultures was published. The yeast collection from the CBS was moved to a separate department in the Laboratory for Microbiology of the Delft Technical University in 1922. Originally, the CBS was a foundation supported by private means but it gained financial support from the Dutch Government after becoming an Institute of the Royal Netherlands Academy of Arts and Sciences in 1968; it was then assigned the joint role of management of the collection and scientific research.

In the USA, the North American Agricultural Research Station (ARS) collection can be traced back to 1904 when it consisted of a collection of strains of fungi for cheese production. The collection was formally established when the Northern Regional Research Laboratory (now the National Center For Agricultural Utilization Research) opened in 1940 in Peoria, Illinois (USA). Strains deposited in the NRRL collection during the 1940s included many isolates relevant for industrial production, such as citric acid-producing aspergilli, antibiotic-producing actinomycetes and various other bacterial strains.

The American Type Culture Collection (ATCC, USA) was established in 1925 as a non-profit, non-governmental organization. It later became the biggest of all culture collections in terms of earnings, total budget and scope of holdings and activities (Cypess, 1996).

In Japan, the Collection of the Institute for Fermentation Osaka (IFO) was established in 1944 with the aims of studying and conserving strains useful in the production of aviation fuel, food and drugs.

The United Kingdom was one of the first countries to recognize the need for an integrated policy for *ex-situ* conservation of microorganisms. At a specialist conference held in 1947, a decision was made to establish a government-supported decentralized system of complementary culture collections. Under this framework, several 'National Collections' were established (Office of Science and Technology, 1996). The International Mycological Institute (IMI), founded in 1920, was given the responsibility to house the National Collection of Fungus Cultures.

In 1962, the Canadian Committee on Culture Collections convened a conference on culture collections which was attended by 266 scientists from 28 countries. After this meeting it was clear that the scattered collections would greatly benefit from the development of an organization that could promote their interests and encourage collaboration and technical developments. In 1968, at the International Conference on Culture Collections held in Tokyo, it was resolved that an international federation should be established.

In 1969, the DSMZ was founded as the National Culture Collection in Germany. It was originally set up as a system of federated collections but was later consolidated in Braunschweig to become a department within the German Biotechnology Institute (GBF).

The World Federation for Culture Collections (WFCC) was established in 1970 at the X International Congress for Microbiology, Mexico City, to support the interest of Culture Collections and their users. Established as a committee of the International Union of Microbiological Societies (IUMS), the WFCC became a highly active and valuable organization affecting many spheres of microbiology throughout the decades (Sly, 1996).

In 1972, the WFCC World Data Center for Microorganisms (WDCM) was established in the University of Queensland, Australia, under the sponsorship of the United Nations Environmental Programme (UNEP) and the United Nations Educational, Scientific and Cultural Organization (UNESCO). The WFCC data center was instrumental in the development of an international directory of culture collections with associated information.

By the 1980s, with the development of biotechnology in industrialized countries, policy makers became aware of the importance of microbial resources centers. Those working in culture collections needed to become skilled in marketing, promoting and fund raising, as well as remaining experts in microbial taxonomy, conservation, and physiological and biochemical testing for research and educational purposes (Hawksworth, 1996).

The Japan Collection of Microorganisms (JCM) was established in 1980 at the Institute of Physical and Chemical Research (RIKEN), a semi-governmental research institute supported by the Science and Technology Agency. Shortly after (1981), the European Culture Collection Organization (ECCO) was established to support improvement in the scientific and technical standards of european collections. Accordingly, culture collection staff computerized information, attempted to coordinate this for the benefit of users, produced attractive and useful catalogues, set up trade stands at conferences, developed databases and, later, with developments in electronic communication, began to put all this information online for the benefit of international science. They succeeded in entering the market economy, but at some cost to the basic scientific skills that they were established to provide. It was difficult to add
new skills without additional financial resources.

In 1986, the WDCM was relocated to Japan and housed at the Institute of Physical and Chemical Research (RIKEN) as a centralized database on culture collections worldwide. In 1997, the WDCM was transferred to the National Institute of Genetics in Japan.

Developments in information technology in the 1980s were instrumental in the establishment of national, regional and international databases and communication networks. The goal was to provide online mechanisms for locating microorganisms and cultured cells with specific properties. The efforts included the international Microbial Strain Data Network (MSDN), the Microbial Information Network Europe (MINE), the British Microbial Culture Information Service (MiCIS), the Microbial Germplasm Database (MGD), in North America, and the Tropical Database (BDT) in Brazil. These, and many other databases are listed in Table 3.

In 1988, the DSMZ became an independent body, financially supported by the Federal Ministry of Research and Technology and the State Ministries of Germany. In the same year, the Tropical Culture Collection (CCT) was established as a private non-profit service collection with support from the Brazilian Federal Government.

The Belgian Coordinated Collections of Microorganisms (BCCM) was established as a consortium of four complementary research-based culture collections. The decentralized activities of the BCCM collections were organized under a central co-ordination team at the Belgian Office for Scientific, Technical and Cultural Affairs (OSTC), the first series of complete catalogues were published in 1989.

In the 1990s, with the implementation of the Convention on Biological Diversity (CBD), collections were required to develop new strategies to fulfill the requirements of the Convention. There are now obligations to track distribution of strains, develop material transfer agreements (MTA) and prior informed consent (PIC) forms for the purpose of benefit sharing (Glowka, 1996).

With the turn of the century, genomic information technology and huge advances in microbial phylogeny and molecular ecology are changing the perceptions and needs of end users. Collections are becoming 'microbial genetic resources centers' and the provision/demand for new specialized services will soon follow. Culture collections around the world are irreplaceable as repositories of the products of over one century of microbial research. They are a fundamental asset, the capital for the development of basic and applied microbiology. Today, hundreds of service and research culture collections have been established around the world for the support of science and technology. However, the underlying principle of centers of expertise providing reliable, authenticated cultures and associated information remains unchanged.

TABLE 3.   List of Internet sites of databases, federations and information networks

Information Networks and Federations

ANMD	Asian Network on Microbial Research. Information network on microbial collections from 8 countries in Asia (China, Indonesia, Japan, Korea, Malaysia, Philippines, Singapore and Thailand).	www.jcm.riken.go.jp/anmrhp.html
CBD	Internet site of the Convention on Biological Diversity.	www.biodiv.org/ (also at www.unep. ch/bio/conv-e.html)
BCCM	Belgian Co-Ordinated Collections of Micro-organisms, Belgium. Consortium of four complementary research-based culture collections: LMG Bacteria Collection, MUCL (Agro)Industrial Fungi and Yeasts Collection, IHEM Biomedical Fungi and Yeast Collection and LMBP Plasmid and cDNA Collection.	www.belspo.be/bccm/
ECCO	European Culture Collection Organization.	www.uia.org/uiademo/org/d4656.htm
EFB	European Federation of Biotechnology. Non-profit-making European scientific and technical societies active in promoting biotechnology. Links 81 member societies from 25 European and 5 non-European countries.	www.dechema.de/englisch/europa/ biotec/pages/biotec1.htm
FEMS	Federation of European Microbiological Societies. Links some 38 microbiological societies in Austria, Bulgaria, Croatia, Czech Republic, Denmark, Estonia, Finland, France, Germany, Greece, Hungary, Iceland, Israel, Italy, Latvia, Lithuania, Macedonia, The Netherlands, Norway, Poland, Portugal, Russia, Slovak Republic, Slovenia, Spain, Sweden, Switzerland, Turkey, the United Kingdom and Yugoslavia.	www.elsevier.com:80/inca/homepage/ sah/fems/menu.htm
GRIN	The Germplasm Resources Information Network, USA. Database with information on the national animal (NAGP), invertebrate	www.ars-grin.gov/index.html

TABLE 3.   continued

Information Networks and Federations

GRIN	(NIGP), microbial (NMGP) and plant (NPGS) germplasm programs from USDA/ARS research programs.	
IUMS	International Union of Microbiological Societies.	www.iums.rdg.ac.uk/
JFCC	Japanese Federation of Culture Collections database. Holdings of bacteria, filamentous fungi and yeasts, microalgae, protozoa and viruses (bacteriophage, invertebrate, plant and vertebrate viruses from several affiliated collections.	wdcm.nig.ac.jp/wdcm/JFCC.html
MICRO-NET	Microbial Information Network of China. List of Chinese Research Institutes and Microbial Culture Collection. Link to centralized catalogue of bacteria, actinomycetes, filamentous fungi, yeasts and viruses of the China Committee for Culture Collection of Microorganisms (CCCCM).	www.im.ac.cn/
MIRCEN	UNESCO Microbial Resource Centers. Global network in environmental, applied micro biological and biotechnological research (biological nitrogen fixation, culture collection.	www.unesco.org/general/eng/
MSDN	Microbial Strain Data Network, UK. Links and database with holdings of several affiliated culture collections.	panizzi.shef.ac.uk/msdn/
UKNCC	United Kingdom National Culture Collection. Co-ordinates the activities, marketing and research of the UK national service collections, with links to strain databases and affiliated collections.	www.ukncc.co.uk/
USFCC	United States Federation of Culture Collections, USA.	nagual.ese.ogi.edu/usfcc/default.html
WDCM	World Data Centre for Microorganisms, Japan. Comprehensive worldwide directory of culture collections and holdings, and links to databases on microorganisms, biodiversity, molecular biology and genome projects.	wdcm.nig.ac.jp/

TABLE 3.   continued

Information Networks and Federations

WFCC	World Federation of Culture Collections, Network of specialist microbiologists and culture collections.	wdcm.nig.ac.jp/wfcc/index.html

Databases and Virtual Libraries

ALGAE	WDCM world catalogue of algal collections.	wdcm.nig.ac.jp/cgi-bin/ALGAE.pl
Bioguide	Web page with information on microbial Culture Collections in the UK.	dtiinfo1.dti.gov.uk/bioguide/culture.htm
CBS Databases	Databases with information on strains from the CBS Collections (Baarn and Delft), collaborating collections in the NCC (Netherlands Culture Collections of Microorganisms) and the IGC Collection (Oeiras, Portugal); taxonomic and nomenclature data on *Aphyllophorales* and *Fusarium* and genetic strains from the Phabagen Collection (Utrecht).	www.cbs.knaw.nl/database.html
MGD	Microbial Germplasm Database, USA. Shared database of information on collections of microbial germplasm maintained primarily for research purposes in laboratories, universities, private industry, USDA research field stations, and in NSF supported collections. Holds data on several small non-catalogued collections (searchable online database).	mgd.nacse.org/cgi-bin/mgd
SBML-Db	Systematic Botany and Mycology Laboratory Fungal Databases (ARS-USDA, Beltsville), USA. Holdings of the U.S. National Fungus Collections.	nt.ars-grin.gov/fungaldatabases/ DatabaseFrame3.cfm
Virus Databases Online	Site developed by the Research School of Biological Sciences (Australia). Holds databases on viral taxonomy, descriptions and links to other viral databases.	life.anu.edu.au/viruses/welcome.html
WWW VL: Mycology	Virtual Library of Mycological Resources on the Internet. Distributed and comprehensive catalogue of internet resources.	www.keil.ukans.edu/~fungi/

4 The World Federation for Culture Collections

Since its establishment, the WFCC has played a catalytic role in encouraging the exchange of ideas and collaboration among scientists, collections and international scientific organizations. Throughout the years a number of committees have been set up to carry out specific tasks in different aspects of culture collection work. An example was the role of the WFCC Patents Committee (wdcm.nig.ac.jp/wfcc/patent_committee.html) in the development of practical regulations and procedures for the operation of the International Depository Authorities established in accordance with the Budapest Treaty (Ilardi, 1996). Training courses, international workshops and eight International Congresses have been organized. Books, videos, information documents and issues of the WFCC Newsletter have been published. The Guidelines for the Establishment and Operation of Culture Collections, published in 1990, is a reference in this area (Hawksworth *et al.*, 1990), and a new revised version will be published in 1999.

The WFCC is very active in promoting the debated of issues relevant to the implementation of the Convention on Biological Diversity (CBD), and has been responsable for the publication of information documents on the role of resource centers on the *ex-situ* conservation of microorganisms (Kirsop & Hawksworth, 1994), access to microbial genetic resources (Kirsop, 1996b), biosafety regulations (Smith, 1996) and the value of microbial genetic resources (Kirsop & Canhos, 1998).

A number of databases and network activities established in the 1980's under WFCC sponsorship evolved to become important biological information resources and biodiversity networks. A key instrument for the development of WFCC activities was the World Data Center for Microorganisms (WDCM). The WDCM holds information on 498 microbial culture collections located in 55 countries, informaticus on these collections are available through the Internet (wdcm.nig.ac.jp/). Many of the Collections registered at the WDCM have developed their own information systems available through the Internet. The Internet addresses of collections with online information are listed Table 4.

TABLE 4. List of Internet sites of culture collections with online information

Acronym	Description	Holdings	URL addresses
ACAM	Australian Collection of Antarctic Microorganisms, Australia (searchable online text catalogue).	Microorganisms isolated from the Antarctic continent as well as from subantarctic islands and the Southern Ocean.	www.antcrc.utas.edu.au/antcrc/micropro/acaminfo.html
ACT	Arbuscular Mycorrhizal Culture Collection Center in Taiwan (searchable online catalogue).	Specialized collection of mycorrhizal fungi.	www.tari.gov.tw/.
ARSEF	ARS Collection of Entomopathogenic Fungi (USDA), USA (not searchable online,	Enthomopathogenic fungi.	www.ppru.cornell.edu/Mycology/ARSEF_Culture_Collection.html

TABLE 4.   continued

Acronym	Description	Holdings	URL addresses
	downloadable catalogues on pdf format).		
ATCC	American Type Culture Collection, USA (searchable online database).	Archaea, bacteria, animal, human and plant cell lines and tissues, filamentous fungi and yeasts, microalgae, protozoa, seeds, viruses, antisera, purified DNA (bacteria, human cell lines, yeasts, cloned viruses), recombinant materials (genome project clones, hosts, libraries, expressed sequence tag cDNA, vectors).	www.atcc.org/
CBS	Centraalbureau voor Schimmelcultures, The Netherlands (searchable online database).	Bacteria, filamentous fungi and yeasts, phages and genetic materials (host strains, plasmids, gene libraries, vectors).	www.cbs.knaw.nl/
CCALA	Czechoslovak Database of Algae and Cyanobacteria, Czech Republic (searchable online database).	Microalgae and cyanobacteria.	panizzi.shef.ac.uk/msdn/ccala/
CCAP	The Culture Collection of Algae and Protozoa, UK (searchable online database).	Microalgae (freshwater and marine) and protozoa	www.ife.ac.uk/ccap/
CCBAS	Culture Collection of Basidiomycetes, Czech Republic (downloadable catalogue on txt format).	Basidiomycetic fungi.	www.biomed.cas.cz/ccbas/ fungi.htm
CCCCM	China Committee of Culture Collection of Microorganisms. Consortium of seven Chinese culture collections (include holdings from general, agricultural, industrial (food, fermentation and antibiotics), medical, veterinary and forestry collections; searchable online database).	Bacteria (and actinomycetes), filamentous fungi, yeasts and viruses.	www.im.ac.cn/database/ catalogs.html (list of affiliated collections at www.im.ac.cn/database/ aboutccccm.html
CCF	Culture Collection of Fungi, Charles University, Czech Republic (searchable online database).	Filamentous fungi and yeasts.	panizzi.shef.ac.uk/msdn/ccf/

TABLE 4. continued

Acronym	Description	Holdings	URL addresses
CCFC	Canadian Collection of Fungal Cultures, Canada.	Filamentous fungi and yeasts	res.agr.ca/brd/ccc/
CC-Kemijski Institute	Culture Collection at the Kemijski Institute, Slovenia (searchable online database).	Filamentous fungi.	panizzi.shef.ac.uk/msdn/slov/
CCM	Czech Collection of Microorganisms, Masaryk University, Czech Republic (searchable online database).	Bacteria and filamentous fungi.	panizzi.shef.ac.uk/msdn/ccm/
CCMP	Provasoli-Guillard National Center for Culture of Marine Phytoplankton, USA (searchable online database).	Marine phytoplankton, benthic algae, marine macrophytes and fresh-water microalgae.	ccmp.bigelow.org/
CCT	Tropical Culture Collection, Brazil (searchable online database).	Bacteria, filamentous fungi and yeasts.	www.cct.org.br/
CCUG	Culture Collection, University of Göteborg, Sweden (searchable online database).	Bacteria (some filamentous fungi and yeasts).	www.ccug.gu.se/
CIP	Collection of Bacterial Strains of Institut Pasteur, France (searchable online database).	Bacteria (mainly pathogenic strains to animals and humans).	www.pasteur.fr/CIP/
CNPAB	Culture Collection of Diazotrophic Bacteria from EMBRAPA Agrobiologia, Brazil (searchable online database).	Nitrogen fixing bacteria.	www.cnps.embrapa.br/cnpab/colecao.html
DAVFP	Pacific Forestry Centre's Forest Pathology Herbarium, Canada (searchable online database).	Forest fungi and disease specimens (herbarium).	www.pfc.cfs.nrcan.gc.ca/biodiversity/herbarium/index.html
DSMZ	Deutsche Sammlung von Mikroorganismen und Zell-kulturen GmbH (German Collection of Microorganisms and Cell Cultures), Germany (searchable online database).	Archaea, bacteria, filamentous fungi and yeasts, animal, human and plant cell lines and tissue cultures, phages, plant viruses and plasmids.	www.dsmz.de/
ECAC	European Collection of Cell Cultures, UK (searchable online database).	Animal and human cell lines (patent deposit of viruses, plant cell lines, bacteria and DNA probes).	www.camr.org.uk/ecacc.htm
GPDATA	Collection of soil-borne fungi of the Plant Pathology Department (Institute of	Filamentous fungi (some yeasts).	www.res.bbsrc.ac.uk/plantpath/cultures/

TABLE 4.   continued

Acronym	Description	Holdings	URL addresses
	Arable Crops Research - Rothamsted), UK (search able online database).		
IAM	IAM Culture Collection at the Institute of Molecular and Cellular Biosciences, University of Tokyo, Japan (not searchable).	Bacteria, filamentous fungi, yeasts and microalgae.	www.iam.u-tokyo.ac.jp/ misyst/ColleBOX/IAM collection.html
IBSBF	Collection of the "Seção de Bacteriologia Fitopatológica do Instituto Biológico" (Campinas), Brazil (search-able online database).	Bacteria (mainly phy-topathogenic).	www.bdt.org.br/bdt/ibsbf/
IBSO	Culture Collection of Luminous Bacteria of the Institute of Biophysics (Siberian Branch, Russian Academy of Sciences), Russia (searchable online database).	Luminescent bacteria.	panizzi.shef.ac.uk/msdn/ibso/
IHEM	Scientific Institute of Public Health - Louis Pasteur, Mycology Section, Biomedical Fungi and Yeasts Collection, Belgium (searchable online database).	Filamentous fungi and yeasts (public and animal health isolates).	www.belspo.be/bccm/bccm col.htm
IEGM	Institute of Ecology and Genetics of Microorganisms, Russia (searchable online database).	Regional specialized col-lection of alkanitrophic microorganisms (mostly actinomycetes).	www.ecology.psu.ru/iegmcol/ index.html
IMI	International Mycological Institute Culture Collection, UK (searchable online database).	Filamentous fungi, yeasts and plant pathogenic bacteria.	www.cabi.org/institut/imi/ grc.htm
INVAM	International Culture Collection of Arbuscular and Vesicular-Arbuscular Mycorrhizal Fungi, USA (searchable online database).	Specialized collection of mycorrhizal fungi.	invam.caf.wvu.edu/
IPPAS	Culture collection of microalgae at the K.A. Timiryazev Institute of Plant Physiology, Russia (searchable online database).	Microalgae (including cyanobacteria).	panizzi.shef.ac.uk/msdn/ ippas.html
ITCC	Indian Type Culture Collection, India (online text catalogue).	Fungi.	molbiol.soton.ac.uk/msdn/ temp/itcc.html

TABLE 4.   continued

Acronym	Description	Holdings	URL addresses
JCM	Japan Collection of Microorganisms, Japan (searchable online database).	Archaea, bacteria, filamentous fungi and yeasts.	www.jcm.riken.go.jp/
KCTC	Korean Collection for Type Cultures, Korea.	Archaea, bacteria, animal and human cell lines, filamentous fungi and yeasts, recombinant genetic materials.	kctc.kribb.re.kr
KMMGU	Russian Bacteria Database with holdings of the collection at the Department of Microbiology, Lomonosov State University, Russia (searchable online database)	Bacteria.	panizzi.shef.ac.uk/msdn/ kmmgu/
LAM (UNCOR)	Culture Collection at the Laboratorio de Microbiologia, Universidad Nacional de Córdoba, Argentina (search-able online database).	Bacteria and filamentous fungi (strains for applied microbiology research).	panizzi.shef.ac.uk/msdn/uncor/
(LE)BIN	Russian Basidiomycetes Collection at the Komarov Botanical Institute, Russia (searchable online database).	Basidiomycetic fungi.	panizzi.shef.ac.uk/msdn/lebin/
LMBP	Laboratorium voor Moleculaire Biologie (Universiteit Gent), Plasmid and cDNA Collection, Belgium (searchable online database).	Plasmids and cloned cDNA libraries (fungal and human).	www.belspo.be/bccm/bccm col.htm
LMG	Laboratorium voor Microbiologie (Universiteit Gent) Bacteria Collection, Belgium (searchable online database).	Bacteria (specialized collection of phytopathogenic and plant associated).	www.belspo.be/bccm/bccm col.htm
MICH	University of Michigan Herbarium Fungus Collection, USA (searchable online database).	Filamentous fungi.	www.herb.lsa.umich.edu/ index.html
MUCL	Mycothèque de l'Université Catholique de Louvain (Agro) Industrial Fungi and Yeasts Collection, Belgium (search-able online database)	Filamentous fungi and yeasts.	www.belspo.be/bccm/bccm col.htm
MSU	Russian Yeasts Database, with holdings of the collection at the Department of Soil Biology, Moscow State	Yeasts.	panizzi.shef.ac.uk/msdn/msu/

TABLE 4.   continued

Acronym	Description	Holdings	URL addresses
	University, Russia (search-able online database).		
NBIMCC	National Bank for Industrial Microorganisms and Cell Cultures, Bulgaria (searchable online database).	Bacteria, filamentous fungi and yeasts, animal and plant cell lines, animal and plant viruses, plasmids.	bioinfo.ernet.in/cgibin/asearch/ msdn/nbimcc/read (also at panizzi.shef.ac.uk/ msdn/nbimcc/nbimcci.html)
NCAIM	National Collection of Agricultural and Industrial Microorganisms at the University of Horticulture and Food Industry, Hungary (searchable online database).	Bacteria, filamentous fungi and yeasts.	ncaim.kee.hu/ (also at panizzi. shef.ac.uk/msdn/ncaim/)
NCIMB	National Collections of Industrial and Marine Bac-teria, UK (searchable online database under construction).	Non-medical bacteria.	www.abdn.ac.uk/~aur014/ ncimb.htm
NCS	National Catalogue of Strains, Brazil. Holdings of 36 brazilian culture collections (searchable online database).	Archaea, bacteria, fila-mentous fungi and yeasts, microalgae, animal and human cell lines, proto-zoa and viruses.	www.bdt.org.br/bdt/index/ biotecnologia/micropage/ micromenu.catalogue/catbr
NCYC	National Collection of Yeast Cultures, UK (search-able online database).	Yeasts.	www.ifrn.bbsrc.ac.uk/ncyc/
NEPCC	North East Pacific Culture Collection, Canada.	Marine microalgae.	www.ocgy.ubc.ca/projects/ nepcc/index.html
NRRL	Agricultural Research Service Culture Collection (NRRL Culture Collection), USA (searchable online database).	Bacteria, filamentous fungi and yeasts. Patent deposits.	nrrl.ncaur.usda.gov/
OSU	Oregon State University Herbarium Mycological Collection, USA (search-able online database).	Filamentous fungi (herbarium types).	mgd.nacse.org/myco_coll.html
PCC	Pasteur Culture Collection of Cyanobacteria, France (searchable online database)	Microalgae (axenic cyano-bacteria).	www.pasteur.fr/Bio/PCC/
PCM	Polish Collection of Micro-organisms, Poland. Holdings from eight associated collections (searchable online database).	Bacteria, filamentous fungi and yeasts.	immuno.pan.wroc.pl/pcm/

TABLE 4.   continued

Acronym	Description	Holdings	URL addresses
PGC	Peterhof Genetic Collection of Yeasts at the Biotechnological Centre, St. Petersburg State University, Russia (searchable online database).	Yeasts (strains for genetics and biotechnology).	panizzi.shef.ac.uk/msdn/peter/
RIAM	Research Institute of Applied Microbiology, Russia (searchable online database).	Filamentous fungi.	panizzi.shef.ac.uk/msdn/riam/
SBML	Systematic Botany and Mycology Laboratory Fungal Culture Collection and Databases (ARS-USDA, Beltsville), USA.	Filamentous fungi.	nt.ars-grin.gov/sbmlweb/cul tures/cultureframe.htm
TIMM	Institute of Medical Mycology Culture Collection at Teikyo University, Japan (searchable online database).	Clinically isolated patho genic filamentous fungi and yeasts.	timm.main.teikyo-u.ac.jp/cul ture.html
TISTR	Thailand Institute of Scientific and Technological Research Culture Collection, Thailand (searchable online database).	Bacteria, filamentous fungi and yeasts.	www.jcm.riken.go.jp/cgi-bin/TISTR.pl
UNSW	The University of New South Wales, School of Microbiology and Immunology Culture Collection, Australia (online text catalogue).	Algae, bacteria, filamentous fungi, yeasts and protozoa.	www.unsw.edu.au/clients/ microbiology/culture.html
UPSC	The Uppsala University Culture Collection (Mycoteket), Sweden (searchable online database).	Filamentous fungi and yeasts.	ups.fyto.uu.se/mykotek/ index.html
USDA/ARS-Rhizobium	USDA ARS National *Rhizobium* Germplasm Resource Center, USA.	Specialized collection of nitrogen-fixing bacteria.	bldg6.arsusda.gov/pberkum/ Public/cc1a.html
UTCC	The University of Toronto Culture Collection of Algae and Cyanobacteria, Canada (online text catalogue).	Cyanobacteria, and microalgae.	www.botany.utoronto.ca/utcc/
UTEX	The Culture Collection of Algae at the University of Texas at Austin, USA (searchable online data-base and species list).	Microalgae.	www.botany.utexas.edu/ infores/utex/
UTMB	University of Texas Medical Branch (Galveston, USA)	Medical filamentous fungi and yeasts.	fungus.utmb.edu/myco.htm

TABLE 4.   continued

Acronym	Description	Holdings	URL addresses
VKM	Herbarium and Fungus Culture and Image Collection. Collection at the Institute of Biochemistry and Physiology of Micro-organisms, Russia (searchable online database).	Bacteria, filamentous fungi, yeasts and geneti-cally modified organisms.	www.vkm.ru/ (also at panizzi. shef.ac.uk/msdn/vkmst/)
UAMH	University of Alberta Microfungus Collection & Herbarium, Canada (search-able online database).	Filamentous fungi (special-ized in ascomycetous and hyphomycetous fungi).	www.devonian.ualberta.ca/ uamh/

5   New Roles for Microbial Resource Centers

In the last decade major forces have changed the environment in which culture collec-tions operate. These include political changes, the effect of regulations on activities related to microorganisms, the emergence of new areas of research, the impact of information technology on the operation of resource centers and the trend towards a decrease in public funding for microbial culture collections.

The political challenges are mainly associated with the implementation of the Convention on Biological Diversity, particularly the issues related to inventories of microbial diver-sity, capacity building in taxonomy, ownership and access to genetic resources, and the regulations on the trans-border movement of genetically modified biological material and invasive species (Glowka, 1996; Stackebrandt, 1996, 1997).

The aims of the Convention on Biological Diversity are the conservation of biological diversity, the sustainable use of its components and the fair and equitable sharing of benefits arising from the use of genetic resources. The CBD, now ratified by more than 160 countries, is a framework Convention in the sense that it has outlined broad principles and objectives. There is a need for further detailing to establish specific goals and work programs. Therefore, the CBD is in constant progress, coordinated by the negotiations at the Conference of the Parties (COP) which is supported by the Secretariat, the Subsidiary Body on Scientific, Technical and Technological Advise (SBSTTA), the Financial Mechanism, the Clearing-House Mechanism, Protocol-negotiation processes, inter-session activities and inputs from Parties, observer countries, international organizations and non-governmental organizations. To comply with the obligations of the Convention, each Party must organize and maintain its own data and develop strategies to meet the Convention goals.

Article 15 of the CBD states that the access to genetic resources is determined by national laws, and that parties should facilitate access, provided that it is done on Mutually Agreed Terms and subject the Prior Informed Consent. Therefore, the new legal framework governing the access to genetic resources is country driven and provides an opportunity for the parties to the Convention to assert sovereignty over their genetic resources. Scientific collaboration, operation and management of biological collections, bioprospecting efforts and commercial utilization of genetic resources, including microorganisms, must adapt to this new order.

With the aim of providing guidance to the member culture collections on the issue of access and benefit sharing, the WFCC has published the 'Information Document on Access to *ex-situ* Microbial Genetic Resource within the Framework of the Convention on Biological Diversity' (available online at wdcm.nig.ac.jp/wfcc/InfoDoc.html).

Specific consideration has been given to microorganisms under the theme agricultural biological diversity under the priorities of the CBD. On agricultural biological diversity, the third meeting of the COP, held in Buenos Aires (1996), adopted decision III/11, paragraph 16, which "... *encourages Parties to develop national strategies, programs and plans which should focus* inter *alia on* ... *micro-organisms of interest for agriculture.*" Paragraph 11 of the same decision encourages interested Parties and international organizations to conduct case studies on the two initial issues identified by SBSTTA in recommendation 11/7, namely pollinators and soil micro-organisms in agriculture. The latter was expanded to cover soil biota by decision IV/6 adopted at the fourth meeting of the COP in Bratislava (1998). This is a very good opportunity for the development of an international program involving microbial resource centers with expertise in the study, conservation and utilization of microorganisms in sustainable agriculture.

Changes in biosafety regulations for the transport of biological material are having a tremendous impact on the operation of microbial resource centers (see Chapter 16; Kirsop, 1996a). As consequence, there is a trend of increasing operational costs associated with the distribution of biological material, a matter of concern for the development of taxonomy, systematics and microbiological research as a whole. The IATA (International Air Transport Association) Dangerous Goods Regulations (1998) require that packaging used for the transport of hazard groups 2, 3 or 4 must meet defined standards (IATA packing instruction 602, class 6.2) and shippers of microorganisms assigned to these hazard groups must also be trained by instructors certified and approved by IATA. The 'WFCC Committee on Postal, Quarantine and Safety Regulations' constantly monitors changes in regulations concerning these issues and how they affect culture collection management and procedures worldwide. Summarised guidelines and reports, including information on changes in legislation in quarantine, transport, packaging and postal regulations, and safety in the handling of biological agents are given in Chapter 16 and can be found at the WFCC site (wdcm.nig.ac.jp/wfcc/postal_committee.html). Information on IATA Dangerous Goods Regulations is available at IATA's DGIO site (www.iata.org/cargo/dg/index.htm). Additional issues subject to regulation include the handling and distribution of genetically modified microorganisms (GMOs) and transgenic animal and plant material (including cell lines) and the handling and distribution of biohazard and infectious agents. Other online sources of information on biosafety and biohazard are listed in Table 5.

The emergence of new areas of research such as molecular systematics, microbial phylogeny and functional genomics is bringing challenges and opportunities for microbial resource centers. The molecular characterization of unicellular life will continue to affect research and development in medicine, agriculture, and environmental sciences. These changes will impact significantly on the nature and holdings of microbial resource centers and are already reflected in the growing number of specialized 'genetic resource centers' and Internet sites with information on functional genome projects (Table 6).

Large funds are being invested in functional genomics and metabolic reconstruction (Overbeek, 1996), the microbial resource centers must be prepared to be active partners in this process. The genomes of several microorganisms have already been completely sequenced, including representatives from the Domains Archaea (*Aquifex aeolicus, Archaeoglobus fulgidus, Methanobacterium thermoautotrophicum, Methanococcus jannaschii* and *Pyrococcus horikoshii*), Bacteria (*Bacillus subtilis, Borrelia burgdorferi, Chlamydia trachomatis, Escherichia coli, Haemophilus influenzae, Helicobacter pylori, Mycobacterium tuberculosis, Mycoplasma genitalium, Mycoplasma pneumoniae, Rickettsia prowazekii, Synechocystis* sp. and *Treponema pallidum*) and Eukarya (*Saccharomyces cerevisiae*). A comprehensive list of Internet sites with details on concluded and on-going genome projects can be found at The Institute for Genome Research (TIGR, USA) Microbial Database (www.tigr.org/tdb/tdb.html).

TABLE 5.   List of Internet sites with information on regulatory issues, biosafety and biohazards

Acronym	Description	URL address
ACDP	Advisory Committee on Dangerous Pathogens	www.doh.gov.uk/bioinfo.htm
CBD Biosafety Homepage	Convention on Biological Diversity documents and links to biodiversity information.	www.biodiv.org/biosafe/
BINAS	Biosafety Information Network and Advisory Service. Service of the United Nations Industrial Development Organization (UNIDO); monitors global developments in regulatory issues in biotechnology.	binas.unido.org/binas/binas.html
IRB	United Nations Environmental Programme (UNEP) International Register on Biosafety. Focuses on information useful in establishing a regulatory framework for the safe development, transfer, and application of biotechnology.	irptc.unep.ch/biodiv/
NIEHS	National Institute of Environmental Health Sciences (USA): Health and Safety Manual (guidelines on handling of hazardous biological agents and recombinant DNA research).	www.niehs.nih.gov/odhsb/ manual/home.htm
OECD Bio-track online	Databases and documents of the OECD's program on harmonization of regulatory oversight in biotechnology.	www.oecd.org/ehs/service.htm
ICGEB	International Centre for Genetic Engineering and and Biotechnology (Italy) Biosafety WebPages.	www.icgeb.trieste.it/biosafety/
DGIO	International Air Transport Association Dangerous Good Information Online www pages.	www.iata.org/cargo/dg/

TABLE 6. List of Internet sites with information on genetic resource centers and genome projects

CGC	The *Chlamydomonas* Genetics Center, at the Developmental, Cell and Molecular Biology Group (Duke University, USA) holds nuclear and chloroplast mutants of *Chlamydomonas reinhardtii* (and other *Chlamydomonas* species) and genomic and cDNA clones of nuclear, chloroplast and mitochondrial genes (searchable online database).	www.botany.duke.edu/DCMB/ chlamy.htm
CGSC	*E. coli* Genetic Stock Center Collection (Yale, USA). Holds *E. coli* strains and data on genotypes, gene names, properties, linkage map, gene products and specific mutations (searchable online database).	cgsc.biology.yale.edu/
FGSC	Fungal Genetics Stock Center, USA. Database on genetically characterized strains from several fungi genera and other genetic information, such as RFLP and genetic maps, cloned genes and gene libraries (searchable online databases)	www.kumc.edu/research/fgsc/
MIPS	Munich Information Center for Protein Sequences (Germany) Yeast Genome Project (*Saccharomyces cerevisiae* genome database).	speedy.mips.biochem.mpg.de/ mips/yeast/index.htmlx
MycDB	The Integrated Mycobacterial Database, Sweeden. Searchable nucleotide and protein sequences of mycobacteria (searchable online database).	www.biochem.kth.se/MycDB.html
PGSC	*Pseudomonas* Genetic Stock Center (USA). Collection of strains derived from the prototrophic *Pseudomonas aeruginosa* strain PAO1. Also includes some *P. putida* strains (searchable online database).	www.pseudomonas.med.ecu.edu/
SGD	*Saccharomyces* Genome Database, USA. Scientific database of the molecular biology and genetics of the yeast *Saccharomyces cerevisiae*. (searchable online databases).	genome-www.stanford.edu/ Saccharomyces/
SGSC	*Salmonella* Genetic Stock Centre, Canada. Maintains genetic stocks of *Salmonella,* plasmids, phages, transposons, gene libraries and data on genetic maps and genomic sequences (searchable online database and downloadable data files).	www.acs.ucalgary.ca/~kesander/

The revolution in information technology is increasing the user's expectations on information content associated with biological material. Biological collections must be prepared to handle the integration of massive amounts of information (Canhos, 1996; Tiedje *et al.*, 1996; Canhos *et al.*, 1997; Beach, 1998). This will ultimately lead to the development of resource centers as knowledge institutions, especially in the fields of taxonomy, phylogeny and functional genomics. It is important to stress that knowledge-based goods and services comprise 60% of OECD's wealth production. Microbial resource centers must reexamine their roles as information centers and knowledge institutions if they are to remain viable in the 21st century.

6   References

Amann, R.I., Ludwig, W. & Schleifer, K.H. (1995). Phylogenetic identification and in-situ detection of individual microbial cells without cultivation. *Microbiological Reviews* **59**, 143-169.

Beach, J.H. (1998). Employing metadata for biological collections. *NFAIS Newsletter* **40**, 117-120.

Bull, A.T., Goodfellow, M. & Slater, J.H. (1992). Biodiversity as a source of innovation in biotechnology. *Annual Review of Microbiology* **46**, 219-252.

Brussaard, L., Behan-Pelletier, V.M., Bignell, D.E. et al. (1997). Biodiversity and ecosystem functioning in soil. *Ambio* **26**, 563-570.

Canhos, V.P. (1996). Networking the knowledge base on microbial diversity: a role for culture collections. In *Culture Collections to Improve the Quality of Life*, pp. 71-73. Edited by R.A. Samson, J.A. Stalpers, D. van der Mei *et al.* Baarn, The Netherlands: Centraalbureau voor Schimmelcultures.

Canhos, V.P., Manfio, G.P. & Brefe, C.A.F. (1997). Data bases on microbial diversity: needs and prospects. In *Progress in Microbial Ecology*, pp. 29-35. Edited by M.T. Martins, M.I.Z. Sato, J.M. Tiedje. São Paulo: Brazilian Society for Microbiology and International Committee on Microbial Ecology.

Colwell, R.R. (1997). Microbial diversity: the importance of exploration and conservation. *Journal of Industrial Microbiology and Biotechnology* **18**, 302-307.

Cypess, R.H. (1996). ATCC, a diverse culture collection. In *Culture Collections to Improve the Quality of Life*, pp. 186-192. Edited by R.A. Samson, J.A. Stalpers, D. van der Mei *et al.* Baarn, The Netherlands: Centraalbureau voor Schimmelcultures.

Ehrlich, P.R. & Ehrlich, A.H. (1981). *Extinction: the Causes and Consequences of the Disappearance of Species* New York: Random House.

Freckman, D.W., Blackburn, T.H., Brussaard, L. et al. (1997). Linking biodiversity and ecosystem functioning of soils and sediments. *Ambio* **26**, 556-562.

Garrity, G.M., Tiedje, J. & Searles, D. (1998). Report on a workshop on the phylogeny of prokaryotes based upon sequence similarity of the small ribosomal subunit (Unpublished report: personal communication).

Glowka, L. (1996). The Convention on Biological Diversity: issues of interest to the microbial scientist and microbial culture collections. In *Culture Collections to Improve the Quality of Life*, pp. 36-60. Edited by R.A. Samson, J.A. Stalpers, D. van der Mei *et al.* Baarn, The Netherlands: Centraalbureau voor Schimmelcultures.

Glowka, L., Burhenne-Guilmin, F, Synge, H. et al. (1994). *A Guide to the Convention on Biological Diversity* (IUCN Environmental Policy and Law Paper 30), IUCN, Gland.

Harper, J.L. & Hawksworth, D.L. (1994). Biodiversity: measurement and estimation. Preface. *Philosophical Transactions of the Royal Society of London, Series B* **345**, 5-12.

Harvey, C.A. & Pimentel, D. (1996). Effects of soil and wood depletion on biodiversity. *Biodiversity and Conservation* **5**, 1121-1130.

Hawksworth, D.L. (ed.) (1991a). *The Biodiversity of Microorganisms and Invertebrates: Its Role in Sustainable Agriculture.* Wallingford, UK: C.A.B. International.

Hawksworth, D.L. (1991b). The fungal dimension of biodiversity: magnitude, significance, and conservation. *Mycological Research* **95**, 641-55.

Hawksworth, D.L. (1996). Microbial collections as a tool in biodiversity and biosystematic research. In *Culture Collections to Improve the Quality of Life*, pp. 26-35. Edited by R.A. Samson, J.A. Stalpers, D. van der Mei *et al.* Baarn, The Netherlands: Centraalbureau voor Schimmelcultures.

Hawksworth, D.L. & Colwell, R.R. (1992). Biodiversity amongst microorganisms and its relevance. *Biology International* **24**, 11-15.

Hawksworth, D.L., Sastramihardja, I., Kokke, R. et al. (1990). *Guidelines for the Establishment of*

Collections of Cultures of Microrganisms. Surrey, UK: WFCC.

Hugenholtz, P., Goebel, B.M. & Pace, N.R. (1998a). Impact of culture-independent studies on the emerging phylogenetic view of bacterial diversity. *Journal of Bacteriology* 180, 4765-4774.

Hugenholtz, P., Pitulle, K. L. & Pace, N. R. (1998b). Novel division level bacterial diversity in a Yellowstone hot spring. *Journal of Bacteriology* 180, 366-376.

Hunter-Cevera, J.C. (1996). The importance of culture collections in industrial microbiology and biotechnology. In *Culture Collections to Improve the Quality of Life*, pp. 158-162. Edited by R.A. Samson, J.A. Stalpers, D. van der Mei *et al.* Baarn, The Netherlands: Centraalbureau voor Schimmelcultures.

Hunter-Cevera, J.C. (1998). The value of microbial diversity. *Current Opinion in Microbiology* 1, 278-285.

Ilardi, A. (1996). The system of deposit of microorganisms under the Budapest Treaty. In *Culture Collections to Improve the Quality of Life*, pp. 234-248. Edited by R.A. Samson, J.A. Stalpers, D. van der Mei *et al.* Baarn, The Netherlands: Centraalbureau voor Schimmelcultures.

Jaenike, J. (1991). Mass extinction of European fungi. *Trends in Ecology and Evolution* 6, 174-175.

Kirsop, B. (1996a). Biosafety regulations and standards affecting culture collections.In *Culture Collections to Improve the Quality of Life*, pp. 229-234. Edited by R.A. Samson, J.A. Stalpers, D. van der Mei *et al.* Baarn, The Netherlands: Centraalbureau voor Schimmelcultures.

Kirsop, B. (1996b). *Access to Ex-Situ Microbial Genetic Resources within the Framework of the Convention on Biological Diversity.* Campinas, Brazil: WFCC.

Kirsop, B. & Canhos, V.P. (1998). *The Economic Value of Microbial Genetic Resources.* Campinas, Brazil: WFCC.

Kirsop, B. & Hawksworth, D.L. (1994). The Biodiversity of Microorganisms and the Role of Microbial Resource Centers. London: WFCC.

Lovejoy, T.E. (1994). The quantification of biodiversity: an esoteric quest or a vital component of sustainable development? *Philosophical Transactions of the Royal Society of London, Series B* 345, 89-99.

Lovelock, J.M. (1988). *The Ages of Gaia.* Oxford: Oxford University Press.

Marggraf, R. & Birner, R. (1998). The conservation of biological diversity from an economic point of view. *Theory in Biosciences* 117, 289-306.

McNeely, G.A., Miller, K.R., Reid, W.V. *et al.* (1990). *Conserving the World's Biological Diversity.* Gland: International Union for Conservation of Nature and Natural Resources.

Myers, N. (1979). *The Sinking Ark: A New Look at the Problem of Disappearing Species.* Oxford: Pergamon Press.

National Science Foundation USA (1991). *Loss of Biological Diversity: A Global Crisis Requiring International Solutions.* Washington, D.C.: National Science Foundation.

Nisbet, L.J. & Fox, F.M. (1991). The importance of microbial biodiversity to biotechnology. In The *Biodiversity of Microorganisms and Invertebrates: Its Role in Sustainable Agriculture.* pp. 229-44. Edited by D.L. Hawksworth. Wallingford, UK: C.A.B. International.

Norse, E.A. Rosenbaum, K.L.; Wilcove, D.S. *et al.* (1986). *Conserving Biological Diversity in our National Forests.* Washington, D.C.: The Wilderness Society.

O'Donnell, A.G. Goodfellow, M. & Hawksworth, D.L. (1994). Theoretical and practical aspects of the quantification of biodiversity among microorganisms. *Philosophical Transactions of the Royal Society of London, Series B* 345, 65-73.

Office of Science & Technology (1996). *A New Strategy for the UK Microbial Culture Collections: Government Response to the Independent Review of UK Microbial Culture Collections.* London: Office of Science and Technology. URL htp://www.dti.gov.uk/ost/ukmcc/ukmcc.htm.

Overbeek, R. (1996). Possible outcomes of metabolic reconstruction: the characterization of microbial diversity via sequence data; comments in response to a request by Niels Larsen. URL http://www.cme.msu.edu/WIT/Doc/outcomes.html

Pace, N.R. (1997). A molecular view of microbial diversity and the biosphere. *Science* **276**, 734-740.

Pace, N.R., Stahl, D.A., Lane, D.J. *et al.* **(1985).** Analyzing natural microbial populations by rRNA sequences. *ASM News* **51**, 4-12.

Palleroni, N.J. (1996). Microbial diversity and the importance of culturing.In *Culture Collections to Improve the Quality of Life*, pp. 111-114. Edited by R.A. Samson, J.A. Stalpers, D. van der Mei *et al.* Baarn, The Netherlands: Centraalbureau voor Schimmelcultures.

Rifai, M.A. (1989). Astounding fungal phenomena as manifestations between tropical plants and microorganisms. In *Interactions Between Plants and Microorganisms*, pp. 1-8. Edited by G. Lim and K. Katsya, Singapore: National University of Singapore.

Schlegel, H.G. & Jannasch, H.W. (1992). Prokaryotes and their habitats. In *The Prokaryotes*. Vol. I, pp. 75-125. Edited by A. Balows, H.G. Trüper, M. Dworkin *et al.* New York: Springer Verlag.

Sly, L.I. (1996). WFCC: aims and achievements revisited. In *Culture Collections to Improve the Quality of Life*, pp. 3-16. Edited by R.A. Samson, J.A. Stalpers, D. van der Mei *et al.* Baarn, The Netherlands: Centraalbureau voor Schimmelcultures.

Sly, L.I., Iijima, T. & Kirsop, B. (1990). 100 Years of Culture Collections. In *Proceedings of the Král Symposium to Commemorate the Centenary of the First Recorded Service Culture Collection.* Edited by WFCC. Osaka: Institute for Fermentation.

Smith, D. (1996). *Postal, Quarentine and Safety Regulations: Status and Concerns.* London: WFCC.

Stackebrandt, E. (1996). Are culture collections prepared to meet the microbe-specific demands of the articles of the Convention on Biological Diversity? In *Culture Collections to Improve the Quality of Life*, pp. 74-78. Edited by R.A. Samson, J.A. Stalpers, D. van der Mei *et al.* Baarn, The Netherlands: Centraalbureau voor Schimmelcultures.

Stackebrandt, E. (1997). The Biodiversity Convention and its consequence for the inventory of prokaryotes. In *Progress in Microbial Ecology*, pp. 3-9. Edited by M.T. Martins, M.I.Z. Sato, J.M. Tiedje *et al.* São Paulo: Brazilian Society for Microbiology and International Committee on Microbial Ecology.

Staley, J.T., Castenholz, R.W., Colwell, R.R. *et al.* **(1997).** The microbial world: foundation of the biosphere. In *American Academy of Microbiology Critical Issues Colloquia.* Edited by American Society for Microbiology. URL: http://www.asmusa.org/acasrc/pdfs/microb.pdf.

Stolz, J.F. Botkin, D.B. & Dastoor, M.N. (1989). The integral biosphere. In *Global Ecology*, pp. 31-49. Edited by M.B. Rambler, L. Margulis and R. Fester. San Diego: Academic Press.

Stork, N.E. (1988). Insect diversity: facts, fiction and speculation. *Biological Journal of the Linnean Society* **35**, 321-337.

Tiedje, J., Urbance, J., Larsen, N. *et al.* **(1996).** Towards an integrated microbial database. In *Culture Collections to Improve the Quality of Life*, pp. 63-68. Edited by R.A. Samson, J.A. Stalpers, D. van der Mei *et al.* Baarn, The Netherlands: Centraalbureau voor Schimmelcultures.

Trüper, H.G. (1992). Prokaryotes: an overview with respect to biodiversity and environmental importance. *Biodiversity and Conservation* **1**, 227-236.

Wilson, E.O. (1988). The current state of biological diversity. In *Biodiversity*, pp. 3-18. Edited by E.O. Wilson and F.M. Peter. Washington, D.C.: National Academic Press.

Wilson, E.O. (1992). *The Diversity of Life.* Harmondsworth, UK: Penguin Books.

Wilson, E.O. and Peter, F.M. (eds..) (1988). *Biodiversity.* Washington, D.C.: National Academic Press.

Woese, C.R. (1987). Bacterial evolution. *Microbiological Reviews* **51**, 221-271.

World Conservation Monitoring Centre (ed.) (1992). *Global Biodiversity: Status of the Earth's Living Resources.* London: Chapman & Hall.

Zedan, H. (1993). The economic value of microbial diversity. *Society for Industrial Microbiology Newsletter* **43**, 178-185.

16   SYSTEMATICS AND LEGISLATION

DAGMAR FRITZE *and* VERA WEIHS

*DSMZ-Deutsche Sammlung von Mikroorganismen
und Zellkulturen GmbH, Mascheroder Weg 1b, D-38124 Braunschweig,
Germany*

1   Introduction

'It is of no interest to me what it is - I only want to know what it does!' This and similar
statements are often heard in groups of microbiologists active in research or in the applied
areas of the field when it comes to the question: what kind of microorganism are you
working with? Such scientists like to stress their wish to avoid the burden of time-consuming
taxonomic work and to have a rather more stream-lined, straightforward research, development and production programme. This attitude is not new. Even the great Louis Pasteur
was reported to have said when his coworker informed him that an organism which he
had thought to be a coccus in fact was a short rod, 'I wish I could convey to you how little
this information excites me' (Duclaux, 1896).

It may indeed be of no interest what a given microorganism 'really is', this knowledge
may well lie beyond the abilities of the human brain. We can only depict the image as
we see it and in the way we are able to record it. But what we can do, and what we need
to do, is to define what we are talking about. Order is basic to the human way of thinking
and this is why some kind of systematics underly any human activity. Looking into nature,
a vast diversity stretches before us; we have to separate, classify, and to put things into
small boxes to attempt to come to an overview. On the other hand, of course we need the
opposing approach that joins the details and makes visible the larger totality.

Developments in modern biological science have produced an explosion in the description
of new species of microorganisms so that we depend on comprehensive frameworks and
experts for the reliable identification of isolates and their allocation to taxonomic groups.
A few examples will illustrate how intimately the discipline of bacterial systematics -
the backbone for the identification of microorganisms- is interconnected with everyday
microbiological work in the applied sciences and even with the economic and industrial
side of research.

A large proportion of the world's production of antibiotics, enzymes and pharmaceuticals
is derived from microorganisms. Rising numbers of patents on microorganisms or
production processes using microorganisms show the growing industrial interest in living
biological material. Today, "newly constructed", that is, genetically engineered organisms,
as well as those isolated from nature or available from culture collections are considered

F. G. Priest and M. Goodfellow (eds.), Applied Microbial Systematics, 447-469 patentable or usable in patentable processes. Reliable identification of these organisms is a basic requirement.

Imagine a company A that holds a patent for the production of a certain compound by a microorganism of taxon X. Company B wants to enter this lucrative market without infringing the patent of company A. To be able to present something new, company B tries to find a microorganism that is able to produce the desired compound but belongs to another taxon, Y. Finally an organism exhibiting the required ability is isolated and identification reveals that fortunately the organism indeed belongs to a different taxon. The patent is filed, a lot of work and money is spent, and only then it is found out that the identification was wrong and that the organism, after all, belongs to the same taxon as the organism of company A. The patent is, of course, lost.

A second example concerns a laboratory engaged in recombinant DNA technology. A patent is filed describing a production process in which a certain microorganism carrying a constructed plasmid is used to catabolize a certain reaction. Later, a cooperating laboratory finds that the taxonomic allocation of the organism is incorrect, and instead of a registered host strain for safe genetic engineering a hazard group 2 organism is carrying the plasmid. How could this happen? The original laboratory has to confess that, whereas they had always checked for the presence of the plasmid in the organisms they used, they never felt it necessary to recheck the identity of the host strain which, at the beginning of the work, had been the prescribed safety host strain. So, the researchers involved had to carefully track back the sequence of subcultures of the strain in order to eliminate all of the wrong cultures.

In both cases the patent was not correctly described and hence the applications were flawed. Moreover, in the second case, the prescribed safety measures for production had not been met and an unintended exposure of personnel to a possible hazard had occurred. Obviously, in both examples proper attention to sound taxonomic practices would have saved the laboratories considerable time and expense.

In the following account a number of laws and regulations will be cited with which the typical research scientist may not be totally familiar. Short descriptions of their contents are therefore given in the appendix.

2   WHO Hazard Classification System

Microorganisms have different hazard potentials and are classified according to the harm they might cause to humans or animals. 'Harm' -as defined by the International Standard Organisation (ISO, 1997)- is the physical injury or damage to the health of people or to the environment, whereas the term 'hazard' is used to define a potential source of harm (like fire or biological hazard). The combination of the probability of occurrence of harm and the severity of that harm is called 'risk'. The classification of microorganisms according to the risk they pose is used as the background for existing restrictions on who is allowed to import, export or work with which organisms. However, before microorganisms can be correctly assigned to hazard groups, they must have been identified and allocated to an already existing taxon, usually at species level. This practise is necessary to avoid the danger of working with a wrong microorganism, a situation which could prove time-wasting, expensive, and lead to the publication of invalid results. As indicated before, without reliable identification dangerous organisms could be used inadvertently in production processes or unwittingly supplied to laboratories with the risk of harm to employees.

The World Health Organization (WHO, 1983, 1993) established a classification of four hazard groups for microorganisms (1, 2, 3, and 4) applicable for work in a laboratory. Organisms are allocated to these groups according to the increasing risk they pose for human beings or animals. Similarly, respective laboratory (L1, L2, L3 and L4) and production containment levels (P1, P2, P3 and P4) have been thoroughly described (Table 1).

TABLE 1. Hazard groups and corresponding precautionary measures for laboratories and production facilities

Hazard groups and corresponding precautionary measures			
		Containment levels	
		Production facilities	Production facilities
Hazard group	Laboratories	Europe	OECD
1	L1	P1	GILSP
2	L2	P2	C1
3	L3	P3	C2
4	L4	P4	C3

GILSP: Good industrial large scale practice

National and international laws, directives, guidelines and administrative provisions have been brought into force to comply with the WHO system and reinforce it with more details for daily practice. The representative definitions and examples (Table 2) given below have been compiled from a selection of such international and national legal systems.

TABLE 2. Classification of biological material according to its pathogenic potential

Risk assessment for employees, the community and the overall risk			
Hazard group	Risk for employees	Risk for community	Overall risk
1	none	none	none
2	low to moderate	low	low
3	moderate to high	low to moderate	moderate
4	high	high	high

International - WHO Laboratory Biosafety Manual (1993)
Europe   - Council Directives 90/679/EEC and 93/88/EEC on the protection of
                 workers from risks related to exposure to biological agents at work
                 (European Community, 1990c, 1993a)
National   - **Canada**: Laboratory Biosafety Guidelines, Health Canada
                 (Laboratory Center for Disease Control, 1996)
              - **Germany**: BG-Chemie leaflets 'Safe biotechnology, classification of
                 biological agents' (BG-Chemie, 1990-1998)
              - **Japan**: Additional requirements for facilities transferring or receiving
                 particular infectious agents, Department of Health and Human
                 Services, Public Health Service, Proposed Rules (1996)
              - **United Kingdom**: Advisory Committee on Dangerous Pathogens,
                 categorisation of pathogens according to hazard and categories of con-
                 tainment (1996)
              - **United States of America**: Communicable Disease Center-National
                 Institutes of Health - Biosafety in microbiological and biomedical lab-
                 oratories (1993)

The assignment of known organisms to hazard groups is based on long-term experience.
Whereas the potentially hazardous organisms, rated according to their rising pathogenic
potential, are allocated to hazard groups 2, 3, or 4, those organisms which are generally
considered to be non-hazardous are grouped under level 1.

a) non-hazardous organisms
Hazard group 1: organisms and viruses with no risk or which are most unlikely to
(= level 1)   cause diseases in man and other vertebrates
   Examples are:
   - obligately psychrophilic organisms (growth only below 20°C)
   - obligately thermophilic organisms (growth only above 50°C)
   - obligately acidophilic organisms (growth only below pH 4)
   - obligately alcaliphilic organisms (growth only above pH 8,5)
   - obligately phototrophic organisms (no growth without light)
   - obligately chemolithotrophic organisms (no growth with organic matter)
   - saprophytically living organisms; no infectious potential for humans or animals
   - individual strains of organisms of hazard group 2 or higher which have definitely
      lost their pathogenicity (e.g. *Escherichia coli* K12, *Salmonella typhi* 21a) or which
      had been in use for a long period of time (e.g. for production purposes) without
      having ever caused diseases (e.g. certain strains of *Lactobacillus rhamnosus*)

b) potentially hazardous organisms
Hazard group 2: organisms and viruses which may cause diseases in animals and in
employees but are unlikely to spread in the community; the risk to the laboratory worker is
low to medium and minimal for the general population as prophylactic and/or therapeutic
measures exist.

Examples are:
- *Streptococcus mutans*: lives on the mucous membrane of the human mouth and is involved in the development of caries; facultatively causative agent for endocarditis.
- *Vibrio cholerae*: causative agent of cholera; high dose needed for infection; infection mainly *via* drinking water spoilt with faeces; infection by air excluded.
- *Candida albicans*: lives on the mucous membrane of the human mouth and intestines; causes mucosal mycosis.
- rabies virus: invasion of the central nervous system; possibility of active vaccination after infection; incubation time one week up to one year; lethal without medical attention; immunoprophylaxis possible.

Hazard group 3: organisms and viruses which may cause severe and life-threatening diseases; regarding infectiousness, pathogenicity and the existence of prophylactic and/or therapeutic measures the risk is medium to high for the laboratory worker whereas the risk to the community is low to medium.
Examples are:
- *Mycobacterium tuberculosis*: causative agent of tuberculosis; extremely infectious, spreads by air; severe disease with lengthy therapy; protection by vaccination possible
- *Yersinia pestis*: causative agent of plague; as few as five cells are sufficient for an infection; acutely life-threatening disease; therapy with antibiotics successful only directly after infection; protection by vaccination possible.
- *Coccidioides immitis*: causes severe inner mycosis, initial infection of the lungs, spreads to other organs of the body; existing medicines do not guarantee complete recovery.
- Human immunodeficiency virus (HIV): causative agent of AIDS; no reliable therapy or prophylactic measures available: almost 100% manifestation and lethality; virus, however, is extremely unstable.

Hazard group 4: biological agents causing severe diseases and representing a serious risk to employees and the general population; regarding infectiousness, pathogenicity and the lack of effective prophylactic and/or therapeutic measures the risk to the laboratory worker and the community is high.
To date, only viruses have been classified in hazard group 4.
Examples are:
- Herpes-B-virus: laboratory infections known; spreads by air; no therapy or prophylactic measures possible; up to 100 % lethality.
- Lassa fever virus: systemic disease with inflamed and ulcerated throat, fever and convulsions; death within 10-14 days in 50% of cases.
- Poxvirus (Variola): causative agent of pocks; vaccination led to eradication worldwide.

While the existing hazard classification systems of different countries may vary with respect to the grouping of a certain species with one or other of the different risk groups, it can be said that up to now it has been shown that organisms exhibiting no or only low hazard by far outnumber those with a high pathogenic potential. For instance, out of the current 4500 validly described bacterial species, nearly 80% are classified in hazard category group 1; about 20% are considered to present a moderate hazard and are classified in hazard group 2. Only about 1% of the known bacterial species are considered to pose a high potential risk to laboratory workers and the general public and are hence allocated to hazard group 3 (Fig. 1).

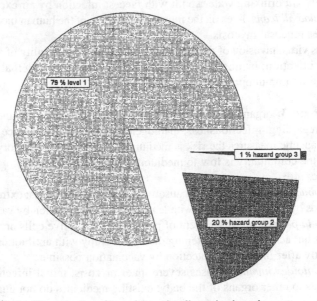

Figure 1. The proportion of bacterial species allocated to hazard groups.

3 Laws and Regulations which Rely on the Hazard and Taxonomic Classification of Microorganisms

The allocation of organisms to hazard groupings in connection with risk assessments plays an important role in various regulatory fields. Here, the main interest usually lies in the hazard potential of microorganisms and consequently in the safeguarding of human beings, animals or plants -as the case might be- against the risk of infection. Testing in living animals has to be performed to unambiguously determine the pathogenicity of a microorganism. In this regard, the hazard classification system, as well as the systematic classification of organisms, helps to avoid excessive use of animals and thereby reduces costs as basic estimations of a pathogenic potential may be deducted from the reliable identification of an organism. A number of laws and regulations in connection with the handling of microorganisms seem to take it for granted that the microorganisms in question are correctly identified, but few explicitly require an actual identification of a given strain before the start of a project. However, it is clear that all kinds of permits to work with certain organisms, to allow the import or export of cultures of microorganisms, or to use organisms for production purposes rely on the availability of a comprehensive systematic framework and the ability of researchers to identify the strains.

3.1 THE HANDLING OF MICROORGANISMS

Before starting any microbiological work various regulations may have to be considered. In some countries, the laboratory may have to register with an appropriate national authority. In general, the laboratory has to be equipped so as to match the hazard level of the organisms which are going to be handled. In addition, the handling of pathogenic microorganisms may be restricted to employees who have been granted special licenses. In Germany, for example, organisms classified in hazard group 2 (or higher) may only be handled by persons who have permission under the German Infectious Diseases Act (for human pathogens) (1990), or the German Infectious Diseases of Animals Enactment (1992). Permits to handle certain plant pathogenic microorganisms are regulated through the German Plant Protection Act (1990) and in the case of genetically manipulated organisms specific permits need to be obtained according to the German Law Regulating Genetic Engineering (1993).

All national and international regulations stress that the handling of hazard group 3 and 4 organisms needs special attention and permits which emphasize the risks that these organisms pose to employees and the public.

In special cases, working permission needs to be obtained even for some organisms classified in hazard group 1. *Paenibacillus larvae* subsp. *larvae* (formerly *Bacillus larvae*, see Chapter 7), for example, is classified in group 1, but because of its high pathogenicity to bees the German authorities require the prospective researcher to obtain a permit under the German Infectious Diseases of Animals Enactment (1992).

Two useful publications have been compiled which compare national and international regulations dealing with safety and quarantine aspects of work with microorganisms, as well as regulations for the transport of living biological material (Smith, 1996; Rhode & Claus, 1998).

3.2 REFERENCE AND QUALITY CONTROL STRAINS

In a number of Pharmacopoeia, and other national and international compendia, procedures are described for testing material for certain features using microorganisms (e.g. the Regulations of the International Standardization Organization [ISO], the Regulations of the European Committee for Standardization [CEN: Comité Européen de Normalisation], or the US-Pharmacopoeia and the European Pharmacopoeia). The required microorganisms are usually available from service culture collections or other reference laboratories as so called test strains, assay strains or quality control strains.

The scope of applications is broad. Probably one of the most widely known procedures is sterility testing, that is, testing for correct functioning of autoclaves (steam sterilization) and other sterilizers (using dry heat, ethylene oxide, H_2O_2, ozone or irradiation). For each kind of sterilization, specific strains of different species have been shown to be useful according to their individual resistances. *Bacillus stearothermophilus* strains, for example, produce endospores with exceptional resistance against moist heat. So, if spores of certain strains of this organism are killed during a steam sterilization process one can be confident that all other organisms have also been killed. Similarly, certain strains of *Bacillus subtilis* have been approved as standards for dry heat sterilization. Other examples of the use of particular microbial strains for quality assessment include testing the performance of media, tests for the presence of antimicrobial agents in cosmetics, and of course, testing for antibiotic residues in substances such as meat or milk.

In the cases described, the allocation of a strain to a specific taxon is needed in order to comply with the regulations, but even more important is the correct activity of the organisms in the given test. It is necessary, therefore, not just to use any strain of *B. stearothermophilus* for steam sterilization control, but to use the specific strain designated as the test strain. Susceptibilities or resistances vary between strains of the same species, while taxonomic developments may require the division of species into new taxa and the renaming or invalidation of species. The strain in question may end up in any of the new allocations, but will still bear the strain characteristics that are vital for the test in question.

An example where identification according to the International Code of Nomenclature is explicitely required is given in the European Commission Directive 95/11/EEC: *fixing guidelines for the assessment of additives in animal nutrition* (European Community, 1995). These additives may be living microorganisms and the directive requests that an identified culture of the strain used be deposited in a recognized culture collection. '*Bacillus toyoi*' or '*Bacillus cereus* subsp. *toyoi* 'is an example of such an additive used as a probiotic in animal feeds. This strain, which belongs to a species classified in hazard group 2, is described as one which does not produce toxic substances. However, recently cases have been reported where toxin producing strains of *B. cereus* have been isolated from feeds inoculated with '*B. toyoi* '. It seems likely in such cases that after inoculation with the -pure and non-toxigenic- starter culture no further examination of the animal feed had been undertaken during processing and hence a spoilt product had been sold on the market. Whether another strain of *B. cereus* had contaminated the feed, or whether the original strain had reverted to toxigenicity is not clear but it seems obvious from such cases that identification of a microorganism -even below species level- is essential not only at the beginning of a process but also throughout production.

3.3   GENETIC ENGINEERING

Genetic engineering covers the alteration of the genetic material of organisms in a way that does not occur naturally by mating and/or natural recombination; be it by recombinant DNA techniques using host-vector systems, techniques involving direct introduction of genetical material (micro-injection, macro-injection, micro-encapsulation) or cell fusion and hybridization techniques.

Special facilities are needed to meet legal requirements when working with genetically manipulated organisms is intended. Existing legislation regulating genetic engineering are (among other international or national laws), for example, OECD guidelines (1992), and EEC-council directive 98/81/EC amending directive 90/219/EEC *on the contained use of genetically modified microorganisms* (European Community, 1998) and directive 90/220/EEC *on the deliberate release into the environment of genetically modified microorganisms* (European Community, 1990b). These regulations cover the actual research work in recombinant DNA technology as well as any other work where such tailored organisms are used. Comparable to the specifications for work with natural microorganisms, the work itself, as well as the laboratories, are categorized according to the potential risk involved (e.g. safety levels S1, S2, S3 and S4 in Germany or, for the European Community, Class 1,2,3 and 4 respectively).

When applying genetic engineering only well characterized organisms can be used to minimize possible negative consequences for human health and the environment. Risk assessments have to be performed taking into consideration the individual potential hazards of the recipient organism, the donor organism, the cloned material, and the vector. An example of a non-hazardous genetically modified organism would be a strain of *Escherichia coli* K12, e.g. DH5alpha, carrying a vector like pUC19 with an insert coding for the alcohol-dehydrogenase of *Saccharomyces cerevisiae*. Strains of *E. coli* K12 have been shown to have lost all pathogenicity factors typical of *E. coli* wild type strains which belong to hazard group 2. Thus donor, recipient and the resulting organism are harmless. On the other hand, an *E. coli* K12 strain containing a vector combined with an undefined and uncharacterized DNA-fragment of *Staphylococcus aureus* would have to be categorized as potentially dangerous as the donor is a potential toxin producer allocated to hazard group 2. Consequently, a prior condition for any work of this kind is the proper identification of the test microorganisms before and throughout ongoing work.

3.4 TRANSPORT OF MICROORGANISMS

Living biological material is sent all around the world thereby presenting -at least theoretically- a danger for persons who might be inadvertently exposed to the material, e.g. postal employees and secretaries. To prevent or to reduce the possibility of the inadvertant release of microorganisms and contact of personnel with these organisms due to accidents, a number of national and international laws and regulations for the transport and transborder movement of living biological material exist. These laws and regulations are explicitly built on the classification of microorganisms into hazard groups and hence systematics, as the basis for identification, is the determining factor for the measures to be taken when microorganisms are to be transported. As a general rule, microorganisms -at least those of hazard group 2 and higher- must not be sent to private persons. Normally, only adequately trained staff in appropriately equipped laboratories should receive microorganisms.

Domestic shipments may be subject to national permit and license requirements. In the USA, for example, shipment permits are necessary in the case of certain human, animal and plant pathogens, as well as for some genetically engineered organisms (American Type Culture Collection, 1996).

Internationally, the individual national postal services are part of the Universal Postal Union (UPU). The Universal Postal Convention (Universal Postal Union, 1996) includes guidelines for the postal transport and the packaging of non-infectious perishable biological substances (NPBS) and infectious perishable biological substances (IPBS):

NPBS: expression coined by UPU; organisms without hazardous potential (corresponding to hazard group 1)

IPBS: expression given by UPU, equivalent to the term infectious substances used by
        the International Air Transport Organisation and to the term etiological agents as
        used in the USA; organisms with hazardous potential for humans and/or animals
        (corresponding to hazard group 2 or higher)

Generally, the exchange of perishable biological substances is restricted to countries
whose postal administrations agree to receive biological material. Further restrictions may
occur when a country does not accept NPBS or IPBS by air mail. For example, the national
postal services of Bolivia, Canada, Egypt and Greece do not allow the shipment (import
or export) of perishable biological substances by mail, be it surface or air. Other postal
services will accept only non-infectious material by air or surface mail (e.g. the Belgian,
Chinese or Spanish postal services) (Rohde, 1999). Here the possibility of sending the
material by air freight should be chosen.

The Universal Postal Convention includes details on how biological material has to be
packaged though procedures differ for non-infectious and infectious substances, that is,
between organisms of hazard group 1 on the one hand and organisms of hazard groups 2, 3
and 4 on the other. The European Norm EN 829: *In vitro diagnostic systems - transport
packages for medical and biological specimens - requirements, tests* (CEN, 1996) deals
exclusively with the transport and packaging of specimens without potential hazard, that is,
those of hazard group 1.

Two organizations have established guidelines for the transport of biological material
by air: the International Civil Aviation Organisation (ICAO) and the International Air
Transport Association (IATA). The IATA Dangerous Goods Regulations (1999) are concerned
with the shipping of infectious biological material whereas the transport of non-infectious
biological material is not considered. It is demanded that a *'shipper's declaration for
dangerous goods'* accompanies the shipment, it also has to be indicated precisely which
kind of infectious material is to be transported.

a) export.   Export of microorganisms is usually governed by national laws.
Organisms classified in hazard group 1 (no risk) generally are admitted for export
without restrictions. However, if an embargo is imposed on certain countries even this
kind of organism is subject to export restrictions.

More general restrictions are imposed through laws and directives designed to control
the export of goods with dual use, that is, use for civil as well as military purposes.
Regulative examples are the EEC council directive 3381/94/EEC *setting up a Community
regime for the control of exports of goods with dual use* (European Community, 1994b)
or the German Law Controlling War Weapons (1991). A validated export license is needed
in the United States of America for organisms or toxins listed with an export control
classification number according to the U.S. Department of Commerce (American Type
Culture Collection, 1996). Comparable laws are in preparation in Australia and Japan.
These regulations prohibit the export, among other items, of some pathogenic microorganisms and viruses and their toxins which potentially could be used as biological weapons.
Permission for the export of such goods may be sought from special governmental agencies.
Organisms categorized as possible biological weapons include the causative agents of
anthrax *(Bacillus anthracis)*, of plague *(Yersinia pestis)*, of legionnaires' disease

(*Legionella pneumophila*) or toxin producing strains of organisms such as *Aspergillus fumigatus, Clostridium botulinum,* and *Staphylococcus aureus.*

b) import. Import of biological material may be subject to the quarantine regulations of a given recipient country. These often depend on the special situation of a certain country which, for example, might be free from a certain pathogen and tries to remain so, or, where it is expected, because of the climate or missing predators, that a newly-introduced microorganism could spread excessively. Equally, the knowledge of the existence of endangered target organisms would lead to restrictions of the import of specific harmful organisms. In the USA, for example, a license is needed for the import of animal cell cultures; the latter are also subjected to a period of quarantine to avoid the possible introduction of animal pathogens of, for example, a viral nature. In some countries only officially recognized laboratories are authorized to accept infectious substances; in France, for example, seven institutions are nominated for this task (Smith, 1996). Independent of the hazardous potential of the microorganisms, some countries, such as Australia and New Zealand, require permission for the import of all microorganisms.

The EC Commission Directive 92/103/EEC (European Community, 1992) restricts the import of organisms harmful to plants and plant products and their spread within the Community. In Germany, comparable restrictions concerning, for example, *Synchytrium endobioticum,* the causative agent of potato cancer and *Erwinia amylovora,* which causes fire blight in fruit trees are governed by the German Plant Protection Act (1990). Similarly, the German Infectious Diseases of Animals Imports Enactment (1992) restricts the import of *Chlamydia psittaci,* which causes psittacosis, and *Zymonema farcinimosus,* which is responsible for lymphatic vessel inflammation of one-hoofed animals. Similar laws exist in other countries.

It is evident that the rules and regulations governing the transport of microorganisms is quite a complex subject. National as well as international regulations apply and within individual countries different regulatory bodies may be involved depending on the microorganism concerned and the means of transport chosen. Some regulations are even contradictory, this underlines the need to have regulations harmonized internationally.

4   The Patenting of Living Biological Material

The patent system involves a two-sided process: an inventor is granted exclusive rights to exploit his/her invention for a limited period of time if he/she, in turn, provides to the public a description of the invention that would enable 'a person skilled in the art' to rework the described invention. With biotechnological inventions, where living biological material is involved, the deposit of the propagatable material (patent deposit) is either a demanded or internationally accepted practice for supplementing the written disclosure of a patent application.

Patent deposits are regulated internationally through the *Budapest Treaty on the International Recognition of the Deposit of Microorganisms for the Purposes of Patent Procedure* (WIPO, 1989). The Treaty was enacted in 1980 and, as a consequence, single deposits made with any International Depositary Authority (IDA) are recognized as sufficient and valid for all signatories of the Budapest Union. Non-member countries may also accept such deposits as sufficient, but usually these countries require an additional deposit within the purview of their own patent laws. Currently (June '99), 45 states worldwide are party to the Treaty and 31 culture collections have acquired IDA status (Table 3). The general deposition procedure is described in detail in the *Guide to the Deposit of Microorganisms under the Budapest Treaty* (WIPO, 1998); an overview and exemplary descriptions are available (Weihs & Fritze, 1995).

TABLE 3.   List of International Depositary Authorities (= IDA's) as of 1-06-1999

STATE	IDA
Australia	Australian Government Analytical Laboratories (AGAL)
Belgium	Belgian Coordinated Collections of Microorganisms (BCCM)
Bulgaria	National Bank for Industrial Microorganisms and Cell Cultures (NBIMCC)
Canada	Bureau of Microbiology of Health Canada (BMHC)
China	China Center for Type Culture Collections (CCTCC)
	China General Microbiological Culture Collection Center (CGMCC)
Czech Republic	Czech Collection of Microorganisms (CCM)
France	Collection Nationale de Cultures de Micro-organismes (CNCM)
Germany	DSMZ-Deutsche Sammlung von Mikroorganismen und Zellkulturen GmbH (DSMZ)
Great Britain	Culture Collection of Algae and Protozoa (CCAP)
	European Collection of Cell Cultures (ECACC)
	International Mycological Institute (IMI)
	National Collection of Type Cultures (NCTC)
	National Collection of Yeast Cultures (NCYC)
	National Collections of Industrial and Marine Bacteria Ltd. (NCIMB)
Hungary	National Collection of Agricultural and Industrial Microorganisms (NCAIM)
Italy	Advanced Biotechnology Center (ABC)
	Collection of Industrial Yeasts (DBVPG)
Japan	National Institute of Bioscience and Human-Technology (NIBH)
Latvia	Microbial Strain Collection of Latvia (MSCL)
Republic of Korea	Korean Cell Line Research Foundation (KCLRF)
	Korean Culture Center of Microorganisms (KCCM)
	Korean Collection for type cultures (KCTC)
Russian Federation	All Russian Scientific Centre of Antibiotics (VNIIA)
	Russian Collection of Microorganisms (VKM)
	Russian National Collection of Industrial Microorganisms (VKPM), GNII Genetika
Slovakia	Culture Collection of Yeasts (CCY)
Spain	Colleción Española de Cultivos Tipo (CECT)
The Netherlands	Centraalbureau voor Schimmelcultures (CBS)
USA	Agricultural Research Service Culture Collection (NRRL)
	American Type Culture Collection (ATCC)

When granting patent rights on biotechnologial inventions, the patent regulations usually do not demand the identification of the relevant microorganisms. The European Patent Convention, for example, requires only that a strain be described according to the information on the relevant characteristics of the microorganism as available to the patent applicant, and that strains be identifiable by a designation. In the extreme, this could be reduced to no more than a random combination of numbers and letters, but the examples given earlier suggest that thorough and up-to-date taxonomic identification may prove vital for the defense of patent rights.

However, it is usually necessary to indicate the taxonomic designation of the strain when depositing microorganisms with a culture collection for patent purposes. While this is not mandatory as far as patent regulations are concerned, each IDA may have to observe regulations which imply knowledge of the taxonomic allocation of the culture. Each IDA decides individually which kinds of microorganisms (e.g. bacteria, fungi, viruses, plasmids, animal cell lines) it will accept for deposit under the Treaty. Following national and international regulations, an IDA has to decide on the degree of pathogenicity of the biological material it will be able to handle (e.g. restriction to hazard groups 1 and 2) and take appropriate measures. These decisions are published in the WIPO Journal *Intellectual Property Laws and Treaties* in which updated compilations appear annually. The individual IDA is bound to its statements so long as no other decision has officially been taken. Kinds of organisms not covered by an IDA, or which exceed its hazard limits have to be refused by the IDA. Thus, material sent to an IDA has at least to be characterized to a degree that allows its taxonomic and hazard classifications according to the applicable laws, for example in Europe the Council Directive 93/88/EEC *on the protection of workers from risks related to exposure to biological agents at work* (European Community, 1990b; 1993).

4.1 NOVELTY OF MICROORGANISMS

Another prerequisite for obtaining a patent for an invention is that the process must be new. The definition provided by the European Patent Office states that it should not have been available to the public before, neither by means of a written or oral description, nor by use or in any other way, before the date of filing of the patent application, thus representing more than the state of art. This is probably why, in the past, many researchers felt obliged -to be on the safe side- to describe 'new' species in their patent applications. This led, as can be seen in the case of the genus *Streptomyces*, members of which produce many important antibiotic compounds, to the description of enormous numbers of 'new' species many of which were subsequently shown to be ill-described and synonymous with existing species (Williams *et al.*, 1983). Later the 'creation' of 'new' organisms by recombinant techniques seemed to be sufficiently well defined to avoid the description of new species. However, current practice shows that even isolates from nature may be used in patents when their isolation has been achieved by new techniques or when they exhibit unexpected properties.

An important conflict of interests arises from the wish of researchers to obtain both patent protection on their microorganism and scientific merit. This conflict is based on the need for free availability of taxonomic reference strains versus the patent practice of restricting the distribution of patent deposits. As many of the patent microorganisms currently isolated from unusual habitats are not only of industrial interest but also new to systematists, the desire of the scientists to describe validly a new species is understandably strong. The unfortunate consequence is that in these cases the type strain, which is the strain to which the name of the new species is attached, is usually the strain which is covered by patent protection. From a scientific point of view, organisms designated as type or reference reference strains should be readily accessible to any interested party and no restrictions should impede availability when such strains are requested. However, in most cases the release of a culture deposited under patent law is a laborious and time consuming process.

4.2   AVAILABILITY OF DEPOSITED PATENT CULTURES

According to patent legislation, organisms which have been deposited for patent purposes are available to entitled third parties for examination and trials. However, this accessibility depends on the status of a patent and -under certain circumstances- may be conceded only to selected experts as listed with the patent office. When a culture of an organism deposited under the Budapest Treaty is requested from an IDA, the depositary will provide the relevant official forms (request for the release of a patent strain) which must be completed and sent to the relevant patent office for confirmation. According to rule 11 of the Budapest Treaty samples may be furnished to interested industrial property offices, to the depositor or with the authorization of the depositor to third parties, and to legally entitled parties (this last point usually refers to the ordinary scientist). The entitlement to receive the requested strain is checked by the patent office. After receiving permission from the relevant patent office, the depositary will ship a sample of the deposited strain following the recommendations of other national or international regulators. Only after a patent has been granted, and this fact communicated to the IDA by the patent office, is the strain freely available without further examination through the patent office. The depositor is promptly notified in writing of each furnishing of a sample of the deposited organism (Fig. 2).

It is obvious that such a lengthy procedure is not acceptable from a taxonomist's view. A type strain of a species must be readily available for comparative studies. This provision is clearly covered by the *International Code of Nomenclature of Bacteria* (Sneath, 1992): 'Type strains that are also deposited in connection with patent applications must be released to the public, or the validation of the taxon will not be accepted until the patent is issued and the strain becomes available.' (International Committee of Systematic Bacteriology, 1997). Meanwhile Tindall (1999) proposed to change rule 30 of the code to even better take account of the requirement of availability of type strains. This point is also stressed in the *Guidelines for Authors* (1999) of the *International Journal of Systematic Bacteriology*. One solution is to avoid the description of a new species when the strain is used in a patent application. Alternatively, a parallel subculture of the patent strain under a different accession number could be lodged in the public part of a collection from where it would be available to anyone without delay, but of course, this may compromise the patent protection.

Action of the Requesting Party

Request is sent to the IDA ⟶

Action of the Depositary Authority

The IDA determines whether the strain may be released according to patent law. (It also checks whether other rules have to be observed, ie. public health or export regulations)

If no information is available:

A formal letter is sent to the requestor stating that the IDA has no information but asks for information, if available.

Eventually information is sent to ⟵ the IDA (ie. patent document)

If information is available or copies of documents are received, an official form for the request of cultures to be used with the relevant patent offices can be supplied.

The form must be filled in and ⟵ sent by the requestor to the appropriate patent office.

The patent office examines the entitlement of the requestor and sends a letter of confirmation to the IDA. ⟶

The IDA is now authorised to release a sample of the deposited strain together with a copy of the statement of the original deposit.

A notification on the furnishing of a sample is sent to the depositor together with a copy of the confirmation of the patent office.

Figure 2. Request and release of a patent strain.

5   Convention on Biological Diversity

The United Nations Convention on Biological Diversity (CBD) (United Nations, 1993) came into force in December 1993 after having received a sufficient number of ratifications. The contracting parties now have the responsibility to initiate a series of actions at the national level to fulfil the aims and demands of the convention.

The principal aim of the Convention is to provide a comprehensive and international strategy for the preservation of biodiversity by adopting three objectives: '..the conservation of biological diversity, the sustaining use of its components and the fair and equitable sharing of the benefits arising out of the utilization of genetic resources..' (Glowka *et al.*, 1994). The Convention on Biological Diversity is applicable to microorganisms as well as to plants and animals and the demand for 'fair and equitable sharing' implies unimpeded access to the holdings of *ex-situ* collections. At the same time, the important issues of ownership and repatriation of cultures and information thereof must be clarified.

The provisions of the Convention imply that thorough inventories have to be established on the existing biodiversity. This not only includes the material that is already held in *ex-situ* collections, together with information about its origin, but equally the implementation of projects designed to investigate and document undiscovered populations. To date, most inventory and monitoring programmes for microorganisms have been confined to organisms associated with human, animal and plant diseases and, to some extent, organisms of agricultural significance. It is now timely to adopt a broader approach and to encompass microbial populations with no known associations or pathogenic properties. It is estimated that the microorganisms isolated, cultured, described and named to date represent only a small fraction of the total numbers of existing species; about 95 to 99% of microbial species have yet to be discovered and described (Hawksworth, 1997; Hawksworth & Rossman, 1997).

The importance of microbial systematics is clearly stated in the Convention on Biological Diversity as it is evident that little progress can be made in assessing the degree and value of biodiversity without knowing the individual species involved. A description of an organism and its resulting name are the keys to the various sets of data that are available on the properties of known species. The central issues of biodiversity are systematic biology and taxonomy and both are vital for promoting research in microbial diversity.

6   Conclusions

It is evident from what has been said that regulatory bodies presuppose and demand an unequivocal and clear identification of the microorganisms with which a person is working. This applies not only to clinical isolates but to all other strains isolated from nature or obtained from culture collections. It is the systematist/taxonomist who is central to a number of vital tasks. However, it should not to be forgotten that once a microbial culture has been identified the strain needs to be periodically checked to ensure that contamination has not occurred. To be able to avoid losses due to contaminated production lines, or due to erroneous use of wrong cultures, and to be able to protect personnel adequately well trained systematists are needed. However, there is evidence of a growing shortage of well trained systematists worlwide. Statements such as those issued by IUMS/IUBS (Action Statement Microbial Diversity 21) as early as 1991 (IUMS/IUBS, 1991) and Resolutions ICCC-7 and ICCC-8 issued by the World Federation for Culture Collections (1992; 1996) address these deficiencies and define the needs around which future activities should be built.

7   Appendix

The objectives of relevant legal works are briefly described:

BIOLOGICAL SAFETY

WHO Laboratory Biosafety Manual (1993).
'Safety is good technique': guidelines on basic laboratory design, equipment, its use, and good microbiological techniques for the different biosafety levels (including animal facilities) are presented; special attention is also given to chemical, fire and electrical safety, safety organization and training. A safety checklist is useful for reviewing existing arrangements and helps planning new facilities.

Advisory Committee on Dangerous Pathogens, Categorisation of Pathogens According to Hazard and Categories of Containment, UK (1996).
Bactcrial, fungal, parasitic and viral pathogens are categorized into four hazard groups and listed accordingly; the different containment levels (also for animals) are defined and described. In the appendices the handling of specific pathogenic biological material (e.g. latently infected animals, invertebrates, rabies virus, hepatitis B virus) is covered.

BG-Chemie leaflets 'Safe Biotechnology, Classification of Biological Agents - Bacteria', Germany (1998).
Presents -for the time being- the most extensive and most up-to-date list of bacteria categorized according to their hazardous potential for man. Indications of the possible pathogenic potential for plants, invertebrate and vertebrate animals are also given. The list comprises all validly described bacterial species and subspecies (about 4500) and is updated regularly. (Other leaflets have been produced for fungi, parasites, viruses and cell cultures).

CDC-NIH - Biosafety in Microbiological and Biomedical Laboratories, USA (1993).
Provides specific descriptions of combinations of microbiological practices, laboratory facilities and safety equipment, as well as recommendations for handling of selected parasitic, fungal, bacterial and viral agents infectious for man in four categories or biology safety levels.

EEC Council Directive 90/679/EEC on the protection of workers from risks related to exposure to biological agents at work (1990b).

The purpose of the directive is to encourage improvements, especially in the working environment, to guarantee better protection of the health and safety of laboratory workers; containment measures and containment levels are defined, biological agents are classified in four groups according to their hazardous potential and risk assessments have to be performed.

EEC Council Directive 93/88/EEC on the protection of workers from risks related to exposure to biological agents at work (1993a).

Amending the Directive 90/679/EEC, a list is provided reflecting the classification of biological agents by the European Community (bacteria, viruses, parasites, fungi) according to the hazardous potential. Only agents known to infect humans, classified in hazard group 2, 3 or 4, are listed.

GENETIC ENGINEERING

OECD Recombinant DNA Safety Considerations (1992).

Sets out the first international safety guidelines for the application of genetic engineering in agriculture, environment and industry; risk assessment methods are presented taking into consideration the properties of the donor and recipient organisms as well as the resulting genetically modified organism. The principles of biological and physical containment are deseribed. Step-by-step assessment during research and development is recommended to minimize the potential hazards. The conditions of Good Industrial Large-Scale Practice (GILSP) for the handling of genetically modified organisms in industrial processes are described.

EEC Council Directive 98/81/EC amending Directive 90/219/EEC on the contained use of genetically modified microorganisms (1990a).

Guidance of the European Community for the handling of genetically engineered microorganisms in laboratories and for the protection of human health and the environment, for the prevention of accidents and the control of wastes. The directive takes into account the fact that national frontiers do not exist for unwittingly released microorganisms, hence a common classification system for genetically manipulated microorganisms for the EU member states has been established. The principles of good microbiological practice, occupational safety and hygiene are described as well as the administrative requirements for work with genetically modified organisms.

EEC Council Directive 90/220/EEC on the deliberate release into the environment of genetically modified microorganisms (1990).

Guidance of the European Community for the conditions of the planned release of genetically engineered microorganisms into the environment with the aim of protecting human health and the environment. It is taken as a fact that national frontiers do not exist for released microorganisms. The effects of releases might be irreversible. The 'step by step' principle has to be applied. The placing on the market of products containing or consisting of genetically modified microorganisms has to be notified to a Community Authority and should be carried out only after its consent. Procedures for the involvement of the public are included in the directive.

TRANSPORT, EXPORT, IMPORT

IATA Dangerous Goods Regulations (1999).

The International Air Transport Association established guidelines for transport of dangerous goods by air freight. IATA's special regulations for the transport of such goods are presented, including explosives, gases, inflammable liquids, as well as toxic and infectious substances. The transport of non-infectious material is not regulated here. Consequently no restrictions exist for the transport of non-infectious material by air freight (unless other restrictions have effect).

A classification system for the different dangerous goods has to be followed consisting of identification number, class specification, corresponding label, packing instruction and packing group.

Universal Postal Union - Official Compendium of Information of General Interest Concerning the Implementation of the Universal Postal Union's Convention and its Detailed Regulations (1996).

National postal services are internationally joined in the Universal Postal Union (UPU). The convention serves as a guideline for the postal transport of non-infectious and infectious, perishable biological substances. The exchange of such material in general is restricted to states allowing the receipt of such material.

EEC Council Regulation 3381/94 setting up a Community regime for the control of exports of goods with dual use (1994b).

By establishing an internal market with the free movement of goods within the European Community it must be ensured that goods which can be used for both civil and military purposes ('dual-use goods') are subject to effective control when exported from Member States. According to this regulation an authorization through the competent authorities of the Member States is required for the export of such goods.

EEC Council Decision 94/942/CFSP on the joint action adopted by the Council on the basis of Article J.3 of the Treaty on the European Union concerning the control of exports of dual-use goods (1994c).

Contains the European Community's list of goods which can be used for both civil and military purposes and which are subject to control when exported from the European Community; the list comprises the areas of nuclear materials as well as chemicals, microorganisms (including viruses, bacteria and fungi), toxins, electronics, computers, telecommunications, sensors and lasers, navigation and avionics and the respective technology transfer; 60 microorganisms (including viruses, bacteria and fungi) and 10 toxins are listed.

Rohde, C.: Shipping of infectious, non-infectious and genetically modified biological materials, international regulations (1999).

A complex compendium designed to help resolve problems relating to packaging and shipping of the various kinds of biological material; it is based on the international shipping regulations as given by the Universal Postal Union, the International Air Transport Association and the relevant EC regulations. Step-by-step instructions help to avoid serious mistakes. Additionally helpful is the appendix indicating names and addresses of suppliers of packaging materials, a model of a shipper's declaration, as well as EC lists classifying biological material according to its hazardous potential.

PATENTS

World Intellectual Property Organization - Budapest Treaty on the International Recognition of the Deposit of Microorganisms for the Purposes of Patent Procedure (1989).

Regulates patent deposits in an international frame. The most important result of the Treaty is that a deposit made with one international depositary authority (= IDA) is sufficient and valid for all other member states of the Budapest Union. Any contracting state recognizes the deposit made with any IDA. The deposition procedure, the handling of requests for the release of deposited samples and the administrative work related to a patent deposit are described in detail.

World Intellectual Property Organization - Guide to the Deposit of Microorganisms under the Budapest Treaty (1997).

The aim of the guide is to give information on the procedures and requirements concerning the deposit of microorganisms in more detail than in the Treaty itself. Practical advice is presented for the deposit of microorganisms for patent purposes and for the procedure which has to be followed when obtaining samples of deposited microorganisms. In addition to general explanations, the specific requirements of the different international depositary authorities worldwide are described; an appended checklist summarizes the actions to be taken when considering the points mentioned above.

OTHER RELEVANT REGULATIONS

United Nations - Convention on Biological Diversity (1993).

The scope of the Convention is a comprehensive and international strategy for the three objectives it embraces: the preservation of biological diversity, the sustainable use of its components, and the fair and equitable sharing of the benefits arising out of the utilization of genetic resources. The provisions of the Convention imply the establishment of inventories on biodiversity and the clarification of issues like ownership and repatriation of cultures and information thereof.

Guide to the Convention on Biological Diversity (Glowka et al., 1994).

The Guide has been designed as a reference document to the text of the *Convention on Biological Diversity*. It also provides information on the possible steps for its implementation and a comprehensive view of what could be involved when fulfilling the Convention or its individual articles. An extensive bibliography is included.

EEC Council Directive 70/524/EEC concerning additives in animal feeds (1970).
Lays down the principles relating to the authorization for the use of additives, i.e. enzymes and microorganisms, in animal feeds; it has to be ensured that the products marketed are innocuous to the environment, workers, animal owners and consumers of animal products. It is amended by the EC Council Directive 93/114/EC which regulates the use and marketing of additives in animal feeds in case the additives are produced by or contain genetically modified organisms; a specific environmental risk assessment has to be carried out.

EEC Council Directive 94/40/EEC concerning additives in animal feeds (1994a).
Based on EEC Directive 70/524/EEC and takes into account recent developments in biotechnology, especially the possibility of the deliberate release of genetically engineered microorganisms into the environment. To comply with the EEC Council Directive 90/220/EEC the requested dossier has to contain data -in addition to the requirements specified in EEC Directive 70/524/EEC- concerning the risk assessment of the genetically modified organism for man, animals and the environment.

EEC Council Directive 93/114/EC concerning the use of enzymes, micro-organisms and their preparations in animal nutrition (1993b).
Applies to the use and marketing of enzymes, microorganisms and their preparations in animal nutrition; the conditions and the prerequisites to obtain a permit are presented. Products may not be dangerous to human or animal health. In case of microorganisms or mixtures containing microorganisms the identification of the strain(s) according to the international code of nomenclature and the deposit number of the organism with a recognized culture collection (preferably according to the Budapest Treaty and located in the European Union) are requested.

EEC Commission Directive 95/11/EEC amending Council Directive 87/153/EEC fixing guidelines for the assessment of additives in animal nutrition (1987, 1995).
This annex to the Council Directive 87/153/EEC points out the special requirements for the use of microbial strains as additives in animal nutrition: the microorganism has to be deposited with a recognized culture collection (preferably according to the Budapest Treaty and located in the European Union). In addition to the copy of the receipt of deposit with the culture collection a taxonomic description following the International Codes of Nomenclature has to be provided.

8   References

Advisory Committee on Dangerous Pathogens (1996). Categorisation of pathogens according to hazard and categories of containment, 4th edn. UK Government publication (HMSO); ISBN 0-11-883761-3.
American Type Culture Collection (1996). ATCC guide to packaging and shipping of biological materials;

ISBN 0-930009-55-X.

BG-Chemie (1990-1998). Leaflets Safe Biotechnology, Classification of Biological Agents (B004, viruses; B005, parasites; B007, fungi; B009, cell cultures). Jedermann-Verlag Dr. Otto Pfeffer oHG.

CDC-NIH (1993). Biosafety in Microbiological and Biomedical Laboratories. U.S. Government Printing Office.

CEN (Comité Européen de Normalisation) (1996). EN 829 *In vitro* diagnostic systems - transport packages for medical and biological specimens - requirements, tests.

Department of Health and Human Services, Public Health Service (1996). Additional requirements for facilities transferring or receiving select infectious agents. Japanese Federal Register Vol. 61 (112) Proposed Rules, 29327-29333.

Duclaux, E. (1896). Pasteur: Histoire d'un esprit. Charaire et Cie, Sceaux.

European Community (1970). Council Directive 70/524/EEC fixing guidelines for the assessment of additives in animal nutrition. OJ L 270 of 14.12.1970.

European Community (1990b). Council Directive 90/220/EEC on the deliberate release into the environment of genetically modified microorganisms. OJ L 117 of 08.05.1990.

European Community (1990c). Council Directive 90/679/EEC on the protection of workers from risks related to exposure to biological agents at work. OJ L 374 of 31.12.1990.

European Community (1992). Commission Directive 92/103/EEC amending Annexes I to IV to Council Directive 77/93/EEC on protective measures against the introduction of organisms harmful to plants and plant products and against their spread within the Community. OJ L 363 of 01.12.1992.

European Community (1993a). Council Directive 93/88/EEC amending Directive 90/679/EEC on the protection of workers from risks related to exposure to biological agents at work. OJ L 268 of 29.10.1993.

European Community (1993b). Council Directive 93/114/EEC amending Council Directive 70/524/EEC fixing guidelines for the assessment of additives in animal nutrition. OJ L 334 of 31.12.1993.

European Community (1994a). Council Directive 94/40/EEC amending Council Directive 87/153/EEC fixing guidelines for the assessment of additives in animal nutrition. OJ L 208 of 11.08.1994.

European Community (1994b). Council Directive 3381/94 setting up a Community regime for the control of exports of goods with dual use. OJ No. L 367 of 31.12.1994.

European Community (1994c). Council Decision 94/942/CFSP on the joint action adopted by the Council on the basis of Article J.3 of the Treaty on the European Union concerning the control of exports of dual-use goods. OJ L 367 of 31.12.1994.

European Community (1995). Commission Directive 95/11/EEC amending Council Directive 87/153/EEC fixing guidelines for the assessment of additives in animal nutrition. OJ L 106 of 11.05.1995.

European Community (1998). 98/81/EC amending Council Directive 90/219/EEC on the contained use of genetically modified microorganisms. OJ L 117 of 08.05.1990.

German Infectious Diseases Act (1990). BGBl. I, p. 2002 of 12.09.1990.

German Infectious Diseases of Animals Enactment (1992). BGBl. I, p. 1845 of 02.11.1992.

German Infectious Diseases of Animals Imports Enactment (1992). BGBl. I, p. 2467 of 23.12.1992.

German Law Controlling War Weapons (1991). BGBl. I, p. 913 of 19.04.1991.

German Law Regulating Genetic Engineering (1993). BGBl. I, p. 2066 of 22.12.1993.

German Plant Protection Act (1990). BGBl. I, p. 1221 of 28.06.1990.

Glowka, L., Burhenne-Guilmin, F., Synge, H., McNeely, J.A. & Gündling, L. (1994). A guide to the Convention on Biological Diversity. International Union for Conservation of Nature and Natural Resources. Gland and Cambridge; ISBN 2-8317-0222-4.

Hawksworth, D.L. (1997). Fungi and the international biodiversity initiative. *Biodiversity and Conservation* **6**, 661-668.

Hawksworth, D.L. & Rossman, A.Y. (1997). Where are all the undiscovered fungi? *Phytopathology* **87**, 888-891.

International Air Transport Association (1999). Dangerous Goods Regulations, 40th edition; ISBN 92-9035-863-7.

International Committee on Systematic Bacteriology (1997). VIII International Congress of Microbiology and Applied Bacteriology, minutes of the meetings, 17, 18, and 22 August 1996, Jerusalem, Israel. *International Journal of Systematic Bacteriology* **47**, 597-600.

International Journal of Systematic Bacteriology (1999). Guidelines for Authors 49, ix-xviii.

ISO (International Organization for Standardization) (1997). Revision of ISO/IEC guide 51, safety aspects - guidelines for their inclusion in standards (draft, January 1997).

IUMS/IUBS (1991). Action Statement Microbial Diversity 21. *WFCC Newsletter* **17**, 18.

Laboratory Centre for Disease Control, Health Protection Branch, Health Canada Laboratory (1996). Biosafety Guidelines, 2nd ed.

OECD (1992). Recombinant DNA Safety Considerations, ISBN 92-64-12857-3.

Rohde, C. & Claus, D. (1998). Shipping of Infectious, Non-infectious and Genetically Modified Biological Materials, International Regulations. DSMZ-Deutsche Sammlung von Mikroorganismen und Zellkulturen GmbH.

Smith, D., ed. (1996). World Federation for Culture Collections (WFCC) Committee on Postal, Quarantine and Safety Regulations Report 1994-1996.

Sneath, P.H.A. (1992). International Code of Nomenclature (1990 Revision). Washington, D.C.: American Society for Microbiology

Tindall, B.J. (1999). Proposal to change the Rule governing the designation of type strains deposited under culture collection numbers allocated for patent purpose. *International Journal of Systematic Bacteriology* **49**, 1317-1319.

United Nations (1993). Convention on Biological Diversity, UNEP/CBD/94/1, 94-04228.

Universal Postal Union (1996). Official Compendium of Information of General Interest Concerning the Implementation of the Convention and its Detailed Regulations. Berne, International Bureau of the Universal Postal Union.

Weihs, V. & Fritze, D. (1995). Patent protection in biotechnology - deposit of microorganisms and other biological material at the DSM. *Microbiology Europe* **3**, 18-23.

Williams, S.T., Goodfellow, M., Alderson, G., Wellington, E.H.M., Sneath, P.H.A. & Sackin, M.J. (1983). Numerical classification of *Streptomyces* and related genera. *Journal of General Microbiology* **129**, 1743-1813.

World Federation for Culture Collections (1992). Resolutions ICCC-7. *WFCC Newsletter* **19**, 3.

World Federation for Culture Collections (1996). Resolutions ICCC-8. *WFCC Newsletter* **25**, 4.

World Health Organization (1983). Special Programme on Safety Measures in Microbiology: Report by the Working Group on the Development of Emergency Services. World Health Organization, Geneva, ISBN 92-4-154167-9.

World Health Organization (1993). Laboratory Biosafety Manual, 2nd edn. World Health Organization, Geneva, ISBN 92- 4-154450-3.

World Intellectual Property Organization (1989). Budapest Treaty on the International Recognition of the Deposit of Microorganisms for the Purposes of Patent Procedure. Geneva, WIPO publication no. 277 (E) ISBN 92-805-0744-3.

World Intellectual Property Organization (1998, yearly updates). Guide to the Deposit of Microorganisms under the Budapest Treaty. Geneva, ISBN 92-805-0195-x.

Organism Index